Soldiers and Gentlemen

A History of the University and Public Schools Brigade of the Royal Fusiliers 1914–1918

Colin W. Taylor

Helion & Company

Dedicated to the men who enlisted in the 'Ups' or who served in these battalions.

Also dedicated to my family for their support and patience

Helion & Company Limited
Unit 8 Amherst Business Centre
Budbrooke Road
Warwick
CV34 5WE
England
Tel. 01926 499 619
Email: info@helion.co.uk
Website: www.helion.co.uk
Twitter: @helionbooks
Visit our blog at https://helionbooks.wordpress.com/

Published by Helion & Company 2024
Designed and typeset by Mach 3 Solutions (www.mach3solutions.co.uk)
Cover designed by Paul Hewitt, Battlefield Design (www.battlefield-design.co.uk)

Text © Colin W. Taylor 2024
Images open source unless otherwise credited
Maps drawn by George Anderson © Helion & Company 2024

Every reasonable effort has been made to trace copyright holders and to obtain their permission for the use of copyright material. The author and publisher apologize for any errors or omissions in this work and would be grateful if notified of any corrections that should be incorporated in future reprints or editions of this book.

ISBN 978-1-804514-22-1

British Library Cataloguing-in-Publication Data.
A catalogue record for this book is available from the British Library.

All rights reserved. No part of this publication may be reproduced, stored in a retrieval system, or transmitted, in any form, or by any means, electronic, mechanical, photocopying, recording or otherwise, without the express written consent of Helion & Company Limited.

For details of other military history titles published by Helion & Company Limited contact the above address or visit our website: http://www.helion.co.uk.

We always welcome receipt of book proposals from prospective authors.

Contents

List of Maps		vii
Regimental Abbreviations		viii
Acknowledgements		xi
Introduction		xii
1	Formation, Recruiting and Early Training	15
2	Early Training of the UPS Brigade	32
3	Commissions and Controversy	53
4	Woodcote Park	62
5	Training at Clipstone	77
6	Further Training on Salisbury Plain	92
7	Ex-UPS Men in 1915	105
8	To France	117
9	An Introduction to Trench Warfare	129
10	Christmas on Active Service	150
11	19/R Fus on 2 January 1916	161
12	98th Brigade in Early 1916	170
13	19th Brigade in Early 1916	192
14	Disbandment of 18/R Fus, 19/R Fus and 21/R Fus	204
15	Prelude to the Somme – 20/R Fus in the Spring of 1916	227
16	The UPS and the First Day of the Somme	245
17	20/R Fus on the Somme	252
18	The Fight for High Wood 20th July 1916	268
19	High Wood, the Aftermath	290
20	UPS men in the later Somme fighting, July to August 1916	308
21	The Rest of the Somme	317
22	UPS men later in the Somme fighting – September to November 1916	338
23	20/R Fus and the Winter of 1916–1917	349
24	UPS Men at the Battle of Arras	363
25	20/R Fus and the Arras	374
26	Trenches at Nieuport	397
27	Ex-UPS Men at Ypres, June to November 1917	407
28	20/R Fus and the Third Battle of Ypres	415
29	20/R Fus October–December 1917	421
30	Early 1918 in the Ypres Salient and Disbandment	429

31 *Unternehmen* Michael, March 1918	436
32 Battle of the Lys, April 1918	448
33 UPS Men and the Advance to Victory	459
34 'Side-Show' Theatres	473
35 The University and Public Schools Brigade Legacy	482
Conclusion	493

Appendices
I Victoria Cross Awards to UPS Brigade Members	498
II 20/R Fus. Roll of Honour, High Wood, 20 July 1916	501
III Key Documents	509
IV UPS Brigade Poets	519

Select Bibliography	523
Index	528

List of Maps

1	98th Brigade and 20/R Fus in France and Flanders 1915-1918.	122
2	2 Givenchy Sector.	135
3	Cuinchy Sector.	163
4	Cuinchy Brickstacks.	166
5	High Wood, 20 July 1916: Plan of attack.	271
6	High Wood 20 July 1916: 19th Brigade assault.	275
7	High Wood, 20/R Fus burial map.	293
8	Lesboeufs 5-6 November 1916.	334
9	Concrete Trench, 16 April 1917.	379
10	Tunnel Trench, 20 May 1917.	389
11	Lombartzyde Sector, Nieuport, summer 1917.	402
12	20/R Fus operations, Third Battle of Ypres, September 1917.	419

Regimental Abbreviations

Full 1914 Title	Official Abbreviation	Abbreviation(s) used in text
General Staff	G.S.	GS
Royal Horse Artillery	R.H.A.	RHA
Royal Field Artillery	R.F.A.	RFA
Royal Garrison Artillery	R.G.A.	RGA
Royal Engineers	R.E.	RE
Grenadier Guards	G. Gds.	1/GG
Coldstream Guards	C. Gds.	1/CG
Scots Guards	S. Gds.	1/SG
Irish Guards	I. Gds.	1/IG
Welsh Guards	W. Gds.	1/WG
The Royal Scots (Lothian Regiment)	R. Scots.	1/R Scots
The Queen's (Royal West Surrey Regiment)	R. W. Surr. R.	1/Queen's RWS
The Buffs (East Kent Regiment)	E. Kent. R.	1/E Kents
The King's Own (Royal Lancaster Regiment)	R. Lanc. R.	1/King's Own
The Northumberland Fusiliers	North'd Fus.	1/NF
The Royal Warwickshire Regiment	R. War. R.	1/R Warwicks
The Royal Fusiliers (City of London Regiment)	R. Fus.	1/R Fus
The King's (Liverpool Regiment)	L'pool R.	1/King's
The Norfolk Regiment	Norf. R.	1/Norfolks
The Lincolnshire Regiment	Linc. R.	1/Lincs
The Devonshire Regiment	Devon. R.	1/Devons
The Suffolk Regiment	Suff. R.	1/ Suff R
Prince Albert's (Somerset Light Infantry)	Som. L. I.	1/SomLI
The Prince of Wales' Own (West Yorkshire Regiment)	W. York. R.	1/W Yorks
The East Yorkshire Regiment	E. York. R.	1/E Yorks
The Bedfordshire Regiment	Bedf. R.	1/Bedf R
The Leicestershire Regiment	Leic. R.	1/Leic R
The Royal Irish Regiment	R. Ir. R.	R Irish or 1/RIR
Alexandra, Princess of Wales's Own (Yorkshire Regiment)	York. R.	Green Howards or 1/Yorks
The Lancashire Fusiliers	Lan. Fus.	1/LF

Full 1914 Title	Official Abbreviation	Abbreviation(s) used in text
The Royal Scots Fusiliers	R. Sc. Fus.	1/RSF
The Cheshire Regiment	Ches. R.	1/Ches R
The Royal Welsh Fusiliers	R. W. Fus.	1/RWF
The South Wales Borderers	S. Wales Bord.	1/SWB
The King's Own Scottish Borderers	K. O. Sco. Bord.	1/KOSB
The Cameronians (Scottish Rifles)	Sco. Rif.	1/Cameronians, 5/Scot Rif
The Royal Inniskilling Fusiliers	R. Innis. Fus.	1/R Innis Fus
The Gloucestershire Regiment	Glouc. R.	1/Glosters
The Worcestershire Regiment	Worc. R.	Worcesters or 1/Worc R
The East Lancashire Regiment	E. Lan. R.	1/E Lancs
The East Surrey Regiment	E. Surr. R.	1/E Surrey
The Duke of Cornwall's Light Infantry	D. of Corn. L. I.	1/DCLI
The Duke of Wellington's (West Riding Regiment)	W. Rid. R.	1/DWR
The Border Regiment	Bord. R.	1/Border
The Royal Sussex Regiment	R. Suss. R.	1/R Sussex
The Hampshire Regiment	Hamps. R	1/Hamps R
The South Staffordshire Regiment	S. Staff. R.	1/S Staffs
The Dorsetshire Regiment	Dorset. R.	1/Dorsets
The Prince of Wales's Volunteers (South Lancashire Regiment)	S. Lan.R.	1/S Lancs
The Welsh Regiment	Welsh R.	1/Welsh
The Black Watch (Royal Highlanders)	R. Highrs.	Black Watch or 1/BW
The Oxfordshire and Buckinghamshire Light Infantry	Oxf. & Bucks. L. I.	1/O&BLI
The Essex Regiment	Essex R.	1/Essex
The Sherwood Foresters (Nottinghamshire & Derbyshire Regiment)	Notts. & Derby. R.	Sherwood Foresters or 1/Notts&Derby
The Loyal North Lancashire Regiment	N. Lan. R.	1/LNL
The Northamptonshire Regiment	North'n. R.	1/Northants
Princess Charlotte of Wales's (Royal Berkshire Regiment)	R. Berks. R.	Berkshires or 1/R Berks
The Queen's Own (Royal West Kent Regiment)	R. W. Kent. R.	1/Queen's RWK
The King's Own (Yorkshire Light Infantry)	Yorks. L. I.	1/KOYLI
The King's (Shropshire Light Infantry)	Shrops. L. I.	1/KSLI
The Duke of Cambridge's Own (Middlesex Regiment)	Midd'x R.	Middlesex or 1/Mx
The King's Royal Rifle Corps	K. R. Rif. C.	1/KRRC
The Duke of Edinburgh's (Wiltshire Regiment)	Wilts. R.	Wiltshires or 1/Wilts
The Manchester Regiment	Manch. R.	Manchesters or 1/Manch R

Full 1914 Title	Official Abbreviation	Abbreviation(s) used in text
The Prince of Wales's (North Staffordshire Regiment)	N. Staff. R.	1/N Staffs
The York and Lancaster Regiment	York & Lanc. R.	1/Y&L
The Durham Light Infantry	Durh. L. I.	1/DLI
The Highland Light Infantry	High. L. I.	1/HLI
Seaforth Highlanders (Ross-Shire Buffs, The Duke of Albany's)	Sea. Highrs.	1/Seaforths
The Gordon Highlanders	Gord. Highrs.	1/Gordons
The Queen's Own Cameron Highlanders	Cam'n Highrs.	1/Camerons
The Royal Irish Rifles	R. Ir. Rif.	1/RIRif
Princess Victoria's (Royal Irish Fusiliers)	R. Ir. Fus.	1/RIrFus
The Connaught Rangers	Conn. Rang.	1/Conn R
Princess Louise's (Argyll and Sutherland Highlanders)	Arg.&Suth'd Highrs	Argylls or 1/A&SH
The Prince of Wales's Leinster Regiment (Royal Canadians)	Leins. R.	1/Leinsters
The Royal Munster Fusiliers	R. Muns. Fus.	Munsters or 1/RMF
The Royal Dublin Fusiliers	R. Dub. Fus.	Dublins or 1/RDF
The Rifle Brigade (Prince Consort's Own)	Rif. Brig.	1/RB
The London Regiment	Lond. R.	Londons or 1/Lond R
Machine Gun Corps	M.G.C.	MGC
Army Service Corps	A.S.C.	ASC
Royal Army Medical Corps	R.A.M.C.	RAMC
Labour Corps	Lab. C.	Lab Corps
Army Ordnance Corps	A.O.C.	AOC
Tank Corps	Tank C.	Tank C
Army Cyclist Corps	A.C.C.	ACC

Acknowledgements

Researching and writing a book is a time consuming journey made easier by helpful and supportive people and organisations. This publication could not have been produced without considerable assistance and the author must thank the following people and institutions:

The staff of the reading rooms at the Imperial War Museum for their patience and advice and Anthony Richards the Head of Documents and Sounds. The numerous members of the reading room staff at The National Archives (Kew) without which much of the detail for this book could not have been produced. The reading room staff at the Liddle Collection at the University of Leeds; Dr Beth Wyrill the curator of the Fusilier Museum, London for access to several UPS men papers held in the museum collection.

I am grateful to the following archives: Abigail Hartley at the West Sussex Record Office for the papers of Robert Tudor and Reginald Clements. The Lancashire Infantry Museum, and its curator Garry Smith, for the transcribed letters of E.A. Holden which are held in LIM Archive (origins regrettably unknown). Abby Matthews at the Sutton Archives for help with the photos from the Past on Glass collection by David Knights-Whittome.

Families of members of the UPS have kindly allowed the reproduction of documents and photographs from private collections. This book has relied on their kindness, assistance and patience. I am grateful to Angela Jennings for the unpublished diary and photos of Ernest Brierley; Gillian Häkli for Terence Doherty papers; Pat Isom for permission to use the Cecil Isom papers; Suzannah Schmitt for use of the account by Lewis Jacques with additional thanks Andrew Polkey Assistant Archivist of the Old Derbeian Society; Wendy Stock for the letters of Joseph Leather; Ruth Shield for the use of Frederick Shield's diary; James Skelton for the papers of Godfrey Skelton; Janet Wood for the sketches and account by Ernest Stoneley; Paul Tyson for the letters of Harold Tyson. I regret that I have not been able to publish every photograph or document that I have been generously provided with.

I must also thank Paul Nixon, Peter Reed, Brian Bouchards and Jordan Cassidy who have provided very valuable information; Brian also suggested the title. Andy Pay has also greatly assisted with considerable detail on the UPS men who served with the Rifle Brigade. Likewise, Charles Fair's support and assistance regarding UPS men at Officer Cadet Battalions has been extremely helpful. I am also grateful to Dr Aimée Fox of King's College London for helping me craft the MA dissertation that preceded this book.

This volume would also not have reached publication without the patience, advice and assistance of the Helion & Company team, I must sincerely thank proprietor Duncan Rogers, Dr Michael LoCicero and Victoria Powell. I'm also grateful to George Anderson for creating the excellent maps.

Every effort has been made to obtain permissions from all copyright holders. Finally, I acknowledge that any errors or omissions are mine alone.

Introduction

'Chocolate Soldiers' and 'Rotten Men'

This volume sets out to chronicle the service of four infantry battalions during the Great War of 1914-1918. These were the grandly named University and Public Schools Brigade of the Royal Fusiliers. These units were predominantly recruited from former university students and public schools boys and they were almost unique as a combat organisation in the British Army.[1] The stories of these 'gentleman rankers' and 'temporary gentlemen' requires telling. The Brigade was officially sanctioned as a fighting force by Lord Kitchener instead of employing these men as officer candidates. The UPS Brigade (or 'Ups') started recruiting in September 1914 and formed four battalions; the 18th, 19th, 20th and 21st (Service) Battalions of the Royal Fusiliers (City of London Regiment) (18/R Fus, 19/R Fus, 20/R Fus and 21/R Fus).[2] There were high expectations from battalions of fit public school boys, already with rudimentary school Officers' Training Corps (OTC) training. They could theoretically be trained and sent to France quickly. However, these battalions never achieved their full potential, for various reasons, and none of them survived until November 1918. Not being in existence after the Armistice undoubtedly hindered the UPS Brigade in terms of historiography as their men had been scattered.

The 20th Battalion Royal Fusiliers (20/R Fus) predominantly served with 19th Brigade (part of 33rd Division) which comprised 2nd Battalion Royal Welsh Fusiliers (2/RWF),[3] 1st Battalion The Cameronians (1/Cameronians) and the 5th Battalion Scottish Rifles (5/Scot Rif).[4] 20/R Fus soon acquired the nickname 'Chocolate Soldiers' from 2/RWF[5] and were referred to as; 'rotten men' or 'that rotten crowd' by commentators from the battalion.[6] These comments by

---

1. For the purposes of this book British public schools will refer to those schools listed in the Public Schools Yearbook. H.F.W. Deane (ed), *The Public Schools Year Book* (London: Year Book Press, 1920).
2. A Public Schools Battalion of the Middlesex Regiment was formed on a smaller scale by a different committee.
3. For consistency, the official nomenclature from the Army List of 'Welsh' (not Welch) has been used throughout the text (except for direct quotes).
4. 'Cameronians' is used to refer to the 1st Battalion; 'Scottish Rifles' for the 5th Battalion. On 29 May 1916, 1/5 Scot Rif and 1/6 Scot Rif amalgamated to form 5/6 Scot Rif. See H.H. Story, *History of the Cameronians (Scottish Rifles) 1910–1933* (Aylesbury: Hazell Watson & Viney, 1961), p.109. The 5/Scot Rif war diary first refers to '5/6 Scot Rif' in July 1917. Thus '5/6' is employed after this date).
5. This was the dismissive sobriquet bestowed on 20/RF by the Regulars of 2/RWF. See J.C. Dunn, *The War the Infantry Knew 1914-1919* (London: Jane's, 1938), p.594.
6. Dunn, *The War the Infantry Knew*, p.204; Frank Richards, *Old Soldiers Never Die* (Eastbourne: Anthony Rowe, 1933), pp.212–213; Robert Graves, *Goodbye to All That* (London: Penguin, 1957),

a handful of popular personal accounts, alongside a detailed history of 2/RWF by J.C. Dunn, had a detrimental effect on the perception of the UPS and 20/R Fus and have set the tone for how they have been assessed historically. This assessment is without any detailed examination of what these battalions experienced or achieved and without consultation of the myriad sources available. This book aims to investigate whether the UPS battalions comprised poor officers and soldiers who performed badly, or whether they were just poorly thought of by their peers? The author does not intend to be an apologist for the UPS or the 20th Battalion Royal Fusiliers. Nor is this book an attempt to rehabilitate the reputation of these units. However, closer inspection might suggest that their reputation, certainly at High Wood, might be considered somewhat unjust. This was not to say the Brigade, and 20/R Fus particularly, was not without failings in many aspects of its creation, training and performance.

Due to the negative typecasting of these battalions, and especially 20th Battalion Royal Fusiliers, the UPS Brigade has been under studied. Furthermore, 'gentleman rankers' and 'temporary gentlemen' are both groups that have been generally overlooked in literature on the Great War in favour of regional 'Pals' battalions and archetypal upper class officers. Most of the men of the Brigade were well educated and were literary by education and profession. Therefore, a wealth of source material (both published and unpublished) from the officers and men from within its ranks, though available, has been largely ignored. Over fifty different personal accounts of UPS or ex-UPS men (books, letters and diaries of varying levels of detail) comprise a comprehensive archive of qualitative information; likely more than many other infantry brigades. However, these accounts do not provide uniform coverage as men were killed or departed (wounded, sick or gazetted) and were replaced by more ordinary Tommies, fewer of whom recorded events. This body of material provides limited coverage for 1917 and 1918. These experiences of men of differing ranks and social backgrounds give a unique insight into the war and the lives of the 'footsloggers' of the day and on the recruiting, training and overseas service of a few battalions of 'Kitchener's Army'. Furthermore, the Brigade had two journals published by its soldiers, *The Pow-Wow* and *The Gasper*, which give further insights into the lives of these men and their sense of humour.[7] This abundance of information, combined with numerous official documents and newspaper articles, allows a more detailed examination of both the combat performance of these units but also their wider influence on the Great War. The content of this book has been based on the sources that have been consulted. It is possible that further personal diaries and letters may be found in the future that might cast more light on individuals, actions and events. It is hoped that these will not drastically contradict this narrative. This all provides information to enable the qualitative examination of the experiences of these men and the units they served with. A detailed nominal roll of the UPS has also been compiled by the author which enables further quantitative analysis.

The UPS Brigade was a unique formation, even amongst the many 'Pals' battalions. Few Great War units could match the abundance of information and sources available beyond those whose unit histories were written whilst veterans were still available to be interviewed. In addition, these accounts provide an insight into the 'small scale' activities facing soldiers such as

---

p.185.

7   It would be a great disservice to not highlight the extraordinary journalistic efforts of the contributors and editorial team of *The Pow-Wow* and *The Gasper*. Their talent and wit make these 'trench magazines' exemplars of the genre.

routines and fatigues in camp and in the trenches. Though such procedures likely varied in detail between different units these accounts may give a wider insight into the experience of Great War era soldiers from across the Army.

This work is intended to cover four main facets of both the UPS battalions and the experiences of infantrymen and infantry officers during the Great War. First, the early chapters will describe the formation and initial training of the UPS Brigade. Second, the operational service of the four UPS service battalions in France over the winter of 1915–1916 will be covered, up until three battalions were withdrawn from active service. Third, the experience of 20/R Fus will be examined up until the battalion was disbanded in 1918. Fourth, the history and wider service of many enlisted men and officers from the UPS battalions will be examined throughout the book to determine the wider influence of this formation on the war. How these men served, fought or commanded and how they were wounded, killed, captured or decorated will show how ubiquitous UPS officers and men became within their Regiment and the wider British Army. This book is not only a brigade history but also a battalion history and furthermore covers the subsequent wartime activities of the UPS cohort of officers and men from 1915 to 1918. Finally, the effectiveness and impact of these units will be examined to determine whether this experiment in forming an infantry brigade from potential officers was beneficial or detrimental to the Army and the men themselves. This must be studied both through their combat performance and their success or failure in generating effective officer candidates and future platoon commanders. The Great War also saw an expansion of existing and new 'technical' arms; i.e. the Royal Flying Corps (RFC), Machine Gun Corps (MGC), Tank Corps (Tank C), Special Brigade Royal Engineers (RE) etc. Many ex-UPS men commanded detachments of these corps.

This volume cannot pretend to be a comprehensive account that covers every aspect of the Brigade nor every member of the formation. Almost nothing has been written on the two Reserve UPS Battalions (28th and 29th (Reserve) Battalions Royal Fusiliers) though they played a role in reinforcing the active service battalions. It is regretted that their performance cannot be covered, though several chapters deserve to be devoted to them. Efforts have been made to highlight as much detail on as many of the men of the Brigade as possible. Regrettably, there are thousands whose stories have not been adequately told within these pages. To the memory of these men, and their descendants, who might search this volume for mention of them, and for any errors that have been made, the author can only apologise.

1

Formation, Recruiting and Early Training[1]

The finest body of men I have ever seen

By the end of the month I found myself more than anxious to join up … I wanted to have taken part in the campaign.[2]

More magnificent material I have never seen in my life.[3]
Major General Sir Francis Lloyd

Calm before the storm

The subsequent horrors of war that some men of the University and Public Schools Brigade experienced at High Wood in July 1916 were a long way from the summer of 1914. Though the storm clouds gathered in Europe most people in Great Britain were unconcerned with the events overseas. One future soldier, Vernon Bartlett, spent the last day of peace near Ringwood, in Dorset, fishing with his father. He recalled eating sandwiches by the river, watching the birds, and they prolonged the cycle ride back to Bournemouth for as long as possible. On arrival they found crowds gathering to read the news that Britain was going to war.[4] These bulletins announced the ultimatum to Germany and were followed by the declaration of war at 11:00 p.m. on 4 August 1914. This announced a change that would affect the lives of many young British men. To Arthur Whitten Brown, an engineering apprentice, 'the outbreak of war changed all my plans and hopes and interfered with the career I had mapped out for myself. In

---

1 The origins, recruitment and initial training of the University and Public Schools Brigade is covered in the excellent history of the Brigade. This and some subsequent chapters will overlap with this published account. The intention is to augment the existing narrative to provide a more comprehensive account of the Brigade. This section will additionally explore the motivations, experiences. and backgrounds of some of the men who initially enlisted.
2 IWM: E.D. Shearn Papers, 2033, p.1
3 IWM: Documents 1708, Catalogue date 1985-11-05: Papers of W.B. Medlicott, p.1.
4 Vernon Bartlett, *I Know What I Liked* (London: Chatto and Windus, 1974), p.21.

fact, I was in exactly the same position as many thousands of other young men at the beginning of their careers.'[5] Erroll Shearn recalled:

> At the start of it, it did not appear to be of much concern to me. It was a matter for our professional army, helped, as might be necessary, by the volunteers or territorials. I was having my summer holidays at Frinton-on-Sea. During the month of August more and more of my contemporaries drifted off to join the armed services. By the end of the month I found myself more than anxious to join up. Like most people I thought the war would end by Christmas 1914 in a victory over Germany and I wanted to have taken part in the campaign.[6]

On 7 August Lord Kitchener, the Secretary of State for War, appealed for 100,000 volunteers for the Army between the ages of nineteen and thirty. According to one applicant, Herbert Vinden; 'It came as a stunning blow to the nation. Men flocked to the recruiting offices, falsified their ages upward or downward, learned the opticians' letter card by heart to dissimulate short sight, went to another recruiting office if they had been rejected in their first attempt to enlist.'[7] This was a recruiting boom and men sought to jump on this popular bandwagon. Many were to be frustrated. Young men who had been educated by the eminent public schools and universities desired to serve but initial appeals for junior officers to command the 'first 100,000' were soon filled. Those who volunteered too late for commissions were left in limbo as it was anticipated that the next call for further officers might not happen immediately. These enthusiastic and well educated young men were at a loss as to how to proceed. Serving as an officer was impossible and would also see them potentially arrive too late to take part in a quick war. Alternatively, many had experienced aspects of enlisted service in the ranks whilst members of their school or university OTC. However, such service would not be as enjoyable without being surrounded by like-minded comrades.

On 23 August the British Expeditionary Force first fought the German Army at the Battle of Mons. Three days later, on 26 August, the same day as the Battle of le Cateau, a letter was published in The Times. This was written by eight men who found themselves unable to serve due to the current recruiting environment: 'We have applied for commissions in the new Regulars, but find we are too old. We have offered our services as musketry instructors, and are informed that we are too young, and that none under thirty-five are selected. After endless inquiries there seems only one way in which our services are acceptable, and that is by joining the ranks.'[8] This letter was symptomatic of the recruiting situation, but the plight of these men affected many others. By this date Kitchener's first 'hundred thousand' had been recruited. Many men whose interest was piqued by this letter were not of the same age bracket. However, the idea of a unit in which they could serve as enlisted men, but alongside like-minded comrades, was appealing to both young and old public school and university men. Though six of the 'Eight' had found positions with a London Territorial Force battalion by the date chosen, there was still considerable interest in this meeting.[9]

---

5   Arthur Whitten Brown, *Flying the Atlantic in Sixteen Hours* (New York: Frederick Stokes, 1920), p.7.
6   IWM: E.D. Shearn Papers, 2033, p.1
7   IWM: F.H. Vinden Papers, 5565, p.11.
8   Anon. Author, *University & Public Schools Brigade*, p.15.
9   The two remaining of the 'Eight Unattached' may have been E.J. Stuart and H.F. Fenn.

The meeting on 27 August was a large gathering where much was discussed. The 'official' version of events in the UPS History must be relied upon. Mr Hector Boon made a proposal for raising a force of 5,000 old public school boys. A committee would be formed to recruit this regiment to be part of Kitchener's 'New Army'. Efforts were made to conduct intensive recruiting to reach the 5,000 mark as quickly as possible.[10] Boon's proposal, though ambitious, was seconded by Mr E.J. Stuart and was unanimously agreed. This was beyond the scope that had been anticipated, but the scheme clearly chimed with the mood in the room. A committee was duly selected and sat for the first time on 28 August 1914 and comprised: Chairman, Mr J.P. Thompson; Secretary, Mr H.J. Boon; Members; Mr E.J. Stuart, Captain George Hallett, Mr C.F. Beal, Mr J.W. Henderson, Mr F. Warner-Abbatt, Mr H.F. Fenn, Mr H. Howell, Dr Hele-Shaw, and Mr S.M. Gluckstein. Mr Thompson later resigned to serve as a private and recommended that Mr Boon take over and that Mr Howell be Secretary. Dr Hele-Shaw was made vice-chairman. Several of these men would become officers in the Brigade.

At this meeting it was determined that nothing should be done without having achieved agreement to proceed from the War Office. A letter was drafted (see Appendix III) to be presented to the Secretary of State. After failing to find Lord Kitchener at the War Office, Mr Boon attempted to visit his home and the letter eventually found its way to his desk. The reply from Lord Kitchener was brief; 'Go ahead, and if you can raise 10,000 men I shall be all the better pleased.' When the committee received this response that evening, they sprang into action immediately. A recruiting poster was designed, printed and put up across London. Notices were published in major newspapers and letters were sent to civic leaders around the country to support recruiting for the 'Old Public School and University Men's Force' as the new unit was named. Letters were sent to headmasters of well-known public schools requesting contributions to fund the recruiting of the force. Initial results were promising with recruiting offices being established in over fifty towns and cities and financial contributions covered the early costs of the campaign. As it was impossible to define a public school, beyond the Public Schools' Yearbook, the criteria for recruiting was loosened to any men of the right class. Local recruiting officers were to apply their own discretion.

Alongside this great desire for men to enlist there was also considerable competition between different units and organisations to attract men of the type the UPS desired. The Royal Fusiliers (City of London Regiment) alone, raised different battalions supposed to cater for stockbrokers, bankers, sportsmen, ex-public school boys, men from across the British Empire and frontiersmen. Other regiments also wanted men of a similar class and with similar attributes. There were the smart London Territorial Force battalions such as the London Scottish or Queen Victoria's Rifles or other 'New Army' units like the Public Schools battalion of the Middlesex Regiment; likewise there were battalions for Sportsmen and Footballers. These units were trying to recruit from the same demographic and many potential recruits did not know where to enlist because many units that traditionally accommodated 'Gentleman Rankers', for example the more prestigious Territorial Force battalions, could both demand higher recruit standards and therefore filled up quickly. In many ways the better organised and publicised the unit, the better their chances of recruiting their numbers, succeeding at training, and deploying overseas. Though many men gravitated towards London hoping to enlist, these events were not peculiar to the capitol, as units throughout the UK hoped to attract local men of all classes. Such

10   See Appendix III for the criteria under which this unit was raised.

UNIVERSITY & PUBLIC SCHOOLS BRIGADE

5000 MEN AT ONCE

The Old Public School and University Men's Committee makes an urgent appeal to their fellow Public School and University men to at once enlist in these battalions, thus upholding the glorious traditions of their Public Schools & Universities.

TERMS OF SERVICE.

Age on enlistment 19 to 35, ex-soldiers up to 45, and certain ex-non-commissioned officers up to 50. Height 5 ft. 3 in. and upwards. Chest 34 in. at least. Must be medically fit.

General Service for the War.

Men enlisting for the duration of the War will be discharged with all convenient speed at the conclusion of the War

PAY AT ARMY RATES.

and all married men or widowers with children will be accepted, and will draw separation allowance under Army Conditions.

HOW TO JOIN.

Men wishing to join should apply at once, personally, to the Public Schools & Universities Force, 66, Victoria Street, Westminster, London, S.W., or the nearest Recruiting Office of this Force.

GOD SAVE THE KING!

The first recruiting poster for the UPS Brigade. The brigade would later be re-named 118th Infantry Brigade and was subsequently re-numbered as 98th Brigade. (*University & Public Schools Brigade*) affiliations also encouraged many ex-public school men to enlist locally. All the while, many of the men these battalions hoped to recruit were also looking to secure commissions. The glut of officer applicants meant that the bar could be set high and many apt candidates were turned away. Many recruits were also fickle in their desire to seek commissions. They might enlist whilst awaiting commission applications to be processed or might enlist knowing that they would re-apply for commissions later when the initial 'rush' had passed. Some might enlist for a trial period whilst some were 'die-hard' enlisted men who accepted the perceived privations of enlisted service to get to France faster. Every individual had a different motivation to join the UPS and each one had a different experience.

UPS recruiting begins

Recruiting soon took off once the recruiting offices established themselves and the newspaper notices and posters were published. The publicity created considerable interest, which the Committee and London recruiting staff struggled to meet. There were sometimes several hundred letters arriving every day and sometimes, during the early days, the office barely closed. Some of the recruits assisted with running the enlistment process.

On 1 September, day one of recruiting, 300 men applied but not all were able to be medically examined and attested. Recruits were signed up at 66 Victoria Street, sent for medicals in parties, and returned to be sworn in. Having seen an advert on 1 September Kenneth Norman and a friend, proceeded from Godalming to London to enlist next day. Norman, a dispensing chemist, had initially decided to join the Royal Army Medical Corps (RAMC) but there were no vacancies. He reasoned that if he had to be a private, he should choose the right battalion.[11] They were amongst over 1,200 men who were enrolled, medically examined, and attested on 2 September. Norman found a significant queue. It was three hours before their names were taken and they were marched round in groups to Westminster Hall where Mr Boon explained the aim for the Brigade. Once complete, Norman was able to return home to Godalming to explain his enlistment to his parents.[12] On 4 September 700 men were enlisted. By Friday a steady stream of recruits was arriving. Volunteer doctors were drafted in to conduct the medical examinations and additional premises were needed.

Randolph Arthur Chell was studying science at Reading University at the outbreak of war and was a member of the University OTC. He applied for a commission in the Special Reserve or Territorial Force through the OTC with the answer 'whichever is the quicker'. Having heard nothing for two days Chell travelled to Westminster to join the UPS and was made a recruiting official; 'We were quite careful about who we took; it was supposed to be University or Public School or both … we could be strict…'[13] Chell mentioned the parade at Central Hall Westminster; 'It was a very enthusiastic meeting, I suppose we had got our 4,000 men we wanted then.'[14]

---

11 LA: K.V. Norman Papers, WW1/GS/1188, p.7.
12 Liddle Archives (LA): K.V. Norman Papers WW1/GS/1188, p.8.
13 IWM: R.A. Chell Oral History, 31748, Reel 1.
14 IWM: R.A. Chell Oral History, 31748, Reel 1.

One early recruit was Godfrey Skelton. He had chosen to slightly delay enlisting as he wanted to start a new job for a short period to ensure he would have post-war employment. He left Wolverhampton for London on 2 September having read the UPS advertisement in The Times. He eventually found the UPS recruiting office:

> I was immediately medically examined and filled in a number of papers. About 250 recruits were [at]tested and took the Oath of Allegiance, together, in a room in the basement. Such was the crowd, and hurry, that the Bible was held up at the end of the room and an officer in charge of the Attestation asked if we could all see it, and the shout was "Yes". Only a very privileged few near this officer actually touched or kissed the Book. As soon as we were attested we were bundled out of the room to make space for other eager recruits.[15]

After they had been 'signed up'; 'Orders were issued for us to parade in Hyde Park, near Knightsbridge Barracks, at 11 a.m. the following morning and we were told what it was necessary to bring with us in the way of clothes etc.'[16] Norman went into London a few times over the next two weeks to parade for drill in Hyde Park or to go on route marches (which further aided recruiting).

However, military bureaucracy moved slowly. On the Friday, the Chief Recruiting Officer for Whitehall ordered enlistments to halt because no sanction had been sought to recruit. In fact, it had, but from a much higher power. That approval had been verbal and could not be evidenced. After some lobbying of the 'right people' recruiting commenced again on Wednesday 9 September. In the interim men had been enrolled but had not been medically examined or attested. To manage the sudden influx of recruits the premises of the Institute of Mechanical Engineers was made available to the committee. With this hasty recruitment process many men were determined fit when a more careful examination might have revealed medical weaknesses. Proper checks were not conducted concerning age (at both ends of the spectrum) as men were willing to twist the truth to enlist. These issues likely required considerable administrative effort to iron out later. For example, in late September Godfrey Skelton's platoon had to all be re-attested because something was found to have gone wrong with the original ceremony.[17] Likewise, though Kenneth Norman enlisted on 2 September he had to go back for attestation and a medical on 15 September having already been undergoing training. There were many recruits, and few doctors, leading to long waits to be medically examined and attested. The examination was farcical with little time spent conducting any medical checks; almost every man was passed. The men took their oaths of allegiance in groups. It was midnight before Norman could leave having been accepted as a British soldier.[18]

Men found their way to the UPS from different places and for different reasons. David Kelly was in Spain when war was declared. He landed at Tilbury in August 1914 in time to read The Times account of the retreat from Mons.[19] He found all his friends were joining, or had joined, and though he did not want to be left out of the war he was ignorant in his military knowledge.

---

15   IWM: G. Skelton Papers, 13966, p.25.
16   IWM: G. Skelton Papers, 13966, p.25.
17   IWM: G. Skelton Papers, 13966, p.29.
18   LA: K.V. Norman Papers, WW1/GS/1188, p.10.
19   David Kelly, *The Ruling Few* (London: Hollis and Carter, 1952), p.88.

His problem was solved on seeing the UPS Brigade advert and he enrolled on 15 September. He paraded with a 500-strong group two days later and marched to Hyde Park for training.[20]

In these early days the UPS could be highly selective. Philip Morton Shand, a journalist, was an ideal candidate; being an Old Etonian who had attended King's College, Cambridge, and having served in the Oxfordshire and Buckinghamshire Light Infantry. He enlisted with the 3rd UPS Battalion on 9 September but was discharged on the 24th on medical grounds.[21] Some recruits gave up lucrative or prestigious jobs to enlist. Geoffrey Dunstan (stage name Geoffrey Denys) was acting in 'The Great Adventure' at the Kingsway Theatre and quit to enlist. According to the Brigade History, men returned from Chile, Argentina, the West Indies, and Canada to enlist.[22]

Others were less straightforward when enlisting. George Eyston applied for a commission in the Dorsetshire Regiment. To fill time, he tried to enlist with the Royal Engineers (RE) Searchlight Section, but they were not accepting recruits. Whilst on a 'crammer' course for the Army, Eyston, and two others, enlisted in the UPS in Salisbury on 4 September, knowing that he was awaiting a response to a commission application. Some were turned down by other units. Erroll Shearn applied for the Honourable Artillery Company but their infantry battalion was full; 'This did not seem good enough as I feared that the delay might mean that I should miss the war.'[23] Shearn enlisted in the UPS instead.

Morgan Williams, delayed enlisting until the end of August to ensure that he and some of his university friends could enlist together.[24] Others still had responsibilities or owed allegiance to businesses or employers. Andrew Buxton, a 34-year-old local director of the Westminster branch of Barclay's Bank, wrote on 7 August: 'We live in exciting times in the financial world. A 'moratorium' has been declared … How dearly I should like to enlist at this moment, but it is impossible to leave the Bank at so critical a time. … we have had several amusing incidents in the form of offers from aged … spinsters offering to help if it will enable the [bank] clerks to enlist…'[25] Buxton, though an eligible officer candidate, was lured elsewhere; 'There is a Corps called 'The Artists,' which rather attracts me, as I should not, I think try for a Commission. It would, no doubt, mean three or four months' training, and then choice of volunteering for abroad or not…'[26] He added regarding enlisting; 'This will be all right, provided I get in with nice men…'[27] Buxton was declined by the Artists' Rifles on medical grounds but he joined the UPS on 15 September. Vernon Bartlett, who enlisted with 19/R Fus, was also concerned with joining the 'right' unit; 'Then came the enlistment in the Public Schools Battalion (for we were still very class-conscious) at Epsom…'[28]

Herbert Vinden, another recruit, had wanted to go to Oxford University in the autumn of 1914 and become a barrister. He considered that there was a popular belief that:

---

20 Kelly, *The Ruling Few*, p.88.
21 His granddaughter, Camilla, would become Queen Consort to King Charles III.
22 Anon. Author, *The History of the Royal Fusiliers University & Public Schools Brigade (Formation and Training)*(London: The Times, 1917), p.21.
23 IWM: E.D. Shearn Papers, 2033, p.1
24 Morgan Williams, *From Khaki to Cloth* (Unknown: Western Mail and Echo, 1949), p.44.
25 Edward Woods (ed.), *Andrew Buxton, A Memoir* (London: Robert Scott, 1918), p.38.
26 Woods, *Andrew Buxton*, p.39.
27 Woods, *Andrew Buxton*, p.39.
28 Bartlett, *I Know What I Liked*, p.21.

> Service in the ranks in the Army was not highly regarded in peace time and it must be admitted that in our garrison towns there was a good deal of drunkenness... [in 1914] Groups of people with common professional interests started to raise battalions to counter this attitude and amongst such were battalions formed by members of the Stock Exchange, Footballers ... and the University and Public Schools ... I enlisted in the latter.[29]

Those who enlisted in the provinces, before travelling to London, had a different experience. George Eyston enlisted in Salisbury and was told that he would need to provide his own army shirts and boots. He had to be passed as fit by a doctor but was worried about his eyesight. Luckily, the examiner was called out of the room mid-way through his eye test and he took this opportunity to learn the remaining letters on the board. After passing, his group of recruits went to a local magistrate to be sworn in.[30] Morgan Williams had a harder medical in Cardiff; 'The medical examination proved a pretty thorough one, but I was passed as perfectly fit, without a trace even of dental trouble...'[31] Lewis Namier was an unlikely recruit for the UPS; he was born Ludwik Bernstein Niemirowski and he was of Polish-Jewish extraction. He had already been failed by the RFC on account of his weight, and for army officer selection due to his eyesight. He succeeded in enlisting in the UPS on 5 September after memorising the eyesight testing card before presenting himself at the recruiting office without glasses.

On 17 September the Cardiff Contingent departed for London. Morgan Williams shared a compartment on the train from Cardiff with three others with whom he would share much of his time whilst with the UPS; Stan Tanner, C.H. Watkin and L.B. Greaves. For Arthur Whitten Brown, born in Scotland to American parents, enlisting in the UPS was the first step of a far greater adventure:

> Although, of American parentage and possessing American citizenship, I had not the patience to wait for the entry into the war of the United States. With an English friend I enlisted in the British University and Public Schools battalion, when it was formed in September, 1914. And, although at the time I had no more notion of it than of becoming President of the League of Nations that was my first step towards the transatlantic flight.[32]

Some men had hedged their bets having both sought commissions and enlisted in the UPS as an insurance policy. If they were accepted as officers they could resign from the UPS having received some refresher training. If they were not, they could re-apply later from the UPS. Cleveland Fyfe described his reasons for joining the UPS:

> I enlisted ... Having had previous training in the volunteers and the Territorial Force, it was my purpose to apply for a Commission and I enlisted in order to bring my knowledge of Drill, etc, sufficiently up-to-date to warrant my so doing. With that purpose in view, I devoted a considerable amount of my leisure time to the study of various training manuals.[33]

---

29 IWM: F.H. Vinden Papers, 5565, p.11.
30 George Eyston, *Safety Last* (Spalding: Vincent Publishing, 1975), pp.24-25.
31 Williams, *From Khaki to Cloth*, p.44.
32 Brown, *Flying the Atlantic*, p.8.
33 The National Archives (TNA): WO339/43645: Officer File C. Fyfe.

Hugh Spurrell, an old boy of Llandovery School, was the opposite. He had been offered a commission in the Territorial Force, but, because they were unlikely to deploy overseas, he resigned it having joined the UPS. Likewise, Edmund Page was offered a commission in the Cheshire Regiment based on his having served with the Oxford University OTC. Instead, he and a close friend, Charles Whitley, decided to enlist in the UPS in Liverpool.[34] Page and Whitley would serve together for much of the war. They were mobilised and supposedly joined the UPS at Kennington Park on 2 September, before travelling to Epsom.

Most men had little idea what was happening in France and Belgium beyond what they read in the papers. Nicolas Pease was at a hotel in Neufchatel visiting a friend when war was declared. He rushed back to the UK leaving his brother in Germany, where he was interned. Pease encountered railway wagons of unattended French wounded and conversed with some, 'I came away with a sinking feeling in the pit of my stomach; war was no picnic as far as they were concerned and the Germans were immensely powerful and better armed. I was well aware that it was only a matter of days before I would be taking the King's shilling.'[35] Isaac Hore-Belisha was on holiday in Germany when war broke out; he also rushed back to the UK to enlist in the UPS. Not everyone was an enthusiastic enlistee. Terence Doherty wrote to his father having left school in July 1914; 'I certainly do not want to go into an office, and as far as I can see the only outlet is the Army, and it is the best I therefore decide for the Army....'.[36]

Roderic Hill had not wanted to enlist and had been advised against it by his doctor. On returning to University College London, he was walking along Gower Street observing the University Corps undergoing drill training. He observed the physically weakest member of the university in the ranks. After the parade Hill was approached by the boy who told him that he had enlisted and recommended that Hill did the same. Initially surprised, Hill quickly responded saying; 'I think I will' and went to find the recruiting officer.[37] Namier enlisted in 3rd Public Schools Battalion and was part of C Company. He was an enthusiastic recruit who was popular in his company, but he stood out due to his heavy accent which made him sound like a spy; 'When asked by a friendly N.C.O. what he would do if questioned in a queer spot about his accent, L answered to the enjoyment of everyone including the N.C.O., a Mancunian, "I'll say I cum from *Mahnchester*".[38]

There were also, curiously, several recruits with German names, more than might normally be encountered in a cross section of the populace.[39] It is likely that educated men of German origin felt more comfortable enlisting in the UPS than elsewhere. They might have found commissions harder to achieve despite being otherwise eligible through class and education. Many were second generation, and some anglicised their names to fit in.

* * *

34 Page's diary, written up later, states he enlisted on 20 August. IWM: E. Page Papers, 16924, p.1.
35 IWM: N.A. Pease Papers, 8230, pp.29–30.
36 IWM: Documents.12961, Catalogue date 2003-12-22: Papers of T.O'C. Doherty, p.2.
37 Prudence Hill, *To Know the Sky* (London: William Kimber, 1962), p.20.
38 Julia Namier, *Lewis Namier*, (London: Oxford University Press, 1971), p.117.
39 The nominal roll included the following surnames; Schiller, Frisch, Schenk, Schumann, Haltinstein, Bruhl, Stein, Schur, Schunch, Schwemmer, Oberhoffer, Schwabe, Schurig, Steinthal, Muhlberg, Ladenburg and Ziegler.

During this period Vincent Edwards (formerly of Scarborough College) had endeavoured to get a medical examination and enlist. He finally seemed to achieve this on Saturday 12 September and he spent the next day on a route march to Watford. On Saturday 12 September it was found, after frenzied recruiting, that the number of UPS men exceeded the 5,000 target and recruiting was paused. Another 250 names were quickly taken and added to a waiting list. The administrative nightmare of recruiting 5,000 men in eleven days cannot easily be imagined in the modern age, let alone in the days of pen and ink. Attestation papers had to be filled out and duplicated by hand for all those enlisted. Further premises were required, and the Empire Room at the Trocadero was used to conduct administrative activity. Much was still to be done and some administration caught up later. For example, service numbers (beginning PS/) were not issued until early 1915.

First Parades

Whilst this hurried recruiting was occurring there was a need to commence training to ensure these men were engaged and kept busy. They were paraded for training in Hyde Park each day, in increasing numbers. As there were no rifles or uniforms the only real syllabus was drill or route marching. These parades were taken by Mr James Westwood-Henderson and Mr Frank Warner-Abbatt who were both members of the committee who had previous military experience. The former was an insurance inspector who had formerly served with the London Regiment. The latter was a farmer from Ealing. Both later became majors in the Brigade. Henderson was initially proposed to command a UPS battalion. On 8 September 1914 The Morning Post recorded one of their early parades:

> Over 2,000 mustered at this first parade. They comprised, generally speaking, the London Contingent. Members enrolled in the country had not yet been called up … And an astonishing fine, well-set-up force they are too, if yesterday's muster is typical. Mostly above the average height, and all supple, athletic fellows of manifest strong lung capacity, they made a capital impression…

The London contingent marching together through The City could fill up the length of Whitehall. The physique and bearing of the men, and their proficiency at foot drill, was remarked upon by military observers. However, such attributes were expected when many were both sportsmen and ex-OTC members. Those who were not, would learn quickly. On 14 September, some men, Edwards included, paraded at Knightsbridge and conducted a route march to Barnes and back.

On 10 September Field Marshal Earl Roberts visited the HQ in Victoria and spent twenty minutes talking to members of the committee. He stated he would go straight to the War Office to see about selecting commanding officers for the battalions. Only after this inspection was Edwards able to be formally examined and enlisted.

In the absence or nominated officers and NCOs with recent military experience the value of some of this training might be questioned. It did, however, start to make men realise the enormity of the undertaking they were embarking on and started to build an *esprit de corps*. When Herbert Vinden enlisted in the UPS; 'The recruiting office was a tent in Hyde Park.

On 15 September, the fledgling UPS brigade, 2,000 men strong, was inspected by Major General Sir Francis Lloyd in Hyde Park. He declared them; 'The finest body of men I have ever seen' and 'More magnificent material I have never seen in my life.'[40] The image gives an idea of the different civilian attire that the recruits paraded in which was later described as; '…all kinds of garb of varying picturesqueness …'

With a schoolfellow I joined a queue of 50 or more outside the tent and all around [the tent] were squads of volunteers dressed in blue suits and bowler hats drilling using their umbrellas as rifles.[41] The initial force did not look very martial in appearance; 'There were no uniforms available to us and we had a cardboard badge 'U.P.S.' (Universities & Public Schools). Of this I was extremely proud. Also there were no rifles for a considerable time and no equipment. I remember parading in Hyde Park and doing a march to Barnes Common without any martial trappings …'[42] To distinguish those who had enlisted the cardboard badge was attached to a button hole with a red cord.

Finding officers for the Brigade externally would be difficult; there were few spare Regulars or Territorials available. However, within the men that the UPS had recruited were numerous men of the required quality and with the right motivation. Volunteers were asked for and names were passed to the War Office for ratification and gazetting. Most of the early officers had prior military service or had served in their school or university OTCs. David Kelly recalled that the Brigade 'paraded *ourselves*' in Hyde Park. Once assembled, one of the organisers appealed for volunteers to be officers and NCOs. These volunteers immediately took on the authority of those roles and were given coloured rosettes or arm bands to identify them. Likewise, these

---

40   IWM: Documents 1708, Catalogue date 1985-11-05: Papers of W.B. Medlicott, p.1.
41   IWM: F.H. Vinden Papers, 5565, p.11.
42   IWM: E.D. Shearn Papers, 2033.

appointments were accepted by those on parade.[43] Arthur Whitten Brown remembered the inadequate training of the early officers; 'In knowledge of drill our officers started level with us. Several times I saw a private step from the ranks, produce from his pocket the Infantry Training Manual, and show a lieutenant where he had gone wrong. Doubtful discipline, perhaps, but excellent practice, for most of the original privates of the U.P.S. soon became officers of the New Army.'[44]

Such military peculiarities were likely accepted because everyone was working towards a common goal and many of the 'other ranks' present were of a similar calibre to the officers. On joining, Eyston observed old boys from Oxford and Cambridge amongst his peers as well as celebrated sportsmen and boys from famous public schools. When sorted out, his company had enough renowned rugby players to form a team.[45]

Mobilisation

The committee had the monumental task of getting the Brigade mobilised and established in a permanent location where they could conduct training. Permission to establish a camp at Woodcote Park, near Epsom, was granted. Because it would take too long to erect huts the committee decided to place the battalions in billets in Epsom, Ewell, Ashtead and Leatherhead. Getting the Brigade moved to these towns and accommodated would still be a major task with the only assistance being from the War Office who agreed to provide special trains to carry the brigade.

Mobilisation was in two tranches. The London contingent, along with 300 men from Manchester, and other contingents from near to London, would form the first mobilisation. This comprised approximately two and a half battalions and would mobilise at midday on Friday 18 September in Hyde Park, opposite Knightsbridge Barracks. A second mobilisation contingent would muster in Hyde Park on 24 September and consist of men enlisted from further afield.[46]

First Mobilisation

Each man had been sent instructions for mobilisation which included a label with a specific letter on it. The men, on arrival, deposited their kit bags in the area marked with that letter; these were then loaded on a corresponding bus. The numbered labels corresponded to groups who would form specific companies. Once formed into 'companies' the men met their new officers. On arrival for mobilisation at midday in Hyde Park, Kenneth Norman recalled the area packed with recruits and their parents. Norman considered that the; 'crowd looked like one at a Varsity cricket match'.[47] Having already been divided into companies their bags were stacked in

---

43   Kelly, *The Ruling Few*, pp.88–89.
44   Brown, *Flying the Atlantic*, p.8.
45   Eyston, *Safety Last*, p.25
46   A copy of the mobilisation orders issued to each recruit, and the kit list provided, are in Appendix 3.
47   LA: K.V. Norman Papers, WW1/GS/1188, p.8.

'The crowd looked like one at a Varsity cricket match.' The logistical problems of transporting the first mobilisation of the UPS Brigade from Central London to Epsom. The men marched to Victoria station, their baggage was transported by bus.

company 'piles' and were later loaded on London General Omnibus Company buses for transporting to Epsom. Godfrey Skelton recorded the first mobilisation:

> On the 18th September about 4,000 [sic] men paraded in Hyde Park. There was evidently some organisation, as markers were posted, and men in authority seemed to spring from nowhere. These were officers evidently and wore pieces of red ribbon in their buttonholes. A large convoy of motor omnibuses now appeared on the scene and we loaded our baggage on to these buses.[48]

According to Skelton; 'Our Company Commander was a tall man dressed in loud tweeds and looked like a cross between a bookmaker and a golf professional. His name turned out to be Boyce and he eventually commanded "C" Company, 20th Battalion Royal Fusiliers, which was now the Unit we had joined.'[49]

The 'London' contingent broadly formed the first two UPS Force battalions which would become the 18th and 19th Battalions, Royal Fusiliers. The initial Manchester contingent would form half of the third battalion (20th Battalion). These units would be billeted in Epsom and Ewell. According to Eyston, once the men were ready to leave Hyde Park, they moved to Victoria Station and were seen onto the trains by a large crowd of well-wishers who lined the platforms.[50]

48 IWM: G. Skelton Papers, 13966, p.25.
49 IWM: G. Skelton Papers, 13966, p.26.
50 Eyston, *Safety Last*, pp.25–26.

The Manchester Contingent

Whilst the extensive UPS recruiting process and initial training had occurred in central London the same had occurred in far flung cities and towns in response to the call from the UPS committee. The response in Manchester was extremely strong and almost a whole battalion was raised from men of the city. The Lord Mayor, on receiving the letter from the committee, on 1 September, requested that the headmaster of Manchester Grammar School, Mr L.J. Paton, should investigate the matter. Another story suggests that a Manchester public school battalion was the brainchild of a young lecturer, R.A. Wardle, and Captain J.F.H. Templar, one of his students, who took the idea to Mr Paton.[51] After initial enquiries amongst grammar school boys in the Manchester University OTC, chief amongst them was Captain Templar, a meeting was arranged at the University. Captain Lapage represented the University OTC and Lieutenant E. Moore Mumford represented the School OTC and the Headmaster represented the Grammar School; 150 men attended. At this meeting the UPS proposal was endorsed and recruiting commenced. Other trade groups in Manchester comprising university and public school educated men (accountants, architects and law students) were also meeting to discuss their response to the war. They too adopted the UPS proposal which added further weight to the Manchester numbers. Mirroring events in London the Manchester contingent recruited 1,023 men in eight days. These enthusiastic recruits started to parade at Chetham's Ground near the School and later paraded on Plates Fields. As has already been mentioned, a first contingent of 300 departed at 1:00 a.m. on 17 September ready for the mobilisation on the 18th. On 24 September the remainder were addressed on the Piccadilly flags by the Lord Major. They then marched to Mayfield Station and embarked. The remainder of the Manchester contingent, which would form the rest of the third UPS battalion, travelled direct from Manchester to Leatherhead, where they arrived that evening.

Other recruiting offices around the country had varying levels of success:

Location	Recruits	Notes
Cardiff	250	Half the contingent came from the University College of South Wales and Monmouthshire.
Derby	146	146 were initially recruited rising later to 300.
Exeter	157	Represented all the schools of the West Country; Allhallows (Honiton), Blundell's, Clifton College, Exeter, Kelly, Newton and Sherborne.[52]
Durham	150	Included old boys from northern schools.
Worcester	120	Recruits were from Worcester Cathedral King's School and the Worcester Royal Grammar School
Birmingham	120	
Harrogate	110	Some recruits had travelled from a radius of 40–60 miles
Chelmsford	100	Men from across Suffolk and Essex were enrolled.
Brighton	80–100	

51 IWM: E. Stoneley Papers, 7716, Unknown Newspaper Cutting – 'The Final Agony'.
52 This contingent included a future Victoria Cross winner.

Location	Recruits	Notes
Cheltenham	87	Recruits were from Cheltenham College, St Paul's Training College and Cheltenham Grammar School
Bedford	85	
Plymouth	80	Recruits were from the Plymouth and Mannamead College; Kelly College, Tavistock; Provost Truro Mining College Institution, Cambourne, and The Hoe Grammar School, Plymouth.
Eastbourne	53	
Belfast	51	These men formed a platoon in 20/R Fus
Bristol	30–40	Included recruits from Clifton College
Guildford	40	Some men joined the Sportsmen's Battalion on arrival in London
King's Lynn	30	
Salisbury	20–30	
Portsmouth	20–30	
Worthing	22	
Buxton	15	
Horsham	10–20	
Carmarthen	7	
Aberystwyth	3–6	Only a few men were recruited

For example, the Derby contingent comprised many old boys from Derby School. Two of them were Lewis Jacques and Edwin Wilmshurst who had both been non-commissioned officers (NCOs) in the school OTC. They enjoyed a dinner hosted by local dignitaries before the first batch were sent southwards to join the UPS on mobilisation. However, the recruiting system was inefficient with different regions being represented by multiple recruiting offices. For example, recruits from Plymouth were also enlisted in Exeter. Likewise, men from towns and cities near London had surged into the city to enlist. Some local recruiting centres for the UPS also recruited concurrently for other battalions who were after similar quality candidates. Other organisations and regiments were trying to create similar units; notably the Public School Battalion (16th Battalion Middlesex Regiment (16/Mx); The Welsh Army Corps Public School Battalion (13th Battalion Royal Welsh Fusiliers) and the Royal Naval Division Public School Battalion. There was a Public School Special Corps which was another different organisation training Public School men for commissions.

The mobilisation on 24 September centralised the contingents from around the country and comprised the bulk of the fourth UPS battalion (later 21st Battalion Royal Fusiliers). The same administrative system was in place when Hugh Spurrell reported at Hyde Park having received mobilisation orders:

> I left for London, with a full kit-bag and various odds and ends besides. I spent a night at the Grange and was presented with a new greatcoat by my Grandmother. Next day I went to Hyde Park after lunch and fell in with my kit in a pen marked off by the mysterious letter "P".[53] Here I was first introduced to the Army's administrative system.[54]

---

53 It is probable that 'A' corresponded to A Company, 1st Battalion, and 'P' represented D Company, 4th Battalion.
54 IWM: H.W. Spurrell 2138, p.4.

According to Spurrell:

> Men of all shapes and sizes were there, in every conceivable sort of dress. There were men in bowler and spats; men in check trousers and caps; some had spectacles, some without. A few real militarists were in putties already. The gentleman who was detailed to march us to the station presented a most singular appearance. He wore a military cap and sword (though not an officer); the rest of his clothes were those of a civilian. Of course he was at once seized by the authorities and made a Field Marshal on the spot.[55]

Spurrell's 'company' was addressed by:

> … a gentleman with a face like a communist flag. He wore a ribbon to match it; this denoted that he was an officer. He turned out to be one of the company promoters who had 'raised' the Corps. He was promptly gazetted Major and, as he had served overseas in the last war in South Africa, he was excused from foreign service during this one.[56]

This was possibly Major H.F. Fenn, though he did serve oversea. Not every new officer was welcomed. Lewis Jacques of the Derby School contingent (4th Battalion) wrote:

> One of the junior officers appointed when the battalion was formed was a chap from school of about my age, who was to have gone to Oxford, but who now put on airs and graces and refused to acknowledge his former schoolmates and became excessively regimental, until we threatened to put him in the village duck pond.[57]

The different contingents went through the same process at Hyde Park before they embarked for Ashtead. This saw the initial phase of the recruiting of the UPS finish. Epsom would commence a new phase which would see training begin in earnest and the character of the different battalions start to develop.

It might be of interest to examine why these ex-public school men did not want to serve in other units. This might, at first glance, appear to be snobbery or elitism, and there may be elements of truth in such suggestions. However, disparities in social standing in Kitchener's New Army might easily be a source of friction. James Hodson, a 23-year-old journalist with the Manchester Daily Mail, decided to enlist. He was told that there were no vacancies in either the Lancers or the Rifle Brigade (RB). He tossed a coin between the King's Royal Rifle Corps (KRRC) and 'the Lanchesters' and he ended up joining the latter.[58] On his first night in barracks at Seaforth he 'lay there feeling like he had dropped from the sky into a filthy slum from which there was no escape.' He was posted to a battalion in Aldershot after a few days. His comrades were from a broad spectrum of jobs and classes:

---

55 IWM: H.W. Spurrell 2138, p.4.
56 IWM: H.W. Spurrell 2138, p.4.
57 This was likely Edwin Wilmshurst, formerly of Derby School, who had planned to attend Keble College, Oxford. Personal account by Lewis Jacques <https://online.flipbuilder.com/zyzh/lsvb/index.html#p=9> (accessed 16 February 2023) Reproduced with the permission of Suzannah Schmitt.
58 Hodson's service record states that the 'Lanchesters' were the 12th Battalion King's Liverpool Regiment.

One man was a shipping clerk, one a doctor's son, one a reporter, a few were decent warehouse youths, and one or two were "old sweats" – that is, soldiers who had served their time long since and had come back to do barracks work. But for the most part they were the labouring and dock-labouring class, trying to live with when sober, and insupportable when drunk.[59]

Being educated, Hodson was given a lance corporals' stripe and had to manage these men in the barrack room. However, as a middle class man he felt very much out of place with men of a different social class; 'Their proximity offended him; probably his annoyed them.'[60] After six months with the 'Lanchesters', Hodson decided to try and transfer to the UPS and used a past contact, who was a company commander with 20th Battalion Royal Fusiliers, to engineer this.

Gordon Jacob also desired to enlist; looking back, he remembered; 'It is easy to dismiss as naïve or even ignoble the love of country, but rightly or wrongly it bore us all along on great waves of excitement and fine feeling.'[61] He had been studying for a BSc in London but whilst on his summer holidays he had found he had failed most of his exams. He and a friend resorted to enlisting; Jacob joined the Royal Field Artillery (RFA) in August 1914 as a private because they wanted 'educated men'; his friend was gazetted and was dead within a year. Jacob initially spent a night in Woolwich Barracks and caught lice whilst there. Next day he moved to Shorncliffe Camp where they started to train despite having no guns or uniforms. He remembered that they 'were a pretty sorry lot'. The men he was with were rough, ill-mannered and some were criminals; he witnessed numerous fights, one of which led to a man being murdered using a tent mallet.[62]

His experiences highlight the potential frictions experienced by university and public school men when joining locally-raised units comprising of men of a different social class. However, his experiences also showed that in such conditions men of all classes could get along; 'But in spite of the roughness of behaviour and language the majority of the chaps were good and kind at heart and there was an excellent feeling of comradeship.'[63] 'Strings' were pulled somewhere and Jacob was able to transfer to the UPS after a couple of months; he joined C Company, of the 1st PS Battalion where he found, 'it was a relief to have more congenial companions who shared the same sort of background and spoke the same sort of language (good and bad!)' Others were similarly delayed. William Bentley, a teacher trainee at St Paul's College, Cheltenham, enlisted whilst there. It was only in November 1914 that he was informed his party had been transferred to the 2nd PS Battalion (later 19/R Fus).

---

59 James Hodson, *Grey Dawn – Red Night* (London: Gollancz, 1929), p.101.
60 Hodson, *Grey Dawn*, p.112.
61 LA: G.P.S. Jacob Papers, WW1/GS/0841, p.98b.
62 LA: G.P.S. Jacob Papers, WW1/GS/0841, pp.101–102.
63 LA: G.P.S. Jacob Papers, WW1/GS/0841, pp.101–102.

2

Early Training of the UPS Brigade

Those were wonderful days for all concerned

For half a year they have drilled us, till we murmur 'How much more?'
And from Woldingham Hill to Woodcote we have trenched till our backs are sore.
We have marched from Ew'll to Dorking amid moors an' farms an' towns,
And we've blobbed on the Derby Race-course till we're sick of those d – d old Downs.[1]

Those were wonderful days for all concerned in the early training of our battalion at Epsom.[2]

The Invasion of Epsom

Before the Brigade could start training properly it needed to be established in Epsom, Leatherhead and Ashtead. The logistics of getting the men and their kit to their destination had been achieved. There was still the problem of getting the men accommodated in small groups, in houses, lodging with members of the population. Godfrey Skelton recollected, 'We arrived at the Race Meeting Station, Epsom, in the early afternoon and were marched down to the town where we found our kit bags in a heap in the roadway. Now came a long wait, after each of us had found his kit bag on which we solemnly sat for about two hours.'[3]

Norman recalled marching about a mile from the station to the market square where they found their bags and settled down to wait. The Police had worked hard to find billets but the sheer numbers of men made this slow and haphazard. That evening George Eyston and his new comrades waited patiently in the square at Epsom. Small groups of men were selected and led away to a prospective billet. The quality of that billet was down to luck. Some men went to live on affluent streets whilst others were sent to poorer parts of the town.[4] Kenneth Norman's B Company was accommodated around Lower Court Road and Hook Road where

---

1   IWM: Documents 10271, Catalogue date 1976-03: The UPS Song Book, by A.W. Lloyd, p.3.
2   Whitten Brown, *Flying the Atlantic*, p.8.
3   IWM: G. Skelton Papers, 13966, p.26.
4   Eyston, *Safety Last*, p.26.

6.0. A.M. "NO LUCK! GET UP; IT'S A FINE MORNING!"

A sketch from *The Pow-Wow* showing the comfort of some billets inhabited by UPS men in Epsom and Leatherhead. Presumably some morning parades were called-off due to inclement weather. (Sketch by Ernest Stoneley; reproduced courtesy of Janet Wood)

poorer families lived. Norman was billeted by 10:00 p.m. in a small dwelling that already housed two families. The next day, they paraded again but were assigned a different house.[5] This change led to discontentment. However, in the re-organisation B Company, 1st Battalion, were accommodated in Church Street and adjacent roads; Norman was assigned to the Vicarage and B Company drilled in Roseberry Park. Edmund Page, No 4 Company, 4th Battalion, recorded that he 'travelled to Epsom – Marched to Ashtead and billeted in a wee maisonette – Mrs Johnston, a very unpleasant woman, who did nothing to make us comfortable. Slept in an attic of absurd dimensions with three of us in the bed – I slept on the floor.'[6] Kelly remembered his first prospective landlady who believed them to be 'ordinary' soldiers and refused them entry. Tired and hungry, Kelly and his fellow recruits pushed past her and were reluctantly fed and

5   LA: K.V. Norman Papers, WW1/GS/1188, p.16, p.18.
6   IWM: E. Page Papers, 16924, p.1.

housed. After a short time she softened to them having realised they were volunteers and not the "brutal and licentious soldiery" that many British civilians believed the Army was composed of. Only then were they welcomed and properly fed.[7]

Skelton and his comrades were led to a billet:

> … we marched off up a road leading to Epsom Downs and the numbers gradually dwindled away as different parties of five, six or seven men were billeted, in the houses bordering the road. Five others, and myself were the last to be billeted and we were sent to a large house, standing in its own grounds, this house belonging to a Mr Tyron. He received us with great cordiality, and we sat down that night to a four-course dinner with wine and port. We all had separate bedrooms and were waited on hand and foot by maids, including early morning tea and hot water for shaving. This, for private soldiers in Kitchener's Army, was not so bad![8]

In his first billets in Epsom Andrew Buxton was quartered with two others in a small room with only two beds; they drew lots, but Buxton ended up having to sleep on the floor.[9] After being left homeless in the street until after midnight Eyston and two comrades were taken by a policeman and billeted in a bare room in a large house on the edge of town. Pease remembered; 'I was billeted in a little villa house in Hook Road, Epsom, with a friendly working-class family.'[10] Erroll Shearn was another of the lucky ones; 'I was allotted a billet with Skilton the vet to the Epsom Race Course. He and his family were extremely hospitable.'[11] Likewise, the food in Donald Addams-Williams' billet was enough to enable him to put on weight during the months he was at Epsom.[12]

Ultimately the Brigade was accommodated as follows; the 18th, 19th and half of the 20th Battalion were billeted in Epsom. About a week later the 21st Battalion was to be billeted at Ashtead and the second half of the 20th Battalion was to be based at Leatherhead. The Brigade HQ office was situated in Waterloo House in Epsom. The derelict Tun Beer public house that was opposite was used as the Brigade Post Office. To make the second tranche of mobilisation go smoothly men from 1st Battalion were sent to meet prospective landladies and examine billets prior to the next units arriving.[13]

The same concerns were demonstrated a week later when the 4th Battalion (21/R Fus) paraded on the common by the station at Ashtead. Here special constables took them away and assigned them billets. Spurrell was billeted at 'The Meade'; 'The inhabitants had been told that soldiers would be billeted on them, but they expected coal-heavers or jailbirds. I think they were glad to find that they were excused such a luxury … the foundations of a lasting friendship were soon laid…'[14] Morgan Williams was billeted in Epsom with 'a bachelor and his spinster sister, both of whom were very kindly, but a little bewildered by this sudden invasion.'[15] He also recorded; 'Our rooms were large and comfortable, suffering from only one serious drawback. The two

7   Kelly, *The Ruling Few*, p.89.
8   IWM: G. Skelton Papers, 13966, p.26.
9   Woods, *Andrew Buxton*, p.40.
10 IWM: N.A. Pease Papers, 8230, p.30.
11 IWM: E.D. Shearn Papers, 2033.
12 LA: WW1/GALL/001: Papers of D.A. Addams-Williams, undated letter.
13 LA: K.V. Norman Papers, WW1/GS/1188, p.22.
14 IWM: H.W. Spurrell 2138, p.4.
15 Williams, *From Khaki to Cloth*, p.46.

beds allotted to us varied so much in size that we changed round every other night to ensure an equal measure of sleep.'[16] Some UPS men did not even start training. The day after they arrived in Epsom, Randolph Chell was ordered to report to Belfast; his commission had come through.[17]

Once everyone was accommodated the further organisation and training of the Brigade could commence. There was another false start according to Spurrell:

> On the day after our [21st Battalion] arrival we had a morning parade. It was a most interesting affair; we stood up from nine o'clock until ten and then sat down until eleven. Meanwhile such members of the Stock Exchange as were present held a conference. At the end of two hours they had made the startling discovery that there were four companies in a battalion. They then hit upon the brilliant idea of dividing the men into four equal parties. This took an hour and we were then sent off for dinner. We came back at 2-30 and again sat down for an hour. The early evening papers then arrived and somebody saw in one of these that there were four platoons in a company. More staff work followed and by five o'clock I found myself in the second platoon of the fourth company. Feeling that our organisation was at last complete, we adjourned for tea. After this we had to do another hour's "work". As there were four men in the company who had served in South Africa, they were made officers for the time being.[18]

Private Hugh William Spurrell (PS/3261) had attended Llandovery School and was 'Gazetted' on 8 September 1914 to be an officer in the RFA. Instead, he enlisted on 2 October 1914 with the UPS and joined Number 14 Platoon, No 4 Company, 21/R Fus. (Reproduced with permission of Sutton Archives)

Spurrell recorded that one of these temporary 'officers' was later promoted to company sanitary man. Training began the next morning with drill under some ex-OTC boys with 'more knowledge of their drill books than of tact' that afternoon they went on a four-mile route march. Edmund Page, by coincidence, was also in No 14 Platoon, No 5 Section, alongside Spurrell.[19]

Page considered that; 'Much enthusiasm was shown, but the ignorance was painful – Officers and NCOs were literally appointed by acclamation with the result that you got the least satisfactory elements to volunteer.'[20]

16 Williams, *From Khaki to Cloth*, p.46–47.
17 IWM: R.A. Chell Oral History, 31748, Reel 1.
18 IWM: H.W. Spurrell 2138, p.5.
19 Neither man mentioned the other in their accounts.
20 IWM: Documents 16924: Papers of E Page, p.1.

The daily routine for the 1st Battalion is summarised in the table below based on Kenneth Norman's memories:

Time	Activity
6:30 – 7:15 a.m.	Physical drill or running
8:00 a.m.	Breakfast back in billets
9:00 – 12:00 a.m.	Squad drill or musketry.
1:00 p.m.	Lunch back in billets
2:00 – 4:00 p.m.	Squad drill or route march
5:00 p.m.	Tea back in billets
5:30 – 6:00 p.m.	Occasionally a lecture
10:00 p.m.	Lights out in billets

George Eyston recalled a different regime in the 2nd Battalion where the day lasted from physical drill at 5:00 a.m. until 6:00 p.m. Though uniforms were non-existent, and only a handful of rifles were provided, it was presumed that the battalion would be ready for the front in six weeks. Early on, the battalion conducted drill on the greens of the RAC golf course at Woodcote Park until they were banned from them.[21]

Parades took place in streets around the town. D Company, 20/R Fus, paraded near the Spread Eagle public house.[22] According to Shearn; 'One of my recollections is of early morning parades in shorts and singlets in bitter winter weather for runs up to the Epsom Downs and back. Doubtless very invigorating but anything but enjoyable!'[23] Albert Knighton found the training physically demanding, but; 'Being so young I could adapt myself to anything'.[24] C Company 18/R Fus paraded and drilled on the Epsom Recreation Ground and went on route marches through the Surrey countryside.

However, though the men had been grouped by companies when they arrived, they needed to be moved around to ensure that each battalion was billeted correctly. As such almost every party moved to a new set of billets. Some were winners; some were losers. Skelton's 'cushy' billets 'only lasted two days, however; eventually someone got to hear of the excellent time we were having and thought it was not sufficiently like war – not for private soldiers anyway – perhaps for officers!'[25] He was moved to the Epsom Town Hall with sixty others. Accommodation was more Spartan, but they were given guest membership of the Epsom Conservative Club which was in the same building. Occupancy rose to eighty with only one basin and one toilet for all to share and catering, provided by the Conservative Club, left something to the imagination. When all the billets were re-arranged Eyston and his colleagues ended up in a coal heaver's house where the three of them were cramped into a single bed.

21 Eyston, *Safety Last*, p.27
22 IWM: G. Skelton Papers, 13966, p.28.
23 IWM: E.D. Shearn Papers, 2033.
24 PS/6779 Private Albert Eric Leslie Knighton enlisted underage in 18/R Fus in April 1915. He was educated at Clifton Hill College, Margate and was from a family of bootmakers. He was posted to 17/R Fus on disbandment of 18/R Fus and was sent home underage shortly after. He later transferred to the HLI. IWM: A.E.L. Knighton Sound Recording, 10263, Reel 1.
25 IWM: G. Skelton Papers, 13966, p.26.

Hugh Spurrell's platoon commander in No 4 Company was Guy Charles Neave Mackarness, aged 21, an Oxford man who was the son of the Archdeacon of Scarborough. Spurrell considered Mackarness 'looked after us well. We used to regard him as the best officer in the company. We also had a liking for the Second in Command, Captain Betts. For the others we had at first no use.'[26] Meanwhile, Edmund Page considered Mackarness 'an empty-headed boy from Charterhouse.'[27] (Reproduced with permission of Sutton Archives)

Often training took place at company level during the first few weeks. Spurrell's early company commander, Pearce Gould, was a former Territorial Force cyclist; 'He wore dark blue breeches to match his khaki puttees and had a thick stick and a thicker skin. On his arrival he announced his intention of carrying out a revolution in the system of training … There was to be no more fooling – the Empire was menaced, and each man had got to make himself efficient…'. Spurrell's company was soon engaged in physical activities, between which Gould read to them passages regarding infantry and musketry training, 'All this, of course, did us a lot of good, but we did not think so at the time.'[28] Page considered Gould was 'an evil-tempered impossible man'.[29]

On 14 October 1914 the Brigade was inspected by King George V at Leatherhead.[30] Terence Doherty (19/R Fus) recounted an inspection by a VIP:

> Today has been awfully exciting. We paraded at 7 a.m. as per usual and were dismissed early and told to shave and look spick and span, because we were going to be inspected by some big nut. We weren't told who. At 9 we marched to the downs and lined up. Two battalions. Lots of people expected it to be Kitchener, some thought Bobs, some Sir E. Wood. No one saw who it was till [sic] he was close by, and it turned out to be the King. We were surprised. He looks so ill, poor man, I suppose it is worry … He has got a fine manly voice for such a short person.[31]

---

26 Guy Charles Neave Mackarness, aged 21, had attended Charterhouse School and was from. Scarborough, Yorkshire. He enlisted on 15 September and presumably commanded No 14 Platoon, No 4 Coy, before being 'gazetted' to become a lieutenant on 27 October 1914. IWM: H.W. Spurrell 2138, p.7.
27 IWM: E. Page Papers, 16924, p.1.
28 IWM: H.W. Spurrell 2138, p.6.
29 IWM: E. Page Papers, 16924, p.1.
30 IWM: G. Skelton Papers, 13966, p.29.
31 IWM: T.O'C. Doherty Papers, 12961, p.3.

Kenneth Norman recorded that the King considered that the men in the 1st Battalion were of the finest physique to be found in Kitchener's New Army.[32]

After a short time at Epsom D Company 20/R Fus had to be re-attested and medically examined again; to pass the medical Skelton had to cheat the eye test in the same manner that Eyston had.[33] According to Skelton the men of the battalion were inoculated during October.[34] When 20/R Fus was vaccinated the men were warned not to imbibe alcohol due to the likely after-effects. Afterwards, Lewis Namier returned to his billets to find his roommate, 'Stottie', had been sick in the wash basin having had a couple of drinks. Stottie was desperate to find a calling card with which to put with the basin, so Namier was not blamed by their landlady.[35]

The New UPS Committee

In agreeing to the formation of the UPS Brigade, the War Office had stated that the committee would need to make the arrangements for it to be clothed, fed, equipped and accommodated during the early stages. Supporting this force of 5,000 men was a difficult financial and commercial feat. The original 'recruiting' committee was replaced by a more business-savvy and influential committee to secure the resources required. The committee secured the assistance of The Honourable Arthur Stanley CB MVO MP to take the role of chairman of the new team in addition to another duty as chair of the British Red Cross Society. Mr Hector Boon was asked to join the new committee and Mr Howell became the secretary. Mr Gordon Selfridge and Mr Samuel Gluckstein, both of whom were involved in catering and hospitality, also joined.[36] Lord Lurgan KCVO accepted the role of vice-chairman. These prominent men and captains of industry were needed to work in the background to keep the money and provisions flowing whilst the Brigade was able to start to train and continue to recruit. Lord Lurgan was able to keep good relationships going between the committee and the Brigade staff and battalion commanders. Mr Selfridge and Mr Gluckstein later left the committee; this may have been because of a conflict of interest as their firms were presumably supplying many of the provisions to the Brigade. They were replaced by Mr Julian Orde.

The War Office was clogged with paperwork and enquiries from numerous units and getting answers and information proved difficult. The Committee had to rely on its own resources, and resourcefulness, to prevail. Whilst the men trained in civilian attire, the committee had to both find, secure and pay for uniforms at a time when every unit, and the War Office, was trying to do the same. This equipment and uniform also needed to satisfy the exacting standards of the Ordnance and Clothing Departments. This required deep pockets from those backing the Brigade as the Committee also had to accommodate 5,000 men and build a camp.

---

32 Norman recalled this visit being in late November 1914. LA: K.V. Norman Papers, WW1/GS/1188, p.32.
33 IWM: G. Skelton Papers, 13966, p.29.
34 IWM: G. Skelton Papers, 13966, p.29.
35 Namier, *Lewis Namier*, p.118.
36 Samuel Montague Gluckstein was the director of Lyons and Co who ran the famous tea shops. Mr Selfridge ran the famous department store which carried his name.

Despite the efforts of the Committee there were playful jibes from the ranks of the Brigade.[37] The Committee also provided some people with a substitute for military service; 'Mysterious people had had a hand in the formation of our brigade. One or two, who seemed anxious to advertise their connection with it, while still themselves remaining out of khaki.'[38] One of these men was Hector Boon or 'Hector the Spectre':

> I'm Hector, the Spectre, the pride of the UPS!
> I drive my two-seater through all you young pups;
> I have an eye-glass, through which I can see
> What a BOON and a blessing to you I must be![39]

Whilst at Epsom the brigade commander and the commanding officers for the four UPS battalions became known to their men. The commanding officers were hand-picked and the selector had almost free-reign to choose candidates.[40] The UPS Brigade was initially commanded by Colonel Robert Gordon Gilmour. He was born in 1857 as Robert Gordon Wolrige and was educated at Eton and Christ Church College, Oxford. He was first gazetted in 1878 and served in the Zulu War in 1879; the Nile campaign and in South Africa. During the latter campaign he was Mentioned in Despatches (MiD) twice, made CB and appointed to the Distinguished Service Order (DSO). He later commanded the Grenadier Guards (1908–1910).[41] By 1914 he was aged 57 and was, though highly experienced, quite old for a field command. He was seemingly popular with the men of the Brigade:

Colonel, later Brigadier General Robert Gordon Gilmour, a popular and experienced officer, commanded the UPS Brigade during its training. He was not destined to command it in action.

> ... the UPS early realized that they possessed in him a good soldier and moreover a friend. He quickly gained the highest esteem of all serving under him. Owing to the peculiar difficulties associated with the Brigade, General

---

37 IWM: A.W. Lloyd, 10271, The UPS Song Book, p.12.
38 Ashley Gibson, *Postscript to Adventure* (London: Dent & Sons, 1930), p.135.
39 Gibson, *Postscript to Adventure*, p.136.
40 TNA: WO 374/5742: Officer File Lieutenant Colonel Bennett.
41 <https://thepeerage.com/p2960.htm#i29592>, accessed 21 September 2023.

Gilmour had a by no means enviable position to fill. He was dependent upon the Committee for practically everything that was needed by the force ... However, the Committee at once gained his hearty co-operation, help, and advice in their heavy undertaking.[42]

Another man described him as 'a fine soldierly figure.'[43]

The man selected to lead the 1st Battalion (later 18/R Fus) was Colonel Lord Henry Francis Montagu-Douglas Scott, aged 46, from Edinburgh. He was the fourth son of the Sixth Duke of Buccleuch and was educated at Eton and Christ Church College, Oxford. He was a member of the Royal Company of Archers and of the King's Bodyguard for Scotland. During the Boer War he was a major in the 3rd Battalion of the Royal Scots, later commanded the Battalion (1905–1912) and was Honorary Colonel afterwards. He was an experienced soldier who was well-educated and well-connected. However, some of his men were less complimentary. According to Walter Medlicott writing in mid-1915; 'The Colonel is now Brigadier General attached to a cavalry regt. – I don't know who will take his place – it won't be a bad thing – he has very little initiative + no powers of organisation – the 18th is always behind the other Bns in getting ranges + sites for bayonet fighting etc ...'[44]

Colonel Lord Henry Montague-Douglas Scott, the Commanding Officer of the 1st Battalion (later 18/R Fus); he had served in the Volunteers for many years and was a Boer War veteran.

Walter Gordon Wolrige Gordon, aged 53, was appointed to command the 2nd Battalion (19/R Fus). He was a Major with the Reserve of Officers who had formerly served with the Black Watch. Wolrige Gordon was the third son of a Baronet and from an ancient Aberdeenshire family. One of his men considered that Wolrige Gordon did not cut as much of a soldierly figure as the Brigade Commander, Wolrige Gordon's brother. However, his paternalistic interest in his soldiers, his pride in them, and his behaviour as a disciplinarian, made him a fine choice.[45]

---

42 Anon. Author, *University & Public Schools Brigade*, p.80.
43 Liddle Archives: WW1/GS/0125: Papers of W.G. Bentley, p.5.
44 IWM: Documents 1708, Catalogue date 1985-11-05: Papers of W.B. Medlicott, p.16.
45 Liddle Archives: WW1/GS/0125: Papers of W.G. Bentley, p.5.

Lieutenant Colonel Walter Wolrige Gordon, the Commanding Officer of the 2nd Battalion (19/R Fus), was a seasoned campaigner with considerable experience from Victorian era campaigns. Whilst his military knowledge was not recent, the interests and needs of his soldiers were foremost in his mind.

Lieutenant Colonel Charles Bennett DSO commanded the 3rd Battalion (20/R Fus). He had only recently retired and had been decorated during the Boer War.

Wolrige Gordon was educated at Eton College and was commissioned via the Inverness-shire Militia. He had seen active service in several campaigns in Suakim, the Sudan, Matabeleland and during the Boer War.

In early October 1914 Major C.H. Bennett DSO of the Worcestershire Regiment was on retired pay and was employed as a Railway Transport Officer (RTO) in London District. He was selected by the General Officer Commanding London District to command the 3rd Battalion (20/R Fus) but was the second choice because another major was unavailable. Charles Bennett had been an officer since 1888 and had been awarded the DSO during the Boer War but had retired in 1914. Bennett, aged 47, was described as 'a fire eater' by one of his men.[46]

Lieutenant Colonel John 'Jack' Stuart-Wortley, aged 34, late of the Scottish Rifles, was selected to command the 4th Battalion (21/R Fus). In 1914 he was *Aide-de-Camp* to Lieutenant General Franklin. Stuart-Wortley supposedly served as a trooper in the Boer War before commissioning and was one of the youngest captains in the British Army in 1902; both were achievements to

46 IWM: Vinden, p.14; TNA: WO 374/5742: Officer File Lieutenant Colonel Bennett.

which many UPS men aspired. John Buchan later wrote of him, 'Jack was the ideal officer to train young men of that class. He created a most efficient battalion…'[47] One of Stuart-Wortley's acolytes described him; 'That cheerful host was Jack Stuart-Wortley – an uproarious overgrown boy, prone to executive blundering, but gallant and tenacious as a bulldog.'[48] There was a negative view of Stuart-Wortley amongst his own men according to Spurrell: 'We had now seen enough of our Colonel to enable us to sum him up pretty well. He was a member of a well-known military family but [was] not one of its shining lights…'[49] This opinion was reinforced up by Edmund Page who considered Stuart-Wortley 'a bad specimen even for the family …'[50]

These five men would shape the units they commanded over the following months.

Meanwhile, uniforms were still in short supply and most men still paraded in their civilian attire. This eclectic mix of dress was described in the newspapers:

> Epsom is now a town of Norfolk Jackets and grey flannel trousers. Over 3,000 of the University and Public Schools Corps are billeted here, and in the temporary absence of any uniform they look as military as possible in clothes more suggestive of golf than drill … In about a month huts should be ready in Woodcote Park, near the Downs, and there the whole corps will go into winter quarters … if the huts are built in time – which the townspeople here seem to doubt. … Epsom people are proud to have such a fine body of men among them, and many have insisted on putting them up as their guests…[51]

Lieutenant Colonel 'Jack' Stuart-Wortley commanded the 4th Battalion (21/R Fus). Youthful, well-connected and with a track record of bravery and leadership, Stuart-Wortley would seem an obvious choice. However, being younger and less-experienced than his peers, made his a slightly 'riskier' appointment.

---

47 John Buchan, *These for Remembrance* (London: Buchan and Enright, 1919), p.57.
48 Gibson, *Postscript to Adventure*, p.134.
49 IWM: Documents 2138, 1992-11: Papers of H.W. Spurrell, p.6.
50 IWM: Documents 16924: Papers of E. Page, p.1.
51 LA: K.V. Norman Papers, WW1/GS/1188, Un-named newspaper cutting,

Andrew Buxton likened service, and living in billets in Ashtead, to his experiences at school; 'This is a strange life, and I feel I am back at both Harrow and Cambridge mixed up in one, with drills corresponding to 'schools' at various times, and more or less a Cambridge life with others in this house. One sore toe is the extent of my ailments so far, though this becomes a big thing with long road grinds.'[52] James Hodson described feelings of wellbeing whilst undergoing training: 'We are simpler men than we were a month ago ... we have begun to regard easiness and luxury as worthy of naught but contempt, and a brave death as the crown of a good life ...[53] However, many of his comrades were not attracted by a Spartan military lifestyle; they instead continued to enjoy the benefits of being based near London. Norman suggested that discipline was poor with soldiers slipping up to London in the evenings. At one point trips to London were completely forbidden and regimental police were employed at Epsom Station to meet arriving trains. The two imposing members of the regimental police for 19/R Fus were two huge brothers who had attended Cambridge and were distinguished rugby players.[54] The 19/R Fus Provost Sergeant was Lance Corporal Charles Jeffcock (PS/1155), of No 1 Company. He was aged 35, a former boy of Charterhouse School and had an MA from Keble College Oxford. He was an artist and landscape painter; not a normal skillset for a man responsible for maintaining discipline. However, they were easily dodged by those who owned cars or who walked to other railway stations. Others avoided the picquets on the station by escaping out of the carriage furthest from the platform, and after lowering themselves to track-level, they sneaked away down the embankment.[55] As men were not in uniform it was almost impossible to identify UPS-men from the public.[56]

According to Nicholas Pease:

> Life was by no means intolerable. We were all keen, morale was high and when our duties finished in the early afternoon, there was time to slip up to the West End for entertainment, and back by a late train. It was carefree and irresponsible. Of course, we all boasted and pretended that we were longing to get to the front and dreaded the thought that the war might be over before we were in action. It is my belief that each one of us was telling damned lies.[57]

Pease considered that this was bravado and he wished the war would be over before he was within fifty miles of the front. Morgan Williams developed into a conscientious objector in khaki during his training.[58]

The influx of UPS-men was also good for the economy in Epsom as they were generally well-behaved and many were well-off financially. It was said that for the tradesmen in Epsom their routine income was better during the eight months that the UPS were there than it was during Derby Week.[59] Epsom had limited scope for entertainment with only two cinemas,

---

52   Woods, *Andrew Buxton*, p.43.
53   James Hodson, *The Soul of a Soldier* (London: George Routledge, 1918), p.1.
54   LA: WW1/GS/0125: Papers of W.G. Bentley, p.5a.
55   IWM: E.D. Shearn Papers, 2033.
56   LA: K.V. Norman Papers, WW1/GS/1188, p.28.
57   IWM: N.A. Pease Papers, 8230, pp.30–31.
58   Williams, *From Khaki to Cloth*, p.54.
59   LA: K.V. Norman Papers, WW1/GS/1188, p.26.

which were overcrowded nightly, a few billiard halls, a YMCA, and the Town Hall where concerts were sometimes held. Private Donald Addams-Williams (PS/5) recalled in November 1914 that during inclement weather his company was given lectures in the cinema which had been commandeered by their company commander.[60] A bugle band was formed by 18/R Fus and was run by an ex-Dulwich College boy; he invited Gordon Jacob, also from Dulwich, to play the drums. This gave Jacob the chance to miss fatigues and fostered his love of music. A few men were distracted by heavy drinking. Among the men of Shearn's platoon (20/R Fus) were Victor Bass and Arnold Wills whose families were heavily involved with brewing and tobacco respectively:

> Bass was an extraordinary character. He was immensely strong and had gone in for weight-lifting. He also very industriously lifted his elbow … after his death his mother – Lady Bass – wrote to me that she had received a bill for something over £3,000 from a wine and spirit merchant in Epsom for goods supplied to her son. … Victor had, to my knowledge, always been most hospitable to his friends …[61]

Some men who had enlisted were largely unsuitable as infantrymen. One man was Eric Forbes-Robertson, a 48-year-old artist, who was an eccentric 'gentleman ranker'. He was both a concert pianist and the brother of a well-known actor. One soldier remembered that he was tall and gaunt and not a very martial figure, but he got on well with everybody. However, 'Forbie' often went missing during route marches and after searches of the nearby area he was normally found in a nearby house entertaining local ladies with his piano playing. He could not be changed, and his misdemeanours were tolerated by his chain of command and even the Colonel.[62] 'Forbie' was protected by the officers of 21/R Fus and was eventually employed as the battalion post corporal. The UPS Brigade did not comprise 'ordinary' infantry battalions nor 'ordinary' men. Soldiers observed some of the unusual characteristics of the UPS whereby private soldiers arrived for parades in motor cars or wealthy aunts visited them. There were also some well-known people who had enlisted in the UPS who were in the ranks one day and then disappeared, only to take up commissions shortly afterwards. One example was Ivor Andros de la Rue whose father owned the Onoto fountain pen company and who returned as a platoon commander.[63]

This 'different' leadership style was evident elsewhere. John Morton suggested that the UPS was not all about 'standard' NCO leadership styles based around shouting and rigidity. Morton suggested that instructors learnt to curb their sarcasm and employ patience with those who learned slowly whilst they also harried those who were genuinely lazy. Soldiers also helped train and develop one-another as everyone was learning new skills.[64] Private Donald Price remembered that some of the NCOs were regulars and were disciplinarians. The UPS NCOs did not have the same stamp on them; they were more likeable and were less bullying. The disciplinarians were more efficient but the men resented being sworn at or bullied. However, the UPS NCOs were not as efficient at man-management.

60 LA: WW1/GALL/001: Papers of D.A. Addams-Williams, undated letter.
61 IWM: E.D. Shearn Papers, 2033.
62 LA: WW1/GS/0125: Papers of W.G. Bentley, p.5a.
63 LA: WW1/GS/0125: Papers of W.G. Bentley, p.5a.
64 Also, 'The condensed wisdom of the old non-commissioned officers meant rather more.' See J.B. Morton, *The Barber of Putney* (New York: Penguin, 1939), p.13.

Epsom Downs were used for attacks and tactical schemes, but later Walton Downs and Headley Heath were used.[65] The UPS men were taught tactics and how to manoeuvre under fire rather than advance, at the walk, in lines:

> We have often in the afternoon 'extended' drill, which is interesting. It is considered of considerable importance to heave yourself on the ground and get up for the next sprint as quickly as possible as casualties chiefly occur at these times. An afternoon of this sort is splendid exercise. I am glad to find I can usually get up quicker than others and have a considerable lead. I am also glad that my clothes so far need no mending.[66]

To Nicholas Pease, close-order drill training was largely out-dated but more modern field training was taught; 'The extensive Downs and race course provided an excellent area for all kinds of training, mock attacks in extended formation … Extended order was a lesson learned from South Africa, as being the answer to the effectiveness of the modern magazine rifle.'[67]

The following describes the training regime for 21/R Fus:

> We rose at six o'clock (I had a cold bath regularly every morning). We had to walk a mile to the parade ground in time for physical drill at 6-45. This generally took the form of Sandow exercises. We got back to breakfast before 8 and paraded again an hour later. From 9 to 10 we did platoon and company drill – then after a short interval we sallied forth in full marching order and did an attack, an outpost scheme or an advance guard. After dinner we did another hour's drill and an hour's musketry. After tea we had the evening free as a rule. On free evenings we played bridge or read by the fire. Occasionally we had night operations; these were usually a scream…[68]

Night operations were not always taken seriously according to John Morton, 18/R Fus, and there were often comedic occurrences such as a sentry challenging a Colonel using coarse language; soldiers falling into streams; or personnel getting lost and accidentally 'joining' the enemy force.[69] Whilst these activities might be ridiculed, such night training would likely prove useful in the future. Erroll Shearn was clearly proud of his and his unit's physical fitness; 'Among other experiences I remember a prolonged exercise during which, with full pack and rifles, in three days we covered 63 miles and did outposts at night. I don't think anyone fell out.'[70]

Donald Price joined late, and underage, and found himself in No 16 Platoon, D Company, 20/R Fus.[71] He was only at Leatherhead for six weeks but recorded the established training routine:

---

65 LA: K.V. Norman Papers, WW1/GS/1188, p.24.
66 Woods, *Andrew Buxton*, p.44.
67 IWM: N.A. Pease Papers, 8230, p.30.
68 IWM: H.W. Spurrell 2138, p.7
69 Morton, *Barber of Putney*, p.14.
70 IWM: E.D. Shearn Papers, 2033.
71 PS/5473 Private Donald James Price, from Cadishead, Manchester, was a draper's assistant. He enlisted underage in 20/R Fus in December 1914. He went to France in November 1915 with D Company, 20/R Fus. IWM: Donald Price Sound Recording 10168.

Members of the 21st Battalion on parade in 1914 somewhere near Leatherhead. They are still in civilian attire and have a handful of rifles shared between about sixty men. (Author's collection)

We'd get up in the morning, get our shorts on and parade for running. In Leatherhead there was a park there we used to run round at six o'clock in the morning 'til breakfast [back at billets] ... After breakfast we paraded again and we might go out for a walk, a march ... and we probably used to march five or six miles. We'd got into the habit then of marching and drilling. Just forming 'fours' had got to be taught properly. We were raw... Even turning right and left, it all had to be taught properly. ... after a month we pretty much knew what drilling meant.[72]

Price found this 'tiresome' but they were generally back home for lunch; 'Afternoon [training] might have map reading or anything ... or we might be shown the rifle and the parts ... and things like that.' They were finished by 4, had tea back home, and were free for the evening.

During this period there was another inspection of the UPS Brigade by Major General Lloyd on the Downs at Epsom. This occurred on 13 November and coincided with the brigade still being incompletely trained and inadequately equipped.

Christmas in Epsom

The run-up to Christmas was generally quiet with some field days and night operations planned and delivered as training. A charity rugby match was played between the UPS and a team of

72   IWM: Donald Price Sound Recording 10168, Reel 2.

Canadian soldiers on 12 December 1914 in aid of Belgian refugee relief funds. The Public Schools team won by thirteen points to ten but over 4,000 people watched the game including a thousand UPS personnel and, as *The Pow-Wow*, the UPS Brigade journal, noted, a few fashionably dressed ladies.

Half of the Brigade was given leave for the Christmas period; the other half were away over New Year. On Christmas morning Harry Oakley assisted the medical officer (MO), an easy chore, and so was excused church parade. He instead went to church by himself after his duties. Christmas dinner comprised some dry roast beef, plum pudding and mince pies. Luckily one of his comrades had received a box of 'goodies' including dates, figs, cakes and crackers.[73] Oakley had enlisted on 23 November 1914 and had only been delayed by vital war work. As an accomplished cutter of silhouettes, he had helped create one of the best-known recruiting posters which was displayed throughout the country; his contribution to the war effort had already been considerable.[74] Those personnel that remained in billets over Christmas were generally well looked after by the families whose houses they shared.

A further rugby match in early January 1915 saw the UPS Brigade face 11th (Northern) Division which provided a picked team. After a close match, the UPS won by twelve points to ten. In the late Autumn of 1914, the Brigade Commander established a football challenge cup competition between the sixteen companies in the Brigade. Details of some of the matches were recorded in *The Pow-Wow*. The final, played between B Company of the 20th Battalion and C Company of the 21st Battalion on Saturday, 6 February, made the local newspapers. After a hard-fought match the result was that B Company beat C Company by five goals to three. The cup was presented by the Brigadier after the match.

Uniforms – 'You're gentlemen no longer; you're bloody soldiers'

In the early days of formation, the battalions were financed by the deep pockets of several captains of industry including Sir Joseph Lyons and Mr Gluckstein. Uniforms were slow to arrive; an idea of the state of equipment on 13 December 1914 is provided by Andrew Buxton; 'No uniforms yet except cap and puttees, which latter are useful for wet days, and a pair of boots which seem to fit well and to be good ones in spite of simply 'drawing' them by size only.'[75] According to Vinden uniforms were not plentiful; 'We had no uniforms and eventually when a consignment arrived it was found to consist of peace time walking out dress of blue patrol jacket, red striped trousers and forage cap … There was only sufficient for one piece per man. I drew a fore and aft [cap]. We were proud to wear our bit.'[76] Uniforms did not arrive for two months.[77] When they did get uniforms the men were not always smart and soldierly looking. In Andrew Buxton's case; 'As I told you I now possess a uniform as outward and visible sign of being a full

---

73 Jerry Rendell, *Profiles of the First World War, The Silhouettes of Captain H.L. Oakley* (The History Press, 2013), p.36.
74 PS/3058 Private Harry Lawrence Oakley, aged 31, had studied at the Royal College of Art. His work on the recruiting poster earned him an MBE. He served with No 3 Company, 21/R Fus, and in May 1915 he was gazetted to the 11/Yorks and served with 8/Yorks.
75 Woods, *Andrew Buxton*, p.45.
76 IWM: F.H. Vinden Papers, 5565, p.12.
77 IWM: G. Skelton Papers, 13966, p.29.

Sketch entitled 'UNI de FORM ities' by an unknown artist from C Company, 18/R Fus, showing the limited amounts of ill-fitting military clothing that had been issued to the UPS Brigade. This was published on 27 November 1914 in *C's Fire*, a short-lived unit journal.

'private'! I am told that my hat does not fit and I look like a 'bus driver'!'[78] According to Kenneth Norman a few rifles arrived in November and, afterwards, during route marches the men took turns to carry them.[79] Oakley highlighted the lack of equipment as he missed dining with a local family on Christmas afternoon as he had to repair one badly-worn set of army trousers with the remains of a worn out pair.[80]

In about January 1915 the four battalions were taken over by the War Office and funding by the previous donors ceased. Presumably, from this point onwards, equipment started to arrive in increasing quantities. William Bentley joined 19/R Fus in January 1915 and was lavishly equipped with uniform, underwear and durable boots; he had to spend inordinate amounts of time preparing and polishing the latter. The rifle he was issued was of the Boer War pattern and he soon became familiar with its operation, mechanism and method of cleaning it.[81]

By early 1915 leather web equipment and uniform had been issued. Men then carried out route marches of up to 18 miles.[82] Drill for the 4th Battalion (21/R Fus) was taken very seriously: 'All our senior N.C.O.'s being guardsmen, and our own battalion's first Adjutant, Sir Robert Walker, likewise a "guardee" for whom Spit and Polish were the first two gods of war, the UPS' smartness on the square was something to take a proper pride in.'[83] One of the drill-loving ex-Guards 'characters' in 21/R Fus was the Regimental Sergeant Major, called George Fletcher. Though suitably experienced to be a capable RSM, Fletcher was far from popular according to Morgan Williams:

> A gaunt guardsman, he represented the least pleasant side of the British Army. On parade he was a martinet, often foul-mouthed and insulting, and off parade he cringed and crawled

78 Woods, *Andrew Buxton*, p.47.
79 LA: K.V. Norman Papers, WW1/GS/1188, p.28.
80 Rendell, *Profiles of the First World War*, p.36.
81 LA: WW1/GS/0125: Papers of W.G. Bentley, pp.3–4.
82 IWM: G. Skelton Papers, 13966, p.31.
83 Gibson, *Postscript to Adventure*, p.135.

No 2 Company, 21/R Fus looking smart on parade in new uniforms and webbing equipment. On closer inspection several men are still in civilian dress; none have any webbing shoulder straps and none are armed. (Author)

for drinks to men who had more than a private soldier's pay. For weeks he addressed us as "Gentlemen," …. To our amusement he persisted in this practice till the day when we put on the King's uniform for the first time. Then he strutted up and down the ranks, yelling, "You're gentlemen no longer; you're bloody soldiers."[84]

Though reviled, it could be argued that Fletcher helped form the battalion into a cohesive and disciplined unit.

Trench Digging

On many occasions during the training of the UPS battalions detachments were sent to Woldingham and Caterham to dig trenches. Kenneth Norman recollected he and his comrades having to get up at 5:00 a.m. several times a week to catch the train to Caterham to excavate these defences. He considered this 'dismal work'.[85] Norman chose to join the battalion stretcher bearers to employ his medical knowledge but also escaped this mundane labouring as SBs were excused fatigues. In mid-November Skelton's company was sent to Caterham for six weeks of trench digging; 'The girls at Caterham seemed glad to see us and smiles always greeted us from the windows. Even the officer in charge would halt our company occasionally and partake of a cup of coffee with one of his lady friends before resuming the march.'[86] The digging through the

---

84 Williams, *From Khaki to Cloth*, p.47.
85 LA: K.V. Norman Papers, WW1/GS/1188, p.32.
86 IWM: G. Skelton Papers, 13966, p.30.

BETWEEN TRENCH AND TRENCHER.

Sketch depicting UPS men finishing their duties trench digging (presumably at Caterham) and rushing to a nearby public house. Sketch by Ernest Stoneley published in *The Pow-Wow*. (Reproduced with the permission of Janet Wood)

chalk was gruelling. There were also problems caused by troops sloping off to the local pub, The Harrow, and a guard had to be placed on the establishment. However, the guard might let his friends in, or men could use the rear entrance of the pub. One enterprising UPS man, Private Gould, arranged for beer to be delivered by a local tradesman to where the trenches were being dug.

Terence Doherty recorded on 13 December; 'On Friday we went to Warlingham, 5 miles from Purley, to dig trenches. We started at 6.40. It poured most of the time. We are going again tomorrow…'[87] Not everyone disliked this activity which was seemingly accelerated in January 1915. William Bentley, 19/R Fus, rather enjoyed being taken in special trains to dig defences on Sir Walpole Greenwell's estate. At 11:00 a.m. each morning the elegant house maids of this stately home served coffee on the front lawn to the diggers.[88] Buxton was not enthusiastic about travelling to dig trenches on 2 January 1915:

---

87   IWM: T.O'C. Doherty Papers, 12961, letter 13 December 1914.
88   LA: WW1/GS/0125: Papers of W.G. Bentley, p.7.

We had a rotten day going early to W[oldingham] with an inevitable full half-hour wait at Ashtead Station first and about 11.30 knocked off owing to the rain and marched to the station, where we waited from 12.30 till 3.30 for a special train. It was driving rain all the time and a very great number of men had to stand on the uncovered part of the platform. I was very fortunate being under cover and having a Times to read. ... All day was very cold and draughty...[89]

These trench-digging excursions ceased once training started at Woodcote Park.

Inspection by Kitchener

In January there was another inspection, this time by Lord Kitchener and Alexandre Millerand, the French Minister of War. This visit to the UPS helped shape British and French relations going forwards:

It was in order to arrive at some agreement between the two countries upon the manner in which the new armies could most advantageously be used that M. Millerand came to London on the 21st of January, 1915. Escorted by Lord Kitchener to Epsom and Aldershot, he was given a glimpse of a "K" Army, and, in spite of the abominations of deep snow and bitter fog, he was impressed by the numbers and carriage of the men.[90]

Brigadier General Gordon Gilmour and the UPS Brigade on parade in the snow awaiting Lord Kitchener. Godfrey Skelton merely noted; 'We stood for hours awaiting the arrival of these important people during an intense snow storm and many men fainted through sheer cold.' (Reproduced with the permission of James Skelton)

89 Woods, *Andrew Buxton*, pp.46–47.
90 Reginald Viscount Esher, *The Tragedy of Lord Kitchener* (London: John Murray, 1921), p.94.

Spurrell's memories were somewhat critical:

> It had been a very cold day with thick snow of the ground. After a few rehearsals in which Fletcher had been the star turn and Stuart Wortley had cut a sorry figure, we marched up to Epsom Downs. A long wait had followed. K[itchener] had walked along the front line [i.e. the front rank] and had then driven off in most disappointing fashion. I was so intent on looking to my front that I never saw him at all.[91]

William Bentley, despite having only been under training for a few weeks, was surprised to have passed the physical endurance test of Kitchener's visit.

This was a milestone moment in the Brigade's history which was seemingly far closer to being ready to deploy overseas than at any previous time. The training and equipping of the Brigade was going well, and the work to accommodate the battalions was progressing. There was one fly in the ointment, and that was the numerous UPS men who were departing for commissions.

---

91   IWM: H.W. Spurrell 2138, p.11.

3

Commissions and Controversy

I … felt rather like a rat leaving a sinking ship

> *Some of the more ambitious of us are seeking commissions, but most of us are content to shoulder the rifle …*[1]

> *Many of my friends began to drift away with commissions, so I thought to do likewise at the end of January 1915 and felt rather like a rat leaving a sinking ship …*[2]

Much work had been done by the officers and men and the 'new' committee to train and prepare for war. However, some men did not stay with the UPS for long. Many had joined having concurrently applied for commissions. Soon after some began training, their commissions came through. George Eyston was sent to Battalion Headquarters one day; he was told that he was to be a commissioned officer in the Dorsetshire Regiment and was told where he must report.[3] To further exacerbate the situation some of Eyston's comrades asked if he could also get them into the Dorsets as well. Vernon Bartlett later followed Eyston into the Regiment. To some, the UPS was a place-holding activity, whilst they arranged commissions, or used enlisted service to prepare themselves for officer training. Others, who had enlisted with the UPS to get to France quickly, grew impatient having seen no progress in training and also sought commissions. These departures further undermined the whole rationale for having joined the UPS in the first place. When accepted as officers, these men had to be discharged and further replacements needed to be recruited and trained to fill the gaps. There was a constant 'pull' of UPS-men moving away for commissions which retarded the training and readiness of their former battalions and perpetually delayed their deployment to France. By the end of October 1914 hundreds of men had departed; it was further announced that 150 men (if aged between nineteen and twenty-five) would be eligible for a three-month course at Sandhurst if they took permanent commissions.

---

1   *The Marlburian*, Vol. XLIX, No 741, 17 October 1914, p.153.
2   Personal account by Lewis Jacques; <https://online.flipbuilder.com/zyzh/lsvb/index.html#p=9>, accessed 8 October 2023.
3   Eyston, *Safety Last*, p.27.

However, this outflow of officer candidates from the UPS Brigade, and their service as junior officers, ultimately added to the over-all contribution to the war effort made by the UPS.

There were appeals by the Brigade to encourage further recruits and to raise the profile of the formation. The following appeared in the Marlborough College magazine:

> Some of the more ambitious of us are seeking commissions, but most of us are content to shoulder the rifle – which by the way we have not yet got – as ordinary "Tommies." Training is going forward with all speed and we are to be sent abroad as soon as we are ready. ... A camp consisting of a thousand huts (five men to each), is in process of erection in Woodcote Park. At present we are billeted in the town and the surrounding neighbourhood ... We have no uniforms as yet, and men may be seen wandering about in all kinds of garb of varying picturesqueness... Marlburians are very plentiful here. ... Indeed one cannot walk down the High Street without noting an O.M. tie.[4]

However, a letter published in the Gresham's School magazine exacerbated these problems by recommending their old boys apply for commissions from the ranks:

> I am far from wishing to encourage O.G.'s to remain in the ranks, more especially since I have learnt the great need for officers. I ... have been successful in getting a number of O.G.'s to apply for commissions. Though I sympathize with the commanding officers of the Public School Battalions, when they see their units melting away, I remember that Public School boys were clearly given to understand that, by joining these units, they would increase their chances of obtaining commissions...[5]

Public schools later observed that their old boys, who had enlisted, were prevented from getting commissions:

> ... many well-educated young men, in the first outburst of enthusiasm, when war broke out, enlisted as privates, that they have since had the offer of a commission in some other unit than that to which they were originally attached, but have been unable to induce their OC to give them their release, not because they were not fit to take a commission, but simply because the OC, naturally enough, after training them for months, objected to parting with them?[6]

In mid-November the Brigade numbered about 4,500 men. Around this time another 240 more recruits had arrived and they were split between 20/R Fus (70) and 21/R Fus (170).[7] By January the committee estimated that about 1,000 men had left to take commissions. Hundreds more will have been discharged as unfit or unsuitable. Clarification was needed on the long-term purpose of the Brigade. The following order was promulgated to clarify the situation; 'The G.O.C. is authorized by the Secretary of State for War to state that the Brigade is not to be

---

4   *The Marlburian*, Vol. XLIX, No 741, 17 October 1914, p.153.
5   *The Gresham*, Vol. VI, No 2, 19 December 1914, p.35.
6   *The Stonyhurst Magazine*, June 1915.
7   Anon. Author, *University & Public Schools Brigade*, p.60.

"TO BE OR NOT TO BE."

Sketch by H.C. Mumby highlighting the dilemma for young men who wanted to get to France, but, following their experience in the ranks, also desired gazetted status. (*The Pow-Wow*)

utilized as an O.T.C., and that nothing is further from Lord Kitchener's intention …'[8] More recruits were needed. A Brigade recruiting office was opened at 20 Westminster Palace Gardens in January, run by the 21st Battalion, to help enrol further recruits.

However, not every ranker wanted to be an officer; Andrew Buxton declined the offer to become an NCO: 'I am glad to say I was able to get out of it, though at the time I did not think I could work getting off … Non Commissioned Officers have already had a lot of training and the responsibility of being in charge of fourteen men has duties which I do not consider without previous training that I was able completely to fulfil.'[9] Nor was every applicant allowed to apply or leave. On applying for a commission Edmund Page was rejected by Colonel Stuart Wortley; 'The applications were turned down by the CO – a bad specimen even for the family – presumably because we were useful to him as NCOs.'[10] Once UPS men had decided to apply, and were

8 Anon. Author, *University & Public Schools Brigade*, p.61.
9 Woods, *Andrew Buxton*, p.44.
10 IWM: E. Page Papers, 16924, p.1.

rejected, their view of enlisted service changed. Page recalled; 'All these weeks of discomfort and distasteful companionship and lack of employment commensurate with our abilities was very trying.'[11]

These rejections sparked controversy as men complained to their families of being turned down for commissions or were prevented from leaving. These issues were touched on by the press in Manchester and evidently reached the ears of men in power. Questions were asked in the House of Commons by Mr Edward Goulding MP in March 1915:

> I regret to say that there are certain service battalions who resist very effectively any promotion from the ranks for those commissions, and the colonel absolutely declines to consider applications for commissions. They are notably the 18th, 19th, and 21st Royal Fusiliers (Public School Corps) … Officers commanding these regiments have systematically put obstacles in the way of men applying for commissions; in some regiments they have actually declined altogether to see them. … Some of those who were fortunate enough, as they thought, to get before the colonel, the instant the word "commission" was mentioned, the sergeant-major came to the rescue – "Rightabout turn, march!" They received no consideration, whatsoever … Here is one of several [letters] from colonels … "I beg to inform you that I cannot grant the necessary recommendations for these men to take up commissions in the battalions under your command, in consequence of the general officer commanding the brigade having received verbal instructions from the Secretary of State for War that no further men are to be granted commissions…"[12]

A robust response to criticisms of the committee and the commanding officers was drafted in April 1915 and letters were sent to the major newspapers.[13]

There were further questions specifically concerning the 20th Battalion in relation to the small proportion of commissions offered to them compared to the other three battalions. Sir Clement Kinloch-Cooke MP asked the Under-Secretary of State for War in the House of Commons in May 1915 'whether he can explain how it happens that in the 17th Battalion of the Public Schools Brigade 400 men have been selected for commissions, in the 19th Battalion 200, in the 21st 120, and in the 20th only 25?' Mr Harold Tennant, the Undersecretary of State for War, replied:

> I imagine the hon. Gentleman is referring to the 18th Battalion. … The number of commissions granted to men of the various public school battalions is necessarily regulated by the number of suitable candidates serving in each of those battalions. The numbers quoted by the hon. Member are not accurate in detail. I am not aware that the numbers forwarded to the War Office are the full numbers contemplated by the officers commanding.[14]

---

11 IWM: E. Page Papers, 16924, p.2.
12 *Hansard*, Army (Promotion and Pay), Vol. 70, Thursday 11 March 1915.
13 See Appendix III.
14 *Hansard*, Army (Promotion and Pay), Volume 71: Thursday 17 May 1915.

More than 25 men of 20/R Fus had been gazetted up to this point. By April about 1,700 men had been selected from across the Brigade.[15] The UPS battalions, and 20/R Fus especially, were clearly reluctant to release officer candidates but men were allowed to depart from 20/R Fus. Erroll Shearn recalled a frosty interview with Colonel Bennett in about May 1915:

> I was ordered into his presence when he informed me that my commission 'had come through' and I was to report in due course to a young officers' training centre in Bedford. He did not appear to be at all pleased at my promotion. Looking back I can sympathise with him: he had trained and brought to a pretty high standard a bunch of very raw recruits, and now he had the prospect of his ranks being greatly depleted and having to take in an appreciable number of untrained replacements. He asked me acidly: "Why do you think you are likely to make a good officer?" I replied as deferentially as I could that I had received a reasonably good education and had a professional qualification … and that I anticipated that the Army would provide adequate instruction before I assumed responsibility. My … suave rejoinder elicited a grunt by way of reply.[16]

The tasks the UPS battalions had acquired (trench digging and camp erection) also encouraged some men to seek commissions more actively. David Kelly recalled that his battalion was issued rifles and uniforms, which was a sign that their training was progressing. However, they were also employed labouring to build the hutted camp at Woodcote Park which was not what he had enlisted for. He 'felt that the joke was being carried rather far.'[17] George Gillett, a friend of Kelly's was offered a commission in 6th Battalion Leicestershire Regiment (6/Leics R) by the colonel of that battalion, with whom he played cricket. That colonel asked Gillett to bring any friends to whom he would offer commissions. Kelly took advantage of this offer and received a commission in the Leicesters.[18] For Andrew Buxton, disillusionment started to set in and he actively sought obtaining a commission; 'This letter brings the prospect of an event in my life … [being accepted for a commission in the Rifle Brigade] … Now I am taking steps to get a transfer to the Cambridge O.T.C. as a Private.'[19] In mid-January Andrew Buxton was accepted and departed to an OTC Course at Cambridge before reporting to 6/RB. In April 1915 lobbying from a friend of Edmund Page paid off; Page and his friend, Charles Whitley, departed from 21/R Fus for commissions with the KRRC. Life was not immediately difficult for those gazetted; Vernon Bartlett's commissioned service started with an idyllic period 'in command of a platoon that patrolled the most beautiful stretch of the Dorset coast, from Warbarrow Bay to Lulworth Cove…'. These pleasant times did not last.

Commissions were always being discussed in the background and in letters home. Addams-Williams wrote of both hoping for a commission but being apprehensive about commanding soldiers, many of whom would be older than him.[20] When Addams-Williams departed to a commission in 4th Battalion South Wales Borderers (4/SWB) he had to endure Army bureaucracy spending most of the morning in his battalion headquarters arranging his discharge

---

15 Anon. Author, *University & Public Schools Brigade*, p.62.
16 IWM: E.D. Shearn Papers, 2033.
17 Kelly, *Ruling Few*, p.89.
18 Kelly, *Ruling Few*, p.90.
19 Woods, *Andrew Buxton*, p.46.
20 LA: WW1/GALL/001: Papers of D.A. Addams-Williams, undated letter.

paperwork. He arrived at his new unit at a camp at Chiseldon late at night and in his private's uniform. He was sent into London for a few days to arrange his banking, uniform and equipment as a commissioned officer. He was also aghast to find that he arrived 'absolutely unqualified' but that the battalion was well advanced in training and he was one of the last officers to be recruited.[21] Lewis Jacques departed with a commission in January 1915 and recalled his officer training: 'I left with several others for a three-week course at Cambridge University OTC ... We were now very well fed at Pembroke College where I was billeted. We re-learned everything we had already learned at Ashtead, but with far less efficient instructors.'[22] Men left for different reasons. Lewis Namier, of 20/R Fus, was discharged in February 1915 to work in intelligence. His knowledge of languages and history would far outweigh his minor value as an infantryman.

All these departures meant that the marginal decline in quality was noticed by the men of the UPS Brigade; 'Months before embarkation, our brigade began to lose its advertised character. From the ranks our brightest and best were gazetted out in scores to officer newer units of K.'s [Kitchener's] mushroom army. ... The War Office began to behave as if we were an O.T.C., which hadn't been what we wanted.'[23] As UPS officers left for other units, they needed to be replaced.

John Ashley Gibson returned to the UK from Ceylon to enlist. He found and old friend, Herbert Fenn serving with 21/R Fus; 'Fenn was a great character in our set. A hard-bitten ex-gunner, bored with peace-time service routine he had sent in his papers after the South African War, and in the ensuing years had tried his hand at most things, from pitch and toss to manslaughter.'[24] Fenn was the battalion second-in-command (2IC) of 21/R Fus:

> The UPS Brigade numbered other friends of mine and had just gone into huts at Epsom. I spent a guest night there in the mess of Fenn's battalion, dossing on the colonel's spare truckle-bed ... Six months of training, most of it in the open, had already given him and all his officers bronzed and military countenances that shamed my sickly equatorial buff.[25]

Sketch of Second Lieutenant Ashley Gibson published in his personal account (*Postscript to Adventure*). Gibson was gazetted into 21/R Fus on 15 June 1915 after a short stint as a private soldier.

21 LA: WW1/GALL/001: Papers of D.A. Addams-Williams, undated letter.
22 Personal account by Lewis Jacques; <https://online.flipbuilder.com/zyzh/lsvb/index.html#p=9>, accessed 8 October 2023.
23 Gibson, *Postscript to Adventure*, p.135.
24 Gibson, *Postscript to Adventure*, p.53.
25 Gibson, *Postscript to Adventure*, p.134.

Second Lieutenant Frederick Stewart Modera, aged 30, was educated at Manchester Grammar School and attended University College Oxford where he was awarded an MA. He was a barrister by trade. Modera was gazetted on 20 March 1915 and was later both the Adjutant and the commander of 20/R Fus. (Reproduced with the permission of James Skelton)

Sergeant Godfrey Skelton (PS/5641) enlisted with 20/R Fus on 3 September 1914 and initially served with C Company in 1914. He transferred to D Company when he was appointed lance corporal. He was steadily promoted and was eventually an acting WO2. He returned to Britain on 28 November 1916 and was gazetted into the RE in February 1917. He was employed with 205th Field Company RE and was awarded the Croix de Guerre. (Reproduced with the permission of James Skelton)

Ashley Gibson was presumably vouched for by Major Fenn prior to receiving a commission in 21/R Fus; 'I took another visit to Epsom and had a few words with the Adjutant in his office. When I walked out, it was with the King's shilling in my pocket.'[26] Gibson served for five weeks in the ranks whilst he awaited his commission paperwork, 'I felt ... and indubitably was, the lowliest of all God's creatures, a private soldier. Other people... have told me that they liked it. I didn't. I absolutely hated it.'[27]

New NCOs were also required to replace these who had been gazetted; they, in turn, needed to be trained. Godfrey Skelton was promoted to sergeant after an interview with Lieutenant

---

26 Gibson, *Postscript to Adventure*, p.134.
27 Gibson, *Postscript to Adventure*, p.135.

Colonel Bennett with a view to applying for a commission. Meanwhile, Lieutenant Modera took over Skelton's platoon; Skelton considered him a good officer. Due to the limited training amongst UPS NCOs in February 1915 an NCO training cadre was established whereby UPS senior NCOs were sent to Chelsea Barracks for training under Guards NCOs in drill and infantry skills.[28] The NCOs sent did have the luxury of being honorary members of the Guards Sergeants' Mess. Subsequently, in March 1915, the SNCOs underwent a course of musketry though they were largely self-taught as there were no regular NCOs to instruct. Medlicott was sent to London on 26 April for a three-week course at Chelsea Barracks. He became a corporal on 2 May either due to his performance on the course or because, by this time, about 800 men from 18/R Fus had departed for commissions.[29]

The Next Recruits

More UPS recruits were needed to fill the gaps left by both those who were commissioned but also those who were found to be medically, physically or temperamentally unsuitable as infantrymen. According to Kenneth Norman the replacements were of a lower standard than those whose places they were filling.[30] William Bentley joined B Company 19/R Fus in late 1914 and was separated from the rest of his draft from Cheltenham; he noticed that they formed a 'clique' in D Company. The clique embraced the specialist roles in the battalion (sanitary men, signallers etc) to remain separate but were derided by their UPS peers as 'that lot from Cheltenham'.[31]

Recruit standards lowered over time in terms of the perceived social class and age of recruits. Donald Price was a sixteen-year-old apprentice in the fur trade in Manchester who decided to enlist on 29 December 1914. He recalled, decades later:

> We went down to a hotel in Piccadilly, Manchester, to join. Without any second thought … about the consequences or anything. We were going into the Army and we thought it was going to be a damn good time for about six months and it would all be over and we'll have seen the war, a sort of holiday. We were bored with our work in those days, so tedious.[32]

Price found his way to Epsom and the UPS by accident. During the first week in January about 100 recruits arrived at Leatherhead from Manchester; Price was likely amongst them. He remembered his reception with the battalion alongside fifty others:

> We were called up to line up in front of a church school… There were probably fifty of us and this very smart sergeant major came out. He went right through us all and said, 'now look here, you're all in the Royal Fusiliers … the best regiment in the country'. He came along and examined everybody, questioned everybody. He questioned somebody in the front row in front of me, and said; 'How old are you?' … Dick Parker says, 'Eighteen sir'; he

---

28 IWM: G. Skelton Papers, 13966, p.32.
29 IWM: W.B. Medlicott Papers, 1708, p.2.
30 LA: K.V. Norman Papers, WW1/GS/1188, p.32.
31 LA: WW1/GS/0125: Papers of W.G. Bentley, p.6.
32 IWM: Donald Price Sound Recording 10168, Reel 2.

says, 'You're not eighteen, step out'. He stepped out and was sent home again. ... He never asked me how old I was...[33]

He recalled adapting to the different class system in the UPS:

> I was just shoved in ... They wanted to fill it up quick ... they accepted me and I accepted them. I talked with solicitors... the fellow sleeping next to me was the Honourable F.S. [A.S.] Bligh; he accepted me and I accepted him as 'Joe' ... They were lovely to me and I learned a lot ... They were loaded with money some of them.[34]

He further added; 'It didn't follow that just because they were well-educated that their morals were any better...'. Though many of the men were not prudish, Morgan Williams despised the few men in his number who openly talked about female conquests.[35] Increasingly often, some risks were taken by the recruiters regarding recruit standards. Private Albert Knighton enlisted in April 1915; 'With two pals from Dulwich College we all went to London, to Scotland Yard, and joined the 18th Battalion Royal Fusiliers. When I volunteered to join the Army I was 16 ... they were just glad to get me in.'[36]

However, as the UPS history records, though the decisions over commissions were controversial the ultimate record of the Brigade must be remembered:

> Over 7,000 officers were supplied from the Brigade since it was raised. This is a unique record, which has never been, and probably never will be, equalled. Such a result of untiring work on behalf of the Brigade was far beyond the expectations of the members of both committees. It would have been a sin to have sent the Brigade as originally raised to the firing line.[37]

The 'manning churn' from this cycle of officer candidates leaving, and recruits arriving, affected the individual and collective training for each battalion. The effects of this will be seen whilst examining the training the Brigade underwent whilst in Great Britain.

---

33 IWM: Donald Price Sound Recording 10168, Reel 2.
34 IWM: Donald Price Sound Recording 10168, Reel 2.
35 Williams, *From Khaki to Cloth*, p.51.
36 IWM: A.E.L. Knighton Sound Recording, 10263.
37 Anon. Author, *University & Public Schools Brigade*, p.63.

4

Woodcote Park

Each individual will do his utmost to prepare ... himself for war

---

> *Living in huts was a great change after the cosiness and semi-privacy of good billets...*[1]
>
> *... whoever occupy the huts among the trees will, on fine warm days, think they are having quite a pastoral existence...*[2]
>
> *The Brigadier hopes that all ranks will realize that the proficiency of the whole Brigade may be judged by the bearing and behaviour of the smallest unit ... each individual will do his utmost to prepare and perfect himself for war.*[3]

Building the camp

When the Brigade was first formed there was a consideration of how the formation might be brought together for centralised training. A major coup for the committee was securing areas of the Woodcote Park estate to allow a purpose-built training camp to be erected. That building work would take time and the War Office recommended for the Brigade to be billeted in nearby towns which allowed the UPS Brigade to mobilise. The building of the camp would follow later.

With the UPS Brigade settled in billets, and undergoing training, work commenced on a camp at Woodcote Park in the following month. The intention was to build a hundred huts to accommodate fifty men each. Early on, there was the realisation that such a camp would require heating, running water, drainage, roads and lighting, and catering facilities, ablutions and entertainment facilities. This added significant complexity and likely increased the scope of the project; this required greater time and expense to complete. The contractors were Messrs Humphreys Limited, of Knightsbridge. The UPS Committee appointed Mr H.B. Longley, from Epsom, to be the architect.

---

1 Williams, *From Khaki to Cloth*, p.51.
2 *The Epsom Herald*, 13 February 1915.
3 Anon. Author, *University & Public Schools Brigade*, pp.79–80.

The preliminary work began in the Autumn of 1914 and when The King visited the different battalions on 13 October, he saw two companies erecting huts at Woodcote.[4] Early on, the contractor employed 150 men, but this rose over time to a staff of two general foremen, six sub-foremen and between 300 and 400 workers. Due to the slowness of the contractor getting started on the camp, and the onset of winter, it was decided to employ some of the men in constructing the camp. Even during the early stages of the building work about 400–500 UPS personnel, under UPS NCOs with practical engineering experience, were roped-in. Many of them were skilled in engineering, surveying, and draughtsmanship, and this accelerated the work. This arrangement worked well though Mr Boon had to intervene when there were threats of striking from the UPS men regarding having to work on weekends.[5] Kenneth Norman recalled that the camp took several months to build and required the men of the brigade to dig the ditches and construct the huts as cheap labour. He noted that the camp was so good that it was later turned into a convalescent hospital.[6]

Private Erroll David Shearn, aged 23, from Sidcup, Kent, was an articled clerk. He enlisted in 20/R Fus in September 1914 and was gazetted into the Hampshire Regiment in May 1915. Shearn was promoted Lieutenant in January 1916 and went to France in May 1916. (From the papers of E. Shearn)

The work on drainage ditches may have been the result of a rumour, heard by Terence Doherty, about the previous ditches having been condemned.[7] According to Erroll Shearn:

> Someone in authority conceived the idea that the gallant troops – the U.P.S. – might well be employed in helping to construct the camp ... Accordingly we were paraded and asked to fall into groups comprising 'skilled, semi-skilled and unskilled' workmen. I calculated ... it would be folly for me to designate myself as 'skilled'. On the other hand, 'unskilled' labour seemed to be likely to be required for heavy and burdensome jobs ... I promoted myself to the ranks of the 'semi-skilled' ... I have a distinct recollection of finding myself on the roofs of a number of huts hammering extremely long nails into boards of which the roofs consisted.[8]

---

4   The UPS History suggests work started by 13 October 1914. This contradicts a February 1915 article printed in *The Epsom Herald* which stated work commenced on 25 November. See Anon. Author, *University & Public Schools Brigade*, p.56, p.75.
5   IWM: G. Skelton Papers, 13966, p.27.
6   LA: K.V. Norman Papers, WW1/GS/1188, p.26.
7   IWM: T.O'C. Doherty Papers, 12961, undated letter.
8   IWM: E.D. Shearn Papers, 2033.

The construction of Woodcote Park Camp with the labour being provided by men of the UPS Brigade. Their civilian clothing suggests this was taken in late 1914 (Author's Collection)

According to Spurrell, 'Not being under proper supervision we made a rare mess of things … After a few days the contractors decided that it would be better to pay for their labour than use us …'[9] Meantime, Andrew Buxton enjoyed the work:

> When I was in Canada my great grief was to feel incapable of attempting to put up a simple wooden 'shack' which everyone seemed able to do for himself, though I tried hard to learn the system of it, and so you can imagine I delighted in this chance of experience in what is just similar work. The site is a splendid one, high up and looking on to the Grandstand of the Racecourse.[10]

In January there was an examination of the progress of the camp at Woodcote Park. The site of the Farm Camp which was intended for 18/R Fus was nearer completion, but more work was still needed before it could be occupied. There was a focus on the huts for 20/R Fus, presumably to allow the battalion to centralise having previously been split between two locations. Work on the huts was delayed due to poor winter weather and the completion date was difficult to ascertain.

Despite this 'captive labour', the cost of the camp would have been astronomical. However, in the meantime the cost of billeting soldiers in private dwellings had mounted; 'At 3s. 4d. per man per day the money spent on billeting alone worked out for 5,000 men for thirty weeks at over £177,000. Large sums of money, moreover, were spent with the tradesmen of the locality, and this, in the practical absence of racing, formed a considerable source of income for the

---

9   IWM: H.W. Spurrell 2138, p.8.
10 Woods, *Andrew Buxton*, p.44.

town.'[11] There were other demands on labour, such as for trench-digging. Meanwhile, training pressed ahead. Hugh Spurrell recalled that; 'During January and February 1915 training was very strenuous. We went in for a large number of tactical exercises. These, though more tiring, were more interesting than square work [i.e. drill]. We had many amusing battles on Banstead Downs, Walton Heath and the race course at Epsom. We felt very fit as a result of all this.'[12]

Occupation

By late February sufficient parts of the camp were ready for occupation. On 20 February the advance party of 20/R Fus moved into Woodcote Park Camp to prepare the way for the four companies. By the following Thursday the rest of the battalion had taken over. By 6 March two companies of 21/R Fus had taken the number in camp to 1,600.[13] Kenneth Norman recalled that 20/R Fus and 21/R Fus moved into Woodcote Park Camp first; presumably due to their separated locations at Leatherhead and Ashtead. A month later, 18/R Fus and 19/R Fus moved into the camp. Many were reluctant to leave cushy billets and good friends in various houses. On about 30 March 1915 D Company 19/R Fus, including Terence Doherty, moved into the hutted camp.[14] Around this time the battalion underwent a musketry course which was welcome news as it meant they had to stop digging trenches for a fortnight.

By 2 April 1915 there were two-and-a-half battalions in camp. The drum and fife band of 21/R Fus provided an attraction, it marched up and down the 'street' in the camp playing tunes in the evening. The entire Brigade was in Woodcote Park camp by 15 April.

Camp Quarters

Hugh Spurrell described the hutted camp for 21/R Fus:

> Woodcote Park was an irregular sort of place, straggling out along the side of a hill. The 21st Battalion lay furthest from Epsom and nearest to Ashtead and the racecourse. As the ground was uneven, one end of the hut was level with it, but to reach the other you had to climb up a dozen steps. Each battalion had it's canteen, divided into the orthodox 'wet' and 'dry' sections. Then there was the regimental institute, where there were a few lounge chairs and where you could read or write in peace. There was one Y.M.C.A. for the whole camp but it was not in working order when we left Woodcote. The Officers' and Sergeants' messes occupied other corners.[15]

A description of the new huts was furnished by Lance Corporal Reginald Piper (PS/3096, a Derby journalist) to The Derby Times:

---

11 Anon. Author, *University & Public Schools Brigade*, p.84.
12 IWM: H.W. Spurrell Papers, 2138, p.10.
13 Anon. Author, *University & Public Schools Brigade*, p.75.
14 IWM: T.O'C. Doherty Papers, 12961, p.9.
15 IWM: H.W. Spurrell 2138, p.11.

Our camp is composed of long structures of wood and corrugated iron, and we take an especial pride in them from the fact that we have taken no mean share in the work of construction. There is an average of forty-five or fifty in a hut, and every man has a spring bed, mattress, pillow, two sheets, and four blankets...[16]

The most detailed description of the camp was published in *The Epsom Herald*:

It was on November 25 that a start was made with the erection of the camp in that portion of Woodcote Park not laid out by the owners for golf, etc... The 18th Battalion quarters are among the trees near the farm, some little distance away from the quarters of the other three Battalions. Trees overhang many of the buildings, which comprise the huts in which will sleep, eat, read, and write the members of the 18th Battalion ...whoever occupy the huts among the trees will, on fine warm days, think they are having quite a pastoral existence... The huts of the 19th, 20th, and 21st Battalions are in two long lines with their fronts facing each other. Between the two lines is a space about 70 feet wide. ... A portion of this space is being utilized for the purposes of a road... The 19th Battalion

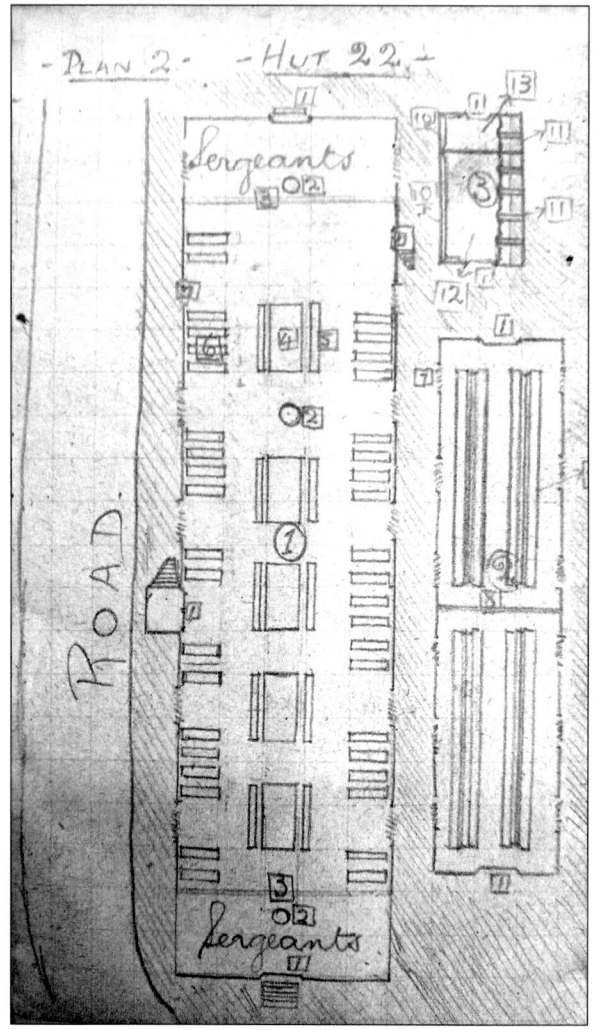

Once the UPS battalions took over Woodcote Park Camp the men lived together in these confined conditions. James Dykes' hut in the 20/R Fus lines was typical. The entrance was at the top and curtains separated the Sergeants from the remaining men. Over forty beds lined the side walls and tables and benches ran along the centre. Outside the hut on the top right are the latrines and lower right the washhouses. (From the papers of J.N. Dykes)

---

16 PS/3096 Lance Corporal Reginald George Piper, aged 34, served with No 2 Company, 21/R Fus, having enlisted in September 1914. He stayed in England with 29/R Fus and was discharged in January 1916 due to myopia, general debility and having had a nervous breakdown. Anon. Author, *University & Public Schools Brigade*, pp.78–79.

ANIMATED SCENE IN A U.P.S. HUT HALF-AN-HOUR AFTER REVEILLE.

Sometimes the warmth and relative comfort of the beds in the huts at Epsom made rousing the soldiers a difficult task. Piper recalled that the NCO in charge of a hut often had to resort to ungentlemanly language to get everyone out of bed at reveille. (*The Pow-Wow*, sketch by H.C. Mumby)

quarters are those nearest the main entrance from Headley Road, then come those of the 20th Battalion, the one which is going in first and then those of the 21st. The ground is undulating, and consequently few of the buildings are exactly on the same level … There are twenty-four huts to each Battalion, and each hut is 120 feet long and 20 feet wide. It has a timber frame, built on wooden supports, which prevent any part of the floor resting on the ground. The sides of the hut are constructed externally of corrugated iron, and lined internally with match-boarding … For heating purposes each hut has three slow-combustion stoves, and the lighting is by electric lamps… Each hut is to accommodate fifty men, for sleeping and for meals…[17]

The officers and sergeants of each battalion were accommodated in separate huts with reading rooms and ante rooms for their respective messes. There were three cookhouses; 18/R Fus and 19/R Fus had their own and 20/R Fus and 21/R Fus shared a double cookhouse. Ablutions were provided with hot and cold showers.

17   *The Epsom Herald*, 13 February 1915.

Camp Life

With movement into Woodcote Park the dynamics between UPS men changed: 'Living in huts was a great change after the cosiness and semi-privacy of good billets. Henceforth we ate and drank, slept and woke, dressed and undressed, in the company of forty of fifty other men.'[18] The author J.B. Morton, though writing fiction, recalled the morning routine of his protagonist.[19] Morton would likely have experienced the same at Woodcote Park. On reveille being called by the bugler, the soldiers got up, donned their trousers and socks and went to get washed and shaved at the washhouse. Afterwards, they dressed in their hut, folded their blankets according to regulations and swept around and under their bed. The sweepings were centralised and removed by the hut orderly. Around this time the mess orderly brought in a dixie of tea which they dipped their mugs into to drink. This was followed by a whistle or shout to get onto parade outside.[20] In addition, boots and buttons needed to be cleaned before the men went on parade at about 7:00 a.m. Reginald Piper suggested that many men may have found sympathy with the servants and chambermaids who may have waited on them in the past.[21]

James Hodson described a scene from the huts: 'Two tables away, a field telegraph is croaking and buzzing as a signaller taps out his practice message. A yard away is a little group telling stories, smoking, and playing bridge. A stretcher bearer has just returned from a twenty-five mile tramp... Men are beginning to come in, bringing cakes, jam, treacle, cocoa, biscuits, and so on, for supper.'[22]

The change in accommodation also led to a change in catering as Godfrey Skelton remembered:

> The food at this time was supplied by a contractor working in the camp kitchens, as we had no trained cooks of our own, and the food was extremely bad. Tea was almost undrinkable, made in big vats, and the milk was added by throwing in the tins of condensed milk after jabbing a hole in the top, the labels then soaked off and mingled with the other "foreign bodies" in the tea. We frequently did a hard day's marching or took part in mock battles on a cup of such tea, one slice of bread and butter and possibly a cold pickled mackerel or one rasher of smelly bacon.[23]

Some men were less fussy than others:

> We were in huts there ... they were quite good; everybody got a bed ... there was a big table down the centre and that was for feeding... the food was brought in great big dixies and we'd feed along these big tables and we used to use them to play cards... It was big helpings of army food, stews every day and maybe you'd get a lump of bacon in the morning, plenty of bread and plenty of butter, plenty of cheese about; more often than not for lunch

---

18 Williams, *From Khaki to Cloth*, p.51.
19 PS/1847 Private John Cameron Andrew Bingham Morton had attended Harrow School and Worcester College, Oxford, and was a journalist and author by profession. Morton enlisted on 15 September 1914, aged 21, and was from London.
20 J.B. Morton, *The Barber of Putney* (London: Penguin, 1939), pp.11–12.
21 Anon. Author, *University & Public Schools Brigade*, p.78–79.
22 Hodson, *Soul of a Soldier*, p.20.
23 IWM: G. Skelton Papers, 13966, p.31.

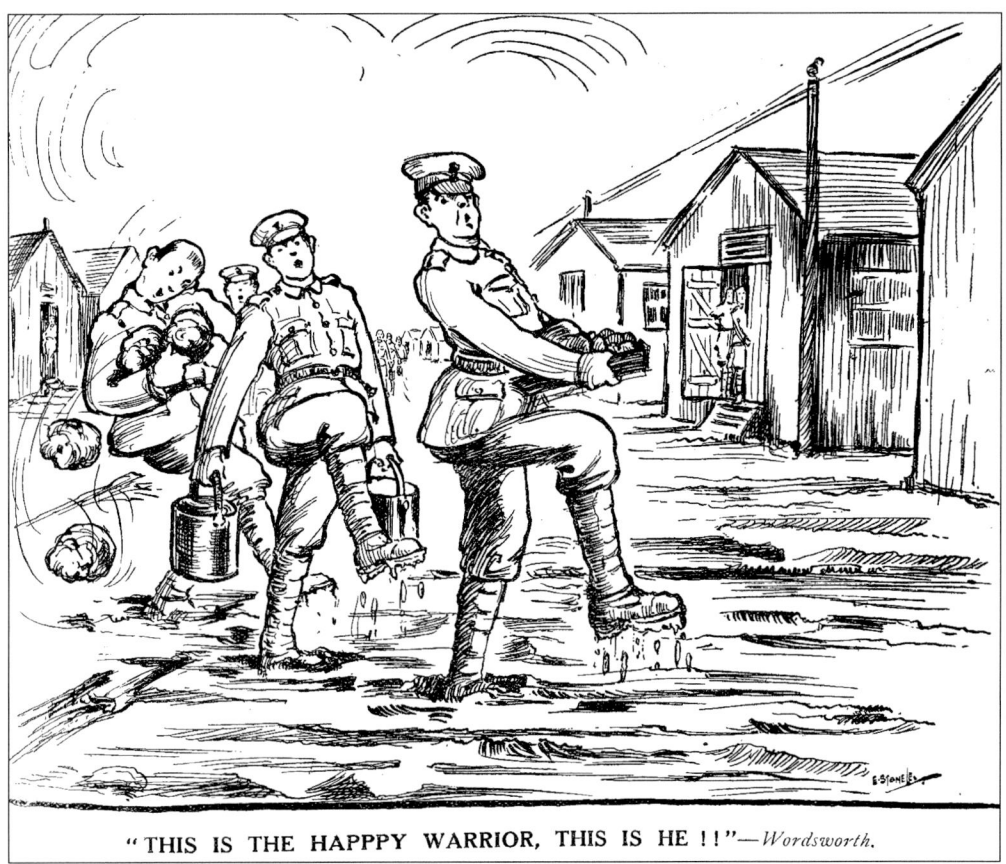

"THIS IS THE HAPPPY WARRIOR, THIS IS HE !!"—*Wordsworth*.

UPS mess orderlies carry a cooked meal across camp to their hut. This shows the standard of rations at Woodcote Park, the muddy conditions under foot, and the spacing between the huts on the main avenue through camp. Part of the same route is now a private road for luxury homes. (Sketch by Ernest Stoneley: reproduced courtesy of Janet Wood)

a great big stack of stew, but sometimes a chunk of beef. Wholesome, and plenty of it, but it was crude. You might eventually get a lump of pudding or a basin of rice as a treat… No complaints about quality or quantity but no finesse.[24]

Despite the good work of the UPS Committee there were still negative comments regarding the quality and quantity of rations; 'Six men to a loaf today. Salmon and Gluckstein are making a good thing out of this.'[25]

---

24 IWM: Donald Price Sound Recording 10168, Reel 2.
25 IWM: H.W. Spurrell 2138, p.10.

A group of seemingly happy UPS men enjoy a meal outside their huts at Woodcote Park Camp. The relaxed attitude and variations in dress suggest that they are enjoying afternoon tea after a day of training. (Author)

Ongoing training

Training continued and equipment and uniform arrived periodically. By early 1915, leather web equipment and uniform had been issued. Men then carried out route marches of up to 18 miles according to Godfrey Skelton. The leather webbing was not well-liked as it was fiddly to take apart and adjust.

How the brigade was equipped and looked on parade might appear to be cosmetic, but it mattered when external visitors inspected the Brigade to see how ready they were for service overseas. Another major inspection occurred after arriving at Woodcote Park:

> On March 27 General Sir Archibald Murray, K.C.B., C.V.O., D.S.O., Deputy Chief of the Imperial General Staff [CIGS], inspected the Brigade on the Downs. This inspection was a most minute and careful one, and such as would enable an accurate judgment to be formed of the forces. The Brigadier was later instructed to express his satisfaction with what the General had seen, and to say that the General would recommend certain steps with a view to the Brigade being included in one of the early armies going to the Front. The Brigadier then added the following: "The Brigadier hopes that all ranks will realize that the proficiency of the whole Brigade may be judged by the bearing and behaviour of the smallest unit, and that each individual will do his utmost to prepare and perfect himself for war." These words of the Brigadier, coupled with the General's message, were welcomed by the Brigade, as they were the first official intimation that the Brigade was likely to get a "move on" in the near future…[26]

26 Anon. Author, *University & Public Schools Brigade*, pp.79–80.

HOC LABOR HIC OPUS EST.

Hoc labor hic opus est ('this is the problem, this is the hard task'). Illustration depicting the difficulty of assembling the P14 leather infantry equipment issued to the UPS and the consequent discomfort of wearing it. The comparative simplicity of an officer's Sam Browne belt may have added to the frustration for men considering commissions. (Sketch by Ernest Stoneley, reproduced courtesy of Janet Wood)

Hugh Spurrell provided an un-official summary:

> The second inspection was by Sir Archibald Murray and it was understood that a limited number of commissions would be given after the inspection and that afterwards commissions would close down until the battalion had been overseas. I did not apply, as this wound have upset my whole plan … The inspection was more interesting than the previous one had been. The General walked along the lines and we marched past in column.[27]

A few men departed to commissions shortly after. According to the UPS chronicler; 'The question of commissions, moreover, was an exceedingly delicate one, and had to be handled with considerable tact.'[28] More UPS men were needed to fill these gaps. When Leonard Salter, a

27   IWM: H.W. Spurrell 2138, p.11.
28   Anon. Author, *University & Public Schools Brigade*, p.80.

21-year-old farmer from Aldershot, joined 19/R Fus in late March. He was assigned to a recruit drill squad after a week in Epsom; whilst he awaited this training, he was issued his equipment and covered the duty of post orderly.[29] After parading with different companies, including E Company, he passed out as a recruit and was assigned to No 13 Platoon, D Company, as a scout. Leslie Woodhouse Cubitt Ireland arrived in Epsom as a recruit on 21 April having walked the 2.5 miles to camp; 'I continued loafing until 5.15 because the adjutant and half the camp were at the Races. He came in eventually and asked me if I was Irish and drafted me off to where I am [E Company, 19/R Fus]. I sleep in a hut with about 50 others on a bed with a spring mattress …[30] Leslie Ireland noticed that a lot of other new men were arriving; 'Recruits are joining daily from all over the country. In our hut we have one fellow who came from Mexico City to join, another from Valpariso, another from Queensland. We have about 5 or 6 fellows from Manchester – quite a colony…'[31]

However, though the recruiting for the brigade had changed, there were high standards expected for soldiers within 21/R Fus. The men of the Battalion were forced to submit to a hut inspection every Saturday in the same manner that was inflicted on the Brigade of Guards. This involved extensive washing of hut floors, displaying of equipment and holding up shirts and socks to inspecting officers. This was not a popular activity for the men but was clearly enjoyed by George Fletcher, the RSM. In fact, Fletcher was so much in his element that he was lent to 19/R Fus to lick both soldiers and officers into shape at battalion drill. William Bentley described him as being over 6ft tall, having a very soldierly physique and was without any additional fat. He was dressed immaculately and had a 'tremendous' moustache. His eyes captured even the smallest movement and he responded with a bellow that could be heard anywhere on the drill square. He had the whole battalion marching up and down the parade square for long periods until the men were exhausted. He was unafraid of drilling and critiquing the officers in their sword drill. Bentley considered that this was unforgettable and of invaluable benefit to his battalion.[32]

Bentley considered men of 19/R Fus to be similar in physique to guardsmen. However, he also recalled there being two misfits in B Company; one was a rough Irishman and the other a coarse Yorkshireman who came from Hebden Bridge.'[33] Though no-one knew how they came to join the UPS they became stalwarts and were loved by the UPS-men for being kindly and down-to-earth.

Leisure time

There were plenty of facilities in the new camp to enjoy, as well as activities for the men in the surrounding area. The camp boasted the following for training, relaxation and amusement for the men:

---

29 IWM: L.E. Salter Papers 21103.
30 IWM: Documents.14441, date 2006-01-31: Papers of L.W.C. Ireland, letter 21 April 1915.
31 IWM: L.W.C. Ireland Papers, 14441, letter 28 April 1915.
32 LA: WW1/GS/0125: Papers of W.G. Bentley, p.7a.
33 LA: WW1/GS/0125: Papers of W.G. Bentley, p.7.

Men of 19/R Fus on parade, likely at Woodcote Park. Compared to the two previous pictures of 21/R Fus on parade, more of the men appear to be fully equipped and many are armed. The lack of web equipment, high turnout and number of officers and NCOs around suggests that this was part of a larger inspection rather than a precursor to conducting daily training. (From the papers of L.E. Salter)

Is it necessary to say that a canteen, fitted with "wet" and "dry" bars, will form part of the Battalions' equipment? But the biggest building … is the recreation hall … It is 130 feet long and 60 feet wide, and the distance between the floor and the eaves is 22 feet … This hall is to be fitted with a stage, and will be very suitable for lectures, concerts, picture displays, and other forms of entertainment … Standing isolated from the other parts of the camp is the miniature rifle-range, a closed-in place, having skylights. There are five miniature ranges in it, fitted with Solano and bull's-eye targets. The length of the range is 100 feet. The camp is being fitted up with postal and telephone facilities, and anyone who wants a newspaper or a shave will find that his wants can be readily attended to. A power station, to generate electricity for the 19th, 20th, and 21st Battalions has been erected at one corner of the camp. The 18th Battalion will be supplied with current from Epsom electricity main, which goes as far as Woodcote House … The Epsom water supply is laid on to the camp, and, as a safeguard in case of fire, hydrants have been fixed in various parts. There are also [anti-fire] hand-grenades provided. The camp has been connected up to Epsom's system of drainage, having regard to the designs prepared by the War Office for typical camps. … A Y.M.C.A. building has yet to be erected. The Brigadier General and his staff will be quartered in the Woodcote Park House.[34]

Soon after the Spring race meeting at Epsom Downs took place and many UPS men went to visit, some to view the spectacle, some had a 'flutter':

The crowd was immense and the noise they made, combined with the shouts of the bookmakers, created a perfect babel of sound. I got a place opposite the Grandstand – free to all

---

34 Anon. Author, *University & Public Schools Brigade*, pp.77–78.

Photograph depicting the scale of Woodcote Park camp and the effort required to build it. On the right are accommodation huts stretching into the distance. The centre shows the recreation hall and on the left is the rifle range. (Author)

soldiers in uniform. So little, however, could be seen that I got tired and went back to camp. On that evening we had an excellent concert in the Cinema Hall – a huge place in the centre of the camp. It was a variety show with a vengeance – a Salome dance was followed by an address from Bishop Wheldon…[35]

Leonard Salter's diary recorded that he had the afternoon off on account of the race meeting and was able to attend the first day on 20 April.[36] He lost a bit of money on bets and recalled that the concert occurred on the 21st and that there was an orchestra on the 25th. Moving to Woodcote Park provided an added attraction to keen golfers who could play on some of the nearby golf courses. Whilst at Epsom several UPS soldiers got married in nearby churches and this continued whilst at Woodcote Park.

'Active service conditions'

During their time at Woodcote 21/R Fus, under Lieutenant Colonel Stuart Wortley, took part in a route march, under active service conditions, down to Stuart Wortley's ancestral home. After parading at 6:00 a.m. on the first day they marched over Headley Heath and through Dorking and fell out to eat some stew for lunch. At about 2:00 p.m. they moved off again and were billeted overnight in Rusper village. Here they were supposed to be treated to a Spartan existence. However, as they were the first marching troops to pass through since war was declared the people of the village provided every assistance and treat to the Fusiliers. The Colonel's instructions to the village not to do so were ignored and the evening stew was ignored

---

35 IWM: H.W. Spurrell 2138, p.8.
36 IWM: L.E. Salter Papers 21103.

in favour of the villagers' hospitality. The vicar even arranged a concert in the church after the pub was put out of bounds.

After parading at 6:00 a.m. next day the battalion marched off and passed Ockley, Leith Hill, Holmbury St Mary, Gomshall and Shere, where they ate lunch. After climbing the hill out of the village they stepped out to complete the march at Ockham having gone through Horsley. They covered about 24 miles in a day. On arrival, Colonel Stuart Wortley retired to his Aunt's stately home. Spurrell's No 4 Company was the last to arrive and their evening meal consisted of undrinkable tea and the buns that had been provided were insufficient in number. The men retired to a loft above the coach house to sleep but were turfed-out to provide outposts to guard the main house just as it started raining. A piquet was set up at the lodge gate. Spurrell recorded the comedy scene that followed:

> At about ten o'clock Otto, who was blotto, came round the posts. The thing that worried him most was the direction of the supposed enemy. The sentries, he assured the corporal, were watching the wrong front. He pointed to the house from which he had just come and told them that they must watch that … he congratulated them [the sentries] on their good fortune in having their heads under a tarpaulin. Were not many men in France devoid of such shelter? This reduced the piquet to a spirit of resignation…[37]

Next morning the outpost positions were collapsed having had little sleep. After a quick breakfast the march home commenced. On arrival back at Woodcote Park the Colonel kept the men fallen-in for a congratulatory talk. They had covered 54 miles in two-and-a-half days. Other battalions went on longer toughening marches. Salter's diary recorded that on 3 May he covered 22 miles between 9:30 a.m. and 5:00 p.m. going via Kingston, Surbiton and Esher.[38]

The amenities of Woodcote Park would not be enjoyed by the UPS Brigade for very long. In the words of Hugh Spurrell:

> Towards the end of March it became pretty evident that our days at Woodcote were numbered. Rumour said that we were going to move, but whether to France, Egypt, Leicestershire or the Isle of Man, nobody knew. The camp at Woodcote was to be turned into a convalescent depot…[39]

The camp at Epsom had to be thoroughly 'spring cleaned' before leaving. That the battalion was going somewhere was certain. Where it was going, in both the short and long terms, was the subject of rumour:

> The Egypt rumour seems the most consistent, as it was positively attested that the Secretary had ordered helmets. This was only "knocked out" when the Secretary, on being asked if it were true, stated "Oh! yes, and I have ordered 5,000 camels." France, Egypt, Aldershot, and so forth were mentioned by turns, and various others are chronicled in The Pow-Wow.[40]

---

37 IWM: H.W. Spurrell 2138, p.13.
38 IWM: L.E. Salter Papers 21103.
39 IWM: H.W. Spurrell 2138, p.14.
40 Anon. Author, University & Public Schools Brigade, p.83.

Departure

According to Kenneth Norman, Epsom was one of the most enjoyable places he based at during his military service.[41] The people of Epsom were equally sad to see the UPS Brigade go. A local lady wrote to the family of one of the soldiers billeted with her that she was so sorry to see him and his friend depart after eight months.[42] According to one officer; 'We missed the dear people of Epsom, Ashtead, and Leatherhead, who had nearly killed us all with kindness.'[43] The Brigade history recorded the departure from Epsom:

> On Tuesday at the camp there was all the activity and excitement natural under the circumstances … The men who were to leave the Woodcote Park camp felt that they would not leave it without some regrets, and without doubts as to whether they would ever live again in such a comfortable and prettily situated camp … The men, in the daylight before the evening shadows fell, took a last look round.
>
> Epsom Downs Station at the hour of midnight is usually wrapped in silence, but this was not the case on Tuesday night. Eight trains, each made up of carriages, luggage-vans, horseboxes and trucks for horses, were despatched … between 12 and 8 a.m. The first train left at 12, and there was just an hour's interval between the departures of the others. Both the inhabitants of Epsom and the UPS men, too, would have liked the departure to have taken place in the daytime so that there could have been a "regular send-off." … The entraining of the men was carried out in perfect order, and as each train passed out cheers were raised by the handful of people standing on the platform … there was no fuss or ceremony. Everything was conducted with quietness and precision, and as the men left Epsom and went out into the night, away from where their Brigade has been since last September, they were carefully obeying instructions to pull down all the blinds. Their action symbolized the pulling down of the curtain at the close of the work and play performance by the UPS Brigade on the stage of Epsom.[44]

Epsom was considered to have been renamed 'Upsom'.

---

41 LA: K.V. Norman Papers, WW1/GS/1188, p.42.
42 LA: K.V. Norman Papers, WW1/GS/1188, Letter by Mrs Elsworthy 11 May 1915.
43 Gibson, *Postscript to Adventure*, p.136.
44 Anon. Author, *University & Public Schools Brigade*, pp.85–86.

5

Training at Clipstone

We shall be 100 miles from anywhere

Training at Clipstone was of a less systematic nature than it had been before ....our training was very scrappy, and ... we felt that we had learned very little there.[1]

It is awful, we shall be 100 miles from anywhere and I don't see how I can get home for weekends ...[2]

There was allegedly some confusion during the communication of the destination of the Brigade that confused the location between 'Clipston' near Market Harborough and 'Clipstone', near Mansfield (the correct destination).[3] Luckily this was straightened-out before any troops departed. According to Private Leonard Salter's diary, some of the first UPS men to depart Woodcote Park for Clipstone were the battalion scouts of 19/R Fus. The UPS Brigade history summarised their departure:

> The exodus really began on Sunday (May 9, 1915) when about thirty members of the Brigade, belonging to the Scout contingents, started to walk all the distance to Clipstone. They had a great journey before them, and the men at the camp turned out of their huts to give the little party a rousing cheer as it set forth on its long pilgrimage.[4]

Terence Doherty sent a postcard home which was posted at the first overnight stop at Feltham: 'I joined the Scouts, it is very interesting work. We started this morning, about 30 of us to march to Clipstone in Notts. We are taking as long as we like as it is going to be an instructional march. It will be luck[y] if we get a roof over our head every night... From Epsom to Clipstone is about 155 miles.'[5] The scouts marched between fifteen and eighteen miles a day, sometimes through heavy rain and did several sections as night marches or across country. One day they marched in the heat and Doherty got a touch of sun stroke. They received a great reception on

---

1   IWM: H.W. Spurrell 2138, p.16.
2   IWM: T.O'C. Doherty Papers, 12961, letter 4 May 1915.
3   Anon. Author, *University & Public Schools Brigade*, pp.83–84.
4   Anon. Author, *University & Public Schools Brigade*, pp.84–85.
5   IWM: T.O'C. Doherty Papers, 12961, letter 10 May 1915.

arriving in Clipstone at 2:00 p.m. on Thursday 20 May. The scouts were given a long weekend of leave to recuperate; Salter promptly took a train back to London to see his family. It should be noted that due to Clipstone's distance from many locations, weekend leave was often extended until Monday nights to allow for UPS soldiers to travel home.

On Monday 17 May advance parties of 100 men from each battalion departed for Clipstone. The 18/R Fus party consisted of part of B Company which included Kenneth Norman. They went on a special 'through-train' to Edwinstowe which was packed with soldiers. After marching for four miles some huts became visible in the distance. Clipstone Camp seemed to be and made Woodcote Park look like a palace by comparison.[6] These men spent two days of hard graft to get the camp ready. No food had been laid on for them.

The rest of the Brigade would depart on Wednesday the 19th for Clipstone. Private Spurrell was in the 21/R Fus main body:

> People began to come round and check the stores and at last definite news arrived… We spent a whole night in packing up and storing our bedding. Eventually we paraded at four o'clock in the morning and marched off to Epsom Downs Station. By seven we were passing through London.[7]

Leslie Ireland was part of the 19/R Fus 'main body':

> Tuesday reveille was at 5 and it was the day we moved. Of course there was a terrific amount to do; such as packing up beds, mattresses, blankets, loading vans and trolleys with kit bags and all such like baggage, cleaning all the huts, cookhouses etc. At 6 o'clock that evening we had some tea consisting of bread, butter, jam and tea. We finished work about then and had to parade with full pack at 7 o'clock for inspection and roll call. We were dismissed about 7.45 and we [were] told to parade again at 12 o'clock midnight. … we paraded … at 11.45 for letters and started from Epsom at 12.30 a.m. Wed[nesday] morn[ing]. We marched to Epsom Downs Station some 2–3 miles and then loafed about waiting for a train to turn up. At length we started from Epsom about 3 a.m.[8]

On arriving at Edwinstowe the men of the main body detrained and marched to Clipstone. Some arrived at about 4:00 p.m. Leslie Ireland's party marched from the station in drizzle and a cold wind and arrived at 10:15 p.m.

Despite the advanced party having worked hard, the camp was incomplete. The area allotted for the parade square was covered with bracken and the huts were surrounded by gorse bushes. Hut roofs were not covered in roofing felt and were not weatherproof; though electric lights were to be fitted no bulbs had arrived. Spurrell mused; 'Rome was not built in a day. Once more we were to be distinguished members of the Camp Improvement Society.'[9] The troops stuffed palliasses with straw and settled down for the night. *The Pow-Wow* quipped that the UPS were 'the official Fatigue Party to the Army.'

6 LA: K.V. Norman Papers, WW1/GS/1188, p.46.
7 IWM: H.W. Spurrell 2138, p.14.
8 IWM: L.W.C. Ireland Papers, 14441, letter 16 May 1915.
9 IWM: H.W. Spurrell Papers, 2138, p.14.

Terence Doherty, like many, was reluctant to move away from Epsom:

Dear Mother,
News!

We are leaving Epsom on Wednesday next. We are going to Clipstone … It is awful, we shall be 100 miles from anywhere and I don't see how I can get home for weekends. I suppose it is awfully stupid to grumble, but it is a bit thick. We are going to march to St Pancras – at least rumour says so …[10]

The arrival at Mansfield was dreary; Godfrey Skelton recalled:

It was raining and bitterly cold, the huts were only half finished and there was nothing to eat of any description. It appeared that the Army did not intend feeding us, so, after a "whip round", I [was] sent to the village for a dozen packets of Quaker Oats. We scrounged some wood and, making the porridge in tea buckets … the Platoon was satisfied …[11]

Leslie Ireland added regarding the 'rations' they received on arrival, unfed, after an 18-hour journey; 'There was enough for 6 men and we were 30 to mess …we were in a state varying from utter exhaustion to extreme rebellion.'[12] The huts were poorly built with huge cracks between boards, no lights and doors that would not shut. Walter Medlicott stated; 'It is a depressingly barren spot, stumps of black heather in all directions, coal mines in the near distance and a big heather moor towards Edwinstowe. It is all part of Sherwood Forest…'[13] Ashley Gibson recalled that; 'Mansfield, our nearest and, indeed, our only town, seemed to us the last place God could possibly have made.'[14] Morgan Williams observed that Clipstone was a 'Shocking hole, miles from anywhere.'[15]

There was no cooked food available from the kitchens and rations were distributed for the first few days; 'For the next two days we feasted on biscuits and cheese and unsweetened tea; after that the new system got into working order and we lived like kings.'[16] According to Spurrell; 'No longer were we to be rationed by Gordon Selfridge and his committee. From now on we were to be fed by the ASC'[17] Medlicott was pleased with the new catering:

Food is much better than at Epsom. With Government rations you get very little variety – meat, bread, potatoes, tea and oddments – but each battalion manages its diet sheet and by reducing bread ration they get flour for pies etc. Our cooks have done well and will improve

---

10 IWM: T.O'C. Doherty Papers, 12961, letter 4 May 1915.
11 IWM: G. Skelton Papers, 13966, p.33.
12 IWM: L.W.C. Ireland Papers, 14441, letter 16 May 1915.
13 IWM: W.B. Medlicott Papers, 1708, pp.2–3.
14 Gibson, *Postscript to Adventure*, p.136.
15 Williams, *From Khaki to Cloth*, p.52.
16 IWM: H.W. Spurrell Papers, 2138, p.14.
17 IWM: H.W. Spurrell Papers, 2138, p.14.

Cartoon highlighting the absence of food on arrival at Clipstone Camp. (*The Pow-Wow* sketch by Ernest Stoneley, reproduced with the permission of Janet Wood)

I expect. We buy extras like marmalade and potted meats as before. I could do with a cake shortly, it goes well with our nightly cocoa.[18]

As the Army made a special messing allowance of 5½d a day per man there was plenty of money for the battalion cooks to spend on extra vegetables and higher quality foodstuffs. The food was inconsistent in quality and quantity as Terence Doherty wrote in letters home, 'the food here is fair but there is not enough …'[19] 'Today was a good example of a bad dinner. We had enough

---

18   IWM: W.B. Medlicott Papers, 1708, p.4.
19   IWM: T.O'C. Doherty Papers, 12961, 1 Jun 1915.

meat, but it was bad and smelt. There were not enough potatoes to go round and no pudding, my dinner consisted of bread & marmalade.'[20] Another man wrote home that food quickly improved with roast beef, potatoes, pears, apricots and bread for lunch.[21]

Talk of food might seem trivial but it's quality clearly related to the morale of the men and demonstrated the unpreparedness of military institutions for managing and training Kitchener's Army.

A local newspaper described the place:

> … the camp is scarcely an inspiring sight. The wooden rectangular structures, which differ from each other only in the matter of size, are painted an inconspicuous grey, and the great piles of timber, the toiling engines, the uneven ground, and the clang of hammers suggest some great industrial undertaking rather than the panoply of war. Even the lighting installation is not complete, and candles are the chief form of illuminant.[22]

All the UPS men were initially employed making the camp habitable. It would have been disconcerting to know that work on the camp had begun in December 1914. The parade ground needed to be cleared and levelled, a rifle range and bayonet fighting area built. Early on, time was also spent in making roads. The parade grounds were so uneven that men from 21/R Fus toiled for two weeks moving earth from the higher ground in coal tubs to enable a flat surface to be made. Furthermore, a breastwork was built along the front of their camp which was anointed with the regimental crest. In the words of Godfrey Skelton; 'Serious training was at a minimum and really we were used as a large working party to help complete the new camp.'[23] A range was constructed at Rufford.[24] Laying out of ranges at Clipstone was still taking place in early July.[25] For the range creation, PS/5062 Private Ernest Holden's party of 30 men moved an estimated 15 tons of sand in a day as they excavated a 7ft deep trench. *The Pow-Wow* recorded: 'Extensive excavations some three miles from camp, at Rufford, did at least show some semblance to a range; and the levelling of parade grounds and digging of trenches went on apace.'[26] Once this hard work was done some soldiers created gardens around their huts in their spare time.

The camp was on a road running between Mansfield and Edwinstowe and was laid out for twelve sets of lines; each would accommodate a battalion of about 1,000 men. The first four camps, consisting of No 5 to No 8 lines were occupied by the UPS Brigade.[27] Hugh Spurrell described the surroundings:

> In front of the camp the ground was broken. It stretched downwards gradually to a small lake, known as Vicar Water. The heather reached the water's edge. On the other side were a few fields and then a branch line of the Great Central Railway. This was at the time under

---

20   IWM: T.O'C. Doherty Papers, 12961, 19 June 1915.
21   IWM: L.W.C. Ireland Papers, 14441, letter 16 May 1915.
22   Anon. Author, *University & Public Schools Brigade*, p.88.
23   IWM: G. Skelton Papers, 13966, p.33.
24   IWM: H.W. Spurrell 2138, p.17.
25   IWM: W.B. Medlicott Papers, 1708, p.22.
26   Anon. Author, *University & Public Schools Brigade*, p.93.
27   Some source state 19/R Fus was in Camp No. 9.

construction. Beyond the railway the heath stretched for miles. The headgear of several collieries could be seen in various parts of the moor.[28]

According to Holden 20/R Fus was in No. 7 Lines:

> This place is simply awful. We came here from Epsom, from scorching hot weather with everything at the height of bloom … and we arrive here in a cold, wet drizzle … and march up to a great bleak moor with the trees almost bare … and a poor-looking miserable town in the valley below … if there is a worse-looking town in the Potteries I shall be surprised. Of course, it has poured down since we arrived. The huts are damp and draughty, and it is so cold that I am wearing all my flannels and cardigan and am still cold.[29]

Despite being May, and chilly, the Army did not provide fuel for the huts. To avoid the cold some personnel by 'acquired' coal from near the railway. There were other hazards than the cold; German measles spread and Leslie Ireland was prevented from returning home on leave due to catching a dose. He got to experience the well-appointed camp hospital for five days.

A bigger town than Epsom, Mansfield had a theatre and a music hall. Though it seemed far away by distance several taxis and hire cars soon appeared to cater for journeys by some of the better-off UPS men. Trips to Mansfield cost two shillings by car, but, according to Holden, 'there is absolutely nothing at all to do in Camp.'[30] The reception of the UPS in Mansfield was cool to start with allegedly because a bad reputation had followed the Brigade from Epsom. The situation soon thawed as the men of the battalions got to know the area and the townsfolk. This may have been helped by a critical article published in *The Pow-Wow* bemoaning the departure from Epsom and indirectly criticising the reception in the Midlands. A branch of the YMCA was later opened in the town run by local volunteers.

Several men preferred to walk the beautiful countryside nearby rather than visit Mansfield; especially as the summer weather in June and July 1915 was quite good. Terence Doherty recalled; 'The woods are just carpeted with Bluebells, quite a carpet of blue.'[31] Some men watched birds on the nearby moor. A keen ornithologist, Captain Cecil Mears was out bird watching during free time at Clipstone. Those who were enthusiastic for cricket and tennis often had opportunities to play. Hugh Spurrell sent for his bicycle and either walked or rode around the surrounding area as far as Lincoln, Southwell, Worksop, Chesterfield and Bolsover on free weekends. Ernest Holden sent for his motorbike. Kenneth Norman made two female friends whilst at the music hall in Mansfield and they invited him to tea and took him round some of the local sights in their motor car. Godfrey Skelton also preferred to travel to Nottingham: 'Altogether Clipstone Camp was pretty good, with our visits to Mansfield, and the boating on the river at Nottingham, especially when accompanied by some of the local girls.'[32] William Bentley was also complimentary of the beauty of the local girls but because he could not easily get home to Stoke-on-Trent he ended up playing cards, gambling and getting into debt.[33]

28 IWM: H.W. Spurrell 2138, pp.14–15.
29 LIM: Letters of E.A. Holden, courtesy LIM, p.32.
30 LIM: Letters of E.A. Holden, courtesy LIM, p.34.
31 IWM: T.O'C. Doherty Papers, 12961, undated.
32 IWM: G. Skelton Papers, 13966, p.33.
33 LA: WW1/GS/0125: Papers of W.G. Bentley, p.8.

Some men hated their time at Clipstone for other reasons. Lance Corporal Joseph Leather's private correspondence highlights that he was not jingoistic and disliked both the Army and his fellow soldiers:

> The Army is undeniably the most dismal experience I have had yet … but the happiest moment of my life lies before me, the moment, I mean, when I throw my khaki back at the Quartermaster Sergeant & get back into my own comfortable clothes. I haven't got the faintest enthusiasm for this war; and I feel pretty sure now that the question is not whether we shall win or lose, but whether we shall lose or just manage to save our bacon. The idea of beating the Germans is so much nonsense, and you only have to be in the Army to see the mess and muddle of everything, the lack of training in the men, and the thick headedness of the officers – nine out of every ten of whom I wouldn't have touched with a yardstick in civilian life … I compare the foolish, ill-educated, stupid officer here with the German officers, many of them my own dear friends indeed…[34]

Joseph Leather was aged 31 and from Moss Side, Manchester. He was the son of a colliery winding engine operator who had not attended a public school. Through his own energy he had secured employment as a journalist for *The Manchester Guardian* (1910–1914). The nature of his experience of living in Germany is unknown but Leather was clearly troubled with the idea of fighting his erstwhile friends.

The Bishop of Southwell came to conduct a service for the men of the Brigade. To the joy of some 'he preached for exactly four minutes, a practice that might be copied by Army padres. It was a curious change after the half hour denunciations … to which Capt. Clarke usually subjected us.'[35] Clarke's services were often conducted in the open air and Kenneth Norman considered them 'impressive'. Likewise, Terence Doherty initially considered; 'Our Chaplain is a very nice man…'[36] He soon changed his tune; 'Clarke is so dismal and dreary.'[37]

According to Morgan Williams the time at Clipstone was generally unexciting; 'Experience now became a series of meaningless happenings. We dug trenches only to fill them up again. We received orders which were cancelled before we had begun to carry them out and in imagination we frizzled in Egypt or froze on the Russian Front. The two things that broke the monotony were fire and water.'[38]

Fire, Water and Farming

Beyond the training that was conducted, two activities were commonly remembered by UPS-men who stayed at Clipstone in the summer of 1915. On several occasions the UPS battalions had to be sent to extinguish brush fires on the moors. Morgan Williams recalled; 'All around us was a great heath and in the blistering weather we often turned out in trousers and puttees, shirts and

---

34 PS/5204 Lance Corporal Joseph Henry Leather. IWM: Documents 2783: Papers of J.H. Leather, Undated Letter.
35 IWM: H.W. Spurrell 2138, p.16.
36 IWM: T.O'C. Doherty Papers, 12961, letter 13 December 1914.
37 IWM: T.O'C. Doherty Papers, 12961, undated letter.
38 Williams, *From Khaki to Cloth*, p.54.

boots, to fight immense fires with spades and buckets of sand. We returned to camp aching in every limb, our faces blackened with smoke, like miners coming from their pit.'[39] However, this took on a slightly comical tone when; 'After making ourselves filthy and getting blinded with smoke on several successive days we received a polite intimation from the Rufford estate agent that they would be obliged if we allowed the fires to take their course. They had been lighted on purpose. Every year the heath was burned out as soon as it was dry enough.'[40] Once the moors were burnt and the weather grew warmer the ash and dust from the roads could be quite severe and almost choking.[41]

The other activity was bathing in Vicar Water, the nearby lake which was excellent for relaxing or swimming after a hot day of work.[42] There were perks to life at Clipstone, especially for those men able to swim strongly. Morgan Williams liked part of his duties:

> There was also the lake, half-a-mile from our hutments. It was a lovely stretch of water, a little choked with weed, and in virtue of being able to swim well I formed part of a bathing picket which paraded with commendable regularity. To arrive on duty clad in bathing costume, greatcoat and gym-shoes, and to patrol the lake all day long in a boat was the best job I ever had in the army. We were supposed to assist swimmers who got into difficulty, but much of our time went in practicing for the water-polo matches which we played in the evenings against other companies and battalions.[43]

Leslie Ireland recalled; 'I had my first route march – 23 miles. I was surprisingly fresh after it, feet a wee bit sore … Immediately on return I went for a swim in a pond a short distance from the camp which proceeding I have repeated several times and intend to continue.'[44]

Whilst at Clipstone the War Office decided to release men temporarily from training to help gather in the harvest. This was forced on the UPS battalions and likely further retarded their training. In Morgan Williams' case he and another soldier were sent to his uncle's farm at Llantrisant for two weeks. Some who remained on camp bemoaned their lack of connections to agriculture.[45]

Training

Kenneth Norman recalled that the training at Clipstone was like Woodcote Park and Epsom but that the tactical field exercises in Sherwood Forrest were on a bigger scale.[46] According to Hugh Spurrell:

---

39 Williams, *From Khaki to Cloth*, p.54.
40 IWM: H.W. Spurrell 2138, p.17.
41 LIM Archive: Letters of E.A. Holden, courtesy LIM, p.35.
42 LA: K.V. Norman Papers, WW1/GS/1188, p.52.
43 Williams, *From Khaki to Cloth*, p.54.
44 IWM: L.W.C. Ireland Papers, 14441, letter 12 June 1915.
45 IWM: T.O'C. Doherty Papers, 12961, undated letter.
46 LA: K.V. Norman Papers, WW1/GS/1188, p.48, p.50.

Sketch of UPS men using Vicar's water for bathing and boating. The lake was visited so often that a lifeguard was provided from within the ranks of the UPS.
(*The Pow-Wow* sketch by Wilfred Phythian)

Training at Clipstone was of a less systematic nature than it had been before. We all went through a course in the theory of Musketry and a number of tactical schemes were done. There were numerous route marches and what was known as "observation" was in great favour. This included sketching, writing reports, map and compass work and the use of field glasses – in short, intelligence work of sorts. I have seldom seen so serious an attempt to train the rank and file in this branch of training as was made by our battalion … Apart from the above mentioned forms, our training was very scrappy, and … we felt that we had learned very little there.[47]

47 IWM: H.W. Spurrell 2138, p.16.

Likewise; 'Most of the tactical schemes which we tried proved farces. The officers and N.C.O.'s kept all the information to themselves and we did not gain much knowledge in consequence.'[48] Not everyone took manoeuvres seriously. PS/6648 Private Frederick Eaves of No 1 Company, 21/R Fus, was a 24-year-old valet who fell foul of military discipline whilst at Clipstone, for drunkenness, absence, being 'unsteady on night operations' and the loss of his hat.[49] Salter recorded one field day: 'Beautiful day. Reveille 5 a.m. Paraded 7.15 a.m. Brigade field day in Birklands Wood. Marched to Ollerton. Enemy were S[outh] Notts Hussars. [I] Was killed about a dozen times. Back at 2.30 p.m. Pay day.'[50] In some ways training at Epsom had been more productive; 'We carried out a great deal of wood fighting in Edwinstowe Forest but did not learn much. Our operations in Oxshott Wood, near Leatherhead, had taught us far more. Many of the shows in Edwinstowe ended in utter confusion; the only benefit we got was that derived from the [physical] exercise.'[51] Field days also included long marches that built up physical stamina; 20/R Fus completed a 25-mile march with full kit in late June.[52]

Rifle ranges were needed because; 'Musketry at this period was badly neglected for lack of ranges'. *The Pow-Wow* mentioned that 'Besides the manoeuvres, good work was done at musketry and bayonet-fighting, and a certain amount of firing practice even, on a miniature range below the camp.'[53] Not every member of the 21st Battalion was steady under fire. PS/4300 Private Harold Morgan Bowen, aged 27, and a groom, was presumably exercising one of the officers' mounts when the horse was startled by rifle fire whilst near one of the ranges on 1 July 1915. Bowen was thrown off and suffered a concussion. However, though the UPS helped build a new range nearby at Rufford it was still unfinished when the battalions left. Training varied for specialists in each battalion. For example, the scouts in 19/R Fus did physical drill, route marching, sketching, surveying and signalling amongst other activities.

There were other drains on manpower and distractions that affected or retarded training. On 20 July 1915 the 33rd Division Cyclist Company was formed and was up to strength by 26 July.[54] This was a unit of the newly formed Army Cyclist Corps (ACC) which was intended to act as divisional scouts for use in open warfare. The company was formed from infantrymen provided by different battalions of 33rd Division. 21/R Fus contributed two officers. Captain Russell Pearce Gould was in command of No 4 Company, 21/R Fus, having formerly served in the Volunteers pre-war including being a captain with 6th (Cyclist) Battalion, Norfolk Regiment (6/Norfolk). According to Spurrell; 'Before he left [No 4 Company] he carried out a very exhaustive inspection of the company by platoons.' This training scheme included physical drill, a river crossing, platoon drill, digging, bayonet fighting and signalling. Gould was chosen to command 33rd Divisional Cyclist Company. 21/R Fus also provided the newly-joined Second Lieutenant Herbert West who had been gazetted with Ashley Gibson on 15 June 1915.[55] 19/R Fus

---

48 IWM: H.W. Spurrell 2138, p.16.
49 Though Eaves went to France with 21/R Fus he was transferred to a trench mortar course within weeks of arriving and later left the battalion.
50 IWM: L.E. Salter Papers 21103, 28 May 1915.
51 IWM: H.W. Spurrell 2138, p.17.
52 LIM Archive: Letters of E.A. Holden, courtesy LIM, p.35.
53 Anon. Author, *University & Public Schools Brigade*, p.93.
54 TNA: WO95/2405: 33rd Division GS War Diary.
55 West went to France with the company and later transferred to the ACC before joining the Intelligence Department in GHQ.

The manoeuvres and field days conducted around Sherwood Forest by the UPS Brigade were not always well-planned and executed. Indeed, they were often confusing for the infantrymen who 'fought' in them. (*The Gasper* sketch by T. Kelly)

"Hullo! What Battalion are you?"

contributed Second Lieutenant Harold Douglas Collis, another recently-gazetted officer, to make up the numbers. One of the soldiers transferred across was PS/8663 Private Reginald Sleight Dearden who was an articled clerk from Harrogate.[56]

There were other distractions from training. On Sunday 26 May the Brigade heard that the General commanding Northern Command would inspect the troops. Many officers were all away, presumably for the weekend, as was half the brigade. Even the cooks and grooms were mustered for parade to make up the numbers and the general was supposedly happy with the turn out.[57]

More visits and inspections followed. On 1 June General Drummond was present to observe field operations including outposts and picquets and an attack. The General was, according to one account, very pleased with proceedings.[58] Another senior officer, Major General Lawson, also visited. On 19 June the Brigade was inspected again; 'Last Saturday we had a very important inspection by General Sir Bruce Hamilton who is in supreme command of all the troops in this country. I think it was quite successful, but it is very tiring work. We did not get away until 4 in the afternoon …'[59] Leslie Ireland considered the inspection a 'beastly shame to spoil our Sat. afternoon and of course all leave was stopped.'[60] According to *The Pow-Wow* these visits,

56   Dearden later transferred to 1/Cameronians as a bomber in May 1916 and was killed in action on 15/16 July 1916 and is commemorated on the Thiepval Memorial.
57   IWM: W.B. Medlicott Papers, 1708, p.5.
58   IWM: L.E. Salter Papers 21103, 1 June 1915.
59   LIM Archive: Letters of E.A. Holden, courtesy LIM, p.34.
60   IWM: L.W.C. Ireland Papers, 14441, letter 20 June 1915.

The 20/R Fus on the parade ground at Clipstone. As the men are formed up by platoons and companies with the band and specialists (e.g. stretcher bearers) in the rear, this may have been a rehearsal for one of the 98th Brigade inspections. Comparing this photo to the previous ones demonstrates how UPS battalions had developed from the initial parades in Hyde Park in terms of small arms, uniforms, equipment and training. (Reproduced with the permission of James Skelton)

though a distraction; '…served to put a new energy into the general training, if they had no other result.'[61] Another distraction was on 13 June when the camp was thrown open to visitors and apparently thousands came.[62]

On 8 July there was a 'rather persistent rumour about India at present' according to Holden. The same had been stated by Leslie Ireland in mid-June. Not knowing the future made it difficult for UPS men to know the right thing to do. Despite preparing for deployment there were further departures for commissions in early July:

> Tom Reed, Walton, Stanley Brown and a man called Baldwin were all passed by the Colonel and are awaiting gazette … At the present time the Colonel is obliged to allow anyone to go [for a commission], who is applied for by another Colonel … Of course, all the Burnley boys know Col Porritt of the East Lancs or have friends at home who have influence in one way or another and so long as there are vacancies they have had no difficulty in getting nomination. I was fairly content here as long as the Burnley boys stayed, but it is rather disheartening to see them all leaving.[63]

61 Anon. Author, *University & Public Schools Brigade*, p.92.
62 IWM: L.E. Salter Papers 21103, 13 June 1915.
63 LIM: Photocopy of transcribed letters of E.A. Holden, courtesy LIM, p.37–38.

Holden meanwhile did not have sufficient influence with a colonel he knew who had already applied for a dozen officers from 20/R Fus and was already pushing his luck. Holden would remain in the ranks. Presumably Colonel Bennett of 20/R Fus was reluctant to lose men to commissions unless he was forced to. Likewise, on 8 July Holden reported that with 20,000 men at Clipstone, divisional manoeuvres would begin shortly and that might see a further cessation of commissions. In a letter on 27 July Leslie Ireland stated that the divisional commander issued a notice stating the division would go overseas on about 15 September.[64] This solid news stopped him applying for an Indian Army commission and likely put paid to others doing the same.

Recruit training

However, the same problems remained; outflow of soldiers to commissions meant that recruits were still needed. The changing character of the Brigade was noticed by Terence Doherty; 'They are taking miners and Navvies into this Brigade, we are all awfully wild about it. We are no longer the U.P.S. as it would be a positive fib to call us University & Public Schools. We are only the 118th Infantry Brigade…'[65] Kenneth Norman also recalled that many of these recruits were not public school boys; moreover, they needed to be trained which further retarded the readiness of the battalion. Norman estimated that about 700 men of those who had originally paraded at Hyde Park had left for commissions, presumably from his battalion alone.[66] On 20 May 18/R Fus was short about 300 men. Some huts only contained a dozen men. Recruits were being sent to Clipstone in batches of 20–30 a day; men of 'all sorts and conditions' according to Medlicott.[67] Recruits were given separate packages of 'basic' training:

> I have been turned on to training recruits for good apparently. It's arduous work; 6.45–7.30, 9.25–1 and 2.25–5. We have 3 squads. I take the newcomers. They get ¾ hr easy in the morning and ½ hr in the afternoon when we have to lecture them. Heaven knows what I shall talk about! I shall have very little time to myself. We have a course to get through in 5 weeks including night training, route marching and lectures … and a deal to do with them in off [duty] hours. I have just been showing them how to put on equipment.[68]

Medlicott described it as important work but a dog's life.

In addition to recruit training new NCOs were needed to replace gaps. 19/R Fus ran a regimental course for NCOs whilst at Clipstone. One of those who was taught was PS/931 Corporal William Benjamin Stimson of D Company. He was aged 23 and had been educated at Bedford School and Emmanuel College, Cambridge, where he served with the OTC and was a science undergraduate. He had already attended a signalling course at Epsom.

---

64   IWM: L.W.C. Ireland Papers, 14441, letter 27 July 1915.
65   IWM: T.O'C. Doherty Papers, 12961, letter 4 May 1915.
66   LA: K.V. Norman Papers, WW1/GS/1188, p.54.
67   IWM: W.B. Medlicott Papers, 1708, p.22.
68   IWM: W.B. Medlicott Papers, 1708, p.5.

HQ 33rd Division was also located at Clipstone and was growing as more officers were posted in. Brigadier General Gordon Gilmour was still commanding 98th Brigade and Major R.H. Hermon Hodge was his brigade major.[69] In addition to 98th Brigade; 99th Brigade arrived in July which consisted of 17th (Empire), 22nd (Kensington), 23rd (1st Sportsmen) and 24th (2nd Sportsmen) Battalions of the Royal Fusiliers. According to Salter, the Sportsmen of 23 and 24/R Fus arrived on 25 June. On 4 July units of 100th Brigade also started to arrive and were complete by 13 July. This brigade consisted of the 17th Battalion Essex Regiment (17/Essex), 16th Battalion Middlesex Regiment (Public Schools)(16/Mx), 17th Middlesex Regiment (1st Football)(17/Mx) and 16th Battalion King's Royal Rifle Corps (16/KRRC)(Church Lads Brigade). The 18th Battalion Middlesex Regiment (1st Public Works) (the divisional pioneer battalion or 'Navvies Battalion') also joined. On 7 July the Divisional Train also turned up as did the 212th and 222nd Field Companies RE and the division appeared to be starting to form into an all-arms formation. At some point the UPS Brigade was, re-christened the 98th Brigade. The initials UPS had increasingly become a misnomer but many clung to this former title.

As the division formed up, so the successive commanders wanted to review their troops. Major General F.S. Maude CB CMG DSO took over as divisional commander on 9 July having commanded an infantry brigade in France. According to Leonard Salter, the GOC saw an attack by 19/R Fus during a field day on 9 July and was pleased with the performance. Three generals watched manoeuvres on 16 July according to Salter though he was unfortunately 'captured' by 21/R Fus. On 22 July General Maude visited to inspect the division. On 28 July 98th Brigade took part in a tactical exercise at Birklands which was watched by the divisional GOC.

Meanwhile, on 21 July 1915 33rd Division received orders to move to Salisbury Plain. The rumour of a move to Salisbury had been circulating around the ranks for a week already.[70] Advanced parties were sent south to start the transition of locations. The 19/R Fus advanced guard left on 26 July.

On 4 August the men of 18/R Fus also rose early and marched seven miles to Warsop to entrain; the march was typically blighted by rain. Hugh Spurrell remembered entraining at Edwinstowe one evening and arrived after midnight. After a mile-long tramp along a dusty road arrived in their new barracks at Perham Down.

Though the UPS Brigade had not had high expectations of life near Mansfield on initial arrival they had grown to like the place. The commander of 33rd Division General Maude, and the Mayor of Mansfield exchanged letters of appreciation to highlight the mutual regret about the formation departing. A letter of thanks from the Brigade to the Y.M.C.A. facilities was also reproduced in *The Pow-Wow*:

> The dusty wayfarer, who wants refreshing, finds all that is requisite to his needs with equal convenience. Now where am I? … Of course it's the Y.M.[C.A.] Theologically speaking, I have no business here. Nor, probably, have many of the others. Yet never once … has one single word designed to discover or alter my beliefs been uttered or implied. … I am leaving room for gratitude to the Mansfield ladies and gentlemen, who so unselfishly devote their time and trouble to our welfare and happiness … We are apt to take kindness rather for

---

69   The UPS Brigade was originally numbered 118th but was later renumbered 98th.
70   IWM: L.E. Salter Papers 21103.

granted. Nevertheless, these devoted philanthropists may, I am confident, be assured that, even if we don't show it, we are grateful …

Moving to Clipstone had enabled the Brigade to conduct further training and to absorb more replacements. The finishing touches to that training would be completed in a new location near Salisbury.

6

Further Training on Salisbury Plain

Marching or mock-fighting in the grey light of dawn

Beat of the drums that rises from the Plain,
And skirling pipes that echo hill to hill,
Winging their plaintive melodies until
The valleys ring with War's sad songs again!
Flashing of steel; avenger of the slain,
And glint of swords that never shall be still
While there be man to strike and foe to kill,
And all to sacrifice and all to gain!

Sarum! Whose very soil of old has bled
That tyranny might die and freedom live,
That England conquering might be England free!
Since thou dost link the living with the dead,
Ask now for all that men of heart can give,
And this, and this alone, their gift shall be.[1]
Salisbury Plain, by R.F. Clements

There were days when they marched wearily, with much halting, on a route which always seemed to circle round Sidbury Hill ... There were other days when they did not march in at dusk, but bivouacked in cold and wind-swept fields, to sleep uneasily and continue the marching or mock-fighting in the grey light of dawn. More utterly miserable than these were the days and nights spent in the cold, chalk trench-system dug for training in the slopes round Bedlam Buildings. These exercises were presumably intended to accustom the men to active service conditions; if so, they erred on the side of thoroughness.[2]
Anonymous – from *The History of the UPS*.

---

1 R.F. Clements, *Salisbury Plain and other poems* (Salisbury: Bennett Brothers, 1917), p.5.
2 Anon. Author, *University & Public Schools Brigade*, p.95.

The main body of the UPS Brigade departed for Tidworth on 3 August. For example, 19/R Fus left Clipstone at 12:45 p.m. and arrived at Tidworth for 8:15 p.m.; the battalion was accommodated in Jellalabad Barracks and nearby married quarters. Leonard Salter and six of his comrades were sharing one room in one of these houses. Kenneth Norman and others were grateful that they were in the barracks as many troops on Salisbury Plain were in tents or huts. Unfortunately for A Company, 18/R Fus, they were under canvas nearby.[3] There was insufficient space in the barracks and the men of 21/R Fus had to be accommodated in huts. Having arrived after midnight, the men 'tumbled into a hut and were soon fast asleep on the bare boards…'. They were accommodated in No 3 Camp, Perham Down.

Holden left Worksop at 3:00 p.m. and arrived along with two companies at Tidworth at 11:30 p.m.; they were in barracks by midnight having not been fed all day. His initial impressions were; 'The barracks are like nothing so much as a huge asylum. Of course, there is no town of Tidworth, it is just an accumulation of barracks.'[4] Leslie Ireland observed, 'the barracks are quite nice, very convenient but very dirty. I picked up a flea last night and another man killed a bug. This morning we disinfected the floor and tomorrow we distemper the walls; [I] trust it will rid us of them.'[5] Robert Sturges and his comrades likewise found the barrack rooms filthy and insect-ridden; it required a week of scrubbing and cleaning to make them habitable.[6] The iron bedsteads made the accommodation reasonably comfortable and there were hot baths available and electric lighting. The food was poor and not to the standard they had experienced at Clipstone.[7]

Tidworth was dreary according to Kenneth Norman, but there was some entertainment to be had in this small town.[8] Tidworth boasted a garrison theatre where various revues were put on; a small YMCA hall where concerts took place; a Church Army Institute where meals were provided and billiards could be played; and a garrison church. A garrison cinema was later opened and one of the UPS men, PS/7066 Private John Edward Pickett, 20/R Fus, was assigned to run it.[9] Likewise; 'Perham was a far less interesting centre for exploration than Clipstone had been. Moreover, the nights were getting longer and the hours for riding about [by bicycle] and seeing things were much shorter.'[10] There were opportunities for sporting entertainment and amusement. Leonard Salter watched a game of polo at Tidworth on 25 August and tried his hand playing bicycle polo with a friend a week later. PS/3345 Private Roy Wilson of 21/R Fus managed to get his motorcycle to Tidworth and used it in the local area. He was from Sunderland and had attended Denstone College. He was an apprentice engineer for an engine repairing company.

What were the UPS battalions to do in this 'luxury' accommodation? In the words of the UPS History:

---

3   LA: K.V. Norman Papers, WW1/GS/1188, p.62, p.64.
4   LIM Archive: Letters of E.A. Holden, courtesy LIM, p.39.
5   IWM: L.W.C. Ireland Papers, 14441, letter 4 August 1915.
6   Private No 940 (Robert Sturges), *On the Remainder of Our Front* (London: Harrison and Sons, 1917), p.135.
7   LA: K.V. Norman Papers, WW1/GS/1188, p.62, p.64.
8   LA: K.V. Norman Papers, WW1/GS/1188, p.66.
9   Pickett had attended Shakespeare's School, Stratford-upon-Avon.
10 IWM: H.W. Spurrell 2138, p.25.

The Brigade took up its duties at Salisbury with a new enthusiasm somewhat akin to the old spirit of those far-off first days at Epsom, Leatherhead, and Ashtead. Pessimism was rife; but, after all, Salisbury is Salisbury! No one could deny that! Comfortably housed in barracks ... work proceeded rapidly. Firing on the ranges alternated with bayonet-fighting and other essential forms of military training, and the discomforts of bad food and unwelcome visitors were soon forgotten in the expectation of approaching departure.[11]

According to Godfrey Skelton; 'The training was intensive, aimed to toughen us up, and we marched miles over the Plain with its numerous woods and "tumuli," ancient burial mounds, digging trenches in the hard white chalk, engaged in night operations and firing on the ranges.'[12] According to Kenneth Norman, Tidworth was the perfect place to enable the men to get physically fit before deploying to France through arduous exercises, sleeping out on Salisbury Plain, musketry on the ranges and full-pack route marches.[13]

Major General H.J.S. Landon CB, arrived to take command of 33rd Division on 16 September. He had returned from commanding a New Army division in France due to ill-health. Landon, like Thesiger, had commanded a brigade of 1st Division 1914–1915. He likely arrived with both command and operational experience and brought a fresh eye for efficiency amongst his new units and their commanders.

The whole of 33rd Division had moved from Clipstone. In early August, the divisional artillery consisting of four brigades of the Royal Field Artillery, and a divisional ammunition column, joined the division on Salisbury Plain. Now the division was nearing completion though more training was needed. There were further changes at the top. General Maude moved onward on 16 August to command 13th (Western) Division at Gallipoli. Major General G.H. Thesiger CB, CMG, ADC took over on 27 August; he also departed very shortly afterwards on 5 September to France to command 9th (Scottish) Division. The history of the UPS Brigade might have been very different under either of these two commanders.

11 Anon. Author, *University & Public Schools Brigade*, p.95.
12 IWM: G. Skelton Papers, 13966, p.34.
13 LA: K.V. Norman Papers, WW1/GS/1188, p.59, p.66.

Trench warfare practices

Part of the renewed efforts at training each battalion was to introduce more realistic experiences of trench warfare. Before General Maude left, he led a demonstration for the officers of 33rd Division concerning trench warfare at Bedlam Buildings on 7 August. On 11 August men of 19/R Fus were shown these trenches. Ernest Holden wrote of 20/R Fus trench training:

> We had to turn out on parade at 8 o'clock for trenching. The trenches here are the real article with dug outs and machine gun emplacements as well as cook houses and officers' quarters. I was supposed to see to the building of a sniper's post. Fortunately, it was already built so the section and I made up on some lost sleep until it was time to go home.[14]

Donald Price had a less enjoyable time digging trenches:

> … about six o'clock in the morning we marched out to some place, we were digging in chalk – it was damned hard work. … It must have been 3 or 4 miles away. Funny enough I fell in the trench myself and sprained my ankle. I was *hors de combat* so they send me home and I'd got to walk with a sprained ankle … 4 to 5 miles.[15]

Skelton was less complimentary of the 'trench' training conducted when he wrote later, with hindsight; 'We knew nothing of anything and had only played at trench warfare at a place called "Bedlam Buildings" at Salisbury Plain in England.'[16] Ernest Holden described the first weekend in practice trenches for 20/R Fus:

> We left barracks at 8 a.m. arriving at 9.30. A & B Coys took the first line, C & D were reserves. We had nothing to do until 3 p.m. but wait in a little valley behind the line. At 3 o'clock we relieved the firing line. We entered the trenches by a gradually deepening passage which led in a zig-zag fashion over the crest of the hill. At the top… it reached its full depth, i.e. about 7ft. It led a very erratic course down the full side through dug-outs (dressing stations, Officers Quarters, Signalling stations etc) to the first reserve trenches. Here half the Company was kept until they relieved the first half in the first trench.
>
> There is only room for a single file of men in the trench and when the front lot are relieved, a regular shunting process has to be gone through. … It is not very comfortable, but I have got used to sleeping almost anywhere in any position. At nighttime one sentry for every 10 men in the reserve trenches is posted. The four trenches are far advanced on the trenches we made at Caterham.
>
> We went into the fire trenches the second time from 3 a.m. until 6 a.m. when we finished. We were not sorry either. The Colonel had just come back from the front and was very keen on our doing things properly. We were very dirty when we had finished, chalk and soil everywhere. Of course, we spent most of Sunday in bed…[17]

---

14 LIM Archive: Letters of E.A. Holden, courtesy LIM, p.42.
15 IWM: Donald Price Sound Recording 10168, Reel 4.
16 IWM: G. Skelton Papers, 13966, pp.36–37.
17 LIM Archive: Letters of E.A. Holden, courtesy LIM, p.44.

According to Hugh Spurrell they only visited these trenches once:

> ... once we went up to occupy the trenches at Bedlam Buildings for twenty-four hours. There were two sets of trenches – one for the dastardly Huns and one for ourselves. Each system was organised to allow one company to occupy the forward [system] and one to support it. This enabled the whole battalion to be in action at once. We were issued with blank [ammunition] and marched up in the morning to take over the trenches from another battalion which had occupied them a day earlier. For the first six hours No 4 Company was in the line. Of course a careful reconnaissance of the trenches was made in advance by the officers, but I was not detailed to take part in this. We reached our posts just before the midday meal. No 5 Section was in the support line at first. ... Fortunately for the welfare of the British Army, there were few places in France in which the shelters for resting troops were not better. One piece of iron was stretched across the trench in one place; that was all. As luck would have it, it did not rain. In the front line, which we occupied three hours later, there were no shelters at all. The only dug out in the system had been made for show purposes; it was naturally used as Company Headquarters.[18]

There were several mock raids during the night and the session in these trenches ended in a dawn attack which resulted in a 'battle royale' in No Man's Land. Even before they got to France there were jibes about how officious the Company Quartermaster Sergeant was just behind the lines and how he 'never got as near to the front line as he did on that day. In France the pressure of business always kept him from coming up.'[19] There were some individual lapses in discipline during this period. PS/4444 Lance Corporal Lionel Ballard of C Company, 20/R Fus, a 34-year-old tutor from Bedford, was absent from parade at Tidworth and was also charged with committing a nuisance outside trenches and concealing a disease. He was busted to private and departed to 29/R Fus days before 20/R Fus departed to France.

In a culmination of trench warfare training 100th Brigade launched a two-battalion attack at Bedlam Buildings to capture Pickford Hill; 98th and 99th Brigades later advanced into the attack. This was all planned and controlled by the divisional staff.

Musketry training

Another key training activity was musketry which was to ensure the infantrymen could fire both accurately and rapidly. There were several field firing areas on Salisbury Plain and 21/R Fus had a 900-yard range behind their camp to conduct musketry practices on:

> The musketry course lasted a week, two companies firing in the morning and two in the afternoon. The companies not firing provided butt parties – I was fortunate and did not click for any of these. The markers signalled the scores of those firing and kept registers showing the progress of every man. Firing was done by details – that meant that every practice was completed by all those due to fire before the next practice was commenced

---

18 IWM: H.W. Spurrell 2138, p.24.
19 IWM: H.W. Spurrell 2138, p.24.

… Those not due to fire for some time were employed in [the] rear, in judging distance, physical drill or bayonet fighting. Competition was very keen … At the end of the course I had scored 108 out of 185 and just qualified as a first class shot. … I shot better when firing rapid than when taking deliberate aim…[20]

The battalions fired a range package to prepare them for overseas. 98th Brigade commenced Parts 1 and 2 of their musketry training on 9 August and five days later they were complete.

The men fired from 100 to 600 yards with both snap shooting and rapid fire. The week-long shooting programme that Leonard Salter went through on Bulford Ranges was as follows:

> Monday 9 [August] – Showery. Reveille 5 [a.m.]. Musketry 7.15–9.30. On Bulford Ranges 11.30–6.30. Fired 100 yds grouping. Passed all.
> Tuesday 10 – Hot. Reveille 5. Bulford Ranges 7.15–1.30 p.m. Courses, 200 yds Snap-shooting, Rapid [fire]. Passed in all.
> Wednesday 11 – Hot. Reveille 5. Bedlam Down in morning examining trenches. Ranges 12 – 5.30. Courses Re-fired for failures.
> Thursday 12 – Hot. Reveille 5. Ranges 7–12.30 p.m. Courses, 300 yds kneeling, 300 yds application, 400 yds Rapid, 400 application. Serving out ammunition all morning. Wet through.
> Friday 13 – Hot. Reveille 5.30. Parade 9. Ranges 500–600 yds. Back at 3.[21]

According to one staff officer with 33rd Division; 'The Division was going through musketry under difficulties, as there were only about eighty rifles available for fifteen or sixteen thousand men. Intensive training was being pushed forward and all ranks were fired with the most intense enthusiasm.'[22] With musketry complete the battalions were able to take part in collective training for the division in a series of brigade and divisional exercises.

War games

There were a few field days and mock battles for the UPS battalions whilst they were based in Tidworth. On 6 August, as soon as 98th Brigade arrived, it was out conducting an advance to contact on a field exercise where the brigade operated against the pioneer battalion (18/Mx) and had both cyclists and an artillery battery attached to them. Another exercise followed six days later when 98th and 99th Brigades faced each other in manoeuvres and a mock battle. Hugh Etheridge, of 18/R Fus, wrote home in mid-October; 'We have had such a long week that I have been too tired to write. We went out for 3 days from Monday, came home [on] Wednesday night [and] started again Thursday morning & came home yesterday, so hope to sleep all this afternoon…'[23] An even more ambitious scheme followed on 20–21 October when

---

20   IWM: H.W. Spurrell Papers, 2138, p.23.
21   IWM: L.E. Salter Papers 21103.
22   Robert Blackham, *Scalpel, Sword and Stretcher* (London: Sampson, 1931), p.221.
23   Fusiliers Museum (FML): H.D. Etheridge Documents, RFM.ARC.2482.30, postcard 16 October 1915.

another battalion advanced up a single road and came into 'action' near Everleigh followed by a two-brigade attack next day. Another large-scale operation occurred on 25 October. Gordon Jacob and two comrades got separated from their unit and became lost; they had to sleep on a farm cart in a shed for the night. On arriving back in camp they were sentenced to be confined to barracks for a week by the CO and were warned that they might have been shot for the same offence whilst on active service.[24]

During one three-day battle towards Everleigh 21/R Fus was advancing along the Everleigh-Ludgershall Road with the hostile force a long distance away. According to Hugh Spurrell:

> Our company was in front; we had not even assumed advance guard formation. On reaching the top of a rise, Otto saw before him in the valley some transport marching along. As soon as he saw horses he raised a cry of 'hostile cavalry' and ordered two platoons forward to wipe them out. This was done and the cavalry were exterminated. A staff officer dashed ahead to tell them so and brought back news that they were our own regimental transport. The cookers were riddled with bullets and the animals were all out of action. The news of this reached the Brigadier, who let us stew in our own juice. He ordered the transport back to camp and told Otto to make what arrangements he could in the circumstances. This did not help much, as can be imagined.[25]

When the enemy were encountered the attack was launched in the wrong direction and the umpires called off that part of the operation. Despite this humorous episode the training was arduous and efforts were made to add realism. It is likely that the rigour of some of this training was diminished by the limited experience and paucity of the umpires. Spurrell again recalled:

> I was acting as runner to Mackarness – now second in command of the company. Before the attack was over we captured a machine gun. It was a very easy matter, as we bore charmed lives. Mackarness ordered the gunners to surrender; they replied that they had wiped us out a dozen times. To say this to an officer was, of course, the height of indiscipline; they were told to surrender at once. They did so…[26]

Sadly, lessons over the hazards and effectiveness of machine guns were not learnt at this point.

The division engaged in practice operations in mid-October with 19/R Fus spending five days out camping in fields and engaging in mock attacks against the other brigades on different days. The exercise ended with an early morning attack, according to Salter, the 'whole affair a mess up. Everybody got lost…'[27] Philip Spaull described a tough day of training on Wednesday 25 October 1915:

> We had a bad day Wednesday … Reveille at 5 a.m., breakfast at 6 a.m. We marched off about 7.30 a.m. and marched until 12 noon. We then had our haversack ration consisting of bread and cheese. Off again at 12.30 p.m. and marched to our camping ground and

---

24 LA: G.P.S. Jacob Papers, WW1/GS/0841, pp.117–118.
25 IWM: H.W. Spurrell 2138, p.23.
26 IWM: H.W. Spurrell 2138, p.24.
27 IWM: L.E. Salter Papers 21103.

had some stew at 7 p.m. We had a little hot tea at 10 p.m. and then bed. One blanket and waterproof sheet on a stubble field. It was rather cold for a bedroom. At 5 a.m. reveille. It was perfectly dark so had to walk or sit about until 6 a.m. (Army red tape just to get us up). Breakfast at 6 a.m. consisting of bread, margarine, jam and tea. The ground was covered with white frost as was everything we had. On Thursday, we made an attack in the rain and then back to camping ground for hot stew which was not cooked properly. The rain continued so they took us back to barracks, having made us all thoroughly fed up with everything and, as far as I can see, did no good to further our training.[28]

Ernest Holden recorded some of the hardships; 'We spend most of the week out of barracks on operations now. It is very uncomfortable as we get very little sleep and no wash in the bivouacs. If we get to bed at 9 we are up by 1.30 for an attack at dawn or something similar.'[29] As it grew colder and wetter in the Autumn these divisional manoeuvres became greater feats of endurance; 'You have no idea how miserable it is to be standing, inactive, with icy rain coming down, in wet things, and unable to shout or do anything to relieve ones feelings.'[30]

Part of these field activities were long route marches. There was a twelve-mile route march for the battalions and their first-line transport on 3 September and again on 23 September when 14–15 miles were moved. On both occasions the battalions were inspected by the different divisional commanders.

Specialist training

There was training in grenades and bombing by a visiting officer. At this point bombing was a specialist activity and each battalion had parties of bombers; one officer and one NCO from each battalion were trained on this course. On 24 September the GOC observed the bombers of 21/R Fus undergoing grenade training. In mid-August two officers from each brigade were trained on the Lewis machine gun on a course at Hayling Island. One of the soldiers trained up as a Lewis gunner in August 1915 was PS/2118 Private Charles Ernest Storey of 18/R Fus. He had formerly served for three years with the London Rifle Brigade (LRB) and eight months with the Imperial Light Horse in South Africa. He attended the Grocers' Company School and Cranleigh School and worked in an Argentinian railway stores facility. Likewise, men from the brigades were attached to the Divisional Train from August for training in transport duties. Two company commanders from each brigade were sent to France for an instructional tour of duty in late August; this was presumably for them to experience the conditions in France and pass on good practice to their units. Concurrently, the General Staff Officer, Grade 1 (GSO1), 33rd Division, Lieutenant Colonel Symons, lectured officers of 98th Brigade on how to train their companies and ran regimental exercises for them. On 8 September officers and NCOs of the division were treated to a talk by Major Campbell, from the Inspectorate of Gymnasia, who lectured on the importance of physical fitness and bayonet fighting in modern war. This was a rite of passage for almost every British Army unit.

28   IWM: P.F.F. Spaull papers, 15453, letter 25 Oct 1915.
29   LIM Archive: Letters of E.A. Holden, courtesy LIM, p.51.
30   LIM Archive: Letters of E.A. Holden, courtesy LIM, p.52.

PS/4113 Lance Corporal Robert Jones of 21/R Fus recalled doing a lot of bayonet practice:

> It wasn't until we got to Salisbury Plain and got under Sergeant Major Humphries. He told us that a Regular soldier had to get fifteen shots a minute onto a four-foot square target. We couldn't be able to do that, but he said that anyone [German] advancing at 300 yards you'd possibly hit at least five times … He said [that] when it comes to bayonet fighting the odds are even. If the German was bigger than you then the odds are against you. He said [to] always use your rifle, forget your bayonet. He said that at six feet or ten yards distance that the rifle is absolutely deadly. It makes quite a small hole where it goes in, but you can put your fist in where it comes out the back …[31]

More inspections took place. On 11 August General Sir Arthur Paget inspected the infantry of the division whilst they were at work and observed bombing classes on 24 and 27 August. He also inspected the division again on 15 September; on the second occasion General Paget was recorded to have been very satisfied.[32] On the morning of 18 August 33rd Division was inspected by Lieutenant General Sir Archibald Murray, CIGS, at Seven Barrows on the second of a two-day visit. The men of 19/R Fus had the afternoon off afterwards.[33] Ernest Holden reported; 'General Murray seemed very pleasant with us today'.[34] It must be wondered what the various generals of the 'Old Army' must have thought of this unique brigade of the New Army. As a staff officer in divisional HQ recalled: 'It was a new experience to speak to a man in dirty ducks who was scrubbing out a barracks and receive a reply in the cultured accents of an Oxford don.'[35]

Many men could see that the longed-for departure to France was soon to happen but there was still talk of commissions amongst some men. Ernest Holden was still considering his options in mid-October: 'There does not seem to be much doubt but that we are going to France very shortly and they are bound to use a lot of us as officers as soon as we get out there, so I don't think it will make much difference whether I take a commission here or wait till I get there.'[36]

Some personal accounts glossed over training at Tidworth. Morgan Williams merely recorded; 'After moving to Salisbury Plain for a final course of training and submitting to a host of inspections by generals grave and gay, we received embarkation leave…'[37] Williams spent it with his father in Swansea; 'Dad and I wandered along the seashore talking far less than usual… Dad came with me to the station. "God keep you, son," was all that he could say, and my "Cheerio" meant the same thing.' Embarkation leave likely started during October and soldiers were presumably released in batches. Some men over-stepped the mark on leave; PS/7152 Private William Barrow, aged 23, of No 2 Platoon, A Company, 18/R Fus, got in trouble for being absent for 14 hours on 1 November 1915.[38]

---

31 This was likely PS/2792 Warrant Officer Class 2 Arthur Robert Humphrey. See LA: WW1/TR/03/92: Tape of R.C.B. Jones.
32 IWM: L.E. Salter Papers 21103.
33 IWM: L.E. Salter Papers 21103, 18 August 1915.
34 LIM Archive: Letters of E.A. Holden, courtesy LIM, p.41.
35 Blackham, *Scalpel, Sword and Stretcher*, p.221.
36 LIM Archive: Letters of E.A. Holden, courtesy LIM, p.50.
37 Williams, *Khaki to Cloth*, p.56.
38 Barrow wrote in The Gasper under the name 'Pete'.

An idealised version of the bayonet training that the UPS Brigade conducted at Clipstone and Tidworth. If Sergeant Major Humphries, 21/R Fus, was to be believed, during close combat the men would be better off forgetting their bayonets and firing at close range. (Sketch by Ernest Stoneley reproduced courtesy of Janet Wood)

The Last Fortnight

Training in the last two weeks before departing for France was intense. Only the war diary for 18/R Fus includes details of their training during November 1915. Likewise, no war diary was kept by 98th Brigade before active service commenced. However, an additional idea of the pre-deployment activities can be gleaned from the 100th Brigade War Diary which commenced on 5 November when the brigade received orders to be prepared to deploy to France. The UPS Brigade history recorded:

> Towards the close came final leave, a glorious week away from the Army, followed by farewell dinners in barracks and all the business of mobilization. The men added to their equipment, ammunition, bivouac groundsheets, spare clothing, and significant trifles such as pay-books and identification discs. On the other hand, they gave up unnecessary luxuries

such as spare uniforms and boots. Even then the weight of the outfit seemed intolerable, even after the long training of marches with pack and rifle.[39]

The training programme for 18/R Fus included training in entrenching, including wire entanglements, bayonet fighting, bomb throwing, judging distance and range finding and MG firing. The MG sections fired on Bulford Ranges. At this late stage Royal Fusilier officers of the brigade were assigned to be in charge of MGs. Second Lieutenant Westover was appointed as Brigade MG Officer; Second Lieutenant Dawson was the 18/R Fus Battalion MG Officer and Second Lieutenant Knight was the 'second' MG Officer – a reserve in case of Dawson's incapacitation. Lieutenant John Jervis was the 21/R Fus MG Officer.

Whilst the Brigade was inspected by the Brigade Commander and equipment was checked on 2 November the MG section continued to fire practices on the ranges; this suggested that they had not been formed for long and were doing 11th Hour training. In addition, at this late stage, drafts were still arriving to replenish numbers in 18/R Fus; thirty men arrived from 28/R Fus at Epsom. In addition, the battalion snipers of 18/R Fus had probably been recently established; on 3 November there was a lecture to four officers on telescopic sights and their care by a lieutenant from the Bisley School of Musketry. The snipers were also firing on 'F' Range at Bulford on this day. There were also lectures on MGs to the MG section and further range firing on 'G' Range at Bulford. There was also instruction on bombing for officers. Whilst this specialist training was taking place the men were all inspected in their new equipment and blankets were issued for overseas. Whilst rumours had been going on for months the fact that leave was stopped on 3 November and lights out was brought forward to 9:45 p.m. suggested movement was likely in the short term.

The recently joined draft fired on the ranges on 4 November and other range activities and lessons continued. The rest of the battalion was engaged in trench warfare training at Hare Warren. The next day there was further trench warfare training with divisional manoeuvres at Bedlam Trenches though 18/R Fus was in reserve. Further range firing took place and the battalion snipers practiced field firing at Beeches Barn. There was also an inspection of the battalion horse transport. Training and activities now became centred around the camp as it was presumably expected that the Brigade was to receive imminent movement orders. On 6 November training took the form of wire entanglements and sandbag revetments, breastworks and bayonet fighting. Officers had also been conducting revolver practices over the previous week. Spare military equipment was handed-in to stores; spare personal kit was sent home, and soldiers looked to offload possessions, like bicycles, that were no longer needed.

There were rumours constantly doing the rounds; however, Godfrey Skelton considered their departure imminent when all men were given free from infection (FFI) inspections prior to departure. All forty men in each barrack room paraded in shirts only to be inspected for venereal disease, 'the Medical officer lifting up the front of each man's shirt with the end of his cane, asking "Are you alight?".'[40] Some men were medically unfit to deploy. Major Herbert Townsend injured his knee whilst at Tidworth and later went to hospital in Edinburgh; he could only hobble on crutches by April 1916.

39 Anon. Author, *University & Public Schools Brigade*, p.97.
40 IWM: G. Skelton Papers, 13966, p.34.

Though most men of the division had no idea it existed, one document 'War Office letter No. 121/7109 (M) dated 6 November 1915' would change their lives forever. That order stated that 33rd Division would embark for France shortly.

Whilst Sunday would normally be a quiet day with a church parade on 7 November there was a parade afterwards at 11:30 a.m. at Seven Barrows to practice the 'march past' for a parade the next day.[41] The parade on 8 November was an inspection of 33rd Division, minus divisional artillery and train, by Her Majesty the Queen on Salisbury Plain. This occurred near Sidbury Hill. The Queen and the troops cheered H.M. The King who was absent. Leonard Salter of 19/R Fus was far from awed by a royal inspection according to his brief diary: 'Fine, Parade 8.15, Marched to Figheldean Down for a divisional review and good-bye by the Queen. Bit boring. Handed in all spare kit. Sold bicycle for 10/- in Tidworth.'[42] Kenneth Norman considered that such inspections were becoming quite troublesome.[43] During the afternoon further instructional parades took place. Next day, on 9 November, a farewell message from The King was received in addition to movement orders. The wet weather prevented parades though company mobilisation inspections took place. The battalion transport was also mobilised. The morning of 10 November was devoted to inspecting clothing and equipment. In the afternoon route marches took place and battalion bombers threw live grenades at Hare Warren. Later on, rifle ammunition was distributed to the battalion; another certain sign of departure. As the great day approached the emphasis on turn-out decreased, much to the pleasure of some; 'All our buttons must be dull for which, after 12 months, I am truly thankful.'[44]

On 11 November there was an hour of bayonet fighting and physical drill in the morning followed by a mobilisation inspection of companies by the Colonel. In addition, the transport brakesmen and details accompanying the transport were inspected by the adjutant. There was also the last opportunity for the battalion bombers to throw grenades. On about 11 November the train timetable for deployment was received but soon after a 24-hour delay was announced due to gales in the Channel. Their departure now being near-certain, some men were less security-conscious in their letters. Philip Spaull wrote home; 'We leave, so we are told, on Saturday. I will let you know our port of arrival by commencing the letter with the initial of the Port.'[45] Many men had their hair cut very short before departure and Robert Sturges bemoaned having his hair cut off by the barber. In the week between leave and embarkation Morgan Williams received sheaves of letters and he spent much of this time writing replies.[46] Some men invited their families down to Tidworth for a last visit. Gordon Jacob remembered spending an awkward and gloomy tea with his mother; the conversation was strained as neither knew what to say in the circumstances.[47]

At 7:30 a.m. next morning, 12 November, each battalion of the UPS Brigade despatched an advance party consisting of their transport and some additional men. They left Tidworth Station at 9:15 a.m.. The 18/R Fus party consisted of four officers and 122 men with all battalion animals and consisted of their Transport, MG Section and details. The 19/R Fus party consisted

---

41 IWM: L.E. Salter Papers 21103.
42 IWM: L.E. Salter Papers 21103.
43 LA: K.V. Norman Papers, WW1/GS/1188, p.66.
44 IWM: P.F.F. Spaull papers, 15453, letter 11 November 1915.
45 IWM: P.F.F. Spaull papers, 15453, letter 11 November 1915.
46 Williams, *From Khaki to Cloth*, p.56.
47 LA: G.P.S. Jacob Papers, WW1/GS/0841, p.121.

of three officers and 124 men; the 21/R Fus party consisted of four officers and 122 men. The composition of the 20/R Fus party was likely similar. The transport departed from Southampton and had a rough crossing to le Havre.

The remaining men of 18/R Fus were engaged in fatigues in barracks and being inspected by the CO; they were intended to depart at 9:15 a.m. on 13 November but at 4 p.m. on 12 November this order was postponed for 24 hours due to inclement weather in the channel. With departure so close the 13th was spent in company parades in the morning and the men were given time off in the afternoon. The stage was now set for the Brigade to leave Tidworth and Perham Down to deploy on active service.

Robert Sturges remembered, later, whilst in France, a postscript to their time at Tidworth and the 'Old Army' ways there. During one of the pay parades they were required to sign to say they had been paid 16 francs but were only given 15. When the troops questioned this sum they were told the additional franc was to cover barrack damages from when they were in Tidworth. This was for the same barracks they had deep-cleaned at their own expense to make it remotely habitable.[48] However, as the UPS Brigade readied itself to deploy the stories of the ex-UPS officers and men who had already left, and some of the casualties suffered by the Brigade, must be described.

---

48 Sturges, *On the Remainder of Our Front*, p.135.

7

Ex-UPS Men in 1915

These gallant lads took the 'Ups' Brigade for an Officers' Training Corps!

Some to the Flying Corps have fled,
Some to the R.M.C.;
Some in the London Gazette have read
The names that they've longed to see.
Now in the queue, when our bobs are paid,
Find they a place no more; –
These gallant lads took the 'Ups' Brigade
For an Officers' Training Corps![1]
     UPS Brigade Song

Many of the dead are still out in front … They are for the most part in extended order as they advanced, some wearing gas helmets … It is not a very cheerful sight…[2]

This UPS Brigade song sums up what the remaining men thought of those who had left to become officers. However, the members of the UPS who departed to commissions added to the over-all contribution made by the UPS in the Great War. Many gazetted officers experienced overseas service, saw action, and some made the ultimate sacrifice, before the rest of the Brigade even crossed the Channel. Their contribution to the campaigns in Flanders and Gallipoli must be weighed against the impact of their departure on the readiness of the whole Brigade.

One of the first UPS men to be killed was Reginald Denman Apps who served with 20/R Fus for two months before attending the Royal Military College in November 1914. He was commissioned in February and joined the 1st Battalion Royal Berkshire Regiment (1/R Berks). He was killed in May 1915 during the Battle of Festubert.[3] One of his men noted; 'I saw

---

1 IWM: A.W. Lloyd, 10271, The UPS Song Book.
2 IWM: P.F.F. Spaull papers, 15453, letter 20 January 1916.
3 Apps was aged 21 and was from Watford; he was educated at Cumberland House School, Gravesend, and Dartford School where he was a member of the OTC. He is commemorated on the la Touret Memorial.

Mr Apps ... in the charge at Richebourg. He was in my own Coy and we were at the time in the 1st German trench which we had taken. After this we made no further attacks. There was only a terrific shelling and nearly all the men killed after Sunday morning were killed by shells ... Everyone was astonished at the complete disappearance of Mr Apps.'[4] Another early fatality was John Parker Norfolk Simpson (formerly PS/3201 of 21/R Fus) who was gazetted to a commission on 20 January 1915. He was posted to 3/R Fus and his battalion was engaged during the Second Battle of Ypres. Simpson was believed captured on 24 May and was admitted to a German field hospital with wounds to his left knee, left hand and right leg. He died on 27 May, near Iseghem, of wounds after only ten days in France.[5] The next ex-UPS man to die was Second Lieutenant Leonard Maurice Powell, aged 20, of Chiselhurst. He was educated at Loretto School and Caius College, Cambridge, and was a distinguished athlete, member of Kent County Cricket Club and a member of the MCC. With these sporting and educational credentials he was gazetted into the Gordon Highlanders and joined the 1st Battalion on 20 May 1915. The Germans bombarded two companies of the battalion on 17 June near Ypres; Powell and 27 men were killed.

Vernon Bartlett went to France on 5 March 1915 as a second lieutenant and joined the 1st Battalion Dorsetshire Regiment (1/Dorsets). He served in the trenches during the Second Battle of Ypres and spent 17 days in the line because every battalion sent to relieve them was committed elsewhere during the heavy fighting. Later, the battalion arrived at billets at Kruistraat and Bartlett recorded that he had never been so tired. Whilst in the street a shell exploded nearby and he suffered minor fragmentation wounds and was evacuated. Shortly afterwards an order arrived committing the battalion to the battle; it suffered heavy casualties. When Bartlett returned, he was appointed as a company commander; '...there was nobody else left. Nearly all my friends had been killed or wounded while I was having a pleasant time in a hospital ward at Rouen ...[6] Bartlett spent his 21st birthday on an ambulance train being evacuated. He was later admitted for treatment suffering from neurasthenia on 2 August 1915 and returned to the UK. He was invalided out of the Army by early 1916.[7]

Ex-UPS officers were also starting to arrive in France and Flanders with New Army battalions or as replacements. 7/KRRC had departed to France before Whitley and Page could join it. Before they arrived, the battalion was decimated during the first German use of flamethrowers at Hooge on 31 July 1915. Page recalled; 'The Bn was relieved that night and came back into rest at Watou where we joined them. Everyone was very depressed particularly at the loss of officers, 21 in the Bn.'[8] They helped counterattack on 23 August and Page recalled half of his company being killed by shelling just reaching the trenches. Charles Whitley was wounded by shell fragments and was evacuated; he did not rejoin for three months. Not every early arrival was killed or wounded. Many had a mundane time; David Kelly was attached to HQ 110th

---

4   TNA: WO 339/3652: R.D. Apps officer file. A. Clutterbuck and others, *Bond of Sacrifice, August 1914 – June 1915*, (Dallington: N&M Press, 1992), p.12.
5   Simpson, from Alnwick, Northumberland, was originally buried in a German cemetery but was reburied in Harlebeke New British Cemetery.
6   Vernon Bartlett, *This is My Life* (London: Evergreen Books, 1941), pp.49–51.
7   Bartlett was later made CBE and died in 1983.
8   IWM: Documents 16924: Papers of E Page,

Brigade which endured a quiet ten months, with few casualties, holding a peaceful section of trenches near Arras.⁹

Closer to home

These deaths on active service of men who had once served with the UPS may have seemed remote for the members of these battalion. There were other deaths of UPS men far nearer at hand as several died of illness or injury in the UK. These deaths were equally tragic and deserve to be recorded here. PS/3753 Private Geoffrey Lennard Davis was the son of a solicitor and had enlisted with 18/R Fus in November 1914 on his nineteenth birthday. He was known as 'Geoff' and was nominated for a commission in 9/Mx to serve alongside his brother. He contracted influenza and succumbed on 19 January 1915 after complications. PS/1851 Private Leslie Cecil Munns, aged 28, was educated at Haileybury (1900–1904) and the Imperial Service College. He had served in D Company, 18/R Fus, and died on 21 January 1915 of influenza. Known as 'Long John' to his friends, Private John (Jack) Percy Dalzell Clarke served with No 1 Company, 18/R Fus. He was from Bishop's Stortford, Hertfordshire and was educated at St Edmund's School, Canterbury, and Christchurch College Cambridge. He enlisted in September 1914, was gazetted on 27 November 1914 into the Royal Fusiliers and died in an accident on 21 February 1915, aged 23. PS/1365 Private George Edward Cohen, aged 18, was from Didsbury and his father was born in Hamburg but was a naturalised British citizen. Cohen was educated at Bedales School and the Ecole de Commerce, Neufchatel. He fell ill with meningitis, temporarily recovered, but relapsed, and died a month short of his 19th Birthday on 3 April 1915. Educated at the High School, Croydon, PS/1476 Private Hubert Hope Flecker was aged 22 and died on 17 June 1915 of pneumonia. PS/2144 Private Arthur George Taylor had formerly served in the Royal Naval Volunteer Reserve and died on 8 August 1915. This might seem a lot of fatalities from illness from a population of largely young men. Unfortunately, these men only represent 18/R Fus.

PS/1148 Warrant Officer Class 2 Thomas Peter Reed was the Company Sergeant Major of E Company, 19/R Fus. He was aged 57, from Widdington, Essex, and was a clerk in patent agency. Reed was formerly a colour sergeant in the Royal Marines who had served in Egypt in the 1880s. He died on 11 April 1915 of pneumonia. Lieutenant Colonel Wolrige Gordon wrote to Reed's widow to offer his condolences and state his appreciation for an NCO who had assisted in the training of his battalion.¹⁰ PS/4356 Private Roland Claude O'Kelly (likely of 19/R Fus) was from Felling on Tyne. The following appeared in his school magazine; 'At school he had been a delicate boy, and the heavy strain entailed by the long hours of severe physical exercise in a training corps for officers had proved too much for his strength. ... He died of septic peritonitis, due to an internal injury accidentally received, on October 1st, after receiving the Last Sacraments.'¹¹ A third man from 19/R Fus, Private Lionel Dempster Miller, aged 22, died on 12 June 1915.

Two men died each from 20/R Fus and 21/R Fus. PS/5137 Private Ian Montgomery Jones, was born in Cirencester but was from Chorlton, Manchester. He was educated at Bromsgrove

---

9 Kelly, *Ruling Few*, p.91.
10 *Thomas Peter Reed* <https://eehe.org.uk/?p=28340#ReedTP> (accessed 6 November 2022).
11 *The Stonyhurst Magazine*, October 1915, p.1356.

School where he was a member of the OTC and played rugby. He was gazetted into 5th Battalion Leinster Regiment (5/Leinsters) on 29 September 1914 but as he was aged 17 his commission was delayed. He died of meningitis on 20 March 1915. PS/5985 Private E.C. Wrigley died on 26 April 1915. PS/4281 Private E.A. Wright, of 21/R Fus, died on 1 March 1915 and PS/2715 Private Louis Robert Garside, aged 25, from Batley, Yorkshire, died on 29 June 1915.

Throughout 1915, as these men sadly passed away, there was an increasing series of ex-UPS men being committed to overseas theatres having been gazetted into different regiments. One of the most voracious theatres was in the Mediterranean.

Gallipoli

The Gallipoli Campaign drew in many UPS men and the landing at Suvla Bay in August 1915 was the first major operation in which officers and men from Kitchener's New Army (and many ex-UPS officers) were engaged in significant numbers. Furthermore, Regular battalions were sent ex-UPS officer replacements such as George Raymond Dallas Moor who enlisted in 21/R Fus on 18 September 1914 and served as an enlisted soldier for a month before he was gazetted into the Hampshire Regiment. Sent to Gallipoli, he was awarded the Victoria Cross during a precarious moment in an attack:

> For most conspicuous bravery and resource on 5th June, 1915, during operations south of Krithia, Dardanelles. When a detachment of a battalion on his left, which had lost all its officers, was rapidly retiring before a heavy Turkish attack, Second Lieutenant Moor, immediately grasping the danger to the remainder of the line, dashed back some 200 yards, stemmed the retirement, led back the men, and recaptured the lost trench. This young officer … saved a dangerous situation.[12]

Moor had already been wounded on 28 April and went sick with exhaustion in early July 1915 before being evacuated to the UK.

Brian and Frederick Melland were two brothers from Altrincham, who enlisted and served with No 5 Platoon, No 1 Company, 21/R Fus. They soldiered for a few weeks before being gazetted as sub lieutenants in the Royal Naval Voluntary Reserve. On 17 October 1914 they joined the Anson Battalion RNVR and went to Gallipoli. Frederick was wounded on 26 April 1915 and was invalided to the UK in July.[13] Brian Melland took part in an attack on 6 May and after he had advanced with his platoon to 'The White House' he was shot dead by a sniper.[14] Brian Melland was possibly the first man who had enlisted with the UPS to be killed.[15]

Cuthbert Taunton was educated at Stonyhurst (1904) and Corpus Christi, Oxford, where he was an exceptional classics and Latin scholar; he also played football for his school and was in the OTC. Taunton was the epitome of what the UPS had hoped to recruit and he enlisted with

---

12   *The London Gazette*, 23 July 1915.
13   Frederick Bernard Melland, age 20, was killed in France as a company commander with the Hawke Battalion on 24 April 1917. He is commemorated on the Arras Memorial.
14   Rev H.C. Foster, *At Antwerp and The Dardanelles* (London: Mills & Boon, 1918), p.104.
15   Melland was buried near the White House but is commemorated on the Helles Memorial.

20/R Fus in September 1914. He was gazetted into 7th Battalion South Staffordshire Regiment (11th Division) in January 1915. His battalion was heavily engaged during the Suvla Bay landings and Taunton was killed on 9 August 1915. An account of his death was provided to his school magazine:

> While I was attacking the enemy on Chocolate Hill, Mr. Taunton was sent up to reinforce me. I … had to fall back on him … It must have been five minutes after this that he was killed. At the time he was looking in a periscope from some little head cover and was hit just behind the temple, probably by a sniper's bullet … he had only been fighting an hour, and was only 25 yards from the Turks … Mr. Taunton was one of the best of our officers, and was most popular both among officers and men, always having a cheery word for all …[16]

Donald Arthur Addams-Williams had enlisted in 19/R Fus in September 1914 and in November he was gazetted to 4/SWB (13th (Western) Division). He was one of the best long-distance runners in the battalion and trained its cross-country team. He landed on the Gallipoli Peninsular on 12 July and on 3 August he sent a letter home describing the boiling heat, the swarms of flies and the large numbers of unburied bodies on the battlefield.[17] Addams-Williams and other ex-UPS men were involved in one of the most important battles of the campaign, the struggle to take and hold Chunuk Bair; a hill that dominated the Peninsular. 13th Division was commanded by General Maude (who had formerly commanded 33rd Division) and some units were charged with holding the hill once it was captured. Addams-William was killed in action there on 11 August, aged only 19, having been in the Dardanelles for less than a month.[18] Another officer wrote that his platoon was forming a covering party whilst new trenches were being dug. As the enemy were in greater numbers than predicted the party suffered heavy casualties. Addams-Williams advanced, and was hit, but he continued forward before he was hit again. He kept commanding his platoon before he was felled by a third bullet. His body was left on the battlefield.[19] Gilbert Gamman was gazetted into 5th Battalion Wiltshire Regiment (5/Wilts)(13th Division) in November 1914 and served at Gallipoli. He was educated at Marlborough College and was from Andover. He was last seen leading a counterattack up a hill to retake a lost position on 10 August 1915. This was undoubtedly Chunuk Bair where the battalion was over-run during a massed Turkish counterattack. Second Lieutenant John Philip Morton Carpenter, educated alongside Gamman, was also with 5/Wilts and was wounded on 11 August.

Though some of these men had not served for long with the UPS, the Gallipoli campaign serves to demonstrate the calibre of the enlisted men and their potential qualities of leadership.

---

16   Cuthbert Andre Patmore Taunton was aged 20 and from Liverpool. He was buried in Green Hill Cemetery. *The Stonyhurst Magazine*, October 1915, p.1338.
17   LA: WW1/GALL/001: D.A. Addams-Williams Papers, letter 3 August 1915.
18   Donald Arthur Addams-Williams, aged 19, from Llangibby, Newport, was educated at Winchester and Marlborough College (1910–1912); in 1914 he was a businessman. He is buried in 7th Field Ambulance Cemetery.
19   LA: WW1/GALL/001: D.A. Addams-Williams Papers, press cutting.

'… an operation more suited to a plumber than a University graduate' – The UPS and Gas Warfare

During the summer of 1915 the UPS experienced another drain in manpower. On 22 April 1915 the Germans emitted chlorine gas from pressurised cylinders opposite two French divisions in the Ypres Salient. The impact of this first deployment of poison gas in terms of casualties and lost ground was taken very seriously and there was a dual need to better-protect the BEF against gas warfare and to harness this offensive capability. The RE was chosen to form 'Special' companies for this purpose which required analytical chemists and chemistry students. The Army in France and the UK was searched for chemists to become 'Chemical Corporals'. 21/R Fus possessed many potential candidates who were 'sick of the repetition of drill and spit and polish and feared that the war would end before they reached France…'.[20] However, those who applied were prevented from leaving; '[21/R Fus] … had its numbers seriously depleted by many hundreds of the rank and file being given commissions. As it appeared that the battalion would soon become a skeleton, the C.O. [Stuart-Wortley] used his best endeavours to prevent further wastage.'[21] The Army trawled universities for lists of chemistry graduates and potential candidates were extracted from their battalions and the objections of their COs were ignored. Several UPS men volunteered for the fledgling Special Brigade in July 1915 and went to France shortly afterwards. Over twenty-five men from 21/R Fus departed to become chemical corporals, more than from the other UPS battalions:

> Their first reaction was a feeling of immense relief at escaping from a battalion with a Guards adjutant and a Guards battalion sergeant-major. They crossed the Channel too late to participate in the Battle of Loos … No doubt they applied a little of their chemical knowledge, but found … that the discharging of gas from cylinders was often an operation more suited for a plumber than a university graduate …[22]

20/R Fus contributed at least seven men to the RE Special Brigade; 19/R Fus three men and 18/R Fus at least one. The Special Companies RE first released chlorine gas in support of the British attacks at Loos. William Harrison Brindley, Reginald Frank Dalton, Thomas Joyce, Wilfred Stanley Pheasey and Edward Charles Varley all joined 186th Company RE from 21/R Fus during the late summer and Autumn of 1915. That company was on the northern periphery of the Battle of Loos around Givenchy and Cuinchy; an area the UPS would later come to know well. Despite misgivings over the wind the company released gas at 5:50 a.m. on 25 September; in many cases the chlorine gas collected in the British trenches and the gas was turned off. When the infantry attack commenced it attracted retaliation from German artillery on the British trenches.

---

20   Martin Fox, *With the Special Brigade RE* (Toronto: Edgar Cross, 1957), p.9.
21   Fox, *With the Special Brigade*, p.9.
22   For some reason 21/R Fus commonly referred to their RSM as the Battalion Sergeant Major (BSM). Fox, *With the Special Brigade*, p.10.

The UPS and the Battle of Loos

If Gallipoli was the first major engagement for the New Armies, then the Battle of Loos was the first occasion where they fought in France *en masse*. Several ex-UPS officers were serving with different New Army, Territorial Force and Regular divisions. The following case studies give a snapshot of their employment and survival rate but avoids detailed accounts of every officer and every battalion.

Second Lieutenant Patrick Hamilton Forrester attended Loretto School and enlisted in September 1914; he was gazetted to 8th Battalion Black Watch (8/BW)(9th (Scottish) Division) in April 1915 and served in France. Forrester was the MG officer and was left out of the battle on 25 September but went up to replace officer casualties. He was mortally wounded on 26 September when he was leading some bombers. He died of wounds on 11 October 1915 and was buried at Colinton Parish Churchyard. He was MiD for his gallantry at Loos.

Further south, 1st Division was involved in this great attack. Second Lieutenant Oliver Sutton, with 1/LNL (1st Division), was wounded in the leg by a machine-gun bullet on 25 September and was gassed. He had been educated at Tonbridge School. On recovering he transferred to the RFC. Second Lieutenant Robert Henry Batten had attended the Royal Military College and joined 1st Battalion Northamptonshire Regiment (1/Northants, 1st Division). He was gassed at Loos but later recovered.

47th (2nd London) Division also attacked on 25 September. Second Lieutenant Arthur Keith Sanderson, aged 21, had formerly served with 18/R Fus. He had attended Rugby School (1908–1913) and Pembroke College, Cambridge. He was attached to 1/7 Londons which advanced against the Double Crassier. He was killed in action with B Company which attacked the trenches to the left of the Crassier.[23]

21st Division and 24th Division were in reserve and joined the battle late. Three ex-UPS officers served with 12th Battalion Northumberland Fusiliers (12/NF)(21st Division). PS/3052 Private Robert Oliver of 21/R Fus was gazetted to a commission in 12/NF in January 1915. He was killed during the 12/NF attack on 25 September 1915 and has no known grave; he was MiD after his death for gallantry. Sydney John Armstrong, aged 32, a schoolmaster at Wigan Grammar School, was also with 12/NF.[24] His last moments were on Sunday 26 September:

> This officer I saw myself shot, after he had given the word 'Come on lads!'. He was immediately struck and he dropped … He was quite near the German trench at the time. We took the trench and then retired from it, being bombed out of it. I do not know whether his body has ever been found …[25]

Second Lieutenant Roderic Maxwell Hill, formerly of 18/R Fus, was a subaltern in this battalion. He recalled later that they were expected to capture a hill to their front and when Zero-Hour came he and his men; '…leapt out like a pack of wolves…' to attack. He received a bullet in his side which knocked him over. Hill remembered the thunder of feet passing over

---

23 Sanderson is commemorated on the Loos Memorial.
24 Both Oliver, from Morpeth, and Armstrong, from Lancaster, are commemorated on the Loos Memorial.
25 TNA: WO 339/4021: S.J. Armstrong officer file.

and past him as his men advanced and he heard the fire of the German machine guns until the sound of the fighting ebbed away and the battlefield was left littered with the dead of his battalion.[26] He was MiD for his conduct and later joined the RFC; he became an Air Chief Marshal in the RAF in the Second World War. George Eyston, an artillery officer with 21st Division, came close to being either killed or cashiered. He was ordered to take his guns forward across the battlefield to Loos, which had not yet been captured, up a road in full view of the enemy and amidst heavy shelling. One of his wagons lost a wheel and caused a traffic jam of vehicles that were under heavy fire. Into this chaos appeared the Corps Commander who demanded to know who was in charge. Rather than face his wrath Eyston gave him a wide berth and got on with getting his vehicles moving.[27]

Also at Loos was Second Lieutenant John Easton who served with 12/R Fus (73rd Brigade, 24th Division. Easton had been educated at City of London School, was a member of the OTC, and enlisted in 19/R Fus in London on 2 September 1914. Easton submitted an account of the fighting to the official historian whilst a POW. He and his men had initially relieved a Scottish battalion in the trenches amongst the houses of *Corons de Maroc* and *Corons de Pekin*. The Germans forced out the troops of both battalions on his flanks and he and some of his men escaped by running a gauntlet of German fire. They occupied cramped trenches with other units:

> Fatigue was so great that the men often fell asleep during a conversation and the clogging of rifles had to be remedied by exchanging them for those of the men in the rear. At 3.30 a noise was heard in the rear and a coy. of the 1st [Royal] Berks[hire Regiment] came up to retake the Dump. The 73rd Brigade immediately joined in and a force some 250 strong charged the Dump. The Dump is some 50 to 60 feet high and is very precipitous: the enemy cleared out in face of the advance and the whole Dump was taken. As the men cleared the crest a wave of machine gun fire hit them breast high and casualties were very heavy. … The enemy had manned a trench running immediately round the foot of the whole dump, and would have been completely at the mercy of our bombers, but it was discovered that there was not a single bomb… The situation was now most critical. Both sides were firing at each other at a range of twenty yards with all the advantages in favour of the enemy. In order to fire down practically perpendicularly we had to expose our heads and shoulders… The machine gun fire … was continuous, and their artillery bombarded the dump … casualties were appalling and we seemed to inflict no loss on the enemy. Both the officers of the Berkshires were killed, handing on their orders, "hold on at all costs". … The party lining the SE edge, now reduced to one officer and fifteen men, were captured as about 4.30 a.m. by a strong attack in their rear…[28]

Easton's men were forced back from the edge of the high ground by enemy bombs and machine gun fire. His remaining five men took up positions in shell holes 20 yards back from the lip. Whilst they waited for an attack from the front the Germans surprised them from behind and

---

26 Hill, *To Know the Sky*, p.28.
27 Eyston, *Safety Last*, pp.33–34.
28 TNA: CAB 45/120: Letters to the Official Historian – Loos: Account by J Easton, p.12.

were within ten yards when they were detected. Easton's party was quickly overwhelmed.[29] He was lucky to survive the engagement as a prisoner, unlike many young officers.

As news had spread to the rear that 73rd Brigade (including 12/R Fus) was on the verge of breaking, Major General George Thesiger, commanding 9th Division, went forward to investigate. Whilst doing so he was killed by shellfire. Barely two weeks before Thesiger had commanded 33rd Division.[30]

For some, the Battle of Loos had almost no impact. David Kelly, of 110th Brigade (37th Division) was still in 'quiet' trenches further south. For others, the diversionary operations near Ypres, and German counter-shelling, had caused some battalions, like 7/KRRC (including Edmund Page), heavy casualties. Moreover, a few days later the battalion:

> … returned to the front line with orders to prepare for an enemy counterattack. It came the next evening just as I had started out on patrol to investigate the enemy trench. The preliminary bombardment was appalling and lasted three quarters of an hour. Mercifully our front trench we had dug unexpectedly close to the enemy and as the barrage lifted we managed to shoot them down in clusters as they arrived. The men fired well…[31]

Ex-UPS 'chemical corporals' notably served at Loos. PS/4977 Private Herbert Hewitt, of 20/R Fus, volunteered for the 'Chemical Corps' in the summer of 1915. He became a corporal in 189th Special Company and went to France on 26 July. Gas was also employed on the last day of the Battle of Loos on 13 October; Corporal Hewitt was killed by shrapnel. He was allegedly recommended for the Distinguished Conduct Medal (DCM) because of his tenacity in a previous attack.[32] Second Lieutenant Charles Gordon Jelf was also killed on 13 October at Loos. He was educated at Felsted School, Marlborough College and Exeter College, Oxford. Jelf was an archaeologist (who had excavated at Luxor, Egypt; a schoolmaster at Fonthill School and worked for The Times in Berlin (1911–1914). He had returned to the UK to enlist in October 1914 and spent nine weeks with the UPS. He was gazetted to the 6th Battalion East Kent Regiment (6/E Kents)(12th (Eastern) Division). Jelf was also recognised with a MiD.[33]

An epitaph for the Battle of Loos was written by Private Spaull of the UPS on taking over nearby trenches in January 1916:

> … Our front shows signs of the last advance (25th September last). Many of the dead are still out in front as we are unable to bring them in now after being out there so long. They are for the most part in extended order as they advanced, some wearing gas helmets … It is not a very cheerful sight…[34]

---

29 TNA: CAB 45/120: Letters to the Official Historian – Loos: Account by J Easton, p.12.
30 He is commemorated on the Loos Memorial.
31 IWM: E. Page Papers, 16924, p.4.
32 106297 Corporal Herbert Hewitt (formerly PS/4977) was aged 22 and was from Stockport. He was educated at Portwood Wesleyan School and Stockport Grammar School and was awarded a degree in chemistry in 1914 from Manchester University. He is commemorated on the Loos Memorial. Manchester Evening News, 21 October 1915.
33 Jelf is buried in Vermelles British Cemetery.
34 IWM: P.F.F. Spaull papers, 15453, letter 20 January 1916.

'Some to the Flying Corps have fled …'

The RFC had attracted many ex-UPS men and some reached France during 1915. Second Lieutenant Gilbert Meakin, a former officer of 21/R Fus, transferred to the RFC. He went to France, as an observer, in February 1915 and joined No 7 Squadron RFC. He flew strategic reconnaissance missions with his pilot Lieutenant F.P. Adams. On 3 July they conducted a long-range photo-reconnaissance mission to Ghent for GHQ. According to Meakin, whilst over the target at 7,400ft, and about to take a 'photograph of the engine failed owing to the exhaust valve of one cylinder breaking off and falling into the cylinder resulting in the complete stoppage of the engine'.[35] Adams, recorded in a letter:

PS/1810 Private Gilbert Edward Robinson Meakin enlisted, aged 19, on 3 September 1914 and joined 18/R Fus. He was from Spondon, near Derby, was educated at Oundle School and was employed as a motor engineer. He was gazetted to a commission in 21/R Fus on 27 October and left to join the RFC in December 1914.

> In one second, knowing the symptoms only too well, I had swung round in the direction of the nearest front in Holland, rather more than 15 miles away. A horrible grinding noise proceeded from the engine as we ran for it, rapidly losing height and leaving behind a thick trail of black smoke and steam. The Huns must have noticed our predicament, for they started shelling us. We were not thinking of "Archies" though, and anyhow their shooting wasn't good. It seemed an utterly hopeless task to clear the frontier, a row of trees in the distance, which was painfully distinct as were all the landmarks shown. The fifteen minutes occupied in gliding down seemed like a lifetime. The sharp crack of bullets was a very welcome sound. It showed that we were very near the frontier. We cleared that frontier by about 150ft. and landed only 600 yards in Holland …[36]

The RE5 landed at Sas van Gent; though Meakin escaped being a POW in Germany he was interned in Holland. Another ex-UPS officer who took to the skies was Arthur Whitten Brown

---

35   TNA: WO339/24243: G.E.R. Meakin officer file.
36   *Gilbert Edward Robinson Meakin RFC* <https://www.theauxiliaries.com/men-alphabetical/men-m/meakin-ger/ger-meakin.html> (accessed 4 January 2020).

who was attached to the RFC. On 10 November 1915 he undertook an aerial reconnaissance of Valenciennes:

> … We completed the reconnaissance as far as Valenciennes, flying at times in cloud and snow, which compelled us to descend as low as 800'[sic] in order to make observations. Snow was falling continually and the engine commenced to misfire. … The clouds closed in again, and the engine failed … We glided down and came under heavy machine-gun and rifle fire as soon as we were beneath the clouds. I received a bullet wound in the left foot, and I understand that the controls were damaged. We landed in a field near, I believe, Courville, crashing the machine badly. In the crash, I had my leg broken … we were surrounded and made prisoners.[37]

He was captured by the Germans. However, during his time as a POW he studied aerial navigation which contributed to his later endeavour to fly the Atlantic.[38]

Second Lieutenant Gilbert Insall had enlisted in 18/R Fus on 15 September 1914. He was educated at the Anglo Saxon School, Paris, and was a dental student. He joined the RFC in March 1915 and qualified as a pilot. He arrived in France in July 1915 and joined No 11 Squadron. On 7 November Insall was recommended for the Victoria Cross with the following citation:

> … He was patrolling … with First Class Air Mechanic T.H. Donald as Gunner, when a German machine was sighted, pursued, and attacked near Achiet. The German pilot led the Vickers machine over a rocket battery, but with great skill Lieutenant Insall dived and got to close range, when Donald fired a drum of cartridges into the German machine, stopping its engine. The German pilot then dived through a cloud, followed by Lieutenant Insall. Fire was again opened, and the German machine was brought down heavily in a ploughed field 4 miles south-east of Arras. On seeing the Germans scramble out of their machine and prepare to fire, Lieutenant Insall dived to 500 feet, thus enabling Donald to open heavy fire on them. The Germans then fled … Other Germans then commenced heavy fire, but in spite of this, Lieutenant Insall turned again, and an incendiary bomb was dropped on the German machine, which was last seen wreathed in smoke. Lieutenant Insall then headed west in order to get back over the German trenches, but as he was at only 2,000 feet altitude he dived across them for greater speed, Donald firing into the trenches as he passed over. The German fire, however, damaged the petrol tank, and, with great coolness, Lieutenant Insall landed under cover of a wood 500 yards inside our lines. The Germans fired some 150 shells at our machine on the ground, but without causing material damage. … during the night it was repaired behind screened lights, and at dawn Lieutenant Insall flew his machine home …[39]

Insall was on borrowed time. On 13 December 1915 Insall and Donald were on patrol when they saw a German aircraft near Arras:

---

37 TNA: WO 339/27707: A.W. Brown officer file.
38 Brown, *Flying the Atlantic*, p.10.
39 *The London Gazette*, 28 December 1915, p.1945.

> We crossed the lines to cut him off and, getting within range near Bapaume, opened fire with our machine gun. … A dive at right angles to his line of flight brought us to close quarters, and Corporal Donald opened fire as we turned right-handedly over the tail of the enemy machine. The latter opened fire … a bullet from his machine gun entering the front on the Vickers' nacelle, passing fore and aft through my observer's seat, cutting both right aileron control cables, grazing my right leg, and going through the petrol tank … The engine, being pressure-fed, immediately stopped … We had glided about 500 feet when an A.A. shell, the first fired, burst close under the machine, fragments hitting Corporal Donald in the legs and myself in the hip. A.A. shells were then fairly plentiful … I felt dizzy for a few minutes after being hit but regained complete consciousness until we landed. … We continued gliding and came to earth the other side of the wood, in a large field …[40]

Insall spent the rest of the war as a PoW.

The men who left the UPS Brigade in 1914 and 1915 for gazetting, predominantly into the infantry, had got to the war faster than their enlisted compatriots. Some had suffered boredom, others terror, and many had been killed, endured wounds or were captured. Similar fates would await the enlisted men of the UPS when they finally arrived in France.

---

40 TNA: WO 339/53916: G.S.M. Insall Officer File.

8

To France

No more playing at soldiers

> *… a lot of very young schoolboys going home for the holidays …*
>
> *No more playing at soldiers for the 21st Battalion Royal Fusiliers!*

Frenetic activity

The time for departure to France rapidly approached. According to Morgan Williams:

> The night before we sailed for France was marvellous. Gone was the brooding and the apprehension, giving way to a strange exhilaration which took the queerest forms. Two brothers, delightful fellows and magnificent in their strength – namesakes of mine – challenged each other to race a hundred lengths of the hut … At last they dropped exhausted, dead beat in a dead heat.[1]

In the hours before departing men made difficult decisions on what to pack and what to leave. Philip Spaull recorded the activity before departing; 'You never saw such a move as this – everything turned upside down. Various fellow doing various thing, washing clothes, mending same, cleaning rifles, putting packs together. I shall have a lot to talk about when I come home… I have been given a confidential book of rules and regulations for billeting and requisitioning in France and Belgium…'[2] Ernest Holden described the last night before departing:

> I tried to get a little sleep on the floor in front of the fire, but I am afraid there was too much subdued excitement for much sleep. I got to sleep however about 11.15 and was turned out again at 12 p.m. I arose … as I wanted to attend Communion at 12.15. It was a strange little service. The soldiers in all states of dress, yawning and blinking, and shivering in the big garrison church. We got breakfast at 1 a.m. and that was our last meal … The

---

1 Williams, *From Khaki to Cloth*, p.56.
2 IWM: P.F.F. Spaull papers, 15453.

A graphic illustration of 98th Brigade departing for France entitled 'Fate cuts the painter'. Though the rumours and uncertainty over their destination were now seemingly decided, there was a clear apprehension as to what the future held. (Sketch by William Beck-Savage from *The Gasper*)

authorities seemed to delight in taking the last 'pound of flesh' from us, as they made us walk to Ludgershall rather than entrain us at Tidworth. However, after that I have nothing to complain of...[3]

At midnight before departing there was a packed service at the Tidworth garrison church which was attended by Lieutenant Colonel Wolrige Gordon. Many took solace in religion at this time; 'I have always thought I should come safely through this War and now feel doubly secure. It [a written prayer] offers protection to all who carry it against capture by the enemy and against sudden death.'[4] However, the chaplain for 98th Brigade, Stuart-Clarke, was not ideal for the role and was 'what "Woodbine Willie" called "a dismal bloke." He had hundreds of keen men to preach to and all he did was to remind us that in three months we might be dead. The statement seemed to contain an element of truth but it grated upon me every time I heard it.'[5] According to Hugh Spurrell, Stuart-Clarke was; '...a bit of a fanatic and a very unsuitable person for the job.'[6] A hurried meal followed the service.

3 LIM Archive: Letters of E.A. Holden, courtesy LIM, p.57.
4 IWM: P.F.F. Spaull papers, 15453, letter 14 November 1915.
5 Williams, *From Khaki to Cloth*, p.54.
6 IWM: H.W. Spurrell 2138.

Departure

By 2:00 a.m. the men of 19/R Fus were on parade and Robert Sturges recalled the battalion moving off after a few delays. They cheered as they left the parade ground and everyone was excited and laughing at jokes; '…like nothing so much as a lot of very young schoolboys going home for the holidays …' The band and singing likely awakened some inhabitants of Tidworth and those who were roused waved from their windows.[7] The UPS Brigade left Tidworth early on the morning of 14 November 1915 and proceeded by train to Folkestone. Lieutenant Modera, 20/R Fus, arranged for his wife visit Tidworth and she marched alongside him from the camp to Tidworth Station. Private Donald Price of 20/R Fus recalled not being allowed to talk or smoke as the battalion marched to Ludgershall which took place at night. The history of the Brigade recorded the secrecy of their departure, 'During a short halt in a street in Ludgershall a window of a house, close to where one of the platoons was standing, was suddenly opened and a woman's voice was heard: "Are you the Fussies?" "Yes," said some-one in a hoarse whisper. "Good-bye and good luck," came the answer. "Stop that talking," snapped out some bad-tempered officer or N.C.O.'[8]

Once they were on the train the blinds had to be pulled down when they departed at midnight. Price remembered their destination:

> It was dawn before we got to Dover. There was a case of excitement; here we were, the sea was there, the ship was there. We were all de-trained and marched on board this ship. A wonderful feeling for us after being cooped up for a year training to go abroad and all of a sudden here we are, we're going. We didn't know what we were in for. … Keen, keen to get there.[9]

18/R Fus departed Tidworth by half-battalions at 2:30 and 2:50 a.m. on 14 November. They arrived at Folkestone at 7:45 a.m. and 8:15 a.m. respectively. The battalion travelled on the *SS Princess Victoria* and enjoyed a good crossing; Kenneth Norman recalled that the ship was waiting at the docks for them when they disembarked from the train. The vessel was a cargo boat with three decks; accommodation on board was uncomfortable with Norman's company in the bowels of the ship in a dirty hold. These conditions made many men seasick. Lifebelts were worn in case of an emergency.[10] Such precautions were advisable; due to heavy weather some of the floating sea mines near Boulogne had come adrift. One such mine had been rumoured to have blown up the last ship to enter the harbour. Disembarkation occurred in Calais instead, and a short train journey to Boulogne was needed. 18/R Fus recorded arriving in France at 11:15 a.m. and after entraining they arrived at Boulogne at 3:00 p.m.

Private Harold Wilson, 19/R Fus, recalled the journey:

> Left Tidworth 3 a.m. arrived Folkestone via Red Hill at 10 a.m. Embarked 1 Batt. Per boat, some crush. Topping crossing – arr. Calais 12.30 – entrained 1.30 for Boulogne,

---

7   Sturges, *On the Remainder of Our Front*, pp.13–14.
8   Anon. Author, *University & Public Schools Brigade*, p.102.
9   IWM: Donald Price Sound Recording 10168.
10 LA: K.V. Norman Papers, WW1/GS/1188, p.88.

arriving at B. station about 5 o'c – some fun ragging the Frenchies all along the line. French trains pretty awful. … Marching up very steep hill 3 miles to St Martin's, where the first Expeditionary Force camped. Rations for 24 hrs – 1 tin bully & 3 biscuits – not bad tuck. 12 men in the tent… Sleeping in clothes with one blanket – quite comfy.[11]

PS/757 Private Nicholas Pease recorded 'in the middle of the night and as the troop train passed through Staffhurst Wood at dawn I caught a glimpse of my home in the distance, perhaps, I thought, for the last time …'[12] According to the 19/R Fus War Diary 19/R Fus left Folkestone, numbering 1,024 all ranks, at 9:30 a.m. and arrived at Calais at 12:55 p.m. On disembarking the men were loaded into a train comprising third-class carriages at the nearby station and proceeded by train to Boulogne at 3:00 p.m. and arrived at 5:45 p.m. They then marched to their camp at Boulogne arriving at 7:00 p.m.[13]

Hugh Spurrell of 21/R Fus recalled: 'No more playing at soldiers for the 21st Battalion Royal Fusiliers! From now on they were to do the real thing.'[14] According to PS/6924 Private Philip Noel Wright, No 2 Company, 21/R Fus:

> As the Huns had eagerly awaited 'Der Tag', so we had looked forward to this day. Reveille was at five, so as to give us plenty of time to 'pack up' before moving off. We were on parade by seven o'clock, on the old and by this time 'too' familiar parade ground between the Y.M.C.A. and the camp. Before finally leaving, the Colonel (Stuart Wortley) gave us a short address, a few words of which I can still remember – "When you get over the other side I want you to remember that you're Fusiliers, uphold the traditions of your regiment, – I won my commission on the field of battle. I want you to do the same". And then – "Battalion will move off from the right in columns of route, …", etc, and we were on the way. A short march across the down from the camp at Perham took us to Ludgershall, where we were to entrain for Folkestone…[15]

Hugh Spurrell recalled getting little sleep as the camp was awake by 3:00 a.m. and they breakfasted before 5:00 a.m.. The battalion marched off with gusto with the band playing various patriotic music; '…we paraded and marched to the station singing the crazy songs of the period … and when the band ceased playing a mouth-organ here and there continued the melody. In an hour we reached Dover and with orderly haste boarded a troopship and put on life belts as a precaution against submarine attack.'[16] According to the battalion war diary, that started with this act, 21/R Fus departed No 3 Camp, Perham Down, at 8:00 a.m. on 14 November and consisted of 26 officers and 872 men (thirty officers and 994 men in total). The rail journey was 'tedious and roundabout' though they were able to stretch their legs at Guildford and Redhill. Arriving at Folkestone at midday they boarded the *SS Princess Victoria* but recorded arriving in Calais at 5:00 p.m. where the men were spared travelling in cattle trucks and were again in draughty carriages. They reached Boulogne at 8:00 p.m.:

11 IWM: H. Wilson Papers, 3606, p.1.
12 IWM: N.A. Pease Papers, 8230, p.31.
13   : WO95/2427: 19/R Fus War Diary.
14 IWM: H.W. Spurrell 2138, p.27.
15 FM: P.N. Wright Diary, RFM.2013.8.1/2, p.3.
16 Williams, *From Khaki to Cloth*, p.57.

'We got out, shouldered our kits and paraphernalia, and fell in in the station yard … Then we moved off across the river, made our way through a maze of streets and began to climb a hill. Everybody who has marched to the Boulogne rest camps in full packs knows all about those hills … As we marched along windows flew open and doors opened. Otto thought that he was making a triumphal entry with the first British troops in the country. We should have sung "Tipperary" or something like that… Unfortunately we all had our work cut out to get our 90-pound kits up the hill…'[17]

According to Morgan Williams, 21/R Fus, the battalion came across a welcoming reception by the French populace as they marched up to their camp 'little French girls and boys yelled themselves hoarse or begged for souvenirs. One tiny chap had mastered enough English to shriek, "Ginger you balmy" to Bowes, a red-headed man in my platoon, while the long lines rocked with laughter in which he joined as heartily as anyone.'[18] The battalion was at Ostrohove Rest Camp by 9:40 p.m. Here there was a 'jigsaw puzzle' getting twelve soldiers and their kit into each small tent.

Leslie Ireland did not accompany 19/R Fus having missed training whilst in hospital earlier in the Autumn. He transferred to 28/R Fus shortly after. He mournfully wrote the following in a letter home:

Gone! Gone are the friends of my youth!
Never to return? Nay, not so.
Some, midst trumpets blare and cries loud
With glory great, see the enchanting hour,
Whilst with shame and ignominy
I watch? Save me from that hour
O God![19]

He was kept working hard for several days before moving to Oxford; he remarked; 'If I was bereft of all friends, I verily believe I should desert and reenlist.'

Rest Camps

Life in the 'rest' camp was bleak; on 15 November 21/R Fus woke up and were, for the first time, introduced to cold Maconachie meat stew for breakfast. There were two taps and two washing bowls per company. Kenneth Norman recalled that the rest camps at Boulogne comprised overcrowded tents that were very cold. Next morning the men awoke almost frozen with six inches of snow on the ground. The food was inadequate but a nearby YMCA hut did a roaring trade.[20] Philip Spaull recorded; 'We have all be a jolly party so far and everyone is in fine spirits. Upon

---

17   IWM: H.W. Spurrell 2138.
18   Williams, *From Khaki to Cloth*, p.57.
19   IWM: L.W.C. Ireland Papers, 14441, letter approx. 14 November 1915.
20   LA: K.V. Norman Papers, WW1/GS/1188, p.92.

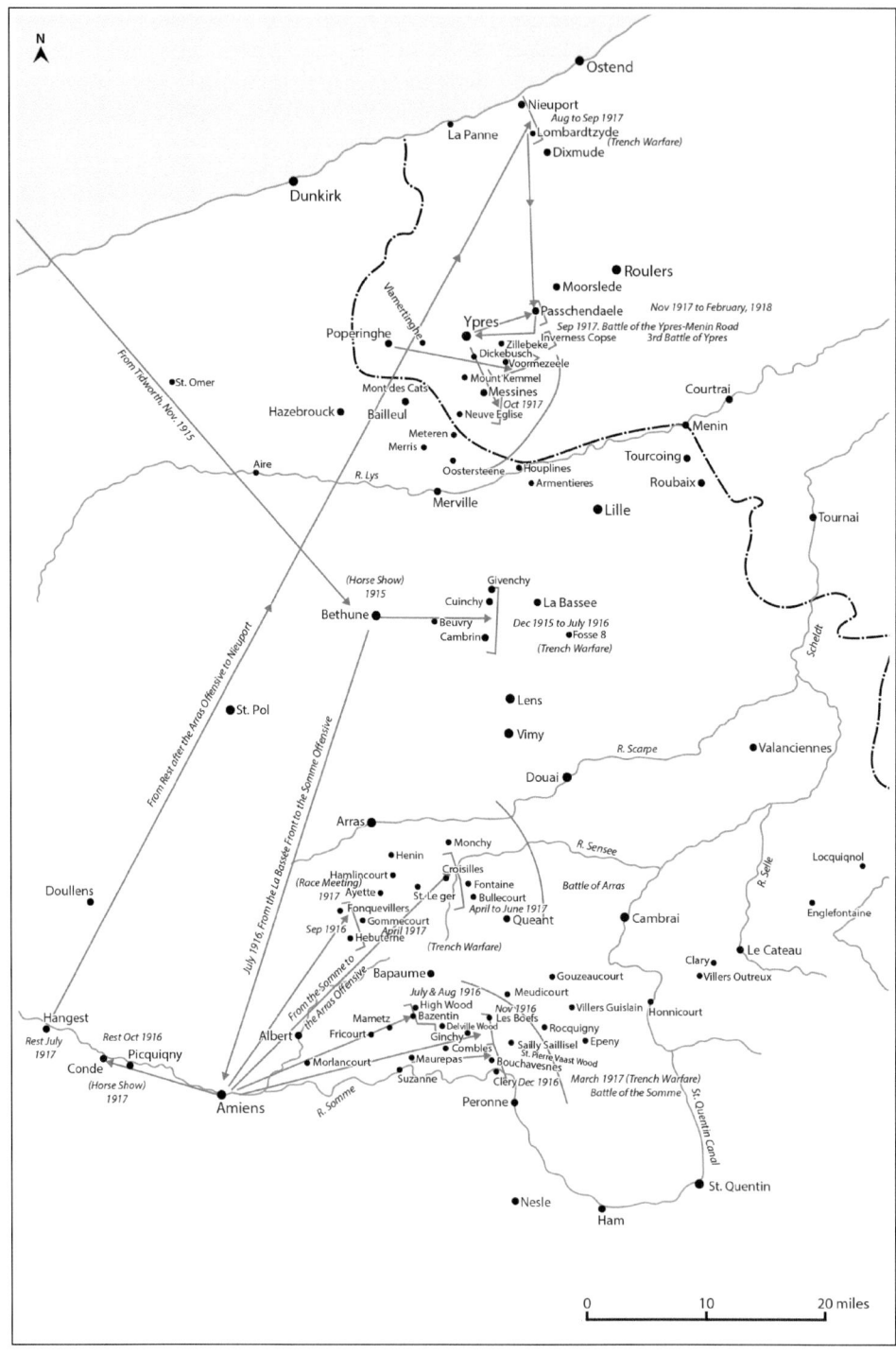

Map 1 98th Brigade and 20/R Fus in France and Flanders 1915-1918.

arriving we slept in tents and in the morning found the ground covered with about 11 inches of snow! It was rather cold for morning ablutions.'[21]

18/R Fus and 19/R Fus spent all day on 15 November in the rest camps but went on a short route march and conducted routine administration. PS/1040 Private Harold Wilson of 19/R Fus concisely recalled the experience of marching through Boulogne 'parade 1.45 for route march through Boulogne – very interesting march – roads cobbles, bad for feet. The Batt. Stopped for half an hour on the sands. Very kind lady gave us 6d packets of chocolate – very few "birds" about – children a nuisance, always begging.'[22] Robert Sturges also remembered this foray by 19/R Fus as they conducted a 'swank' march to show themselves off around the town. This was the first exposure to another country for many men and they saw numerous civilian spectators and several other servicemen. Sturges recollected, 'We were even young and fresh enough to give the once common cry, "Are we downhearted? No," almost asking for the answer, which one of the wounded men gave us, "*You very soon will be.*"'[23] Salter considered the ceremonial march around Boulogne to have been; 'Great fun'.[24] Morale was high for some; Wilson recalled a sing-song that evening in the Y.M.C.A. hut. Due to their ingrained pessimism some men believed they would be staying at the base for a lengthy period; Sturges bet a comrade half a crown that the battalion would not enter the trenches before Christmas.[25]

On 15 November the 20/R Fus war diary recorded nothing of importance. However, Godfrey Skelton recalled 'the Colonel thought that he would like to let the people of Boulogne see what 20th Battalion Royal Fusiliers looked like. A route march around the town was therefore ordered in full pack. This was a gruelling affair, as the town is all hills and the equipment of an infantryman is very heavy.'[26] In the afternoon 21/R Fus did the same; according to Philip Wright; 'We had an opportunity of seeing the town, which, however, didn't impress me very favourably. We passed along the quays and I noticed what a number of hospitals where were on the front, wondering at the time, whether I should ever be lucky or unlucky enough to see the 'inside' of one…'[27]

According to Godfrey Skelton when 20/R Fus paraded in the morning; '…Colonel Bennett … was up betimes, growling as usual and the Adjutant with him …'. Lieutenant Colonel Bennett had likely just demoted his Pioneer Sergeant because he was caught returning to camp drunk having disobeyed orders and gone into town.

Arrivals and departures

The Brigade Commander and Brigade Major had remained in Calais on 14 November and next day drove to Thiennes whilst the brigade remained at Boulogne. No sooner had this occurred than Brigadier General Gilmour was sent back to England. Brigadier General E.P. Strickland

---

21 IWM: P.F.F. Spaull papers, 15453, letter 17 November 1915.
22 IWM: H. Wilson Papers, 3606, p.1.
23 Sturges, *On the Remainder of Our Front*, p.18.
24 IWM: L.E. Salter Papers 21103.
25 Sturges, *On the Remainder of Our Front*, p.16.
26 IWM: G. Skelton Papers, 13966, p.34a.
27 FM: P.N. Wright Diary, RFM.2013.8.1/2, p.3.

CMG DSO took over the Brigade on 16 November.[28] This was a significant change; Strickland was intended as a 'new broom' to adjust the UPS Brigade to conditions in France. He was an ambitious and driven officer and as a divisional commander was known as 'The Hungry One' who potentially put his career before his men.[29] One of Strickland's staff officers at 1st Division recalled his temper and his not welcoming outsiders.[30] Strickland had undoubtedly been appointed to 'shake-up' 98th Brigade.

The brigade commander had not lasted long in France and, similarly, several men from within the ranks did not endure. One character was the paternal figure of William Henry Spershott, whose three sons were serving with him in 19/R Fus.[31] William suffered from severe seasickness and, after a route march next day, was found to be out of breath and unable to play his instrument, the cornet. He went to the MO after four days with stomach cramps. He lasted for five days in France.[32] Men carrying injuries or illnesses had held on long enough to get to France. Donald Price, of 20/R Fus, turned his ankle again climbing up the hill to the camp and went sick on 16 November; he did not re-join until January 1916. Price recalled; 'I felt I was going to miss it. I was urging myself to be fit … Eager, like we all were, to get into the trenches.'[33]

Some other men were in their element in France; 'From the moment his foot touched continental soil Percy Richardson became a foreigner. He adopted a French accent and began a course in the French language for the benefit of ignorant members of the section.'[34] Likewise, Walter Medlicott considered his friend: '[Alan] Beeton especially is a great asset, as he can make himself …understood… by the natives … there are several men in the platoon, and a worthy boy, [Harry] Van der Weyden, who never fail to get all we want out of them…'[35]

28   Brigadier General Edward Peter Strickland CMG DSO was a regimental officer with considerable campaigning and fighting experience. He was born in 1869, was educated at Warwick School and joined the Norfolk Regiment in 1888. He served in Burma, Sudan and was awarded the DSO whilst serving with the Egyptian Army against Ahmed Fedil in 1899. He also served in Nigeria. He was made a lieutenant colonel and commanded 1/Manch R in France from August 1914. He was promoted to brigadier general in 1915 and major general in 1916 to command 1st Division throughout the rest of the war. He was made CBE in 1917, KCB in 1919 and KBE in 1923. He served as a Lieutenant General 1921–1922. He enjoyed hunting, polo and shooting.
29   C.P. Clayton, The Hungry One (London: J.D. Lewis, 1978).
30   Basil Sanderson, *Ships and Sealing Wax* (London: Heinemann, 1967), p.56
31   His three sons were William Frederick Spershott, Henry Percival Spershott Walter John Spershott. PS/8910 Lance Corporal William Frederick Spershott, aged 19, had been an errand boy. In November 1914 he enlisted with 12/R Sussex and was later appointed lance corporal. He transferred to 19/R Fus in November 1915, presumably to go to France with his father and brothers. He returned to the UK in April 1916 and was discharged sick in July. PS/8909 Private Henry Percival Spershott, aged 18, had been employed as a sale room assistant for a tea merchant. He later served with 20/R Fus and was attached to 33rd Division HQ and XIX Corps School. He was discharged in March 1919. PS/8596 Private Walter John Spershott, born in June 1900 was aged only 15. He was discharged, likely underage, in May 1916. He re-enlisted, aged 17, in November 1917 with the Royal Marine Artillery but was discharged in April 1918, unfit.
32   PS/8593 Private William Henry Spershott, aged 45, was an insurance agent and musician. He enlisted on 25 August 1915 in Eastbourne and served as a bandsman. He presumably endured illness only long enough to accompany his sons to France and was sent to the RF Depot five days after arriving. He was discharged in July 1916 due to the strain of playing the cornet in the band.
33   IWM: Donald Price Sound Recording 10168.
34   IWM: H.W. Spurrell 2138, p.28.
35   IWM: W.B. Medlicott Papers, 1708, 87.

Early railway movements by the 98th Brigade in France. Whilst many would travel in cattle wagons, some would be fortunate to return to Blighty in carriages bound for officer training. The unlucky would travel in hospital trains. (Sketch by W.H. Wheatcroft published in *The Gasper*)

On 16 November 1915 18/R Fus, 19/R Fus and 20/R Fus departed Boulogne by rail and arrived at Thiennes; these battalions were billeted there and in surrounding farms. Salter recorded the nature of the journey; 'Left Boulogne for Thiennes by train. Heavy fall of snow. Shoved us [in] vile milk vans, 65 of us in each, no windows. Eight hours journey, wicked. Arr[ived] Thiennes 7[p.m.]. Wretched march to billets, which were farms.'[36] Another commentator described the journey thus; 'The journey in cattle trucks was rather reminiscent of the Black Hole of Calcutta, but by standing up one could see the passing country; and very dull, depressing-looking country it is.'[37] One man of 18/R Fus fell out of the train according to Wilson and was described as being 'some mess'.[38]

Philip Spaull, of 19/R Fus, wrote in a letter home that two mules and four men were injured at a level crossing when a train ran into an Army wagon but presumed they were not suffering from life-threatening injuries.[39] According to Harold Wilson, who must have heard of the incident through rumours, 'nasty accident at level crossing – train caught 3 men, Gore, Archer & Moule – all hurt badly. Gore both legs cut off.'[40] There might have been some exaggerated stories regarding the extent of injuries. According to Robert Sturges the female crossing keeper was distracted by the soldiers passing and had not seen the train. Though the driver of the

36 IWM: L.E. Salter Papers 21103.
37 Sturges, *On the Remainder of Our Front*, p.19.
38 The casualty was presumably badly injured as he relates to no obvious fatality. IWM: H. Wilson Papers, 3606, p.1
39 IWM: P.F.F. Spaull papers, 15453, letter 17 November 1915.
40 These men have been difficult to identify. IWM: H. Wilson Papers, 3606, p.1

wagon jumped clear two men were injured and two mules were dragged along with the wagon and had to be put out of their misery.[41] 19/R Fus found the first time it billeted its soldiers in a French village to be a difficult and lengthy process.

18/R Fus was billeted nearby at Tannay. According to Kenneth Norman, whose section was billeted in a barn with comfortable straws; 'Here we rested for two days and did nothing but eat and drink in the *estaminets* and try our own peculiar French. Here for the first time we heard the boom of the guns in the distance.'[42] Skelton recorded; 'We fed like princes all this time. We all had plenty of money and spent it on French bread, butter, milk, chips, eggs etc. etc. and beer and *vin blanc* and *rouge*.'[43] Walter Medlicott was in a comfy billet:

> I am with Blencowe and 82 men in a fairly cosy barn. Most of the men are on 8ft of straw and occasionally lose a spoon or a cap some foot down. It takes some delving to recover it! We are fed plentifully on biscuits and tinned meat and have had several cups of coffee from the folk nearby … it's not far from the guns, they are noisy enough to be heard anyhow.[44]

21/R Fus arrived at Thiennes later. They went on another route march on 16 November after a short rifle inspection. Some men also did physical training and some bayonet fighting in the snow. During the afternoon they rested as they were to depart camp at 11:50 p.m. that night. According to Philip Wright, 'glad we were too, when the order came to parade, for I distinctly remember how we huddled together in the tent to keep warm all that evening, the weather being intensely cold at the time.'[45] After a four-mile march to Pont-de-Briques Station the battalion had to wait in the rain until 3:55 a.m. to entrain; 'At last one train stopped and the voices of the men of the Machine Gun section were heard inside. They had come from Havre and had been in the train for 28 hours.'[46] This time the battalion travelled in cattle trucks. 21/R Fus reported that its transport departed Southampton on 14 November and rejoined the battalion on this train to Thiennes. However, half of the advance party (two officers and 66 men) was delayed at Southampton and did not rejoin the battalion until 18 November at Thiennes.[47]

The 21/R Fus arrived at Thiennes at 9:50 a.m. on the 17th; which was, according to Philip Wright, 'a small straggling village, containing a few *estaminets*, one or two small stores and a fairly fine church.'[48] Morgan Williams recorded the perceived unpleasantness of French villages:

> … we reached a dismal village, Thiennes by name, in country as dreary as Staffordshire. Its outstanding virtue was the cleanness and warmth of the straw in which we slept. Everything else was ghastly. The doors and windows of our barn opened onto a midden, such as occupies the centre of most French farmyards. It looked like dung, it smelled like dung. It was dung, and the odour of it clung to us for days….[49]

---

41 Sturges, *On the Remainder of Our Front*, pp.20–21.
42 LA: K.V. Norman Papers, WW1/GS/1188, p.96.
43 IWM: G. Skelton Papers, 13966, p.34a.
44 IWM: W.B. Medlicott Papers, 1708, p.87.
45 FM: P.N. Wright Diary, RFM.2013.8.1/2, p.4.
46 IWM: H.W. Spurrell 2138, p.30.
47 Some men were presumably delayed as many disembarked in France on 15 November.
48 FM: P.N. Wright Diary, RFM.2013.8.1/2, p.4.
49 Williams, *From Khaki to Cloth*, p.58.

A sketch showing a typical billet encountered in France. UPS men and officers would encounter similar Spartan accommodation whilst on active service. (Sketch by J.S. Willock published in The Gasper)

Brigadier General Strickland wasted no time getting to know his battalions visiting 19/R Fus at Tannay on 17 November. During 18 November 21/R Fus had a heavy draught mare die having over-exerted itself the previous day. There were further disciplinary problems in 20/R Fus; the Transport Sergeant, Anderson, neglected to post a sentry for the horse lines and was demoted to private after a mule strayed and almost died. He was sent to Skelton's Platoon as a rifleman where he was a welcome addition.[50]

Forced march to Béthune

The brigade marched from Thiennes to villages near Busnes on the 19th and onwards the next day. 18/R Fus departed Tannay at 9:30 a.m. on 19 November and marched via St Venant to 'bad' billets south of the La Bassée Canal at Robecq. Next day, they departed at 9:30 a.m. to Béthune (via Chocques) and stayed at the Tobacco Factory. 19/R Fus left Thiennes on 19 November and after a horrible march arrived in la Miquellerie, near Busnes, though had reasonably comfortable

50 IWM: G. Skelton Papers, 13966, p.34a.

billets.[51] Next day they were awake again at 6:00 a.m. and had a long, weary, march to Béthune covering fifteen miles; they arrived at 1:00 p.m. and were billeted in a school. This was the *Ecole de Jeunes Filles*; '…. a dirty-looking place consisting of bare classrooms with broken windows, round a courtyard. We filled the rooms to overflowing with ourselves and our kit, which was soon strewn all over the floor.'[52] D Company, 19/R Fus, were not well accommodated according to Harold Wilson; 'Our billet in Béthune is pretty awful. 250 men in a school, hardly room to turn.'[53] Salter's succinct summation was 'rotten'.[54]

20/R Fus departed Thiennes at 9:00 a.m. on the 19th and marched to L'Eclème, further east of Lillers. At 7.30 next day six officers and sixteen men went forward to visit the trenches. At 9:30 a.m. the battalion departed for Béthune on foot. On arriving there, 20/R Fus was billeted in Montmorency Barracks.[55] The UPS History recorded: 'Then another day's march and the barracks in Béthune – a dirty hole, with part of the roof removed by a German shell. The U.P.S. were really in touch with the war now.'[56] At 8:45 a.m., on 19 November, 21/R Fus departed Thiennes on foot for Cantraine, a small hamlet east of Lillers, where it was billeted. Next day, at 7:00 a.m., Major Fenn and the seconds-in-command of each company went to the trenches and rejoined the battalion that evening. Meanwhile, at 9:40 a.m., the battalion marched off heading to Béthune. According to Philip Wright:

> The country …was flat, in front of us for miles we could see the road winding in and out … made us feel rather fed-up. We marched for some time without getting any halts, before we realised that we were being treated to a 'forced march.' … A rather amusing incident, I remember, happened on route. For some reason or other the mules harnessed to one of the 18th Battalions 'cookers' took fright, and after some prancing about, eventually deposited the unfortunate cooker in the ditch. As a result, the unlucky company to whom the cooker belonged had a short tea ration that evening…[57]

21/R Fus arrived in Béthune at 1:00 p.m. Billets were found in the Avenue de Bruay, a poor part of the town near the Gare du Nord. Philip Wright's company was quartered in a large, disused building opposite the station. Though the cooker incident had lightened the march for 21/R Fus, it had been a trial and some men could only lie down and sleep in their billets that evening.

In a week the UPS Brigade had rapidly transitioned from training in the UK to active service in France. Despite the accident at Thiennes, the process had gone relatively smoothly. The departure of parties of officers and NCOs to reconnoitre the trenches on 20 November will have been good news to many members in the Brigade. Robert Sturges began to realize 'that things were looking bad for my half-crown …'[58]

---

51   IWM: L.E. Salter Papers 21103.
52   Sturges, *On the Remainder of Our Front*, p.24.
53   IWM: H. Wilson Papers, 3606.
54   IWM: L.E. Salter Papers 21103.
55   According to Skelton his platoon stayed in the girls' school where they had baths. He may have confused their arrival with subsequent visits to Béthune.
56   Anon. Author, *University & Public Schools Brigade*, pp.103–104.
57   FM: P.N. Wright Diary, RFM.2013.8.1/2, p.4.
58   Sturges, *On the Remainder of Our Front*, p.24.

9

An Introduction to Trench Warfare

This was what they had joined the Army for …

> *This was what they had joined the Army for … It takes a few hours for one to realize that there are some men over that pile of sandbags who are trying to kill you.*[1]

> *Great fat rats darted in and out of our shelters or ran along the parapet. Loathsome, foul things they to me the very incarnation of war seeking whom they might devour.*[2]

After the UPS battalions arrived in Béthune, the existing UPS history ceased providing a narrative of events. The story for these units can only be gleaned from the official battalion war diaries and the various personal accounts of individual UPS men. The UPS Brigade history concluded with impressions of the initial move to the trenches:

> On again. Right into the trenches this time. Those eight kilometres seemed absurdly short ones, and still all were keen to be in the fighting. This was what they had joined the Army for. This was what they had done all those attacks in Surrey for. They were right up. It was strange and weird, not a bit as they expected it. What about Headley Heath now? Why, it wasn't much like Bedlam trenches even. It takes a few hours for one to realize that there are some men over that pile of sandbags who are trying to kill you. The realization does come though, first through the graves of the unknown soldiers, and later by losing a comrade. The dreadful, dull monotony of war when there is nothing doing. The same sandbags and mud, the same ping and whizz. The rest has been described often enough. Some of the description is a mere feat of imagination, and, as the truth is never readable, the story may as well end here.[3]

With the arrival of the Brigade in Béthune on 20 November, the men commenced learning to occupy and hold trenches and learning their role of infantrymen from experienced professionals of the BEF. In Béthune, the UPS Brigade was split up for instruction in trench duty. 18/R Fus

---

1   Anon. Author, *University & Public Schools Brigade*, pp.103–104.
2   Williams, *From Khaki to Cloth*, p.59.
3   Anon. Author, *University & Public Schools Brigade*, pp.103–104.

and 20/R Fus were attached to 2nd Division; 19/R Fus and 21/R Fus were attached to 7th Division. These different periods of instruction will be captured battalion by battalion to appreciate their initial indoctrination to trench warfare and the experiences of different UPS soldiers.

Kenneth Norman described Béthune which offered both accommodation (various spacious billets including the Tobacco Factory which housed 2,000 men) and a haven from the war. There were shops and restaurants, cinemas, concert halls and theatres; 'I liked Béthune very much, one seemed to forget about the trenches when here. It was very rarely shelled …'[4]

With so many soldiers around, the prices in Béthune were high, 'things are so awfully expensive; a packet of cigarettes, which in England costs 4d cost 7d here.'[5] There were other options for 'entertainment' according to Harold Wilson; 'The Chaplain warned men this morning in his sermon against visiting resorts of low women, which places abound in all these French towns.'[6]

According to the 18/R Fus War Diary, the Battalion left Robecq at 9:30 a.m. on 20 November and marched, via Chocques, to Béthune; they arrived at 1:00 p.m. and were billeted in the Tobacco Factory. That afternoon six officers and sixteen NCOs from 18/R Fus, under Major Hartley, went into to the trenches for instruction with 6th Brigade. They were back that night. Next day, each company was attached to another battalion; one to 1st Battalion Hertfordshire Regiment; one to 1st Battalion King's Liverpool Regiment (1/King's) and two companies were with 1/KRRC. There were no casualties. The battalion remained in the trenches on 22 November but with companies relieving one another so they each endured a spell in the Front Line. Kenneth Norman's company were 'taught' by the King's Regiment or the Hertfordshire Regiment. He found the trenches were well-made and dry and there were numerous dugouts and shelters. Though the trenches seemed safe from shelling he recalled; 'We could see the Hun barbed wire in the distance with men hanging on [it].[7] However, Nicholas Pease, 19/R Fus, considered that; 'The front here in the Loos, Vermelles area, was inactive and relatively quiet.'[8]

On 23 November they were relived and moved to billets in Annequin-South where the remaining personnel from Béthune, left behind under Major Hartley, rejoined. The ruins of Annequin were, 'another series of mud and hovels on the Béthune Road which we were squeezed into …' according to Walter Medlicott.[9] There was much for the NCOs to do including distributing blankets, issuing rations, posting sentries and getting latrines dug. Medlicott went 'foraging' and purchased some sardines, corned beef, bacon, condensed milk and a French loaf; he and his fellow sergeants had a good dinner. After a night in Annequin, the battalion went into the line, but its companies were attached to other battalions (D Company with 2nd Battalion South Staffordshire Regiment; B and C Companies with the 1/R Berks and A Company with 5/King's). The MG Section was split up between other units. Norman's role as a stretcher bearer excused him from routine fatigues as he had to be both ready to react and rested for the exertions of carrying stretchers. He found that 'When busy the job is not a pleasant one …'[10]

The first casualty in 18/R Fus was on 24 November which was likely PS/8067 Private Henry Clarke, of B Company, who suffered a wound to his scalp which fractured his skull in late

4   LA: K.V. Norman Papers, WW1/GS/1188, p.129.
5   IWM: T.O'C. Doherty Papers, 12961.
6   IWM: H. Wilson Papers, 3606.
7   LA: K.V. Norman Papers, WW1/GS/1188, p.98 and p.100.
8   IWM: N.A. Pease Papers, 8230, p.33.
9   IWM: W.B. Medlicott Papers, 1708, p.92.
10 LA: K.V. Norman Papers, WW1/GS/1188, p.100.

November. After remaining seriously ill for several weeks he spent some time recovering at Woodcote Park which had been converted into a convalescent hospital.[11] According to Kenneth Norman PS/2035 Private Geoffrey Schenk was also wounded and evacuated.[12] Norman's company suffered six casualties and he had a lucky escape as he was knocked down when a whizz-bang exploded near him.[13] On 25 November, PS/1819 Private Arthur George Metcalf, of C Company, was badly wounded and died later that day. Metcalf was aged 21, had been educated at St Lawrence College, Ramsgate, and was an accountant. PS/6518 Private Eric George Frederick Herrmann was killed next day (26 November) with B Company, 18/R Fus.[14] The battalion was relieved at 11:30 a.m. on 26 November and marched to the *L'Ecole de Jeunes Filles* in Béthune. The battalion had lost two men killed and three wounded. The many hazards of the trenches did not claim every life. PS/1798 Corporal Spencer George Mayor was the Acting Sergeant Cook with 18/R Fus; he died of pneumonia, on 26 November, and was buried in Béthune Town Cemetery. This first experience of trench warfare was not as bad as some had expected; PS/8552 Private Charles Rumball wrote to his father:

> Last night we were bombarded for a bit, but it was enough for a start. It occurred just as I was having my supper… All the time, shells and bullets were whizzing above, when, suddenly, there was a bang just in front of me. When I looked up I found I was lying on the floor, but, luckily, not hurt. I had been politely pushed off my perch. It was a "whizz-bang" they fired at us, and it made a nice big hole in the parapet …[15]

Whilst the battalions of 98th Brigade served their apprenticeships in the front line this was the only time when all four battalions were under the same command. On 27 November the UPS battalions were separated. 18/R Fus and 20/RF were moved to 19th Brigade (which had joined 33rd Division). Their place was taken in 98th Brigade by 1/Mx and 2nd Battalion Argyll and Sutherland Highlanders (2/A&SH). The 19th Brigade was commanded by Brigadier General P.R. Robertson CMG and had Major E.K. Twiss as Brigade Major.[16]

---

11 Private Clarke, aged 23, was from Lower Broughton, Manchester where he was employed as a 'maker-up'. He enlisted in Manchester on 2 June 1915 and served with B Company, 18/R Fus. Once his condition was stabilised, he was evacuated to the UK on 20 December 1915. He joined 28/R Fus in late March 1916 but was ultimately discharged in August 1916 due to his wounds.
12 PS/2035 Private Geoffrey Crowhall Schenk, aged 20, from Manchester, had attended Malvern College. He was wounded in late November and was evacuated to the UK on 13 December. He later went to No 1 OCB but was RTU'd in June 1916 and instead went to 7/R Fus in France. He was attached to HQ 190th Brigade and was awarded the MM in April 1918. He was discharged in February 1919.
13 LA: K.V. Norman Papers, WW1/GS/1188, p.104.
14 Metcalf, aged 20, and from Beckenham, Kent, was buried in Béthune Town Cemetery. Herrmann was buried in Cambrin Churchyard. The war diary states Herrmann was killed on 25 November.
15 The Newcastle Daily Chronicle, 9 December 1915.
16 Brigadier General Philip Rynd Robertson was aged 49 and was the son of a general. He was educated at Charterhouse and the Royal Military College, Sandhurst. He served with the Cameronians from 1886 and promoted steadily until he was a lieutenant colonel commanding 1/Cameronians just before the war. He led his battalion until June 1915 when he was promoted to command 19th Brigade. Clearly an able commander he advanced to command 17th Division in the summer of 1916. Major Edward Kemble Twiss was born in Surbiton in 1882. He was a first-class cricketer who played for Oxfordshire before his military career. He was a bugler during the Boer War and served as an

The experiences of 19/R Fus are difficult to accurately determine as the battalion war diary conflates November and December 1915. The battalion likely arrived at Béthune around Midday, 20 November and officers and NCOs of the battalion went to the trenches. They returned that evening and orders were drafted. According to Robert Sturges (B Company) he was about to fall asleep when 'the orderly sergeant came in at last and began reading out his orders by the light of a candle. "The company," he read, "will parade at 8 a.m. tomorrow to proceed to the trenches!" I had lost my half-crown.'[17]

There was a church parade next day for C and D Companies whilst A and B went to the trenches at 'Windy Corner'. Philip Spaull, of B Company, recalled his 'baptism of fire':

> We marched up to our trench … and were paired off with regulars who have been there for some time and are quite used to all the little games of the Huns. We were (our platoon) in the support line situated 100 yards from the Germans. They were fairly quiet when we arrived and did not reply to our artillery who kept them amused for about two hours. A little firing went on by both sides all day … From about 6 p.m. to 7.30 p.m. the Huns sent us over all sorts of things; rifle grenades, trench mortars and wis-bangs [sic]. Our reply was 6 to every 1 of theirs so they soon decided to leave us alone … I quite enjoyed the fun and am looking forward to my next turn in. I quite thought I might feel somewhat nervous but found out that I was just the same as our little affairs at Epsom, Clipstone and Tidworth … as long as one keeps one's head down you are quite safe.[18]

He added that 'everything is very rough. In the trenches, we get into an awful mess. Mud over everywhere and everything. Our most difficult thing is to keep rifles clean.' Terence Doherty recalled; 'We certainly have the superiority in artillery as every time they fire a shell, they get about 8 back at them. We have to thank Lloyd George for that. It is really very silly just sitting behind sandbags and potting at each other.'[19] It is likely that A and B Companies were each attached for tutoring to either 2nd Battalion Yorkshire Regiment (2/Yorks) or 2nd Battalion Wiltshire Regiment (2/Wilts). Leonard Salter and his company presumably took over the factory billets vacated by A and B Companies. On 22 November C and D Companies went up to Windy Corner and relieved A and B (which returned to Béthune). C Company, 19/R Fus, were instructed by 2nd Battalion Bedfordshire Regiment (2/Bedf R) on 26 November and had Bedfordshire Regiment officers attached. D Company was attached to 2nd Battalion Royal Scots Fusiliers (2/RSF) and returned on 27 November. Harold Wilson, D Company, recalled:

> … [we] arrived at Givenchy, passing the village of Beuvry – passed along communication trenches all of which bear names such as Bond St, Shaftesbury Ave, Piccadilly … These trenches are composed chiefly of sandbags in some cases the floors are boarded for

---

Indian Army officer (10th Jats) until he joined 1/Dorsets in France from September 1914. He was a Brevet Major and MiD in June 1916 and was appointed DSO in January 1917. He was a Brigade Major from November 1915 to April 1917. He retired as a lieutenant colonel in 1920. By 1939 he was a hospitalised and disabled ex-officer who died in 1943. IWM: G.K. Twiss Papers, 17109 (the IWM catalogue refers to 'G.K.' Twiss).

17 Sturges, *On the Remainder of our Front*, p.25.
18 IWM: P.F.F. Spaull papers, 15453, letter 23 November 1915.
19 IWM: T.O'C. Doherty Papers, 12961.

drainage. A place we passed this morning called "Windy Corner" was shelled heavily about ¼ hr after we left. Where I am writing the enemy trenches are 80 yards away. Bowller and myself are sitting in a little dug-out. A minute or two ago Robertson was holding a periscope over the trench and a sniper put a bullet clean through it … it is absolutely fatal to show your head for more than a second over the parapet. Shelling goes on all day long.[20]

THE EDGE OF THE WORLD.

A sketch by J.S. Willock depicting the lonely and dangerous experience on sentry in a forward trench or sap for the first time. Whilst such routines had been practiced at the Bedlam Buildings trenches, nothing could prepare a new arrival for such duties. It is worth noting that in late 1915 steel helmets were not standard issue and were often included in trench stores. (*The Gasper*)

20 IWM: H. Wilson Papers, 3606.

Salter experienced the trenches for the first time with his company; 'Occupied front line at Givenchy. Bit noisy, otherwise not too bad. Sentry for four hours. Went in for 24 hours with the 2/R.S.F., relieved by 2/Queens.' Next day; 'Still in trenches. Cold. Saw an R.S.F. [soldier] bowled over. Relieved at mid-day by A and B Coys. Back in same billets by 3.30 p.m.'[21] Wilson also saw that; 'One of the Scots had his brains blown out.'[22] Terence Doherty, of D Company, downplayed his experience of the trenches, 'the bullets whistle and the shells hum and there is a jolly old row going on but it is not really very bad. It is most awfully boring and there is nothing to do but keep your head down. I am very tired, as it is all rather tiring, and nerve-shaking… '[23] The first casualty was a self-inflicted wound on 22 November. 19/R Fus lost one officer and eight men wounded during this tour; six wounded between A and B companies, two with 'C' and none from D Company. The next day (24 November) was spent by C and D Companies cleaning up, though Salter took time to go into town that evening. On 25 November Salter's company went into the line relieving A and B Companies. According to Harold Wilson:

> … we learned that our sappers were to blow up a mine in order to straighten out our line. We got to the bottom of the trench just in time – up went about 120 yards of earth sand bags etc. The row was awful; the column of mud etc went to a height of 50–100ft. We were just far enough away to miss being buried. About 2,000lbs of gun-cotton were used for the mine. It was discovered afterwards that the Germans had been sapping just under our mine, and their mine exploded simultaneously with ours. Parts of our trenches were knocked to blazes, and it took most of the night to repair them.[24]

Wilson and others were called to assist the Wiltshire Regiment to keep a steady fire on the German trenches to prevent a counterattack; they were not released back to the support line until dawn.

Though they were manning a redoubt the next night Salter was on another working party and due to the 'wicked weather' (snow and rain) the sentry duty stints were reduced to two hours. On top of this, Salter was not fed well leading to him; 'Feeling fed up'.[25] Salter's company was relieved on 27 November and marched to billets in Hinges. Nicolas Pease, of 19/R Fus, celebrated his 21st birthday in a muddy farm at Hinges, not the memorable occasion he had imagined. The same was true for Harold Wilson who celebrated his birthday two days later.

The first front line experiences of these UPS battalions demonstrate that every effort was made to set them up for future success. The war diary of 19/R Fus recorded:

> Each Coy had thus been 3 days in the trenches all told … Each Coy has been affiliated to Coys of Regular … Bns which had had experience of the trenches. The experience thus gained was entirely valuable and is likely to stand the Bn in better [stead] than if it had been put into the trenches at once … For a newly arrived Bn the test was a pretty severe one on the whole as there were frequent works and long marches … The spirit of all ranks is admirable.[26]

21 IWM: L.E. Salter Papers 21103.
22 IWM: H. Wilson Papers, 3606.
23 IWM: T.O'C. Doherty Papers, 12961.
24 IWM: H. Wilson Papers, 3606.
25 IWM: L.E. Salter Papers 21103.
26 TNA: WO95/2427: 19/R Fus War Diary.

An Introduction to Trench Warfare 135

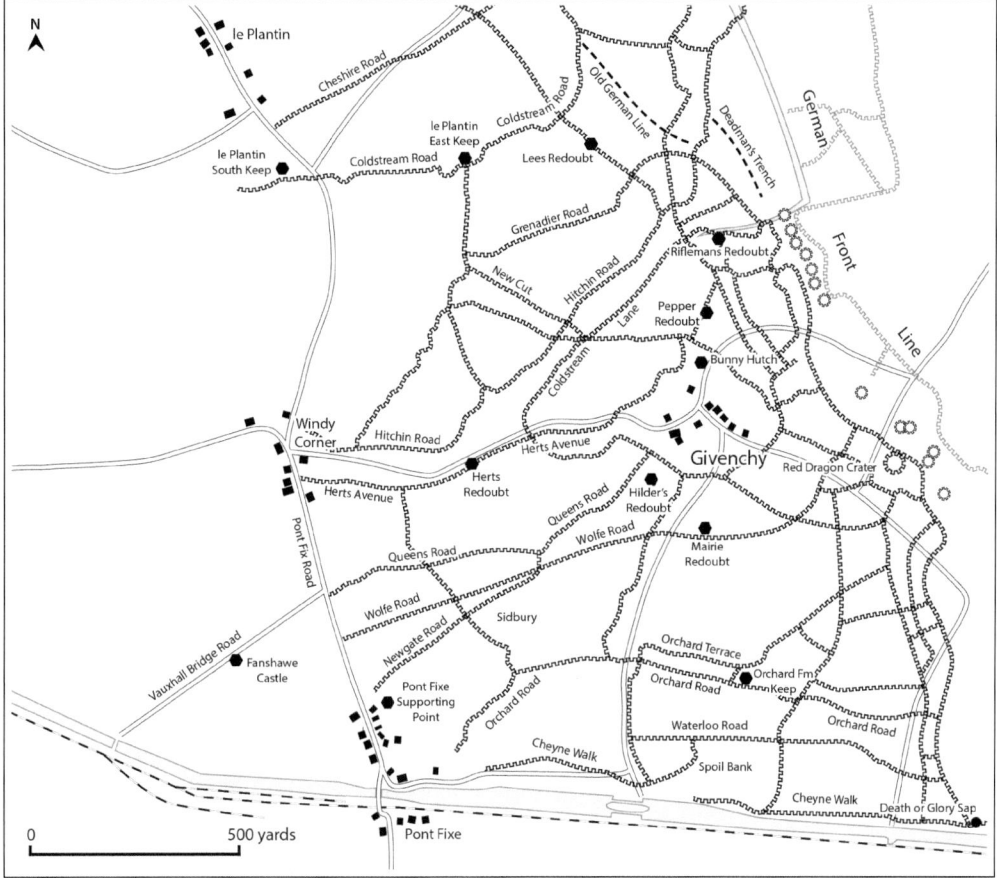

Map 2 2 Givenchy Sector.

As Lance Corporal Bentley of C Company, 19/R Fus, recalled after his first stint in the line '[we were] older and wiser men than when we entered and were we tired, we could have slept marching back.'[27] However, the extent to which the UPS learnt from the Regular and TF battalions is open to interpretations. Certainly, 20/R Fus received lessons of mixed quality from some of their 'instructors'.

* * *

20/R Fus had arrived in Béthune on 20 November and were housed in Montmorency Barracks. However, according to Vinden (of 20/R Fus) he 'marched to Béthune where we were billeted in the tobacco factory with floors of wooden blocks. We heard guns firing – very exciting, we were at last at the war.'[28] They too sent six officers and sixteen men to the trenches for instruction that

27 LA: WW1/GS/0125: Papers of W.G. Bentley, p.12.
28 IWM: F.H. Vinden Papers, 5565, p.13.

day. A and B Companies were sent to the trenches south of the La Bassée Canal. The Fusiliers were attached to 2/A&SH and the 5th Battalion Scottish Rifles (5/Scot Rif).

Next day, at about 6:45 a.m. C and D companies relieved A and B; the latter went to Annequin. Hodson, likely of C Company, described the move into trenches: 'No sound but the splashing of feet … and the slight creaking and rattle of equipment as we march… Sometimes when the starshells rise we halt and stand perfectly still. After a moment or two we move on. Now and then the harsh rattle of a machine-gun strikes the air, vivid and clear. You wonder what the target is and plod on…'[29] C Company were being inducted into trench routine by men of the 'Glasgow Territorials' (likely 5/Scot Rif); '…there are no better soldiers on this side. They adopted us for the day. They wanted to give us half their food and do all our duties. They cheered us by their tales of good times in trenches near Armentieres … and heartened us by their expert ways of cooking meals.'[30] Likewise, the Scottish Rifles recorded they 'made friends' with the two 20/R Fus companies sent to them for induction.[31] However, this easy treatment during their introduction to trench warfare might not prepare 20/R Fus for the hardships of the future nor establish them with a more aggressive ethos. Had Hodson's company been instructed by 2/RWF they might have been imbued with a different outlook. Hodson recalled that they learnt several tips from the Scottish Rifles; 'They shocked me a bit the way they used their bayonets for chopping wood. I asked one about the anti-frostbite grease we are issued with. "Oh!" he said, "you don't want to put that stuff on your feet. … the best thing to do is to put a bit of four-by-two [flannelette for cleaning rifles] in the top and use it for a lamp."'[32] It must be wondered whether such advice might have led to an increase in trench foot for the inexperienced UPS men.

Sergeant Godfrey Skelton, of D Company, recalled his first trip to the trenches:

> [We] … arrived at the entrance of the communication trench, where a guide from the 2nd Battalion of the Argyll and Sutherland Highlanders met us. … It was a daylight 'relief', as the trenches were deep in this part of the line and very good. It was a sunny day although it had been raining earlier … It was arranged that each of the four companies of our battalion was to go into the line with a regular battalion… for twenty hours only, on the first occasion, and we were sandwiched between the other companies of the regular troops. In this way we were instructed in the art of war. … Eventually we arrived in a trench called '*Boyau 19*'. It was nearly dark by this time and the enemy machine guns were beginning to get playful. … Sentries were posted and men detailed off for reliefs.[33]

D Company 20/R Fus did not receive any rations and had to scrounge them from the Argylls in return for lending them their blankets. On the left the Germans had blown a crater the night before and 2/A&SH were trying to recapture it; D Company was to man the fire step at a given time and let loose rapid fire to cover this move or distract the enemy. Skelton recalled:

---

29 Hodson, *The Soul of a Soldier*, pp.52–53.
30 Hodson, *The Soul of a Soldier*, p.31.
31 David Martin (ed.), *The Fifth Battalion Scottish Rifles 1914–1919* (Glasgow: Jackson, 1936), p.61.
32 Hodson, *The Soul of a Soldier*, p.31.
33 IWM: G. Skelton Papers, 13966, pp.36–37.

An Introduction to Trench Warfare 137

NEW USE FOR EVERYDAY IMPLEMENTS.

***Voice inside dug-out*:** I want the tin-opener for the jam. Have you finished with it?
***Voice from outside*:** No! I'm unfixing a jammed bayonet.

The men of 98th Brigade were instructed in the practice of trench warfare by experienced soldiers. It is probable that they also learnt some incorrect practices in the process, such as the misuse of catering utensils for weapon maintenance. (Sketch by unknown contributor published in *The Gasper*)

> ... I ordered my men on to the fire step and they had orders to open rapid fire. The Hun trenches were only about a hundred yards away, if that. Of course, we could see nothing of what was happening but we heard quite a lot very soon afterwards. Evidently the Hun thought an attack was imminent and he retaliated vigorously. The first thing we knew was that the air was full of strange sounds, some quite close at hand. I was walking up and down the trench giving orders, etc. and as I got to our fire bay one of my men fell down into the trench. This was our first casualty. He was not hurt badly, a bullet wound in the wrist.[34]

Skelton's platoon performed well though one of his men did try to run from the trenches and had to be dragged back to his position The parapet was blown in by a shell close to Lieutenant Modera and a sandbag winded him. Meanwhile, A and B Companies lost one man who died of his wounds and another man wounded during this initial tour. PS/5098 Private James Hutchinson, aged 28, from Rawtenstall, Manchester, of C Company, 20/R Fus, was wounded and later died on 23 November.[35] PS/5750 Private Joseph Bernard Troulan, aged 20, from

---

34   IWM: G. Skelton Papers, 13966, p.38.
35   Hutchinson was buried in Béthune Town Cemetery.

Plymouth, was also wounded whilst with 20/R Fus and died of his wounds on 26 November.[36] On 23 November C and D Companies left the trenches. According to Sidney Platt; 'Our liveliest time … was on Wednesday Nov. 24th when we were shelled coming from the Reserve Trenches after having been on a working party. Several of the shells let [sic] only 20 yds away … they were whizzing over our heads dangerously near.'[37]

Meanwhile, 20/R Fus had been attached to 5th Brigade, 2nd Division. On 24 November A and B were attached to 2nd Battalion Oxfordshire and Buckinghamshire Light Infantry (2/OBLI) and 2nd Battalion Highland Light Infantry (2/HLI). Next day, C and D again relieved A and B. On 26 November the Battalion marched back to Béthune and were accommodated in *Rue D'Aire*. Next day, they were moved to 19th Brigade and were relocated to the Tobacco Factory.

Around this time there were few grenades available to use in the trenches. Herbert Vinden recalled fatigues of filling jam-tin bombs early in their time in France which were 'a primitive weapon in comparison to the German stick bomb'.[38]

* * *

Meanwhile, on 20 November, Major Fenn and the 2ICs of each company of 21/R Fus had been taken to the trenches in a motor bus for instruction. Meanwhile, the battalion had arrived and been billeted in the Avenue de Bruay in Béthune. On 21 November two platoons from each company were sent up for 24 hours of instruction in the front line; No 1 and 2 Companies were attached to the 1/S Staffs and 3 and 4 Companies to 2nd Battalion Royal Warwickshire Regiment (2/R Warwicks) 22nd Brigade, 7th Division. Morgan Williams recalled:

> … on Sunday, November 21st, 1915, we went up to the front line for the first time. There the Warwicks and the Staffords, of the famous Seventh Division, made us at home in their shelters and put us through our paces. They instructed us in the mysteries of our craft, of which "bagging a *bougis*" [stealing a candle] seemed the most important … A candle made all the difference in a dug-out. It helped us to see the rats instead of merely hearing or feeling them.[39]

Ashley Gibson of 21/R Fus described the march to the trenches:

> Our hobnails rang on the pavé as we swung into the Beuvry road. For a mile or so the bugles pretended they were a band to encourage us, and when they fell out we sang about Hector the Spectre and other people we knew. By Annequin we had to stop that. There was a kind of Brock's Benefit starting up about a mile ahead. But the Verey lights didn't scare us, nor the noise either. Hadn't we rehearsed all that part on Salisbury Plain? We who had been a whole battalion in columns of fours, then a company, then a platoon, had dwindled to a "blob." Very much aware as usual that a relief was toward, Fritz, intermittently, was lobbing shrapnel at the

---

36   Troulan was an optical student. He was buried in Béthune Town Cemetery.
37   IWM: S. Platt Papers, 17681, p.3.
38   IWM: F.H. Vinden Papers, 5565, p.13.
39   Williams, *From Khaki to Cloth*, p.59.

road. For a few hundred yards the tow-path of the La Bassée canal made part of our route… That absolutely stagnant, deserted waterway made you shiver to look at it. … Shuffling and whispering in the gathering dusk, we dived single file from the path into a communication trench. Good duck-boards under our feet, solid revetting on either hand. Safer this than the *Place de Ville* at Béthune. Half an hour of twisting and turning, then a halt in front, and word came down the line that these are B.H.Q.'s dugouts. At a fork in the trench, B Company's guide detaches my party, we scramble on, in and out, round traverses and past fire-steps now manned by men with fixed bayonets. Lounging and inattentive regarding the matter in hand, they almost seem. But every few yards a couple stand high up on the fire-step, the breast buttons of their tunics level with the parapet. They are not lounging, nor inattentive. Unblinking, their narrowed gaze ranges perpetually the world of shadows that is No Man's Land.[40]

Morgan Williams also described the trenches:

The trenches were long, narrow ditches in the ground, twelve or fourteen feet deep, with communicating alleys branching off here and there. They were, for the most part, one quarter filled with mud, which stank with a stale stench. Great fat rats darted in and out of our shelters or ran along the parapet. Loathsome, foul things they to me the very incarnation of war seeking whom they might devour. Beyond lay the barbed wire – strand upon strand protecting us from sudden attack.[41]

Next day 22nd Brigade was relieved, and the 1st Battalion Queen's Royal West Surrey Regiment (1/Queen's RWS) and the 2/HLI took over tutoring 21/R Fus. At the same time the remaining two platoons of each company relieved the first two. The platoons, once relieved, marched to billets in Beuvry. Hugh Spurrell recalled departing the front line:

To get out of the trenches we had to move some distance along the front line and then down a communication trench, known as Hertford Lane. The battalion had its first casualty during this withdrawal. At one spot there was a sudden rise in the ground, involving a climb up three steps and the turning of a corner. A man standing on the top step could just be seen by a German sniper. One of the fellows in No Three Company was unlucky enough to stop on this top step. Ping! The Hun got him right through the head and killed him at once. When we passed the spot, the body had been removed, but there were ugly signs of the accident. By the time we reached the place, rumour had increased the number of casualties to three. It is true that the sniper had another shot, but he hit nobody. We ran up those three steps like greased lightning.[42]

On 23 November one man of 21/R Fus was killed and one wounded; PS/2689 Private John Felix Frith, aged 24, from Charlwood, Surrey, was likely shot dead under these circumstances.[43]

---

40 Gibson, *Postscript to Adventure*, pp.136–137.
41 Williams, *From Khaki to Cloth*, p.59.
42 IWM: H.W. Spurrell 2138, p.40.
43 Firth had attended Rees School Cumberland and was in business in Sicily. He enlisted in 21/R Fus on 5 November 1914. He was buried in Woburn Abbey Cemetery.

Next day, 24 November, a working party of five officers and 200 men, under Major Fenn, went up to the trenches to conduct repairs. A German rifle grenade exploded in a section of trench where members of the party were working; two men were killed and three were wounded.[44] Spurrell recorded, having heard second-hand, that these two men 'of the working-party had had their heads blown off by a trench mortar...'[45] The UPS men soon became used to death. Morgan Williams recalled the first dead man he saw:

> The first casualty I saw seared my memory. ... He was the son of a clergyman in a town where I had once lived ... He lay very still on his stretcher scarcely scarred by the blow which had struck him. Many such scenes were to haunt me in the months that followed, but the stark horror and suddenness of the first one abides. This was war and it stalked the earth bludgeoning life without sense or reason...[46]

This likely referred to PS/2984 Private William Trevor Morgan, aged 29, the son of the Reverend Edward Morgan of Usk Vicarage, Monmouthshire.[47] The other fatality was PS/6260 Private Donald Morton Bunting, aged 22, from Swaffham, Norfolk. Bunting was educated at Rydal Mount, Colwyn Bay, and was formerly a dental student at Guy's Hospital before he enlisted with 21/R Fus.[48]

The rest of the men of the battalion, in Beuvry, were instructed in the care of feet, the use of gas helmets and gas protection measures. Next day, Major Stanley took a similar-sized working party up to the trenches which returned next morning. On 26 November the battalion moved to Annequin North and again prepared for the trenches. No 1 Company was to be attached to 1/Queen's RWS and No 3 Company to the 2nd Battalion Worcestershire Regiment (2/Worcs R) in the trenches for another 24 hours. According to the Worcestershires' War Diary; 'The enemy's snipers were very active, but the Royal Fusiliers fired continuously through the night and kept the German snipers down.'[49] This suggested that the 21/R Fus acted correctly in the circumstances. Nos 2 and 4 Companies proceeded to the trenches on 27 November and were also tutored for 24 hours in the front line. On the 28th 21/R Fus returned to Béthune and were billeted in the *Ecole de Jeunes Filles*. The two companies from the trenches rejoined before lunch. The lessons learned during this initial period did not come cheaply; during these trench tours 21/R Fus suffered three killed and four wounded. Next day the battalion marched to les Harisoirs, a small village north of Béthune. There was insufficient space to billet the battalion and this was only accomplished with the assistance of the 2/A&SH. The last day of the month was spent cleaning up. 1 December brought the novelty of members of the battalion been given practice at throwing live grenades.

On 29 November the war diary for 2/S Staffs (2nd Division) recorded a company of 21/R Fus attached to them which was split amongst the three front-line companies. Lance Corporal Robert Jones of 21/R Fus recalled:

---

44 The War Diary states three killed and two wounded but only two fatalities could be identified.
45 IWM: H.W. Spurrell 2138, p.41.
46 Williams, *From Khaki to Cloth*, p.60.
47 PS/2984 Private W.T. Morgan, from Newport, Monmouthshire, was killed with 21/R Fus on 24 November 1915. He was buried in Woburn Abbey Cemetery.
48 Bunting was buried in Woburn Abbey Cemetery.
49 TNA: WO 95/1351: 2/Worcs R War Diary.

> We went in to [the trenches of] 2nd Battalion the South Staffs and there were 35 men and one subaltern. They were extraordinarily nice to us. We thought they wouldn't be bothered. But I think holding the line with us was not the same as the normal Kitchener's Army. There was very little difference between NCOs [and men]. Our platoon sergeant was always 'Scotty', he was a bank manager from London. There wasn't the gap between officers, NCOs, and men.[50]

These early periods in the line enabled the UPS battalions to better understand their future role and adapt to trench life. It also enabled the officers and NCOs to learn the capabilities and characters of their men when placed under pressure which could not have been observed during training. However, this was a mutual activity as the men also began to identify the flaws in their leaders. The quality of the NCOs across the battalions, companies and platoons varied considerably. Hugh Spurrell gives an unvarnished account of the perceived failings amongst his platoon staff:

> … we had no officer in charge of the platoon while it was in France. Hassall [sic] was a South African and had been through the Boer War … He was oldish and, on parade, perfectly useless. But he could get things done – more by force of personality than by making a great display. As a leader in action he was far more capable than any officer in the company. Fortunately, they realised this; little was done unless he had said that it was practicable and wise. We had three other sergeants in the platoon. Swears, an over-grown boy, about 6ft. 4in. high, was keen on his work, but had no influence on the men under him, as he lacked self-confidence. Renouf, the third, spent most of his time in preparing cocoa and boiling shaving water for Hassall, whom he insisted on calling "Sir" – much to Hassall's disgust. He was short and stubby as Swears was tall; they were known as "Oil-bottle" and "Pull-through". Atherton was nominally the bombing sergeant, but when he got to France, he lost his nerve, and we were always happy when he was out of the way. If a man ventured to speak above a whisper, Atherton always accused him of carelessness and of trying to draw the enemy's fire. The platoon corporal, Merchant (popularly known as "Happy Fanny") was one of those worrying children who are the bane of a man's existence. In France, however, he proved a good deal better as a N.C.O. than many of his superiors.[51]

As soon as the brigade finished with these trench tours it ceased to be the University and Public Schools Brigade. 19th Brigade was attached to 33rd Division to stiffen it with experienced battalions which were spread throughout the brigades. 18/R Fus and 20/R Fus joined 19th Brigade and were replaced by 2/A&SH and 1/Mx; both were experienced Regular battalions. Therefore, the history of the UPS battalions becomes harder to track as they were split between two brigades. This narrative will shift between the two pairs of battalions to cover their service in France during the winter of 1915–1916.

50 LA: WW1/TR/03/92: Tape of R.C.B. Jones.
51 IWM: H.W. Spurrell 2138, pp.44–45.

98th Brigade

From 27 November 19/R Fus was in billets at Hinges doing routine work. On 30 November the men of 19/R Fus marched to billets in Essars. Next day was spent preparing for the trenches and that evening the battalion took over positions at Festubert. On the night of 1–2 December 98th Brigade took over the line for the first time – 19/R Fus was in the front trenches with 1/Mx; 2/A&SH were in support at Le Touret and 21/R Fus occupied L'Épinette and Festubert in reserve. 19/R Fus relieved 6th Battalion Gordon Highlanders (6/Gordons). According to Leonard Salter the trenches were in a wicked condition. However, he thought he was lucky enough to be attached to battalion HQ as a dispatch orderly which was a relatively comfortable job. After being sent out at 3:30 a.m. next morning for brigade HQ he probably reconsidered this view. 21/R Fus marched to billets in the 'Village Line' on the afternoon of 2 December; they were accommodated with three companies in Épinette and No 3 Company was at Festubert to provide garrisons to several keeps.

On 2 December Brigadier General Strickland visited the trenches; he did the same on each subsequent day. 21/R Fus spent the next few days improving billets and provided a working party for the RE on 3 December which suffered three casualties. A similar party was found for the next two nights. Little occurred for 21/R Fus except for a few German shells landing around No 4 Company's billets which were 'searching' for some nearby British guns; these caused no injuries. Casualties were light; 19/R Fus and 21/R Fus lost a man wounded each on 4 December. However, PS/6580 Private Charles White, D Company, No 16 Platoon, 19/R Fus, was recorded as killed on this date. According to Harold Wilson; 'First casualty this morning. Poor old White got buried in a dugout and died shortly afterwards.'[52] The next day 21/R Fus lost three wounded. On 5 December 19/R Fus lost a man accidentally killed and another was wounded through a self-inflicted wound.[53] 19/R Fus lost a further two killed and one wounded the next day. One of these was PS/6813 Private James Edward Parker of No 16 Platoon, D Company, 19/R Fus. He was, according to Harold Wilson, shot in the head by a sniper but died before he could be evacuated.[54]

On the night of 6–7 December the Brigade conducted a relief in place; 21/R Fus relieved 19/R Fus in the front line. There was a heavy bombardment that morning according to Leonard Salter. 19/R Fus returned to billets in Festubert in reserve.

21/R Fus was in place by 9:00 p.m. having taken over the following positions:

No 1 Company – left of the firing line,
No 3 Company – right of the firing line, and sections of the support line that were considered tenable.
No 4 Company – left of the reserve line,
No 2 Company – right of the reserve line and garrisons for Goldney's Keep.
The MG section remained with two guns at Festubert Central and two at Festubert East Keep.

---

52 PS/6580 Private Charles White, aged 39, was from Great Bentley. He was educated at Haileybury (1891–1892) and was a solicitor from Gateshead. White was buried in Brown's Road Military Cemetery, Festubert.
53 PS/6327 Private Robert Whittle, aged 19, was from Egerton, Bolton and served with 19/R Fus. He died on 5 December and was buried in Brown's Road Military Cemetery.
54 Parker, aged 20, was born in Auckland, New Zealand and lived in Millom, Cumberland. He was buried in Brown's Road Military Cemetery.

During the relief PS/3236 Sergeant Thomas Lancelot William Strother, likely of the MG Section, was wounded by a shell in Festubert East Keep.[55] The trenches were in a deplorable condition with the front line being held by 'islands'; posts that were unable to be reached from behind by daylight due to the communication trenches being flooded with about five feet of water.

Efforts were made to improve communications to some of the more isolated posts by the battalion signallers laying telephone cables to them. The GOC permitted half of No 1 Company to be removed from the line; they went back into billets. On 8 December the companies swapped over with half of No 4 Company, and No 2 Company, taking over the front line. Private Harris, of No 1 Company, was accidentally wounded by Private Metcalfe when the latter was cleaning his rifle. This was, according to the diary, the seventh in No 1 Company and disciplinary action was planned. The company reliefs by 21/R Fus took place on 9 December without any issues or casualties. 10 December passed quietly for 21/R Fus with some repairs being conducted though they were curtailed by the weather. One man was wounded at Goldney's Keep on 10 December. PS/8371 Private Denman Metcalf, reported wounded on 10 December, later recalled:

> It was near Festubert that I was wounded. The battalion had gone into the trenches for five days, and the next morning, just about dawn, while I was standing on a mud platform, I was struck by a German bullet just above the left ankle. The doctors told me afterwards that it must have been an explosive bullet, for part of the bone was blown away. At this spot we were really behind a breastwork and there were no communication trenches. The consequence was that I could not be moved in the daytime. Of course, the others did what they could for me, but I had to wait until night before I could be moved. The following morning, I was taken to a rough and ready dressing station, and they did what they could for me. Afterwards I was taken to the Béthune Temporary Hospital, and later to a casualty clearing station, where I was operated upon. The nature of the wound made it necessary for me to have my leg amputated just above the knee.[56]

Morgan Williams certainly suggested that a 'live and let live' system was present at Festubert by men of 21/R Fus:

> Through a slit in the revetment someone noticed a German boy, cold and half-frozen, swinging a brazier round and round to force it into flame. We could have picked him off more easily than hitting a wicket, yet when one of the men turned to raise a rifle the others bade him desist. "Poor ____'s as cold as we are," was the verdict…[57]

Though the attitude to the enemy was less than hostile some men of 21/R Fus were able to get their own back on their Sergeant Major during their stints in the trenches; 'The day was to come when we should pull the pin out of a Mills bomb, fling it on the parapet behind us, and shriek with

---

55   Strother was from Killinghall, Harrogate. He played rugby for the UPS and was a Full Back for Yorkshire. He was later gazetted to a commission in the Motor MGC and later served in the Tank Corps where he was promoted to captain.
56   PS/8371 Private Denman Metcalf, aged 27, a chemist from Kings Lynn, had enlisted with 29/R Fus in July 1915. Lynn Advertiser, 25 February 1916.
57   Williams, *From Khaki to Cloth*, pp.61–62.

144 Soldiers and Gentlemen

The flooded trenches and breastworks at Festubert required considerable endurance. One way of coping was to make light of the experience. (Sketch by A.S. Palmer published in *The Gasper*)

laughter as he raced to his dug-out, imagining that he was under heavy fire. It cost the Government 7s. and 6d., but it was worth it.'⁵⁸ The situation in reserve remained unchanged until 10 December when 19/R Fus and 1/Mx were withdrawn; 19/R Fus was relieved by 7th Battalion Sussex Regiment (7/R Sussex) and marched to Hingette. Around this time the men of 19/R Fus were issued with leather jerkins and waterproof capes to help with the weather.⁵⁹ On 11 December 98th Brigade was relieved by 36th Brigade; the relief was complete by 12:30 a.m. on 12 December.

21/R Fus handed over to 7/R Sussex in the front line on 11 December. When they were relieved, the men returned to Gorre and le Hamel. One man was reported missing; he was sent to guide a party of 7/R Sussex to the front line and was last seen to return to the rear. The 7/R Sussex history fills in the gaps:

> One man was discovered in the darkness close to the so-called front line after a relief, having slipped off the track and having abandoned his gum boots and trousers in the mud in his efforts to extricate himself. Cold and half-naked, he was warmed by some rum, and as he made his way to the Transport for an issue of trousers he was a strange apparition in the flickering glare of a Verey light.⁶⁰

Even in his 'safe' job as HQ orderly, with the battalion in reserve, Leonard Salter still had a close shave on 9 December; 'Cycling along Festubert Road three shells fell very near. Put the wind

---

58 Williams, *From Khaki to Cloth*, p.44, p.47.
59 IWM: P.F.F. Spaull Papers, 15453, letter 13 December 1915.
60 Owen Rutter (ed), *The History of The Seventh (Service) Battalion of the Royal Sussex Regiment 1914–1919* (London: The Times, 1934), pp.46–47.

up me…'⁶¹ This tour of trenches saw losses to 19/R Fus of three killed and seven wounded; 21/R Fus lost eight wounded.⁶²

Harold Wilson celebrated with his friends the evening after escaping the trenches; 'proceeding with 6 of the boys to a little café and had a feed in celebration of Clarence's birthday. French loaf, 1/2 lb fresh butter, 21 eggs, sardines, beans, apricots, chocolate, dates, biscuits, coffee [and] wine. Got back to billets after having the best tuck in since we left England.'⁶³ Philip Spaull also took advantage of the *estaminets*; 'All the people in the villages are good to us. They should be, for we spend plenty of money. We can always get hot coffee in the "*Estaminets*" … and a warm room to sit in until 8 p.m.'⁶⁴

On 12 December 19/R Fus marched to Gonnehem for fifteen days of rest and 21/R Fus marched to Hingette; next day they moved to rest billets at L'Eclème despite these being reported by the 21/R Fus billeting officer (Captain Musgrave) as being surrounded by inundations. HQ 98th Brigade re-iterated their orders and the infantrymen of 21/R Fus had to wade through a foot of water to reach their billets to allow themselves to 'dry off'. Hugh Spurrell recalled a barn with thick straw which was considered a luxury after Festubert. The next two days were spent by 21/R Fus in cleaning up after a long spell in extremely muddy trenches and baths were enjoyed. After three to four days Spurrell considered that the battalion was 'moderately presentable'. They would be out of the line until Christmas.

19th Brigade

The 5/Scot Rif History recorded the considered detriment of this swapping of battalions to the brigade in which they served; 'The 19th Brigade lost the 1st Middlesex and the 2nd A&SH, who were sent to the 98th Infantry Brigade to stiffen it while we got the 20th Royal Fusiliers in exchange. This was one of the "Public Schools" battalions and we found their ways were very different from those of most soldiers.'⁶⁵

On Sunday 28 November, after morning church services, the commanding officer of 18/R Fus, Lieutenant Colonel Lord Montagu-Douglas Scott, visited 21st Brigade and was guided around the Givenchy trenches they would take over on 30 November. Major Hamilton of 1/Cameronians was attached to provide instruction. 29 November was a day of parades, kit inspections and gas helmet practices. Entering the trenches was postponed for 24 hours so the battalion went on a route march on 30 November instead. Next day, the battalion marched to the Givenchy trenches.

Harry Harvey, of 18/R Fus, recorded an account of the experience for a platoon departing 'up the line' which likely illustrates this experience and starts with them in billets:

---

61 IWM: L.E. Salter Papers, 21103, 9 December 1915.
62 These losses replicated those of 1/Mx which lost three killed and 10 wounded and 2/A&SH two killed and eleven wounded. However, the accidental losses through negligence and error (one killed, two injured) skewed these figures.
63 IWM: H. Wilson Papers, 3606.
64 IWM: P.F.F. Spaull Papers, 15453, letter 13 December 1915.
65 Martin, *The Fifth Battalion, the Cameronians*, p.62.

> A comparative quietude prevailed; to be shortly broken by the bellowed order: "Get dressed!"
> Sergeant … strode into the crowded room.
> "Corporal, detail two men to tidy up."
> This said he was gone.
> Six minutes later, along the stone-paved passage came the impelling roar, "No 12 [Platoon]! On parade!"
> So, at 5-15 on a darkening winter's afternoon, the men of "C" Company, stooping under the sickening, senseless weight of full marching-order approximating a hundred pounds, clumped languidly down the worn stone stairway into the great cobbled Square before the barracks, "falling-in" …[66]

After the platoon sergeant had called the roll the following occurred:

> Rifles, ammunition and gas-helmets [were] duly inspected by a vigilant officer,… [the platoon] turned heavily "to the left in fours," and, "at the slope," with swaying gait beneath bodies cumbrously weighted, and drooping shoulders shrouded with rain-washed, glistening groundsheets, plodded steadily over the irregular worn granite blocks, between which the dismal, soddening downpour trickled away to the flooded, mud-choked gutterings. Beyond the gate the order "March at ease!" – welcome though of seeming irony, brought rifles down to the slung position … eyes were turned to the wide, straight, Napoleonic road ahead…

After a march along roads:

> … in the gathering dusk … [the platoon] wheeled to the left past the finger-post marked "Festubert" – plodding away across the La Bassée Canal, and for a while along its cindered tow path … away to the track of devastation and gruesomeness, the flash of hidden gun batteries, the crack of the sniper's rifle; the slow horrible "tat, ta-ta-tat" of the German machine-guns; the ever-rising-falling fountains of lustrous star-shells which marked the winding, watchful, far-stretched battle-line.

18/R Fus took over positions from Scottish Trench on the right to Fife Road on the left with companies of the 1st Battalion Green Howards (1/Yorks) and 1/4th Battalion Cameron Highlanders (1/4 Camerons) being relieved. Kenneth Norman wrote of their next stint in the line at Festubert; 'Here we often fell in holes up to our knees and occasionally waists full of thick mud. The weather now is very wet and windy and [we] shall be glad to get back to billets and dry. Our fellows were very cheery under these conditions. Except for the Somme these trenches were as bad as any I came across…'[67] The trenches were in a bad condition; poorly drained, wet, weak parapets and uninhabitable or unsafe shelters. Men lost kit and rifles in the mud. On 3 December 18/R Fus suffered another fatal casualty; PS/5827 Private Francis Black, of A Company. Worse was to come.

---

66   H.E. Harvey, *Battle-Line Narratives 1915–1918* (London: Brentano's, 1928), pp.20–21.
67   LA: K.V. Norman Papers, WW1/GS/1188, p.106.

PS/5827 Private Francis Black was educated at Christs Hospital, Horsham, and Reading Grammar School. He was employed by the Canada Bank of Commerce. Black served with A Company 18/R Fus and was appointed lance corporal in March 1915 and corporal in June. He was killed by a sniper after three weeks in France and is commemorated on the Loos Memorial. (Reproduced with permission of the Sutton Archives)

The longer the time the UPS men spent in the trenches the quicker they experienced some of the harsh realities of trench warfare. Harvey, of 18/R Fus, though writing fiction, described a tired soldier, after a fatigue collecting ammunition, being too gentlemanly to wake another man taking his place in a dugout; instead he occupied an unused funk-hole in the side of the trench; 'Herculean – almost frantic digging, at the six feet of sodden earth which had fallen, availed not to restore life to the poor crushed body, which, in a shell-hole to the rear of the trenches, they roughly but reverently buried.'[68] Harvey's semi-fiction mirrored fact; Walter Medlicott recalled on 5 December 1915; 'Two poor fellows were buried in their dugouts in the support trenches last night ... they were smothered before they could be got out. We shall bury them tomorrow in the early morning. There is a peaceful little cemetery around the old farms where there are some 50 graves; French and English.'[69] These two men were PS/6784 Private Howard Clive Thomas, aged 27, from Carisbrooke, Isle of Wight, and PS/6784 Private Victor William Hyatt, aged 18, from Stratford-upon-Avon. Both were serving with D Company, 18/R Fus, when their shelter in George Street collapsed, killing them on 5 December.[70] Work continued to improve the communication trenches and parapet. On 6 December the relief with 20/R Fus was to take place but at 4:30 a.m. a British bombardment invited retaliation; according to Norman; 'When we were billeted in le Plantin a shell fell on a chapel where some of my company were billeted and severely wounded two fellows Sergt Cross and Pte Cooper. We bandaged them up and carried them two miles to the nearest dressing station.'[71] PS/1396 Sergeant Arthur Ernest Selby

68 Harvey, *Battle-Line Narratives*, p.34.
69 IWM: W.B. Medlicott Papers, 1708, p.103
70 Hyatt was the son of a master saddler and was educated at Shakespeare's School, Stratford-upon-Avon (1907–1913). He had enlisted with John Pickett and Ronald Newland. Hyatt and Thomas were both buried in Brown's Road Military Cemetery, Festubert.
71 According to the war diary Sergeant 'Crosse' and two others were wounded in le Plantin. LA: K.V.

Crosse, who had previously attended Keble College, Oxford, later recovered from his injuries and was gazetted to a commission with 13/R Fus. PS/7540 Private Reginald Evans Cooper, a clerk from Great Yarmouth, may have been the other man wounded; he returned to the UK in March 1916 and was discharged sick later that year.

The relief commenced at 7:00 p.m. and was completed by 3:00 a.m. but a Corporal Evans and Private Rason were wounded in George Street, presumably during the relief.[72] 18/R Fus moved back to Essars with A and D Companies billeted in Ferme du Roi; C Company on the Béthune–Essars Road and B Company remained in the Givenchy sector taking over Givenchy Redoubts from 20/R Fus. Cleaning up took place on 7 December and continued next day though 18 'Grenadiers' of C Company were attached to 5/Scot Rif at Givenchy to be taught how to throw grenades. On 9 December 18 men of D Company were switched over to the same instruction. On 11 December the battalion marched to billets in Fontes, about 14 miles away, where further cleaning up and inspection took place.

Meanwhile, on 1 December 20/R Fus, minus A Company, moved to billets, likely east of Rue de Marais where they were the Reserve battalion for 19th Brigade. A Company, 20/R Fus was in support at Givenchy Keep, Hilder, Moat House, Herts Redoubts, Windy Corner Keep and Marais East and South. A Company were relieved by B Company on 4 December. On 6 December 20/R Fus relieved 18/R Fus in the trenches east of Le Plantin. The battalion was disposed as follows:

> Battalion HQ was in the Grouse Butts
> 2 Platoons of C Company were in the front line.
> 2 Platoons of C Company were in George Street.
> 2 Platoons of A Company were in the Grouse Butts.
> 2 Platoons of C Company were in the Old British Line.
> 1 Platoon of D Company was on the right of the front line in touch with 2/RWF on the right
> 1 Platoon of D Company was in the Grouse Butts
> B Company and two platoons of D Company remained at Le Plantin
> On the left was 21/R Fus.

A sergeant was wounded whilst in Le Plantin during the relief. According to Sidney Platt, of D Company:

> Every time we went into the trenches it rained and altogether we had a most uncomfortable experience. At Le Plantin in the firing line we were nearly up to our knees in mud and whilst in the supports we were paddling in water in several places and if your foot slipped off the board walks you went up to your knees in water.[73]

Next day, seven men were wounded by shrapnel in Le Plantin and enemy snipers, likely having realised there were new troops opposite them, were active. Work was conducted to improve

---

Norman Papers, WW1/GS/1188, p.110.
72 Rason and Evans cannot be easily matched to any identified UPS soldiers.
73 IWM: S. Platt Papers, 17681, pp.3–4.

the trenches by building up the parapets, improving shelters, drainage and trench flooring. That evening the platoons in the front line were relieved; conditions in the forward trenches likely necessitated a rest and a chance to get dry. Work continued with the same inter-platoon reliefs until 10 December when the battalion was relieved between 5:00 and 10:00 p.m. by 9th Battalion East Surrey Regiment (9/E Surrey)(24th Division). The battalion marched back to Béthune by platoons and was billeted in the Orphanage. After a day to rest 20/R Fus marched to billets in Ham-en-Artois on 12 December.

This ended the 'trench apprenticeship' and the first few front line stints for the UPS battalions. So began months of endurance of trench conditions. Meanwhile, even in France, talk of commissions continued: 'I think this Corps will open again for commissions. All here are talking about it – you know what I shall do.'[74]

---

74   IWM: P.F.F. Spaull Papers, 15453, letter 11 December 1915.

10

Christmas on Active Service

Oh! We had a fine time

It is not at all Xmasy here …[1]

Oh! We had a fine time … We had a good sing-song and ended up by toasting the dear ones at home, the King, and the Allies.[2]

98th Brigade

98th Brigade was out of the line from mid-December and 19/R Fus and 21/R Fus were both in billets. Having been relieved, 98th Brigade marched back to a rest area at Gonnehem for cleaning up and training. Each battalion was to be inspected by the Brigade Commander. 19/R Fus was in billets at Gonnehem. The 13th commenced with a rifle inspection and the men had the rest of the day off. Next day, the men paraded at 7:00 a.m. for physical drill, went for a bath and were inspected by the brigade commander in the afternoon; that evening 1/Mx R held a concert. Over the next few days there were fatigue parties, parades, inspections and route marches. Some specialists, scouts, bombers etc, paraded separately, presumably to cover some specific training. Free time was spent in Béthune or at the Expeditionary Force Canteen; men went out to eat, find *estaminets* for drinking or just to get their hair cut. There were a few inter-platoon football matches. On 13 December Harold Wilson recorded; 'Tomorrow the Brigadier will inspect the brigade. This is the first time Gen. Strickland has inspected us. Hope he will like our appearance, we are all covered in mud which has dried on our clothes – a somewhat different turn-out to the old Epsom inspection.'[3] On 17 December there was a concert by the battalion, according to Wilson; 'Concert a howling success, hall packed. Piano very tinpot, but always trust Clarence to get music out of anything.'[4] Some men started to prepare for Christmas feasts.

1 IWM: T.O'C. Doherty Papers, 12961, p.50
2 *Halifax Courier*, 8 January 1916, An Ellander Enjoys Himself.
3 IWM: H. Wilson Papers, 3606.
4 IWM: H. Wilson Papers, 3606.

There was some serious training to be done though. 19/R Fus conducted some bombing training including lectures and throwing practices. Harold Wilson and others were kept busy with intensive bombing training in the run up to Christmas whilst many of their comrades were winding down.

Having had a few days of cleaning up in billets, on 16 December 21/R Fus started arms drill, cleaning billets and tidying clothing ready for an inspection by the GOC next day. The Adjutant inspected the companies on parade in the morning and found that considerable work was still needed. In addition, the RE provided instruction to parties of the battalion in the erection of barbed wire and MG emplacements. The GOCs visit on 17 December was a disappointment for the battalion. Brigadier General E P Strickland inspected the battalion on parade and then interviewed the company commanders in the orderly room to state his dissatisfaction with the condition of their men's leather webbing. An officer of the battalion, Ashley Gibson, described the leather webbing equipment and did not miss it from when he was an enlisted man; 'Somebody else's aching fingers and thumbs could fumble with that accursed equipment of mine, of more or less obsolete leather pattern, but as hard and unyielding as a board, which had to be unbuckled and rebuckled in its various degrees of "order" at least a dozen times a day, in the last five minutes before parade.'[5]

Next day, there was an alarm practice to see how quickly the battalion could turn out in marching order, get into formation and march past a chosen point. Nothing went wrong for this test. On the 19th there was further cleaning of equipment and next day there was arms drill, drill bomb throwing and lectures on grenades. The MG section practiced gun drill and were lectured on MG tactics. The CO inspected the battalion on parade that afternoon and was happy with its turn out. Men were already thinking of Christmas; Philip Spaull was looking forward to some turkeys which were being sent to the battalion from home for Christmas day.

After the period in the trenches at Festubert several men of 21/R Fus were suffering from dysentery. The Deputy Assistant Direct Medical Services of 33rd Division visited 'certain units of 33rd Division' where diarrhoea cases had been reported, at L'Eclème; these presumably including 21/R Fus. After reports were sent to First Army the First Army DMS visited the battalion's billets personally. According to Spurrell; 'Our wretched doctor was worked to death and then got sacked. We felt sorry for him, as we could not imagine how he could have improved the Festubert conditions.'[6] Lieutenant David McFarlane RAMC was sent to 99th Field Ambulance and was replaced by a more experienced medical officer, Captain Leslie William Howlett, who had been in France since 1914.

On 20 December 19/R Fus and 21/R Fus were put through gas mask testing. According to the 21/R Fus war diary:

> The Battalion attended a lecture and demonstration of the use of gas-helmets at V.6.c.7.4½ [a farm near Bellerive]… This was most interesting and calculated to give all ranks confidence in their gas helmets, as all had to pass through a chamber filled with gas with their helmets on. The lectures also emphasised the necessity of constantly inspecting these helmets.[7]

---

5   Gibson, *Postscript to Adventure*, p.136.
6   IWM: H.W. Spurrell Papers, 2138, p.54.
7   TNA: WO95/2427: 21/R Fus War Diary.

W.H. Wheatcroft sketch of the realities of trench warfare showing UPS men in their gas masks. This was published around the time that men of the UPS underwent gas mask testing. According to Leonard Salter on 20 December there was a 'Gas helmet practice. Not very nice.' (*The Gasper*)

On 20 December the I Corps commander, Lieutenant General Sir Hubert de la Poer Gough, inspected companies of 19/R Fus and 1/Mx whilst they conducted training. During this period of training the Fusilier battalions needed further attention over the two regular battalions. Also on 20 December the UPS Battalion second in commands and company commanders attended a lecture in the cinema at Gonnehem by Brigadier General Strickland. 1/Mx and 2/A&SH did similar schemes under their respective COs. The brigade commander lectured the officers and NCOs of the brigade, Machine Guns were also fired on the ranges and grenade throwing training took place at a brigade bombing school.

The brigade commander again visited 21/R Fus to inspect No 1 and No 4 Companies whilst they were conducting training. The companies of 21/R Fus each had short route marches on the morning of Christmas Eve; these were conducted with frequent halts to allow for the explanation of tactical situations; this was presumably to assist the officers and men with their tactical understanding for more open warfare, were it to occur.

On Christmas Eve 19/R Fus held a big concert. According to Terence Doherty 19/R Fus had their main functions on 24 December:

> It is not at all Xmasy here, in fact it is raining and quite mild. We have had quite a decent Xmas. Yesterday (Xmas Eve) we did not do any work. At 3.30 p.m., we had a huge feed, of Roast Beef, potatoes, cauliflower & Brussels sprouts. Then some Xmas pudding then, fruit & nuts and champagne. We were so full we could hardly move for an hour afterwards. At

6 p.m. we had a concert, it was very good and finished with a play. … So on the whole we had quite a nice day of it.⁸

The battalion had presumably built up a concert party. By 6 January 1916, 19/R Fus had run two pantomimes in Béthune, Philip Spaull had acted in both.⁹ Harold Wilson added more detail; 'Concert in the evening at Gonnehem Concert Hall. A play written by Pvte Green of D Coy. Actors Lt Skie [Skey], Lt Powell, Pvte Francis (the girl, a perfect darling) LC[pl] Cottam, Sgt Bass and Sgt Hadland. Afforded much amusement.'¹⁰ However, one of these entertainers would be dead just over a week later. Those who desired commissions were allowed to submit paperwork on Christmas Eve.¹¹

According to Philip Spaull the men 'managed to make the best of our various parcels and had quite a good time.'¹² The battalion did receive some turkeys and some Christmas puddings from a fund set up by the *Daily News*. A very jingoistic Christmas account was provided by PS/7896 Private W.H. Iredale of B Company, 19/R Fus:

> I am glad to have had the opportunity of spending Christmas at the front. The fellows in our barn … had arranged for a Christmas dinner, and as the people in the farm are very nice indeed they prepared it for us. Just try and imagine 16 of us sitting down to a Christmas dinner, consisting of two roast ducks, four … rabbits roasted in prunes and raisins, best tinned ham, pudding, cake and dessert, not forgetting bottles and bottles of champagne. There we were in a large bare kitchen all seated at a long table making merry. Oh! We had a fine time, the rabbits being especially nice. We had a good sing-song and ended up by toasting the dear ones at home, the King, and the Allies.¹³

Because there were rumours of a move it was decided that 21/R Fus would have its Christmas dinners on Christmas Eve. Plum puddings 'of a solid nature' in 24lb tins were issued with the rations with each soldier receiving ½ a lb. Spurrell also wrote later; 'In addition to an extra-large ration of beef, money had been raised somehow and a ham and special vegetables were brought. There was also a good supply of fruit, nuts, and liquid refreshment. A programme and menu card were written out in French on a postcard; everybody got a copy.'¹⁴ The feast started a 6:00 p.m. for Spurrell's platoon with them sitting on the straw eating from plates and mess tins. After gorging themselves they had an impromptu concert with Private Garstang, the No 14 Platoon comedian, taking centre stage using two empty biscuit tins and a bell to provide music. The enjoyment went on until late that night.

8   IWM: T.O'C. Doherty Papers, 12961, p.50
9   IWM: P.F.F. Spaull Papers, 15453, 6 January 1916
10 PS/409 Private George Morgan Green, aged 19, he had been at university in London and was a trainee solicitor; he had edited The Gasper. PS/7630 Private Philip Evelyn Clarence Francis; PS/421 Sergeant Harold Hadland, aged 21, from Bedford; PS/6617 Lance Corporal Thomas Wyard Gooch Cottam; PS/6172 Sergeant Eric Bethune Bass; Second Lieutenant Humphrey Crofts Powell and Second Lieutenant Charles Harland Skey all likely made up the concert troupe. IWM: H. Wilson Papers, 3606.
11 IWM: P.F.F. Spaull Papers, 15453, 26 December 1915.
12 IWM: P.F.F. Spaull Papers, 15453, 6 January 1916
13 *Halifax Courier*, 8 January 1916, An Ellander Enjoys Himself.
14 IWM: H.W. Spurrell Papers, 2138, p.53.

PS/3081 Private George Arthur Pagett of Derby enlisted, aged 18, with 21/R Fus. He wrote home on Christmas Day describing the previous night:

> We had our Christmas dinner last night, Christmas Eve, and a jolly success it was, too. You would hardly have thought we could have had such a good do out here under the circumstances. We had a big table rigged up in the barn, and plenty of candles. We started with pork, onions, apple sauce, potatoes, and greens; then beef and plum pudding (supplied by the "Daily Telegraph"), crystallised fruits, nuts, etc, finishing up with wine and cigars for those who indulged in them. When I tell you we had to cart the joints nearly a mile to our barn you will be surprised to hear they were decently hot when we got them. While the French vintages were being sampled, witty speeches were made, songs and recitations given, and peals of merry laughter rang out. As we are resting just now, I did not turn out till nine o'clock this morning…[15]

SANTA CLAUS.

An A.S. Palmer sketch of an over-loaded UPS Tommy on a carrying party or fatigue. Whilst this is made light of by likening him to Santa Claus, these arduous activities sapped morale and meant that soldiers seldom had any rest whilst in support or reserve trenches. It also highlights the cloying mud in the trenches near Béthune. (*The Gasper*)

The aftermath of the revelry was felt on Christmas morning. Hangovers did not feel as bad when members of 21/R Fus learnt that the Battalion Sergeant Major had fallen in a farmyard pond up to his neck, which cannot have done his dignity any good.

For Christmas Day Leonard Salter recorded: 'Early service in Cinema. Not much of a Xmas, usual stew for dinner. Great feed at 4 p.m. with Stimson, the Dingleys and self. Engaged a room at an *Estaminet*. Feeling a bit queer after. Bed early.'[16] According to Wilson it was the; 'Queerest Xmas I have ever spent'. 21/R Fus held no parades and had a quiet day, presumably with the men left to their own devices.

15 *Derby Daily Telegraph*, 31 December 1915.
16 IWM: L.E. Salter Papers 21103.

CHRISTMAS FESTIVITIES!

Though the UPS battalions were not in the trenches over Christmas they must have been able to see the light show of the distant flares at the front. Meanwhile, on New Year's Eve the Germans treated them to shelling and MG fire. (Sketch by William Beck-Savage published in *The Gasper*)

Boxing Day was quiet with some parades and inspections for 19/R Fus. 21/R Fus held some route marches by company. Also on Boxing Day, 21/R Fus held a court of enquiry into how PS/8267 Private D. Metcalfe was wounded on 11 December.[17] On 8 December a Private Metcalfe wounded Private Harris by discharging his weapon accidentally. On 10 December a man was recorded as being wounded in Goldney's Keep which was consistent with Metcalf's account of his wounding. Though no record was made in the diary of a self-inflicted wound it is possible that Metcalf injured himself to avoid any punishment from the previous incident. If so, the plan back-fired tragically with the severity of his wound. However, this episode highlights poor record keeping by the battalion adjutant in the unit war diary.

On 27 December 98th Brigade moved to Béthune again and on 28 December relieved 5th Brigade (2nd Division) in trench subsections A1, A2 and B1. On 27 December 19/R Fus marched to Annequin where it stayed in billets prior to going into the line at Cambrin. 21/R

---

17 This is likely reference to PS/8371 Private Denman Metcalf; he was evacuated from France on 5 January 1916 having been wounded badly in the leg. He was hospitalised again in April 1917 and was discharged due to his wounds on 16 May 1917.

Fus moved to le Preol on 27 December. That afternoon the CO, OCs and MG Officer of 21/R Fus, and presumably those of 19/R Fus, went up to visit the trenches they would take over next day. On 28 December both battalions returned to the trenches. 21/R Fus took over B1 Sub Section from 17/R Fus with the following dispositions:

- No 1 Company were in the right front with a platoon in the front line, a platoon in 'Bayswater', half a platoon in 'Artillery Row' and one and a half platoons in Spoil Bank.
- No 3 Company were in the right front; two companies were in the front line and two were in 'Orchard Redoubt'.
- No 4 Company was in support with half a platoon in 'Sidbury Hill' and the rest of the company at Pont Fixe North.
- No 2 Company was in reserve in billets at Le Preol.

The dispositions of 19/R Fus are unknown but they were split between the front line and supporting keeps. They conducted a company relief to swap the companies in the front line.

According to the 21/R Fus diary, a lot of work was necessary to drain the trenches by clearing and deepening the various drainage channels to the canal. Unlike the trenches at Festubert before, there were sufficient telephones for communication and nothing was needed regarding improvements. There was one man killed and four men wounded on 30 December and next day one man was wounded. On 31 December the reserve company at Le Preol was moved to Pont Fixe to be closer in the event they were needed.

19/R Fus lost two killed and seven wounded during this tour. PS/7602 Private Harry Basil Wick and PS/6116 Private Aubrey Cecil James Coombes, aged 24, from Harrogate, were both killed with 19/R Fus on 28 December. Coombes was killed whilst on duty in reserve at a 'support point'.[18] Coombes was close to Donald Wright who found out about his death in the Daily Mail. Moreover, PS/6082 Private William Ewart Hawkins and PS/178 Private John Chadband both died of their wounds in Béthune and were buried in Béthune Town Cemetery. Meantime, 21/R Fus lost one man killed and three wounded. The former was likely PS/4088 Private Arthur Hutchinson who was killed on 30 December.

Leonard Salter summed up the conditions and events in a few short lines on 31 December having been stood to most of the night on 30–31 December: 'Dull. Very glad for daybreak. Still more straffing [sic]. At Stand to in the afternoon, great doings on our right by the French. Terrible gun fire for an hour. Expected it to continue up the line. Feeling very shaky myself. Very noisy night. Fritz would not let us rest because of it being New Years Eve. Sentry duty 8–11 p.m., 2–6 a.m.'[19] Salter's morale was suffering and there was little to look forward to in the New Year except for more of the same. A New Year would being new beginnings for some, but for others, like Salter, it would bring a definite end.

---

18 Both are buried in Woburn Abbey Cemetery. Donald Wright alleged that Coombes was only 19 suggesting he had lied about his age to enlist.
19 IWM: L.E. Salter Papers 21103.

PS/178 Private John Stanley Chadband, aged 25, from Croydon, was educated at Whitgift Grammar School. He was an articled surveyor and member of the surveyors institute and auctioneers institute. Chadband enlisted in September 1914 and served with C Company 19/R Fus. (Reproduced with permission of Sutton Archives)

PS/4088 Private Arthur Hutchinson was aged 19 and from Walshaw, Bury. He was educated at Bury Grammar School and served with 21/R Fus. Hutchinson is buried in Woburn Abbey Cemetery, Cuinchy. (Reproduced with permission of Sutton Archives)

19th Brigade

18/R Fus spent a period out of the line in the run up to Christmas. They were billeted in a village called Fontes. Whilst here, some training was conducted in physical drill, bayonet fighting, gas mask drill, digging, sandbag filling and wiring. A range was even constructed in a quarry to enable musketry training. Some telescopic sights had also been issued and these were practiced with, presumably by the battalion snipers. The next week was spent in training, route marches and a work party was provided to assist a local airfield with digging drains. There were classes for specialists in the battalion; machine gunners, snipers and bombers. Brigadier General Robertson visited on several occasions and presumably took an interest in the battalion, its range work and its billets.

During this period out of the line, on 14 December, PS/7251 Private Bernard Angelo Parietti, died in hospital of broncho-pneumonia having presumably gone sick during the last tour of trenches.[20] On 19 December the battalion suffered a further casualty as PS/270 Lance Corporal Newman Bruce Dobell, aged 23, from Highbury New Park, died in a nursing home in London. He was buried in Highgate Cemetery.

Like their sister battalions in 98th Brigade, on 22 December 18/R Fus received: 'Instruction at 12 noon by 1st Army gas expert in wearing gas helmets in barn full of gas.'[21] Sergeant Walter Medlicott recalled this incident:

> To-day we had a gas demonstration – very useful to give one confidence in yr [your] helmet, but it was all rather comical, owing to bad management. Lectures in field first + then we filed past the cylinders in a barn – with helmets on – it renders the gas quite innocuous – unfortunately they had forgotten the animals – x2 pigs nearby succumbed + 1 cow is not expected to recover. The wind shifted & laid out 15 men in D Coy before they had got helmets on so they will be laid up for a day or two – Several of us got too much of it before helmets were on + realised the awful torture it must be if not fitted with a helmet. It was most useful practice.[22]

'… a very curious Christmas' – 18/R Fus

On the afternoon of Christmas Eve, Brigadier General Robertson lectured the officers of 18/R Fus and 20/R Fus and afterwards went to a concert by the Divisional Army Service Corps (ASC), according to the Brigade Major, Twiss, who reported that the concert was; 'quite good'.[23] The Christmas enjoyed by 18/R Fus was spent out of the trenches and without any training taking place. Although he and his comrades were living in a barn, Kenneth Norman enjoyed himself. There were numerous parcels of food and some champagne to enjoy. On Boxing Day, the 18th Battalion had to march to the Givenchy trenches.[24] Walter Medlicott's diary recorded the following of Christmas Day for his group of friends. There were early church services at 7:00 a.m. and 8:00 a.m. with a further service at 9:30 a.m. and another at 10:00 a.m. for Catholics. During the day were shooting competitions and inter-company football matches but no formal parades. At 10:00 a.m. some of the sergeants visited Captain Price-Edwards' billet for port and cakes. There was a dinner for the sergeants at 1:30 p.m. and the football final took place at 3:00 p.m. D Company won, 6–4, in a thrilling match with B Company. For lunch, Medlicott's section got hold of five chickens in Aire and got the local *estaminet* owner to kill, pluck and cook four of them. He described the church service on Christmas morning:

---

20 Parietti's family was from Harwich, Essex, and was aged 19. He had presumably enlisted with 18/R Fus whilst under age. He died at 20th General Hospital, Camiers, and was buried in Etaples Military Cemetery.
21 TNA: WO95/2423: 18/R Fus War Diary.
22 IWM: W.B. Medlicott Papers, 1708, p.118.
23 IWM: G.K. Twiss Papers, 17109.
24 LA: K.V. Norman Papers, WW1/GS/1188, p.112.

We had a crowded celebration at 7 [in the morning] in the little village school, the men overflowed into the covered play shed. For some time I was outside but we all joined in the service and hymns. When I got inside I found about 300 men, standing, because there was no room to kneel. The chaplain had arranged the altar under the crucifix at the end of the room and the room was lit by the big stable lanterns issued to the troops for billets. He took the service so reverently under the difficult circumstances and he had no one to help him. Most men had to go before the thanksgiving…[25]

He observed that 'it was a very curious Christmas … but it has all been very happy for everyone.'[26]

On 27 December 18/R Fus marched back towards the Front Line and after a slog of sixteen miles reached Béthune and were billeted in Montmorency Barracks. Harry Harvey's description of the inside of this accommodation in Béthune evokes an impression of many of the other different billets occupied by 18/R Fus:

[The] … platoon, was seated on a dusty floor of the cold, gloomy, prison-like Montmorency Barracks at Béthune, amidst a tangle of infantryman's equipment, rifles, gas-helmets, unrolled putties, grubby blankets of a chocolate hue, mud-caked great-coats, rubber groundsheets outspread and footsoiled, surplus bandoliers, neglected tins of foot-grease, and the discarded wrappings of parcels from Home. Such was the medley which lay around …[27]

20/R Fus at Rest

Meanwhile, Ham-en-Artois hosted 20/R Fus for their next period of training and rest. James Hodson recalled the living conditions; '…Ham-en-Artois was so far back from the line that no shellfire fell near the village. Greatcoats still heavy with mud were floated in a stream behind the farmhouse … Weather was mild, their room was plentifully studded with nails [to hang up kit] … and they had straw for their feet. They read and wrote and smoked and played bridge…'[28] After a day to clean up, the battalion was inspected by Brigadier General Robertson on 14 December. The battalion spent the next ten days conducting company and platoon training, interspersed with route marches and digging some practice trenches. The battalion had a rest day on 25 December with no training. The men of the 20th Battalion likely enjoyed the same festivities as the other three battalions. Sergeant Godfrey Skelton still received packages from home; 'All during my service in France my mother had sent frequent parcels of food and goodies from Fortnum and Mason's, which was much appreciated and shared out amongst my pals.'[29]

Next day, instead of a Bank Holiday, was spent by 20/R Fus in filling in practice trenches and collecting engineering stores. On 27 December 20/R Fus marched towards Béthune and was again billeted in the Orphanage. Unlike the battalions of 98th Brigade, 20/R Fus would not go straight back into the trenches. After lunch on 28 December D Company, 20/R Fus, was

---

25   IWM: W.B. Medlicott Papers, 1708, p.124.
26   IWM: W.B. Medlicott Papers, 1708, p.125.
27   Harvey, *Battle-Line Narratives*, p.20.
28   Hodson, *Grey Dawn*, p.184.
29   IWM: G. Skelton Papers, 13966, p.52.

selected to provide a guard at HQ I Corps at Chocques. According to Private Vincent Platt; 'At the time of writing "D" Coy are in Chocques doing guards and fatigues. The general headquarters are here & we have to turn out perfectly spik [sic] & span. A guard comes around once in 4 days. We may be here for a month. The job was given to us because we are the least criminal company.'[30] Whilst at Chocques 20/R Fus had a canteen which was well stocked with chocolate etc run by Vincent Platt.[31] The remainder of the battalion moved to billets in Annequin South as the reserve battalion for 19th Brigade which was holding the line south of the La Bassée Road. However, being in reserve was not necessarily safe. Major G.K. Twiss, the Brigade Major at 19th Brigade recorded that just before the brigade moved into the line at Cambrin; 'In the afternoon [of 28 December 1915] the Germans shell the Headquarters at Cambrin pretty badly, killing a good many, including a gunner officer. I'm afraid they've got the place 'set' & we shall have to change our Headquarters, a great pity as it is a good one, & there is practically no other.'[32] HQ 19th Brigade would also have to 'rough it' on this tour of the trenches. However, the criticism of the UPS battalion not only originated from 2/RWF. On 30 December Major Twiss recorded; 'It is very trying having these new Regiments [18/R Fus and 20/R Fus], they do no work, they don't seem to understand having a system. By Jove, the work is simply awful, from early morning 'til midnight, I don't get time even to write letter hardly. The trenches are fairly bad, especially on the left where the 18th Fusiliers are…'[33] Many of Twiss' gripes concerned his own workload and he did not state whether these trenches were inherited in this condition. It was not just the UPS men that he resented; 'This [33rd Division] is a pretty trying Division, I wish they would realise we have been out for 16 months & are not just beginning!'[34] He later added after a visit by the divisional commander; 'We had Genl LANDON round, but only talked of trivial details, he is an old 'idiot', if ever I saw one. It was a bad day when we were transferred from the 2nd Division to this crowd.'[35]

This antipathy between the regular battalions in 19th Brigade, and the newly-joined 'service' battalions, would play out to a far greater extent over the coming year.

30 IWM: V. Platt Papers, 17681, p.4.
31 IWM: V. Platt Papers, 17681, p.8.
32 IWM: G.K. Twiss Papers, 17109.
33 IWM: G.K. Twiss Papers, 17109.
34 IWM: G.K. Twiss Papers, 17109.
35 IWM: G.K. Twiss Papers, 17109.

11

19/R Fus on 2 January 1916

Simply hell on earth

... the worst thing I have ever been through ... it was simply hell on earth ...

Fritz was a bit lively ...

As the UPS battalions were split between two different brigades there is a need to examine the fortunes of the two pairs of battalions concurrently. First 19/R Fus and 21/R Fus (with 98th Brigade) and second 18/R Fus and 20/R Fus (19th Brigade). Prior to this, a disaster was to strike 19/R Fus.

New Year's Day saw 19/R Fus in A1 Subsector (Cuinchy Brickstacks) with 2/A&SH in A2 on their right, and 21/R Fus in B1 (North of the Canal) on their left. 1/Mx were in reserve at Annequin. On 30 December PS/4088 Private Arthur Hutchinson, of 21/R Fus, was killed and four men were wounded; another man was wounded on New Year's Eve.[1] On the night 31 December 1915 and 1 January 1916 the Germans 'celebrated' the arrival of 1916 with a bout of heavy shelling, rifle and MG fire from 11:30 p.m. to 12:30 a.m. The damage that this caused to Cheyne Walk was repaired in time for the handover with 2/Worcs R at 10:30 a.m. and was completed by 12:40 p.m. According to Hugh Spurrell:

> The only unusual incident during our spell took place on New Year's Eve. All sorts of warnings had been circulated and shortly after midnight we heard the rattle of musketry in the front line. A sentry with an extra vivid imagination had seen a couple of brigades of Prussian Guards crawling over No Man's Land. Oilbottle (Sergt. Renouf), who was in charge of our post got the wind up with all speed and we had to stand to in the trench below. Here we were up to our knees in muddy water and could not have fired if we had wished, as the trench was nine feet deep. One man got badly told off for not having a full water bottle ... But, after satisfying ourselves that the offensive had been postponed, we became calmer. The Prussian Guards turned out to be the same row of shrubs that had been out in the open every other night. We turned in and slept again.[2]

---

1   The fatality was PS/4088 Private Arthur Hutchinson, aged 19, from Walshaw, Bury. He is buried in Woburn Abbey Cemetery.
2   PS/3183 Lance Sergeant Cyril Percival Renouf, aged 24, from Finsbury. He was gazetted to a

PS/3026 Private Reginald Merryweather was a sniper with 21/R Fus; a role; '....which is good fun, so long as they don't lob a trench mortar over.' He recalled on New Year's Day that; 'At twelve o'clock a single gun was fired by our artillery, which seemed to be the signal for a general cannonade all along the line.'[3]

19/R Fus, at the Cuinchy Brickstacks, had also observed some New Year 'hate' towards the Germans according to Philip Spaull:

> Last night, New Year's Eve, the guns peppered the enemy off and on all the night but at 12 p.m. my word, we did let them have it hot. The roar of the guns was like a rough sea striking the rocks. They must realise what a hopeless task they have before them. We certainly ought to be proud of our artillery.[4]

Such words were potentially prophetic though many had confidence in their defences and shelters according to Philip Spaull on 1 January:

> We have just finished 48 hours in the firing line. We had a fairly quiet time. The line is an old German one. The dug-outs are very well made. Mine had ten steps leading down, so you can tell it was a deep one. We could easily fit ten men in it. I was told the night before we went six shells exploded on the roof ... no damage was done.[5]

On New Year's Day 19/R Fus found the Germans ominously quiet. The artillery observation officer covering the battalion front observed gaps in the German wire and a patrol that night confirmed that the wire had been cut by hand. Expecting a potential German attack neighbouring units were alerted. Whilst 21/R Fus was relieved on New Year's Day, 19/R Fus, 2/A&SH and 1/Mx were to be relieved the next day by 100th Brigade. According to the 19/R Fus war diary; 'The lull of overnight was the precursor to a storm.' What followed on 2 January 1916 was the worst day of the war for 19/R Fus. The Germans somehow discovered that there would be a trench relief taking place. This was later blamed on two men and a woman who lived at an *estaminet* on the corner of Harley Street and the La Bassée–Béthune Road, who were alleged to be spies. On 1 January Leonard Salter had also observed that the barbed wire had been cut opposite their trenches and noted that; 'Fritz was a bit lively'.

The Germans had been largely quiet during the preceding night. The men in the trenches were reminded that it was a Sunday morning by the sound of church bells coming from village behind the lines. Some were slightly buoyed up having received their rum ration and 16/KRRC was to take over the trenches from 19/R Fus.[6] The Germans shelled the 19/R Fus front and support trenches incessantly with 4.2-inch and 5.9-inch guns which also gave Brigade

---

commission in the Motor MGC in April 1916 and later served in the Tank Corps. He ended the war as a major. IWM: H.W. Spurrell Papers, 2138, p.56.
3   PS/3026 Private Reginald Howard Merryweather, aged 22, was from Grimsby and had been educated at Denstone College. In 1914 he returned from Canada where he was farming to enlist in the UPS. He was later gazetted to the Lincolnshire Regiment in August 1916. *The Denstonian*, Vol. XI, March 1916, p.13.
4   IWM: P.F.F. Spaull Papers, 15453, 1 January 1916.
5   IWM: P.F.F. Spaull Papers, 15453, 1 January 1916.
6   *Norwich High School Magazine*, Summer 1916, pp.26–27.

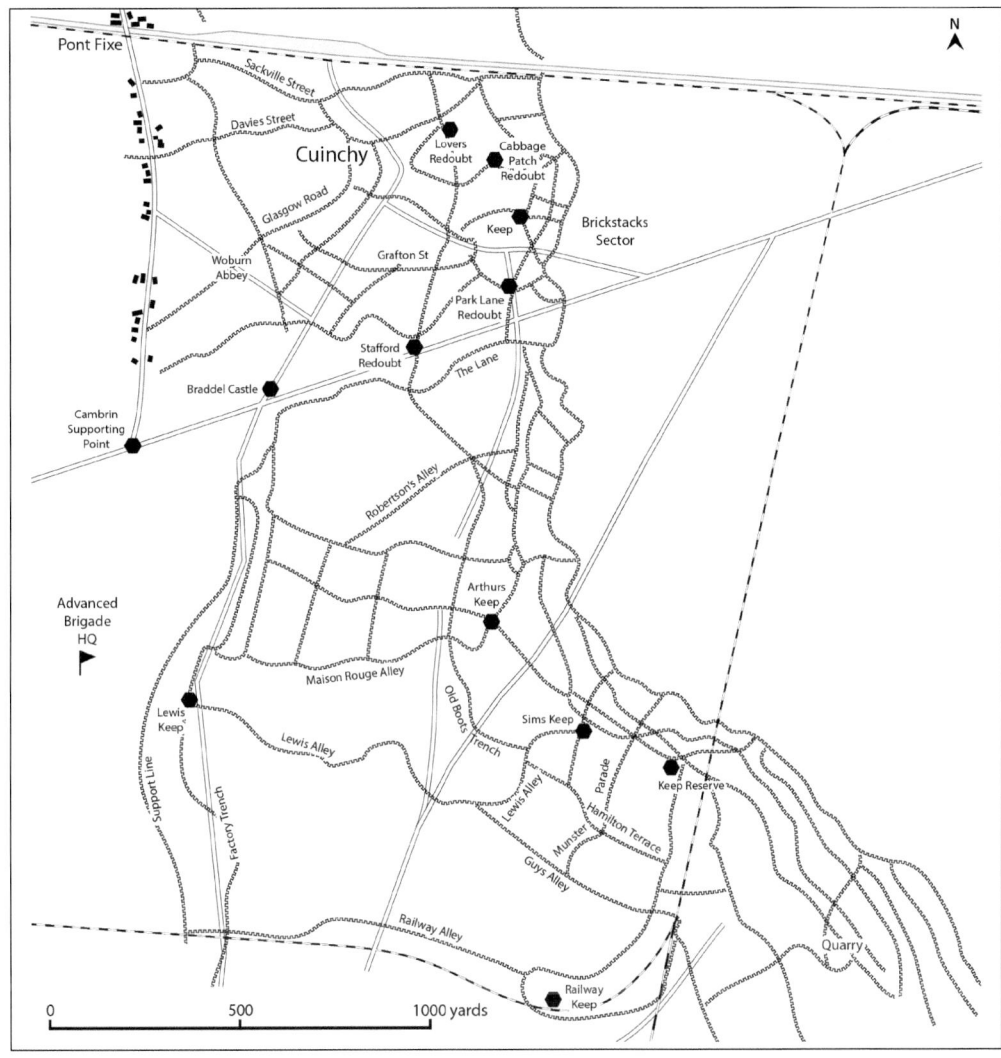

Map 3 Cuinchy Sector.

HQ some unwanted attention. At 11:15 a.m. the Germans detonated a small mine opposite 2/A&SH on the left of 19/R Fus. The bombardment was at its heaviest when the relief was at its height around 11.30 a.m. At the same time the enemy also exploded a mine opposite D Company, 19/R Fus. The parapet was blown in and most casualties occurred here. PS/6701 Private Leonard Salter was a member of this company engulfed by this explosion; 'Just being relieved about 9.30 by the 16th K.R.Rs when Germans exploded a terrific mine under our first line, got wounded badly. Compound fracture right leg due to falling sandbags etc. [I] Lay half buried while Fritz gave us four hours bombardment …'[7] Another man recalled:

7   IWM: L.E. Salter Papers 21103.

… the Germans gave us a terrible strafeing [sic]. The shells were falling all round. In the middle of it all they blew up a mine. It is the worst thing I have ever been through, as for three hours it was simply hell on earth. I am sorry to say that our Company suffered heavily … I have lost several chums in it. It is a miracle how anybody escaped. It was made worse because the regiment that was relieving us had just arrived and we had not left so we were crowded like sardines. I think they [the Germans] had spies at work. On New Year's Eve, they were shouting across that they would "strafe" us…[8]

PS/7010 Private Walter Butterworth wrote home two days later:

We were shelled by the Germans. They kept it up for a few hours. It was hell. They also blew up four mines under our lines, burying our chaps. Some were dug out, but others have not been found. No one can imagine what the blowing up of a mine is like. Tons of earth are blown for yards into the air and it seems an age before it all comes down again. Harry Lord had his arm broken, and the same piece of earth killed one of the KRR men instantly – breaking his back and neck. Our company caught it worst. It is thought that about 40 men were killed and wounded out of about 200, which is terrible. The KRR were worse off than us. … Never have we been face to face with our Maker as were during that terrible time. Our men were crouched and huddled up together – expecting every minute to be the last – helping to dig someone out and calling for stretcher bearers. It was awful. … Our captain was wounded in the head and we lost a good second lieutenant…[9]

The Chaplain of 16/KRRC later wrote:

The men have not had time to settle down in their new surroundings, or even find their bearings, before there comes a series of terrific explosions. The earth trembles violently, and then is blown into the air, mountains high, with a roar that is heard miles away. The falling earth spreads itself like a gigantic pall, and rains down everywhere for some minutes. There is a great gaping hole in the ground; one of the biggest craters seen in the war … a hundred or more brave fellows have been wounded, killed, or buried in the debris. The cries of the maimed mingle with the appeals of those seeking to extricate themselves from the soil that is strangling what remains of life. Feverishly, digging parties bend to their task, and many are got out in time. A few, alas, are beyond help; some die and remain buried for all time where they fell.[10]

Harold Wilson was on a bombing course in Béthune; he was told, second-hand:

… about 11.30 the Hun blew up a big mine between the front line and supports, the usual shower of sandbags etc went sky high and heaps of our fellows were hit and buried. The majority of them were dug out [and] broken legs and arms resulted. It was apparently a

---

8   IWM: T.O'C. Doherty Papers, 12961.
9   *Todmorden and District News*, 14 January 1916.
10 James Duncan, *With the C.L.B. Battalion in France* (London: Skeffington and Son, 1917), pp.37–38.

wonderful though dreadful sight. Everybody said their prayers and waited for their turn. The Coy behaved splendidly.[11]

PS/4536 Private Ernest Edward Brierley was on a 20/R Fus working party; he wrote in his skeleton diary; 'Working party at Cambrin trenches. Mine blown up. Lots of shells. Carrying gas cylinders, rotten job. Dirty trenches. Horrible sights on way back.'[12] Further along the front line, PS/6852 Private Frank Nunn, B Company, of 18/R Fus, experienced the explosion. His trench was 35 yards from the Germans, and he was on sentry scanning No man's land for enemy activity:

> … suddenly I was conscious of the earth giving an awful shake. At first I did not realise what it was … I looked to my left an awful sight met my eyes, for in our lines about one hundred and fifty yards away the earth was being raised by the ton, and it reminded me of a huge fountain, only of earth instead of water. The whole upheaval lasted four minutes, and the mine seemed to creep towards us, and it was then that most of us thought that our last day had come …[13]

Luckily for Nunn the mine explosion did engulf his trench. When the explosion ceased there was relative silence except for the sound of the German bombardment. He likely heard second-hand of the exploits of a sergeant of 19/R Fus who kept his Lewis gun in action and whose fire prevented the German attack. Nunn also heard that the sergeant was to receive the DCM but was killed by a tree falling on him after that tree had been struck by a shell. PS/940 Private Robert Simon Mansfield Sturges of 19/R Fus noted that after a German mine explosion the enemy normally sought to take advantage of the shock effect and launch an attack: 'Knowing this, the machine gun sergeant, wonderfully pre-serving his presence of mind, mounted his gun on the parapet, and sprayed the German trench unceasingly, until a sandbag flung heavenwards by a second explosion fell upon him and broke his neck. The sergeant was afterwards "mentioned," I believe; at any rate he certainly deserved it.'[14] PS/6688 Private William Brislee of the 19/R Fus MG Section – 'the only man left on the [Lewis] gun … propped the gun on the parapet and blazed away.'[15] Whether Brislee's fire was able to deter the Germans, or they did not plan to attack, 19/R Fus was lucky that the Germans did not advance. Brislee was appointed lance corporal the next day but on 2 March he was evacuated to the UK suffering from a gunshot wound hand (self-inflicted) and was later discharged due to this wound.[16] Despite Brislee's bravery, this black mark likely prevented him from being decorated.

11 IWM: H. Wilson Papers, 3606.
12 PS/4536 Private Ernest Edward Brierley was initially a bugler with 20/R Fus and was appointed as a lance corporal on 9 January. He was born in 1895 and lived in Withington, Manchester. Like many in the battalion, he attended Manchester Grammar School and was employed as a railway clerk. Diary of Ernest Brierley, unpublished. With the permission of Angela Jennings. Entry for 2 January 1916.
13 PS/6852 Private Frank Moye Nunn, No. 7 Platoon, B Company, 18/R Fus, was educated at Norwich High School. He was later transferred to the Labour Corps and was discharged in March 1919. Norwich High School Magazine, Summer 1916, pp.26–27.
14 Sturges, *On the Remainder of Our Front*, pp.79–80.
15 IWM: H. Wilson Papers, 3606.
16 PS/6688 Lance Corporal William Harold Brislee, aged 25, was from Bromley Cross, Lancashire. He was an electrical draughtsman and enlisted on 23 March 1915 in Chester. On recovering he was posted to 28/R Fus on 25 April 1916 before being discharged no longer fit in June 1916.

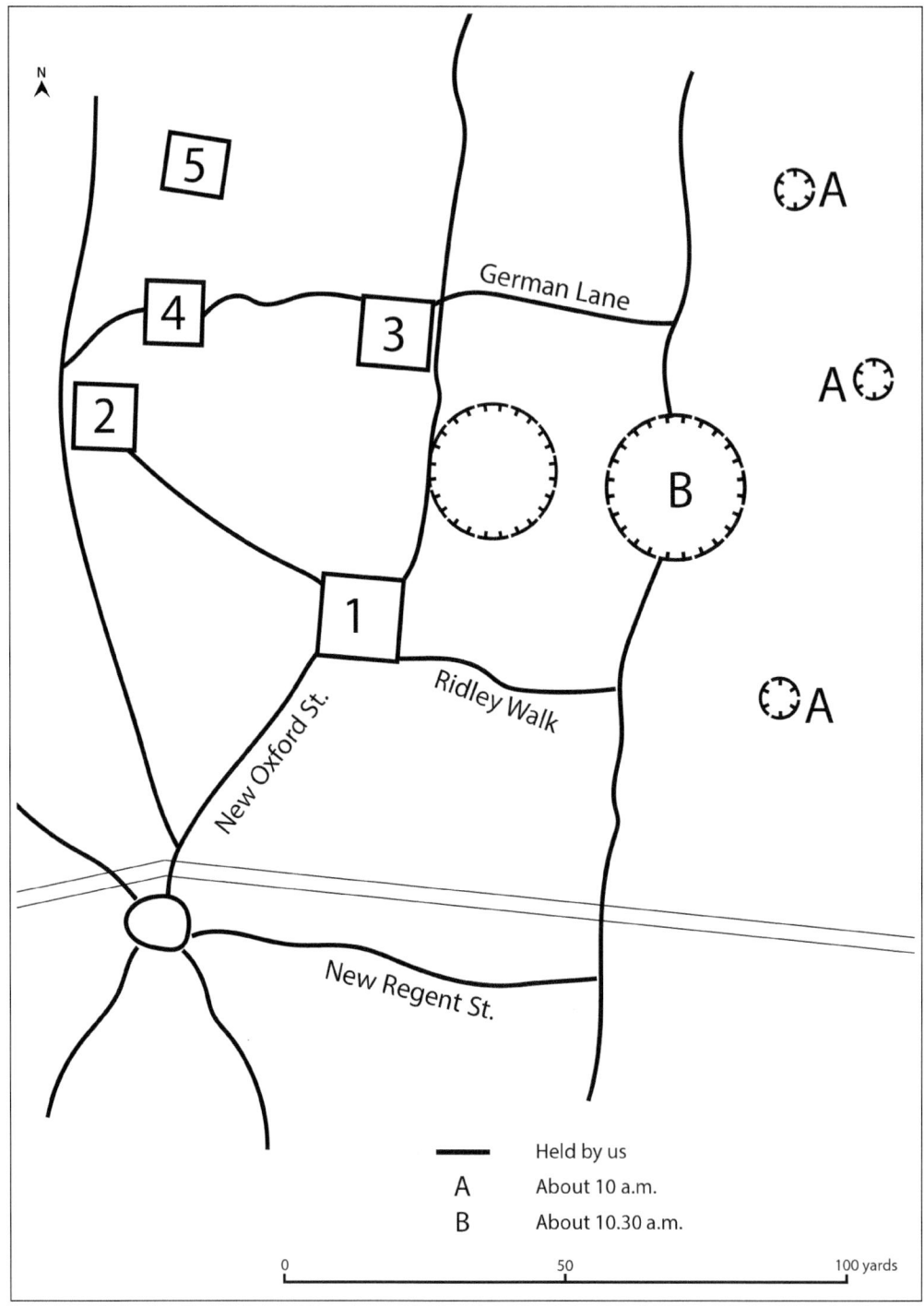

Map 4 Cuinchy Brickstacks.

The relief was completed shortly afterwards. However, the ordeal was not over. Private Butterworth continued, 'our platoon sergeant was killed. After we had left the trench and got about a mile away, the Germans were shelling the main road and he was caught by a shell. He was a good sergeant and he was proud of his Hebden Bridge lads. Our second sergeant is wounded badly.'[17] The last men of 19/R Fus were in Béthune by 3:00 p.m. though many would not return. The corporal in charge of the 19/R Fus stretcher bearers was Vincent Edwards; 'Our battalion ... had to leave behind some bodies of men who had been killed. I considered it my duty to identify the bodies and take back to the battalion the identity discs, pay books... I got the necessary documents into empty sandbags, nine, one for each man killed, and eventually arrived at Battalion HQ loaded with them ...'[18] Edwards was warmly welcomed on his return. He had been gone for so long that he was presumed to have been killed.

Casualties amounted to one officer killed and one wounded, eight men were killed, sixteen wounded and one was missing.[19] Second Lieutenant Henry Wood was one of the first UPS officers to be killed.[20] Captain George Shipster, commanding D Company, was wounded and evacuated.[21] Those killed were PS/2 Sergeant Henry Adams; PS/421 Sergeant Harold Hadland; PS/560 Sergeant Robert Kane; PS/3442 Sergeant Edgar Beale; PS/1036 Lance Corporal Reginald Wilbee; PS/42 Private Walter Banks; PS/747 Private Thomas Pascall and PS/7631 Private Harry Williamson.[22] PS/7014 Private Louis Armstrong died of his wounds.[23] Banks was killed in the trenches by a shell explosion whilst supposedly engaged in bombing operations. Kane was likely killed by a shell whilst helping a wounded man. Armstrong was a sergeant's orderly and cook; he was wounded by a shell whilst going to the aid of an officer.

Leonard Salter, and the other wounded, had a horrific time; he was lucky to survive and was rescued. His ordeal was not over: 'Picked up at 1 p.m. Taken down to 1st Field Dressing Station

17 Todmorden and District News, 14 January 1916.
18 LA: V. Edwards Papers, WW1/GS/0505, p.3.
19 TNA: WO95/2427: 19/R Fus War Diary; TNA: WO95/2424: 98th Brigade War Diary.
20 Second Lieutenant Henry Wood, aged 22, was from Coventry. He was gazetted into 21/R Fus on 15 March 1915. He was buried in Béthune Town Cemetery.
21 Captain George Cecil Shipster, aged 29, was from London. He was educated at Marlborough College and was in business before the war. He was alongside the first batch of UPS officers to be gazetted in October 1914 and was later promoted captain and assigned to command D Company. He was awarded an MC in June 1916, presumably for his handling of this action. He later commanded a draft of 19/R Fus men back to the UK in March 1916. Shipster later joined the MGC, served on the staff and with the Royal Tank Corps.
22 PS/421 Sergeant Harold Hadland, aged 21, was from Beverley, Yorkshire. PS/560 Sergeant Robert Henry Kane, aged 33, from Wexford, Ireland, was educated at Stonyhurst College. He had been turned down for a commission. PS/1036 Lance Corporal Reginald Forrester Wilbee, aged 21, was from Herne Bay. PS/42 Private Walter Ralph Banks, aged 28, was born in Devon but lived in Ealing Common. He was educated at Reading School and was an examiner in the Exchequer and Audit Department of the Admiralty. He enlisted in September 1914. PS/747 Private Thomas Osborne Pascall, aged 21, was from South Norwood. PS/7631 Private Harry Williamson, aged 25, was from Chadderton, Oldham. Banks, Hadland, Pascall, Wilbee and Williamson were buried in Woburn Abbey Cemetery. Kane is buried in Cambrin Military Cemetery.
23 PS/7014 Private Louis Armstrong, aged 26, was one of the 19/R Fus draft that had enlisted in Hebden Bridge. He was educated at the Board School in Hebden Bridge, played football for Hebden Bridge AFC and rugby, and was employed in the wholesale clothing works there. He had previously shown his courage by saving the life of a female employee there. He enlisted in May 1915. Armstrong is buried in Béthune Town Cemetery.

PS/2 Sergeant Henry Neville Adams, aged 19, of the MG Section, was blown to pieces but was responsible for an act of bravery for which he was posthumously MiD in June 1916. Adams was likely the MG sergeant mentioned above. Adams was from Sanderstead, Surrey, and was educated at Whitgift Grammar School, Croydon. He was buried in Woburn Abbey Cemetery, Cuinchy. (Reproduced with permission of Sutton Archives)

PS/3442 Sergeant Edgar Beale, aged 23, was from Frensham, Surrey, and was educated at Mr Poole's Prep School, Lancing College and Peterhouse College, Cambridge. Beale was the platoon sergeant for No 16 Platoon, D Company, and was acting as the platoon commander. Beale is buried in Cambrin Military Cemetery. (Reproduced with permission of Sutton Archives)

on rails. Sent on to Béthune Hospital about 4 p.m. Chloroformed to be dressed there, sent on to No 1 Casualty Clearing Station at Chocques. Arrived about 8 p.m. Feeling a bit rough. Bad night.'[24]

Also amongst the wounded were PS/210 Corporal Archibald Colville who was only slightly injured and PS/6699 Private Edgar Robinson who was shell shocked.[25] They were all luckier than PS/85 Private Samuel Billings who died of his wounds next day, and PS/3639 Sergeant

24 IWM: L.E. Salter Papers 21103.
25 PS/210 Corporal Archibald Ross Colville, aged 27, from Helensburgh, Dumbarton, was educated at Loretto School and Magdalen College, Oxford. He had farmed in South Africa before the war. Colville had enlisted in September 1914 and had attended a trench mortar course. He returned to duty from his wound later in January and returned to the UK for a commission in March 1916; he was gazetted into 3/10 King's and later served in the King's African Rifles in East Africa 1916–1919 and was promoted captain. PS/6699 Private Edgar Whittle Robinson, aged 25, from Crewe, had been a bank clerk.

Louis Colbourne who died on 7 January.²⁶ Colbourne had apparently volunteered, with others, to hold a dangerous forward position. He was buried after the mine explosion and was dug out with difficulty but suffering from severe injuries. He was evacuated to 12th General Hospital, in Rouen, but died five days later. Lieutenant Colonel Wolrige Gordon wrote to Colbourne's family:

> I need hardly say that the death of your son was deeply regretted by the whole Battalion. I knew him personally, and always had a great admiration for him. I did not see him wounded but it was reported to me shortly afterwards, and at the same time I was told how gallantly he stood up against the intense bombardment. I shall always remember him as a gallant comrade. He always carried out his duties in an exemplary manner…²⁷

This was a horrific day for D Company specifically, and 19/R Fus in general, and was the worst event experienced by the UPS battalions over the winter of 1915–16. Despite heavy casualties, the battalion had withstood the German artillery and underground onslaught and their steadfastness in their first action

Having suffered a compound fracture of his tibia and fibula on 2 January, Salter soon had gangrenous wounds due to the earth which had surrounded his leg. On 19 January his leg was amputated below the knee. He had to recover in hospital for three months before he could even return to the UK. Salter's treatment was completed by December 1916 when he was issued with an artificial leg and discharged. (L.E. Salter papers)

does deserve considerable credit. This did not affect the whole battalion. Philip Spaull, writing a few weeks later, possibly recalling this event: 'A mine went off the other night on our left. I was sound asleep and heard nothing although our boys said it shook the ground. I sleep just as well here as I do at home. We are all alike, tired enough to sleep any moment we can get and anywhere.'²⁸

---

26 PS/85 Private Samuel Walter Billings, aged 36, from Cheltenham, was educated at Cheltenham College and was a solicitor by trade. He was buried in Béthune Town Cemetery. PS/3639 Sergeant Louis Colbourne, aged 24, from Brighton, was educated at Eastfield House, Ditchling, Sussex and Tonbridge School where he served in the OTC. He was an accountant in Brighton. Colbourne enlisted in September 1914. Colbourne is buried in St Sever Cemetery, Rouen.
27 Anon. Author, *Tonbridge School and the Great War 1914–1919* (London: Whitefriars, 1923), pp.72–73.
28 IWM: P.F.F. Spaull Papers, 15453, letter 20 January 1916.

12

98th Brigade in Early 1916

The air was alive with shells

> *If we had passed up that road an hour later than we did, we should have got it in the neck ...*[1]

> *I heaved a sigh of relief when I dropped over our parapet again and was once more safe back in the trenches ...*[2]

On 1 January 21/R Fus had withdrawn to billets in Annezin, a small village to the west of Béthune, where they spent 2 January cleaning up. Spurrell was finally able to find and remove the last vestiges of Festubert mud from his greatcoat. The battalion was followed on 2 January by the MG Section commanded by Lieutenant John Jervis the MG Officer.[3] This section brought news of heavy German artillery fire on the trenches that 21/R Fus had recently left and heavy casualties amongst the men of 2/Worcs R.

There were several inspections of 21/R Fus during their time in billets; some of the billets were inspected by the divisional Assistant Director Medical Services on 3 January and the 2IC, Major Fenn, inspected two of the companies on parade. The next day the Divisional Assistant Director Veterinary Services inspected the animals of the battalion transport. Further inspections took place by Lieutenant Colonel Stuart-Wortley and Major Fenn. Visits to baths, arms drill and gas helmet drill took place during the rest of this period away from the line:

> We also investigated the cellars of the school at Béthune, which by this time had become the official divisional baths. The bath rooms were palatial; unfortunately we were only allowed two minutes each under a thin jet of tepid water, which then became cold for another

---

1 Shell Fire Horrors, *Manchester Evening News*, 21 January 1916.
2 Sturges, *Remainder of Our Front*, pp.99–101.
3 Lieutenant John Cedric Jervis, aged 26, from Snitterfield near Stratford-upon-Avon, had been gazetted in 21/R Fus on 17 Mar 1915 and later promoted to lieutenant. With the disbandment of 21/R Fus he was transferred to the RFC as an observer and was killed whilst flying with No 5 Squadron on 26 October 1916. He was buried in Courcelles au Bois Communal Cemetery Extension.

minute. A whistle then blew and our hopes were at an end for that week. A clean change of underclothing was issued with the bath and the old stuff handed in for disinfection.[4]

19/R Fus, having survived their ordeal, went through the *Ecole de Jeunes Filles* baths on their first day out of the trenches on 3 January. They also spent time tidying up their new billets and during this rest period had the opportunity to fire on the rifle range. The battalion was warned again for the trenches on 7 January and were held at higher readiness to move. On 8 January No 2 and No 3 Companies marched to billets in Beuvry ready to take over keeps in Z0 Subsection at 9:30 a.m. next day; this relief was later brought forward to 3:30 a.m. No 3 Company occupied Maison Rouge, Lewis Keep and the trench between them. No 2 Company had platoons in Sims Keep, Railway Keep and a third platoon was sent back to Annequin South due to a mistake. Later that day the remainder of 19/R Fus moved to billets in Annequin South which were exceptionally dirty.

On 7 January the first draft of reinforcements arrived at 21/R Fus; this consisted of 23 NCOs and men; they had embarked on 24 December 1915 and had presumably spent two weeks at an Infantry Base Depot. They went to the trenches and joined No 1 and No 3 Companies. The commanding officer and company commanders departed to reconnoitre the Z Section of trenches whilst the battalion marched to Béthune. Next day 21/R Fus marched to Annequin South to relieve 18/R Fus. Spurrell was assigned as sentry at battalion headquarters and during his stint on the evening of 8 January it snowed non-stop. On the 9th orders arrived for 21/R Fus to take over from 20/R Fus in Z Section at 8:30 a.m. on the 10th but this was cancelled, and the relief commenced at 2:00 a.m. Due to some 20/R Fus guides going astray the relief was not complete until 7:00 a.m. as No 2 Company, 21/R Fus, had to find its own way. The battalion was disposed with all four companies in the line and No 1 Company 19/R Fus in support in Wimpole Street.[5] Each 21/R Fus company had two platoons in the line, one in support and one in the strongpoints behind. According to Spurrell, who was in Russell Keep, life in the strongpoints was quite comfortable:

> … it [the Keep] was to hold out to the last, even if the enemy broke through the rest of the line on both sides of it. We only had to do sentry duty on alternate nights … There were two large dug-outs in the Keep; each had two entrances and was capable of sheltering twenty men. We had plenty of room to lie down comfortably. Of course, the place was rather stuffy … We were not disturbed at all by the Bosche …[6]

1/Mx were on the right from *Boyau 9* southwards and 1/Queen's RWS (100th Brigade) were on the left. The state of the trenches was:

> … On the whole very good – communication trenches being entirely free from water – some work however was found necessary in part of the front line between *Boyaux* 14 and 17. A feature of these sections is the series of craters on the left which practically join our lines

---

4   IWM: H.W. Spurrell 2138, p.57.
5   Likely No. 3 Company 19/R Fus.
6   IWM: H.W. Spurrell 2138, p.58.

with those of the Germans. There are 12 guns (6 Lewis, 6 Machines [i.e. Vickers machine guns]) in this section of the line…[7]

The weather was mild and there was only a little shelling. Whilst there was limited work to do on the trenches the wire entanglements were poor and required work to thicken them into a greater obstacle. Wire was carried into the front line and made into 'footballs' or 'gooseberries':

> Three strands of [barbed] wire were fastened together in the shape of a ball. They were collected together and carried up to the front line on poles at night. The gooseberries were thrown into the main wire entanglements in No Man's Land, in order to make it more difficult for anyone to crawl through. … As I had the only pair of wire-cutting gloves in the platoon, I had to do all the cutting …[8]

These could then be positioned in No man's land at night. Work was also done to improve field telephone communications across this sector by 21/R Fus. There were also efforts to improve fire steps, to clean up old trenches and open disused ones. On 10 January there was a man wounded and on 11 January PS/3299 Lance Corporal Donald Taylor of No 3 Company was killed and PS/3988 Private James Bailey, No 2 Company, was wounded in the scalp.[9] Spurrell recorded that there were only a few casualties which were mostly caused by splinters from trench mortar bombs. There was a lot of wiring work at night which likely caused these losses. As the intensity of German activity increased, through rifle grenades and trench mortars, there was no obvious increase in losses to 21/R Fus and a German sniper's post was claimed to have been destroyed by British artillery fire on 12 January. Three men were wounded. After two nights in Russell Keep Spurrell's platoon moved into the support lines, 'a clammy place devoid of any comfort, most of the shelters consisted of small holes in the sides of the parapet, with a waterproof sheet pegged into the earth by little bits of wood. Some of these palaces held two of us, but as a rule they were only large enough for one.'[10] Here sentries needed to be provided to guard each communication trench (*Boyau*) and men provided carrying and working parties. After one night in support Spurrell's platoon relieved another platoon in the front line:

> The hostile trenches were so close together that the wire met in the middle of No Man's Land. The front trench consisted of all sorts of odd corners an gave the impression that it had been improvised out of the nearest material handy. The Hun line opposite us contained a curious mound, which was reported to contain snipers. … The Huns showered us with rum-jars all day. The trench mortar is a nerve-racking sort of missile, because it takes such a time to come over… Our home in the front line was a tiny little hole under the parapet.[11]

7 TNA: WO 95/2427: 21/R Fus War Diary.
8 IWM: H.W. Spurrell 2138, p.63.
9 PS/3299 Lance Corporal Donald McDonald Taylor, aged 19, from Brynmawr, Breconshire, was buried in Cambrin Churchyard. PS/3988 Private James Edward Bailey, aged 28, from Lower Broughton, Manchester, was a cashier. He had enlisted in 21/R Fus on 20 November 1914 and gone to France in November 1915. He returned to the UK and was discharged due to his wounds in November 1916.
10 IWM: H.W. Spurrell 2138, p.58.
11 IWM: H.W. Spurrell 2138, p.58.

Because of the proximity to the German trenches no work was allowed in No man's land. Sleep was difficult because 'down-time' was disturbed by other changes of sentries and a confined sleeping space. Hour-long stints on sentry were passed by Spurrell and his partner in cooking and eating small meals and occasionally firing their rifles into No man's land to pass the time.

As 21/R Fus grew more assured of the security of their positions they sought to examine the German defences. On the night of 13 January patrols went out at night to check the German wire and an officer-led patrol went around either side of Gibson's Crater but there was no sign of the enemy. On the 15th the Germans shelled Maison Rouge and Arthur's Keep wounding four men. One of those wounded was PS/3017 Sergeant Francis Mills.[12] Another man likely wounded around this time was PS/3204 Lance Corporal John Snowdon.[13] Patrols were curtailed by bright moonlight. On the 16th orders were received for a relief by 2/A&SH though a company and a platoon were to remain manning keeps, due to depleted manpower in the relieving battalion. Spurrell described the Argylls as; '…a very clannish lot of people. They were nearly all coal-miners from Dumbarton and the districts round Glasgow. Their brogue could be cut with a knife…'[14] No 4 Company, commanded by Captain John Betts, including Spurrell and his comrades, garrisoned Wimpole Street and Lewis Keep whilst No 5 Platoon (No 2 Company) manned Arthur's Keep. The remainder of 21/R Fus returned to Annequin South. On the day of this relief PS/4249 Private Frank Taylor was killed.[15]

A Fatigue Party

During this period 19/R Fus provided working parties and endured some shelling around their billets. 20/R Fus also sent working parties forward into the area. An anonymous letter found its way to be published in the Manchester Evening News which recorded what happened after a similar fatigue party had delivered gas cylinders to the trenches:

> At a certain spot the road terminated in a tangle of barbed wire, ruined houses, and broken telegraph posts. Here we branched to the right, across a deep shell hole, or rather a series of shell holes, went down some steps into what had once been a cellar, and, suddenly enough, into the communication trenches. … Finished, we stepped out again along that road, which less than three hours before had been, though certainly a trail of desolation, fairly pleasant in the wintry sunshine. A hundred and fifty yards down, and horror took us by the throat; eight big shell holes in the very centre of the road. Two houses burning furiously …

---

12 PS/3017 Sergeant Francis Richard Mills, from Linden Gardens, London, was a solicitor. He had enlisted with 21/R Fus in October 1914 and had been promoted sergeant in March 1915. He recovered from his wounds and was gazetted to a commission in 13/R Fus in November 1916. He was promoted to Lieutenant in 1918.
13 PS/3204 Lance Corporal John William Snowdon, aged 24, from Winthorpe, Cockfield, Durham, was a colliery surveyor. He enlisted in September 1914 in Durham, appointed lance corporal in May 1915 but was delayed going overseas until 1916. He suffered a wound in the left leg and was evacuated to the UK on 16 January 1916. He was discharged in October 1916.
14 IWM: H.W. Spurrell 2138, p.59.
15 PS/4249 Private Frank Taylor, aged 20, from Middlewich, Cheshire, was buried in Cambrin Churchyard Cemetery Extension.

> One shell had burst in the centre of a mule team ... A motor car with a big piece of high explosive [sic] in the engine lay overturned in the gutter... If we had passed up that road an hour later than we did we should have got it in the neck...[16]

This un-named correspondent was undoubtedly PS/6517 Lance Corporal James Lansdale Hodson who recorded a similar, but less exaggerated, account in his 1918 book, *Soul of a Soldier*. This version records three craters, one shattered house, a car with a broken windscreen and bent bonnet, a grocer's cart and one dead horse.[17] This suggested that the previous account was somewhat exaggerated. In retelling the story in his 1929 novel *Grey Dawn, Red Night* he recorded; 'And on the pavement two huddled, nameless things, wearing equipment, rent in pieces ... foul, unspeakably foul. The sight was an offence. Why weren't they buried, at once, at once, made decent, the place cleansed?'[18] The 19/R Fus war diary recorded on 11 January 1916; 'Enemy dropped a few 5.9" HE [shells] again at Slag Heaps. One of the shells fell on the road, smashing two motor lorries and killing one driver and three civilians, two of whom were children.'[19] Harold Wilson added:

> Breakfast 8. O'clock after which the Hun opened artillery fire on the village ... The shells were falling within a few yards. An ammunition waggon was set on fire, two little French children standing nearby were burned to death. Civilians still live on in this village although not much more than a mile behind the line ...[20]

Another account came from PS/940 Private Robert Simon Mansfield Sturges who recalled a bombardment at this time by heavy German guns on the area of battalion headquarters. A German spotter plane had seen some activity and directed the enemy artillery. Luckily only a few men were killed or injured but one shell scored a direct hit on an ambulance nearby. 'A tangled heap of scrap iron and a broken wheel had been the ambulance. Bits of it had been flung far and wide, fragments of cloth were flapping up in the tree above my head. The driver of the ambulance was simply eliminated; it was only after a long search that they could find enough of him to bury.'[21]

It may have been easier for Hodson to allude to these bodies to be soldiers than children. This whole episode highlights the unreliability of some personal accounts whereby one event might be told entirely differently by different witnesses or by a single witness in three different published works. This demonstrates that Hodson was willing to exaggerate his accounts to suit his audiences but that he did not exaggerate the 'horror' to meet the public demand for 'war books' in the 1920s –1930s. Another casualty in this gruesome tableau may have been Captain Herbert Oakes-Jones of 19/R Fus who was wounded on 11 January.[22]

---

16 Shell Fire Horrors, *Manchester Evening News*, 21 January 1916.
17 Hodson, *Soul of a Soldier*, p.44.
18 Hodson, *Grey Dawn*, p.172.
19 TNA: WO95/2427: 19/R Fus War Diary.
20 IWM: H. Wilson Papers, 3606.
21 Sturges, *On the Remainder of Our Front*, pp.70–71.
22 TNA: WO95/2424 and 2425: 98th Brigade War Diary.

19/R Fus 9–24 January 1916

Meanwhile, for much of this tour in support by 19/R Fus the only enemy involvement was the nightly firing of six shrapnel shells at a given point at 9:00 p.m.; these all burst too high and caused no damage. All was well until 15 January when company commanders from 19/R Fus went forward to reconnoitre the trenches at Cambrin. Next day, 19/R Fus moved into Z0 Subsector and the relief was complete by 7:30 p.m. This trench sector was from Mud Trench to *Boyau 9*. Just to the right of Mud trench was the infamous Hohenzollern Redoubt. The condition of No man's land was reported as good with few craters making it relatively easy to cross for an attack; the German barbed wire was rusty but was likely a suitable obstacle. The British trenches were generally in good condition. The support line was renamed Old Boots Trench from Hamilton Street around this time but this, Guys Alley and Reserve Trench were in poor condition and a company of the divisional pioneer battalion (18/Mx) was to improve them. The Brigade sapping platoon was to work on the Reserve Line. Meanwhile work was conducted to improve the British wire using 'footballs'.

On 17 January there was more activity when German shelling from whizz-bangs caused British retaliation. There was also activity by British MGs and snipers from both sides. At some stage during 17 January PS/94 Private Arthur Blackford was killed and PS/1183 Lance Corporal Edward Mecey was slightly wounded in the head on the 16th.[23] The Germans were generally quiet; the war diary recorded: 'Nothing unusual to report except a most unusual inactivity on the part of the enemy. No working parties were observed, no Very Lights were sent up and the most complete silence prevailed.'[24] This would seem to be the live-and-let-live system of trench warfare in action with neither side desiring to agitate the other to maintain their own seemingly comfortable situation. However, it soon became apparent that the Germans were working hard to improve their positions opposite 19/R Fus; digging saps and strengthening their wire. German wiring attracted the interests of the battalion scouts, and a patrol was decided on to investigate enemy preparations on the night of 18 January. Robert Sturges was approached by one of the battalion scouts and warned that he would be needed to go out on patrol at 2:00 a.m. that night to examine the German barbed wire. He remembered that; 'My blood turned to water and my tongue clave to the roof of my mouth…'; the wire was close to the enemy trenches and there was a full moon. He found it hard to sleep prior to going out as his brain worked hard imagining possible scenarios and worries. At 2:00 a.m.; 'Then I and the other man climbed over the parapet and crawled forward. Almost at once all feeling of nervousness left me. Once we got started, I felt excited of course, but not at all in a funk as I had expected … We got through our own wire easily – it was in a very bad state…' They crawled towards the German positions which were about 130 yards away. Luckily cloud obscured the bright moon. Their movement was cautious, slow and laborious and they passed several dead bodies. Occasionally star shells were fired which required the two men to lie still until the bright light dissipated. A few stray bullets

---

23 PS/94 Private Arthur James Blackford, aged 21, was from Peasenhall, Suffolk. He was educated at The College, Clevedon and Somerset and Bristol University; he intended to become an engineer. He had represented B Company at football. He was buried at Cambrin Churchyard. PS/1183 Lance Corporal Edward John Mecey, aged 23, from Thatcham, was educated at Reading School and was a law student. He enlisted on 15 November 1914 and was appointed lance corporal in July 1915. Though wounded on 16 January he was back at duty on 3 February.

24 TNA: WO 95/2427: 19/R Fus War Diary.

also flew nearby. They crawled for a long period but became lost having not had the trenches or terrain explained to them or without having seen a map. According to Sturges; 'I and the other fellow had a bit of an argument out there in the middle, as to which side the rifle fire on our left came from. He thought it was German, and I thought it was British. However, he was in charge, so we made towards it. But after crawling for some time we never seemed to get to the trenches. So then we had another argument...' The two men then found they were quite close to a German MG that had started firing; '...being under machine gun fire, even at night, when you are in the open gives you a feeling of utter helplessness and insecurity. To listen to the smack, smack of the bullets from the bottom of a trench is a different thing altogether.' They were also overdue returning to their own lines by 30 minutes and they reluctantly decided to turn back. Their route back was far quicker. Sturges: '… heaved a sigh of relief when I dropped over our parapet again and was once more safe back in the trenches … from the greeting accorded to us by everyone we met, we might not have seen them for several years. Our half-hour's overtime had worried them a bit, I think.'[25]

The state of the British wire, that Sturges observed, suggested that more work was needed to make their positions stronger. According to Philip Spaull on 20 January; 'Things are fairly quiet for the infantry. We are, for the most part, occupied with sand bagging and general trench repairs. This helps to pass the time away and is useful work.'[26] Spaull could also observe the state of No man's land beyond the trenches which might highlight the folly of active operations; 'Our front shows signs of the last advance (25th September last). Many of the dead are still out in front as we are unable to bring them in … They are for the most part in extended order as they advanced, some wearing gas helmets.'[27] On 20 January No 2 Platoon of the divisional cyclist company was attached to 19/R Fus for instruction for two days. The platoon commander was Second Lieutenant Harold Collis who had formerly been with 19/R Fus.[28]

On the night of 22 January, a pair of volunteers from No 2 Company, 19/R Fus, went on a patrol to examine the German barbed wire. In the process PS/6330 Private Clifford Calcutt was shot. Robert Sturges, of the same company recalled returning from a ration fatigue and hearing that a British patrol was out in No man's land. Soon after an NCO informed Sturges that one of the men on the patrol had been shot and men went out to bring him in. Eventually a body was carried back into the British trenches. According to Sturges; 'It was the man who had volunteered to go instead of me. He was shot under the arm and died almost at once … it isn't very satisfactory to escape at another man's expense … I'm feeling a bit sick about it. His is the first death in our platoon, quite a nice fellow too and brave as they make them.'[29]

Another man, PS/4357 Private Carl Priest, died of his wounds in Calais on this day; he was presumably mortally wounded a few days before.[30] On 23 January, efforts by British 4.5-inch

---

25 Sturges, *Remainder of Our Front*, pp.99–101.
26 IWM: P.F.F. Spaull papers, 15453, 20 January 1916.
27 IWM: P.F.F. Spaull papers, 15453, 20 January 1916.
28 Second Lieutenant Harold Douglas Collis was gazetted into 19/R Fus on 15 April 1915 and later transferred to 33rd Division Cyclist Company. On 10 May the company joined XI Corps Cyclist Battalion.
29 PS/6330 Private Clifford Percival Calcutt, aged 18, from London, was educated at Manchester Grammar School. He was buried in Cambrin Churchyard. Sturges, *Remainder of Our Front*, p.109.
30 PS/4357 Private Carl Rhodes Priest, aged 29, from Selsey, Sussex, was educated at St Neot's School and Magdalen College Oxford and was a teacher by profession. He enlisted on 4 January 1915 with 19/R Fus. He died in hospital and was buried in Calais Southern Cemetery.

Howitzers against the new German saps were recorded and most of the shells fell near the German trenches and blew away much of his barbed wire. However, three in ten of the fifty shells fired were duds. Robert Sturges' platoon sergeant had a lucky escape during German shelling 'while he was away … the Germans started whizz-banging us … when he returned to his shelter, he found it had disappeared. In its place there is… a heap of earth, under which is his pack, haversack, etc … It's rather discouraging to think that these shelters crumple up so wholeheartedly under the attentions of a whizz-bang.'[31]

On 24 January 19/R Fus was to be relieved by 16/KRRC and return to Billets in Béthune at the Ecole Michelet. Harold Wilson recalled:

> 10 a.m. the air was alive with shells – Enemy started the shelling & our guns got busy & trebled the strafing. Heaps of the Hun shells were duds … Burgon & Bassinet [sic] killed by a shell exploding in their dug-out in which they were sitting. Spohr & Hadfield were blown out of the dug-out. Cecil had the nasty job of digging Burgon & Bassinet out of the debris…[32]

According to the war diary one man was killed and one wounded but both PS/6750 Private William Burgon and PS/7202 Private Thomas Bassnett lost their lives.[33] Private Sturges reported three men in his company wounded, one severely in the lung, by a rifle grenade. Colonel Wolrige Gordon waited for the last platoon to be relieved and walked back with them and carried one of his men's rifles to relieve their load. When the party grew tired, from walking along the Béthune road, the colonel used his rank to flag down two passing lorries to allow the platoon to travel back in style. This was not a long rest though baths were enjoyed at the *Ecole de Jeunes Filles* on 25 January.

On 26 January Private Terence Doherty departed for the UK to take up a place at the Royal Military College Sandhurst. Coincidentally, Harold Wilson, of the same platoon, wrote in his diary around 4:00 p.m. on 26 January 1916; 'Dogwheels [Doherty], the lucky devil, just off to station. He had worked things beautifully & got nominated for Sandhurst & was leaving right away for Blighty.'[34] Whilst his time with the UPS was over, he later referred to an ominous conversation which he had at the RMC; 'The night before last the Assistant Commandant invited me in, as he wanted to know all about the U.P.S. He and others are agitating for the withdrawal of all Public School men, as he thinks it is a waste of good men, keeping them out as privates. He is awfully nice and he is very popular indeed.'[35] The need for suitable officer candidates was growing, and, because of similar conversations, the days of the UPS were numbered.

---

31 Sturges, *Remainder of Our Front*, p.108.
32 IWM: H. Wilson Papers, 3606.
33 PS/6750 Private William Burgon was aged 19 and from Falkland, Fife. PS/7202 Thomas Bassnett, aged 19, was from Chorlton-on-Medlock. He had previously been an errand boy. Both were buried in Cambrin Churchyard. PS/6078 Private Jack Hadfield, from Plymouth, died of his wounds in November 1916 whilst serving with 17/R Fus. Private Ernest Spohr, aged 23, survived the war but was wounded three times and was a sergeant by 1919.
34 IWM: H. Wilson Papers, 3606, entry 26 January 1916.
35 IWM: T.O'C. Doherty Papers, 12961, p.62.

21/R Fus at Rest 17–30 January

Cleaning up took place on 17 January but next day fatigues started in support of the RE. Those not so employed were tasked to improve dug outs. The recent draft, now totalling thirty men, was inspected by the brigade commander.[36] One man was wounded on 18 January. The next few days were quiet with further work on dugouts and shelters. Baths were provided for half of the battalion on 20 January. A further draft of thirty men arrived on 21 January and were initially inspected by Major Fenn and were later inspected by the Brigade Commander on 27 January. On 23 January the Battalion moved to billets in *Rue D'Aire*, Béthune, except for the detached company and platoon. Shelling of the La Bassée–Béthune Road delayed this move and killed PS/4065 Private James Gresty and slightly wounded Second Lieutenant Kenneth Horton.[37] On 24 January No 4 Company and No 5 Platoon were relieved by a company of the Worcestershire Regiment and a platoon of the Queen's Regiment and marched to Béthune to rejoin the battalion. The MG section was also relieved.

On 26 January both 19/R Fus and 21/R Fus were placed at high readiness of an hour notice to move; later in the day two MGs and crews of 21/R Fus were attached to HQ 98th Brigade. This was in anticipation of German offensive operations to mark the Kaiser's birthday on 27 January. High readiness was tiresome as troops were confined to the environs of their billets rather than being free to explore the delights of Béthune. It did relieve the troops from having to conduct works parties and fatigues. 19/R Fus did not stay at high readiness for long as at 9:57 p.m. they were ordered to Annequin Fosse from where they could be deployed into action in the event of the Germans 'celebrating' too vigorously. Donald Wright wrote home around 27 January; 'We had great excitement last night & had to leave our billet (a school) in a fairly big town [Béthune] & march up to this village close behind the line during the night. I think they thought the Gers [Germans] were going to try to celebrate the Kaiser's birthday in some way.'[38] Their new billets on 27 January were not nice; 'This is a filthy place where we are now & the rats swarm the cellar where six of us sleep – but they don't trouble us much now, as we are used to them & when tired you can sleep anywhere although straw is better than bricks…'[39] They were attached to 100th Brigade though the 27th passed quietly with no need for 19/R Fus to move. 21/R Fus remained in Béthune.

The period of rest for 21/R Fus came to an end on 28 January when the CO, company commanders and MGO proceeded to B1 Subsection on reconnaissance. During the stay in Béthune the 33rd Division concert party put on a show in the theatre titled 'The Babes in the Wood':

> Now a pantomime put on by a divisional troupe is at best a compromise. The troupe, consisting of, perhaps, two funny men, a couple of people who can sing and not act and

---

36 This presumably included the 24 December 1915 and 13 January 1916 drafts.
37 PS/4065 Private James Gresty, aged 25, from Manchester was buried in Cambrin Churchyard. Second Lieutenant Kenneth Charles Horton, aged 21, was an articled clerk from Derby; he had enlisted on 15 September 1914 in 21/R Fus as PS/2791. By October 1914 he was a sergeant, and he was gazetted to a commission in 21/R Fus in March 1915. After being wounded he recovered and served with the RE Special Brigade and J Special Company RE; he was awarded the MC and left the Army as a major.
38 IWM: S.D. Wright Papers, 17675.
39 IWM: S.D. Wright Papers, 17675.

half a dozen more who can do neither, but think they can do both, is an awkward thing to manage. The really funny men are, of course, quite in their element, but songs have to be put in for the singers and the plot arranged to suit. All the others have to be dragged in somehow; otherwise they will sulk.[40]

However; 'The show had lasted for two hours; as it had kept us in fits of laughter the whole time, we felt that the actors had justified their existence.'

Meanwhile, 19/R Fus had been aware of German shelling near Annequin which was aimed at a British artillery battery near Annequin Fosse. This shelling started in the morning, continued during the afternoon, and comprised small and large shrapnel and HE shells and a few 5.9-inch shells. According to Jim Peacock, a stretcher bearer, he and others rescued a man wounded by shellfire in a street in Annequin. They bandaged him in a nearby shop:

> … a second shell burst in another building just across the road and Pepperday was badly wounded in the neck. When we got to him blood was pulsating from an artery too deep-seated for us to compress. In spite of our efforts we could not arrest the bleeding and after he cried, "Oh my God" and "bloody hell" he quickly subsided and died in my arms.[41]

The shelling that mortally wounded PS/765 Lance Corporal Gerald Pepperday struck outside a billet; another six men were wounded, one fatally.[42]

On 29 January 19/R Fus moved into A1 Subsection, the Brickstacks between Cambrin and Cuinchy, and relieved 20/R Fus. The relief took from 4:30 p.m. to 7:00 p.m. No 1 and No 4 Companies were in the front line; No 2 and No 3 were in support. Though the trenches were in relatively good condition they required further work which 19/R Fus set out to do. The wire needed improving with footballs being built and rolled out, along with screw stakes and concertina wire. Increased activity likely led to more casualties. On 30 January PS/7405 Private William Bailey was killed. There were some narrow escapes; Second Lieutenant Charles Skey, D Company, was advised not to raise his head above the parapet at a certain point and unwisely did so. He was lucky that a sniper's bullet only grazed his nose.[43] On 31 January Harold Wilson recorded: 'Stand-to at 6 o'clock. Snipers fairly active, we discovered the Huns had no barbed wire in front of their trenches [and] several white flags at intervals along their line – nobody knows what they signify. Scoones wounded slightly in the cheek, went to hospital. Harris killed 11.30 by a sniper. The bullet went in sideways & blew the poor fellow's brains out.'[44] It was PS/3648 Private Joseph Harris who died so graphically.[45] In addition PS/3577 Lance Corporal

---

40 IWM: H.W. Spurrell 2138, p.61.
41 PS/7180 Private James Archibald Peacock, from Newcastle-upon-Tyne was gazetted to a commission in the DLI in September 1916. He recounted his memory of Pepperday's death to his brother in 1965. See Basil Peacock, *Tinkers' Mufti* (London: Seeley Service, 1974), p.202.
42 PS/765 Lance Corporal Gerald Alfred George Pepperday, aged 20, was from Rugby. He had attended Rugby School and was a bank clerk. He was buried in Annequin Communal Cemetery.
43 IWM: H. Wilson Papers, 3606.
44 IWM: H. Wilson Papers, 3606.
45 PS/7405 Private William Bailey, aged 21, from Miles Platting, Manchester is buried in Cambrin Churchyard. PS/3648 Private Joseph Harris, aged 28, from Birmingham, is buried in Cambrin Military Cemetery.

Earl Scoones was wounded in the right cheek but he returned a week later; a lucky escape he would not enjoy again.[46] Work continued on the 31st with limited German and British shellfire. On 2 February the companies switched between the front and support lines; one man was wounded in the process. According to Harold Wilson, 'rifle grenades fairly numerous in the afternoon – nobody hurt except "Gertie" who had a narrow escape, her [sic] hand smashed to blazes & a hole in the head. Gertie was delighted at the prospect of returning to Blighty. "Some girl." St [Sergeant] Jones, standing very near Gertie, but only got 'shock' & was a bit deafened.'[47]

The 3rd was also uneventful but on 4 February there were three casualties due to German rifle grenades and trench mortars, both large and small. On the 5th the Germans shelled the la Bassée Road after dark at around the time rations were carried up. The British guns were more active next day. German shelling was generally confined to the Brickstacks where the Middlesex Regiment suffered several casualties. On 6 February the companies swapped over with D Company replacing C Company in the Front Line. In addition, the battalion practiced a gas alert which was communicated to the men without using gas gongs at 11:00 p.m.; within five minutes all men were masked and standing-to. However, the author of the war diary highlighted some difficulties with their masks; 'It was found difficult to see through the mica of the goggles and the glass of the tube helmets and it is doubtful whether in the event of a real alarm there would be time for both.'[48] This might seem that the battalion was well prepared in anti-gas procedures. However, the fact that they were unused to wearing their helmets and were only finding out about such problems in the front line would suggest they might have fared badly in a real gas attack. Casualties were not referred to during this period. However, one fatality during this period was PS/7184 Private Reginald Lightfoot who was shot through the head by a sniper whilst on sentry duty on 7 February.[49] Also on 7 February a draft of twenty reinforcements arrived; these were likely to replace the losses from 2 January. According to Harold Wilson; 'New draft arrived up in the trenches for the first time. They are a crowd of wash-outs. St[Sergeant] Roberts & Corpl Fenton two fine specimens. All the men very sick that the draft N.C.O.s retain their stripes.'[50]

The next few days were relatively quiet with a thrilling but inconclusive aerial combat on the eighth and the next day there was some shelling by the Germans against the Brickstacks which came close to 19/R Fus positions; one of Wilson's comrades, 'Old Murg', panicked and ran away from the shelling.[51] Several of Wilson's pals had also gone sick or on courses. Sergeant 'Jonah'

---

46   PS/3577 Lance Corporal Earl Foster Scoones, aged 26, from Walworth, was educated at Borough Road Training College for teachers. He enlisted in November 1914 and was appointed lance corporal in October 1915.
47   Sergeant 'Jonah' Jones as likely PS/6482 Sergeant William Edgar Jones, aged 27. He was later gazetted into the R Warwicks in July 1916 and was killed in action on 10 April 1918 with 10/R Warwicks. IWM: H. Wilson Papers, 3606.
48   TNA: WO 95/2427: 19/R Fus War Diary.
49   PS/7184 Private Reginald Lightfoot, aged 18, from Crewe, worked in the electrical department of Crewe Works. He was buried in Cambrin Military Cemetery.
50   IWM: H. Wilson Papers, 3606.
51   This was likely PS/7580 Private Joseph Murgatroyd, aged 37, from Ripon whose trade was described as 'gentleman'. He still went to an OCB but returned to the Royal Fusiliers, presumably having failed the course. He later served with an agricultural company, Labour Corps.

Jones, his section commander, looked ill but refused to go sick; his work was done by another man in the platoon. PS/7363 Private Harry Pointon was wounded in the chest and suffered a severe haemorrhage which proved fatal. He died of his wounds on 10 February.[52]

On 10 February a company of 7/Leinsters (16th (Irish) Division) was attached for 24 hours to 19/R Fus for training in trench warfare. Their arrival allowed D Company to be relieved and return to billets in Annequin. The sound of a German mine near Givenchy may have been heard by members of 19/R Fus. 180th Tunnelling Company RE experienced a German mine being blown which entombed three miners below ground. PS/7025 Private Richard Errington Swan was killed on 11 February 1916 whilst attached to 180 Tunnelling Company and was likely trapped underground with two RE sappers the day before.[53]

On 13 February there was heavy shelling by the Germans but no casualties were caused. Later that day, 19/R Fus was relieved, by 1/Queen's RWS, and returned to Béthune where they were billeted in Montmorency Barracks. This reminded some men of the barracks in Tidworth. The 14th was spent in tidying-up but the troops were not let out into Béthune until after 2:00 p.m. Harold Wilson celebrated another rest period with a good feed on the 14th and a shopping trip around the town. The narrative must now switch back to 21/R Fus's activities in early February.

Farewell Otto!

With the potential excitement of the Kaiser's birthday over, 21/R Fus prepared to go back into the line. On 30 January 21/R Fus marched towards the trenches and relieved 18/R Fus at 4:30 p.m. in B1 Subsection; relief was complete by 6:30 p.m. The battalion was now commanded by Major Fenn whilst Lieutenant Colonel Stuart-Wortley went to England on ten days leave. Spurrell was less complimentary:

> Before we went up [to the trenches], however, we heard that Otto had been entrusted with the care of an important dispatch, which he was to deliver to the War Office in person. After completion of this exhausting duty, he was to be given indefinite leave. In other words, Otto, had got the order of the boot. As he had not yet been in the front line, it was about time for such a denouement.[54]

There were some convoluted reasons for Stuart-Wortley's removal. Major Phipps Stanley had been a very efficient, popular, and well-respected officer during the early training of 21/R Fus but with time his performance had deteriorated. On 19 January two letters were forwarded to HQ 98th Brigade; one from Lieutenant Colonel Stuart-Wortley and another from the Medical Officer for 21/R Fus, Captain Leslie Howlett.[55] Howlett stated; 'I have had Major Stanley under my observation during the time I have been Medical Officer to this battalion knowing from Colonel

---

52 PS/7363 Private Harry Pointon, aged 18, from Manchester, had worked in the offices of a bleach company pre-war. He succumbed to his wounds and is buried in Béthune Town Cemetery.
53 PS/7025 Private Richard Errington Swan, aged 28, from Gateshead, was a drapers shop assistant. He has no known grave and is commemorated on the Loos Memorial.
54 IWM: H.W. Spurrell 2138, p.62.
55 Captain Leslie William Howlett, RAMC, had served with another unit before 21/R Fus having arrived in France on 2 December 1914. He was from Halstead, Essex.

Stuart-Wortley that he [Stanley] was addicted to alcoholism.'[56] Howlett visited Stanley in his dugout and found him suffering from a chilled foot; Stanley was ordered to stay in his dugout to rest. The next morning Stanley had evidently been drinking to excess and Howlett suggested he go sick. After formally examining him Howlett judged Stanley to be medically unfit to command due to his addiction. Stuart-Wortley's letter reported that before August 1915 Stanley's drinking was under control but that: 'During August, September & October he was hardly ever in a fit state to appear on parade, he was usually suffering from chills on the stomach & general nervous breakdown.' Though these issues were temporarily addressed he was unable to conduct his duties in command of a company and he had also become 'unclean in his habits, he hardly ever changes his clothing or has a bath or washed himself …' He was admitted to hospital for trench foot and was also suffering from neurasthenia. Another medical officer recorded: '… I do not consider that any Medical Officer, ignorant of his previous history, could attribute his present condition to alcoholism.' Though Stuart-Wortley tried to back-pedal in a subsequent letter the damage was done to both his and Stanley's reputations. It was Stuart-Wortley's handling of the affair with Stanley, his perceived *laissez-faire* attitude to his subordinate's failings, and contradictory communications, that led to this downfall. Brigadier General Strickland considered:

> The case of Major Stanley has given rise to much doubt as to the power of command and realisation of discipline displayed by Lieut. Col. Stuart-Wortley. … He has no great experience of dealing with a force such as a Battalion… I cannot disguise the fact that an officer of recent considerable experience with a Regular Battalion would be in a position to obtain better results…[57]

That Strickland only realised Stuart-Wortley's failings after several months suggests poor judgement on his part. Even a lowly soldier in 21/R Fus was aware that Strickland disapproved of Stuart-Wortley, 'our old colonel (Stuart-Wortly [sic]) having left us … Rumour had it that he was forced to relinquish his command on the ground of being an incompetent officer. It was quite certain that he never got on well with the Brigadier General [Strickland], who very often, and in our hearing, used to say very 'uncomplimentary things' about him.'[58]

Stuart-Wortley was replaced by Lieutenant Colonel E.B. Denison of the KRRC who amply fitted the description of; '…an officer of recent considerable experience with a Regular Battalion…'. He was described as 'wonderfully energetic, a strict disciplinarian and a believer in hard work. He looked every inch a soldier and the battalion soon began to improve under his guidance.'[59]

21/R Fus in the line – February 1916

Due to previous work to improve the trenches, and the lowering of water in the canal, the trenches were considerably drier than during previous stays in this sector. However, trenches on

---

56 TNA: WO339/5206: P.E. Stanley Officer File.
57 TNA: WO339/5206: P.E. Stanley Officer File.
58 FM: P.N. Wright Diary, RFM.2013.8.1/2, p.24.
59 IWM: H.W. Spurrell 2138, p.62.

the left of the line were still bad. Work was required to improve communication trenches, build up the defences for the gap in the front line (caused by the canal) and better drain the trenches. Occupying this sector was still described by Spurrell as 'a clammy and unpleasant period. … Quite a lot of excitement was crowded into a short time …' Spurrell recalled four days in reserve at Pont Fixe which was 800 yards behind the front line, and over the ridge line, so troops could move around out of cover. They spent plenty of time working on fire bays in the reserve line but when well-managed such labour was not without perks; 'We worked under a good sergeant, who took an individual interest in his men and knew how much to give every man as a fair task. He was always willing to let us knock off a little bit early, if we had put our backs into the task and got through it rapidly.'[60]

Capture the flag – 21/R Fus in No man's land

It could be presumed that there would be little time for anything but labour in the Front Line but on 31 January it was reported that the enemy had planted a German flag in No man's land opposite the right-hand company (probably No. 1 Company) which was holding the line down to the canal. This might be considered a challenge to British mastery of No man's land; the longer the flag was in place the greater German confidence might become whilst degrading the morale of British sentries who would see it through their periscopes or mirrors. A pessimist might see it as a 'come-on' to draw British patrols into a trap.

Major Fenn was in command at this time. According to the War Diary, on the night of 31 January two patrols were despatched into No man's land. The first was an officer's patrol under Captain Richard Whittington consisting of nine NCOs and men. They were to examine the German wire. The second patrol of a sergeant and one man, were to remove the flag. The former patrol lost their way and was surprised by the enemy whilst negotiating their barbed wire defences; Whittington and two men were captured. According to Captain Whittington:

> … on the night of Jan 31st [19]16 I was sent out to patrol the enemy's wire and to discover if possible how strongly their line was held. I had with me nine NCOs & men. Orders were given that no patrols were to be sent out from our lines until I had returned, nor were any lights to be put up. I obtained the information required and was returning as I thought to be our trenches when lights were put up in front of me, this continued as I advanced. Knowing what orders had been given, I thought I had lost direction. I consulted with my NCOs and decided to look for either a road which was on the left of our position or a ditch which ran out from our right. At this time lights were going up on every side. I proceeded to get through the wire, which was extremely thick, when, in the middle I again consulted with my men who thought we were in our own lines. I was the first to get through the wire followed by L/C [Lance Corporal] Bowen, upon seeing the trench I recognised it as being German and saw a number of the enemy waiting for us. I asked L/C Bowen to stay with me and prepare to bomb the trench. We laid down on the parapet and L/C Bowen threatened the Germans immediately in front of us with a bomb. I then gave the orders to the remaining men who were still getting through the wire to return to our trenches.

60   IWM: H.W. Spurrell 2138, p.63.

The enemy on our flanks fired at us without result but those in front of us were prevented from doing so by L/C Bowen. We remained in this position until the rest of the patrol had got clear when we had to give ourselves up. I discovered afterwards that a patrol had been sent out from the company on our right, they had lost direction and made the same mistake as myself. It was this company on our right who were responsible for putting up the light which caused us to lose touch.

When captured I had no letters or papers on me so the enemy were unable to gain any information.[61]

Captured with Whittington was PS/2465 Lance Corporal John Bowen, also of No 3 Company. Three other men, likely of the same patrol, were reported as captured on 1 February 1916 suggesting they hid for a period in No man's land before being taken. They were PS/7076 Lance Sergeant Francis Bird, PS/6296 Lance Corporal John Taggart and PS/7193 Private Max Goodman.[62] PS/2874 Sergeant Leslie Jones led the six remaining men of Whittington's patrol back to British lines and reported on events. He was MiD on 15 June 1916, presumably for this action, as was Lance Corporal Taggart who was gazetted in

Studio portrait of Lieutenant Richard Whittington. As a captain he commanded No. 3 Company, 21/R Fus. Whilst he showed his bravery and daring by leading patrols this was not normally the role of a company commander. (Author)

---

61 TNA: WO339/63530: Captain R.H. Whittington Service File.
62 PS/7076 Lance Sergeant Francis Edward Bird, from Worcester, was captured and held at Munster and Quedlinburg. PS/2465 Lance Corporal John Victor Bowen, aged 33, was born in October 1882 in Maestaf and lived at Merthyr. He was a fitter in a steel works. He was captured on 1 February 1916 and was held in Munster whilst a PoW and was later discharged to the Class Z Reserve in December 1919. He likely hit hard times in the 1930s as in 1938 he had not paid his Old 21st Association subscriptions for 5 years. PS/6296 Lance Corporal John Percival Taggart was born in November 1889 in Liverpool and lived in Tottenham, London. He too was captured on 1 February 1916. Taggart was discharged to the Class Z Reserve in February 1919. PS/7193 Private Max Goodman, aged 26, was from Pontypridd. He was a graduate of Cardiff University in modern languages and was a good German scholar. He had previously served with 29/R Fus, He was captured at Givenchy with Whittington on 31 January 1916.

absentia.⁶³ The five captured men, according to Private Goodman, were held at Lille for a week before travelling to Munster:

> I was not allowed to write from here [Lille]. I saw no cruelty, and was well treated, and have nothing to say of the few days I was there. I was 50 hours in a cattle truck, which was dirty, at Brussels. We got coffee at Cologne. We got a good dinner in the station restaurant… Then we were put in a decent train…⁶⁴

On 31 January 1916 PS/4224 Private Robert Sanderson died of wounds in Béthune.⁶⁵ Lieutenant Wilfred Holland took over No 3 Company after Whittington.

Patrolling had grown more vigorous on both sides. According to Philip Wright:

> It happened that one night the Germans had the cheek to put up a flag on their parapet. The next night two of our fellows determined to get hold of it and set out with that idea in view. By the following morning they had not returned, but the flag had disappeared, as it was assumed that the two fellows had got it down but had been taken prisoners on some Hun sentry giving the alarm … The night after we lost a captain (Whittington, 3 Coy.) who rather foolishly took part in a raid.⁶⁶

Later; 'the Germans put up a board on their parapet with the legend – "Further comrades made welcome"! They also put one up to say that one fellow taken prisoner had died of wounds, the other was still well…'⁶⁷ The identities of the two soldiers caught capturing the flag are unknown, nor the mortally wounded man.

However, these setbacks did not dampen the ardour for patrols. On 4 February Spurrell took part in his first patrol. They observed the ground they would cross from some high ground the day before:

> When the hour arrived, we crawled out from the end of the sap in diamond formation. I was in the centre … and had to follow close behind the sergeant. We had very precise orders about firing; fortunately, I had little cause for anxiety on this point, as common sense told me that I could not fire without hitting one of our own party. An officer, Davis, and Sergeant Hassell were in charge … at the first emergency Hassell took command. In the first hour we advanced not more than 100 yards. After that, progress became even slower. The ground was absolutely sodden. One of the greatest troubles was the difficulty of

---

63 PS/2874 Sergeant Leslie Norman Jones, aged 24, and was from Chester. He was a merchant and manufacturer or bricks and tiles. He enlisted in September and joined No 3 Company; he was quickly promoted to sergeant by March 1915 and shortly after arriving in France attended a course in trench mortars. He left the battalion on 14 March 1916 and joined No 6 OCB. He was gazetted into the RE in July 1916.
64 TNA: WO161/100/579/1811: POW interview of Private Goodman.
65 PS/4224 Private Robert Sanderson, aged 24, from Bredbury, was a clerk with the Northern Assurance Company. He enlisted in November 1914. Sanderson was buried in Béthune Town Cemetery.
66 FM: P.N. Wright Diary, RFM.2013.8.1/2, p.21.
67 FM: P.N. Wright Diary, RFM.2013.8.1/2, p.21.

keeping the muzzles of our rifles from getting clogged with mud. Men were sent forward to scout; these were often away for ten minutes or so. On their return we advanced about ten yards. After nearly four hours of this the excitement began. A voice suddenly growled out in the darkness and the words, though I cannot remember them, were certainly German. One of our men, who could talk German and tried to bluff them, answered. This did not work and a great scrambling noise was heard in front. Davis suddenly jumped up and fired two or three shots with his revolver. He had lost his head completely and quite wrecked Hassell's little plan of working round behind the Bosche and taking them. Before starting Davis had indulged in a lot of hot air about "cold steel" and so on. … his getting up not only made it impossible for us to fire for fear of hitting him, but also made him a good target for the enemy, who hit him in the arm in two places. This finished the whole matter; Davis had to be taken in and Hassell ordered us to go back to the trench.[68]

It had taken the patrol four hours to get 150 yards from their line before this altercation; it took them seconds to get back under the threat of German MG fire. According to the War Diary this patrol went out from the right of the line. They 'encountered 3 Germans who unfortunately escaped: they were pursued and 2Lt Davis was wounded. The patrol were forced by fire to give up pursuit within a few yards of enemy wire and returned without any further casualties.'[69]

According to Philip Wright:

A patrol of our men encountered a Bosche patrol in No Man's Land, and as the Huns did not see our fellows, it would have been a comparatively easy matter to take them prisoners. The officer in charge of our patrol rather thoughtlessly discharged his revolver at the German officer, missed and was himself 'plugged' in the wrist. After this 'exchange of compliments' each side made for its own lines at top speed not wishing for any further civilities.[70]

Davis was evacuated for his wound to be treated.[71]

On 5 February shellfire blew in the parapet where Death or Glory Sap joined the Front Line. Two machine gunners were wounded and trapped beyond this blockage in the trench. Knowing that one, Corporal Nielson, was wounded badly, PS/3259 Private Charles Smith, a stretcher bearer, cut and forced his way through this blockage. Smith treated Neilson and carried him out whilst exposed to enemy fire.[72] For this Smith was awarded a DCM in June 1916. Neilson died an hour after being rescued, despite his efforts. Also killed was PS/6258 Private George Adams and PS/3010 Private William Mann was wounded.[73] No 2 Company, in

---

68 PS/2813 Sergeant Reginald Talbot Clements Hassell, aged 39, from near Blyth, Northumberland, was a ship broker. He had formerly served in the Boer War with the 14th Squadron, 5th Regiment Imperial Yeomanry (1900–1901). He enlisted in September 1914 and was a sergeant by February 1915. IWM: H.W. Spurrell 2138, p.64.
69 TNA: WO 95/2427: 21/R Fus War Diary.
70 FM: P.N. Wright Diary, RFM.2013.8.1/2, p.21.
71 Second Lieutenant Lancelot Samuel Davis, aged 25, from Spondon, was a hosiery manufacturer. He enlisted in the UPS in September 1914 and was gazetted in January 1915. After being evacuated wounded he served with 105th Training Reserve Battalion and joined the RFC in December 1917. He was injured in an aero accident in October 1918.
72 PS/3027 Corporal Nathaniel Clarke Neilson, aged 30, was from Durham.
73 PS/6258 Private George Percy Adams, aged 22, from Leicester, was buried with Neilson in Cambrin

A sketch of a Machine Gun post in Death or Glory Sap (dated 3rd February 1916) where Private Smith, a stretcher bearer with 21/R Fus, earned his DCM. PS/3259 Private Charles Hermon Smith, aged 35, from Haverbreaks, Lancaster, was educated at Lancaster Royal Grammar School. His DCM citation read: 'For conspicuous gallantry. He crawled out, in full view of the enemy, to attend a severely wounded corporal, and later succeeded in bringing him in on his back.' Smith was later gazetted into the MGC in September 1916 and was captured, whilst serving with 19th Battalion MGC, on 24 March 1918.

support, were to repair the sap and a working party under PS/2598 Sergeant Roy Dyer started work during daylight. The Germans detected their work and engaged them with whizz-bangs, MG and sniper fire and they withdrew with Sergeant Dyer having been badly shaken; the job was completed that night. On the night of 7 February, Captain Holland, the newly promoted commander of No 3 Company, was anxious that the Germans were going to raid his front-line trench; No 2 Company reinforced his positions against this eventuality.[74]

These recent patrols were an attempt to wrest control of No man's land from the enemy and were out of character for the battalion. Up to this point there had been little patrolling reported. This may have been a deliberate effort by Major Fenn to show both the battalion, and possibly himself, in a positive light during his caretaking period in command. The next notable event was a planned raid on the German trenches planned for the night of 8–9 February. This was conditional on the British artillery suitably clearing German barbed wire from the site of the planned raid; a feature known as 'The Tortoise'. The arrival of a new commanding officer, Lieutenant Colonel Denison, on 7 February did not alter these plans. Before the raid no obvious effect on the wire had been noticed but British MGs were emplaced to dissuade German working parties from conducting repairs. On the night of the raid the following was recorded:

> Raiding party which went out last night at 11.30 p.m. from Death or Glory Sap returned at 3.20 a.m. this morning: no casualties. Their scouts reported that the wire was not very badly damaged by our artillery but that an enemy working party was engaged upon it: sentries were posted on enemy parapet & were also patrolling behind their wire so that entrance to the trenches was impossible. The party waited for some time & then Sergt. Hassell, who was in command sent some of them back, while he and four others went forward & dispersed the working party with bombs.[75]

The raiders consisted of 35 men each with assigned roles; Spurrell was to clear wire whilst his friend, Thomas Turton, was to kill Germans once entry was made. According to Spurrell the raid was 'more ambitious in the planning than in the event'. The wire was uncut and could not have been penetrated whilst maintaining surprise. However, Spurrell recorded of Hassell's party of bombers; 'A few bombers stayed behind and threw several Mills' bombs into the [German] trench. Judging by the noise that followed, we concluded that some of them had found their mark.'[76] Hassell was later awarded the DCM in June 1916 with the following citation; 'For conspicuous gallantry and good work when in command of patrols. On one occasion, when leading a raid, he ran into an enemy working party, which he dispersed with bombs, and eventually withdrew under heavy fire with no casualties.'[77]

    Military Cemetery. PS/3010 William Edgar Mann aged 20, was born in Madras and lived in Llanfanfechan. He was educated at Kings College Taunton, where he was a member of the OTC, and was a motor and electrical engineer. He enlisted in 21/R Fus in September 1914, presumably serving in the MG Section, and was wounded on 5 February 1916; he presumably remained at duty and on 29 April joined 24/R Fus. He was discharged due to wounds in June 1917.
74 FM: P.N. Wright Diary, RFM.2013.8.1/2, p.22.
75 TNA: WO 95/2427: 21/R Fus War Diary.
76 IWM: H.W. Spurrell 2138, p.65.
77 Hassell left 21/R Fus in the first tranche of officer aspirants on 15 March 1916 and was gazetted to a commission in 13/S Lancs in July 1916. He later served with No 1 Special Company RE, was

The raiders looked forward to a lie-in the next morning but their billets at Pont Fixe were periodically shelled from 8:00 a.m. by a German nine-inch gun which meant they had to spend their time in nearby trenches rather than in their beds:

> The first of them burst outside the company office and compelled the sergeant major to retire to the nearest cellar. Another burst at the point where the sentry was posted and killed him. A third came across the road and burst in the yard just outside our billet. So did the next; this time a shower of splinters came flying in through the door.[78]

The sentry was likely PS/6278 Private Brindley Howells who was reportedly killed on 8 February.[79] On 10 February a company of 7/Leinsters was attached to 21/R Fus for instruction; they took over a sector of the front line and were provided an officer and four sergeants from 21/R Fus to teach them. There were also inter-company reliefs. By this time the Front Line was in a very poor state of repair. Much of the trenches in one section were waterlogged and two posts were established to guard this section; these posts could only be reached at night by travelling overland and were checked by patrols every hour of darkness. During this tour a barrage by British 9.2-inch howitzers was planned against 'The Brickstacks'. The proximity of the respective trenches and the danger of fragmentation meant that the British front line was only manned by sentries whilst this happened. As a result Spurrell, as a sentry, got a grandstand view of this event whilst the rest of his company was safe:

> After a few rounds, the guns got on to some of the brickstacks and made the dust fly merrily. All kinds of things could be seen going up into the air; the aim must have been good, as sandbags were very much in evidence. No splinters came in my direction. The German guns fired on the communication trench on the other side of the canal; I do not suppose they expected a serious attack. In any case their guns did no damage, as there was nobody there.[80]

The Leinster company was relieved on 12 February and next day a reconnaissance party of officers from 16/KRRC appeared to plan the next relief. On Valentine's Day 1916 21/R Fus was relieved by 16/KRRC and withdrew to billets in *Rue d'Aire*, Béthune.

The battalion cleaned itself up again and took advantage of the baths in Béthune. During this rest period the most recent draft, which had arrived on 14 February, was inspected by Colonel Denison and a few days later the acting brigadier (Colonel Rowley, 1/Mx) also inspected them. Denison also inspected the billets and each individual company. Otherwise, fatigues were the order of the day with about 300 men being provided to the RE each day. However, rumours were afoot regarding the future of 21/R Fus whereby the battalion would be sent home and most of the men given commissions. As such there was a lot of optimism about the next spell in the

---

wounded in the toe at Ploegsteert in May 1917, recovered, and was wounded in the lung at Fresnoy in September 1918. He won an MC for both supporting an attack with his mortars and then leading a party of infantrymen to capture several enemy positions. The London Gazette 21 June 1916, p.6144.
78   IWM: H.W. Spurrell Papers, 2138, p.65.
79   PS/6278 Private Brindley Joshua Howells, aged 26, from Llanelly, was buried in Cambrin Churchyard.
80   IWM: H.W. Spurrell Papers, 2138, pp.65–66.

trenches potentially being the last for the battalion. There was still some danger to be endured; Philip Spaull recorded in a letter on 25 February a narrow escape from enemy trench mortar fire whilst on a fatigue party going to the front line to work in a mine.[81]

This relatively uneventful rest period for both 19/R Fus and 21/R Fus came to an end on 21 and 22 February respectively. 19/R Fus went into Z1 Subsector. 21/R Fus went forward to Annequin South and the next night they relieved 2/A&SH in the front line in Z2 Subsection. The weather was very cold with snow over the following days and hard frosts making life difficult.

19/R Fus had some relatively peaceful days with little of note reported, 'The advent of snow and cold winds made the gunners of both sides unenterprising.' By tea-time on 22 February there was four inches of snow. The freezing and desolate conditions were not improved for Wilson's comrades by their sergeant being stingy with the rum ration. There were already rumours flying around about what was happening next. According to Wilson on 23 February; 'Rumours buzzing round that we shall not be coming into the trenches again as a Batt[alion]. Hoorah! Hope for the best. A weekend in the trenches would be ideal, sometimes, in the nice weather – the novelty wears off after a time, though & the cheeriest optimists think to themselves more than they could tell anyone.'[82] Sturges also remembered, 'The story now goes ... that we are to be withdrawn to the base. There we are to become O.T.C., from which batches of men will be taken from time to time for commissions. Everyone is bursting with excitement, as, if it is true it means that this will be our last spell of the trenches...'[83]

On 25 February the Germans sent over some heavy *minenwerfer* shells. The only casualty was one man who was wounded by a sniper. This was likely PS/4335 Private George Hart who died of wounds in Béthune on 26 February 1916.[84] On the 26th 19/R Fus was relieved by 2/A&SH and was billeted in Annequin South. Donald Wright recorded on 28 February 1916, 'The trenches were in an awful mess when the snow all thawed [and] I wish the frost would have lasted for another day or two, but we have all had baths & clean underclothes in the MGS [Machine Gun Section] & feel like lords now.'[85]

Sturges recalled their state as the marched back from the line and alluded to a grudging respect for the commander of 19/R Fus; 'We were still pretty well plastered with trench mud, including the colonel on his horse in front. He had been up in the trenches the previous evening, as usual seeing the last of the battalion out.'[86] On arrival back in camp the rumours were, for once, true and the 19/R Fus war diary recorded; 'The fate we have some time anticipated has now overtaken us, i.e. being turned into an O.T.C.' The next morning 19/R Fus marched to Béthune and got on a train to the next stage of their journey in France.

Though there were no casualties amongst 21/R Fus during their relief on 22 February PS/8349 Private Cyril Thomason, of No 2 Company, was killed on 24 February and another man wounded the next day.[87] On 26 February PS/6872 Private Stanley Pierce was killed and

---

81   IWM: P.F.F. Spaull Papers, 15453, 25 February 1916.
82   IWM: H. Wilson Papers, 3606.
83   Sturges, *Remainder of Our Front*, p.140.
84   PS/4335 Private George James Hart, aged 25, from Brighton, was buried in Béthune Town Cemetery.
85   IWM: S.D. Wright Papers, 17675.
86   Sturges, *Remainder of Our Front*, pp.147–148.
87   PS/8349 Private Cyril Thomason, aged 18, from Runcorn, Cheshire, was buried in Cambrin Churchyard.

two men were wounded; PS/3102 Private William Powell succumbed to his wounds later that day.[88] This must have been a difficult loss bearing in mind the proximity to a potential reprieve for those who wanted commissions and the length of the winter that had already been endured.

Much of this tour was uneventful. According to Spurrell, back guarding Lewis Keep, it was 'short and more or less unexciting'. However, there was a thaw during this tour in the trenches and many trenches collapsed or became mud ridden. After a few days in this Keep, No 4 Company was required to relieve No 2 Company in the front line. The latter had been unable to hold the line for the full period as had been planned due to the poor conditions. According to Spurrell (of No 4 Company):

> We always looked upon this company [No 2 Company] as the "dud" company in the battalion and were very annoyed that they could not stay in the forward line the whole time… we found them holding trenches at the point where the line swings round to the east at the north side of the Loos salient. It was a place of evil repute. In No Man's land were the remains of 17 mine-craters; these showed that the sappers on both sides had been active…[89]

On 27 February the CO and company commanders of 4th Battalion King's (Liverpool) Regiment arrived to plan the next relief and on the 28th 21/R Fus was relieved in the trenches for the last time and moved to trenches in Béthune. This move was 'a somewhat arduous march after the bitter weather we have experienced during the last six days in the trenches'.[90] According to Spurrell; 'It was with a feeling of exhilaration that we tramped off down the La Bassée road for the last time. Good-bye to the trenches of Flanders for many of us. When we came out again, most of us were wanted down on the Somme. Personally, I never went into the line again on the flat Flemish plains.'[91] After a short rest in billets in Béthune 21/R Fus boarded a train for Wardrecques early next morning. Whilst awaiting a railway engine for their train, members of Spurrell's company read some out-of-date French newspapers and first learned of the German attack on Verdun. Though it was the end for three of the UPS battalions the war was reaching a critical phase and momentous events were still to take place.

---

88 PS/6872 Private Stanley Pierce was buried in Cambrin Churchyard. PS/3102 Private William Powell, aged 22, from Buxton, Derbyshire, died of his wounds and was buried in Béthune Town Cemetery.
89 IWM: H.W. Spurrell 2138, p.66.
90 TNA: WO95/2427: 21/R Fus War Diary.
91 IWM: H.W. Spurrell 2138, p.67.

13

19th Brigade in Early 1916

Absolute despairing ruin everywhere

> … *every tree as far as one could see shot to pieces! … Shell holes – broken stone work – torn barbed wire – equipment old and broken – dead Germans – rifles – absolute despairing ruin everywhere …*[1]

Though the activities of 18/R Fus and 20/R Fus with 19th Brigade were similar to their sister units, their stories over the first two months of 1916 were also significantly different. Though for these two months their trench duties were entwined, with successive mutual reliefs, the two battalions deserve separate narratives due to the bifurcation of their existence later in the year. The experience of 20/R Fus will be covered second to allow for a more seamless narrative of this battalion with subsequent chapters.

18/R Fus

New Year's Day found 18/R Fus still in Z2 Subsection with 19/R Fus on the left and 2nd Battalion Royal Welsh Fusiliers (2/RWF) on the right. The battalion had been busy improving and tidying up the trenches with the only activity being minor artillery fire by both sides. Coinciding with the bombardment of 19/R Fus the Germans also shelled the Brigade HQ office. The latter staff moved to avoid the shelling and took over the 18/R Fus battalion HQ which had to make other arrangements. According to E.K. Twiss:

> Everything was quiet until 12.15, and I was working in my room when suddenly 'bang', and a 5.9 [inch shell] pitched in the road just outside. All the glass in my office was broken. The General & I went & took all the others to the cellar, & there we remained, while the Huns pitched 5.9 shells at us for 1 ½ hours at the rate of 1 per minute. Besides this, innumerable 'pip-squeaks'. The first one got about 8 men, one poor beggar having both his legs taken

---

1 *1916 brought into perspective – 1986 – Europeana 1914-1918, Europe – CC BY-SA* <https://www.europeana.eu/item/2020601/https_1914_1918_europeana_eu_contributions_5366> (accessed 11 February 2023).

Photo showing the extent of the damage to buildings in the village of Cuinchy. The surrounding villages like Cambrin, Givenchy, Festubert, and those just behind the lines, would be similarly ruined. (Author)

off. Two of the servants carried him in and there he lay for 1 1/2 hours. ... He must have suffered agonies.[2]

Luckily Twiss' room was at the back and was unaffected by the shelling.

Kenneth Norman, a stretcher bearer with B Company, 18/R Fus, summarised, 'we had a terrific bombardment. Shells seemed to fall everywhere yet we only had a few casualties but a village behind Cambrin suffered very much even soldiers walking casually through the village were killed. Here we had a slight gas attack but our helmets were quite safe ...'[3] The trials of trench warfare were made easier because; 'The comradeship between officers and men being excellent.'[4] As the men gained experience they learned to identify different shells and projectiles going over them and could predict where they might land:

> There were very few bombardments, but there were 'Whizz-Bangs' knocking about all the time. These Whizz-Bangs were fierce; they'd explode before you even heard them coming... they'd usually dip a wee bit and hit the back of the trench. They were pretty lethal and quite frequent. They were the smallest shell and the fiercest. ... The great fifteen-inchers, 'Jack Johnsons' ... a bloody great thing; blew a hole as deep as a bus. ... Some of the heavy shells you could see coming ... some big shells sounded like a tramcar coming,

2 IWM: G.K. Twiss Papers, 17109.
3 LA: K.V. Norman Papers, WW1/GS/1188, p.112, p.114.
4 LA: K.V. Norman Papers, WW1/GS/1188, p.114.

terrifying. … '*Minenwerfers*' were like a great big oil drum, you'd hear them pop and see them going over and over before the dropped… Rifle grenades you could hear the pop and see them coming – the trouble was seeing where they were going to drop.[5]

Having sorted out their trenches 18/R Fus sent out parties at night to improve their barbed wire. Second Lieutenant Henry Cecil Hamilton led one party but was shot in the thigh whilst leaving the trenches on the evening of 3 January.[6] At some stage during this tour Paul Gardner, a Charterhouse-educated twenty-nine year old of B Company, 18/R Fus, was wounded; he died of his injuries on 4 January 1916.[7] This stint in the line ended on 4 January when 20/R Fus relieved 18/R Fus; the exception was D Company 18/R Fus which remained as the reserve company whilst D Company 20/R Fus provided the guard for I Corps HQ at Chocques. The relief was complete by 10:45 a.m. and 18/R Fus moved to billets at Annequin South where baths were provided by companies. Though out of the line, one company of 18/R Fus had to remain at five minutes notice to move during this period. They at least avoided fatigues. Not only were parties sent up the line to work with the Sappers but also carrying parties went to Russell's Keep to transport gas cylinders; other personnel dug further trenches in the lee of their billets to protect from German shellfire.

Whilst at Annequin on 6 January Lieutenant James Anderson RAMC, the MO of 18/R Fus, went sick and was replaced by a Canadian; Lieutenant Hicks RAMC. Other men of the battalion also were sent to the rear; at some point Gordon Jacob of C Company was sent to a base hospital with laryngitis and did not rejoin until later in the year. On 6 and 7 January the CO inspected first C Company, then B Company and the MG Section. That night carrying parties were provided to bring new gas cylinders to Russell's Keep. On 8 January A, B and C Companies marched to Fouquereuil and A Company was inspected the next day. Baths were available for these three companies in Béthune. D Company presumably rejoined on the 10th and were inspected on 11 January. There was time at Fouquereuil for training which took the form of physical drill, bayonet training, platoon drill, musketry, and bombing instruction. Second Lieutenant Francis Bacon left to be the intelligence officer for 19th Brigade.

Further training took place and on the 12th the CO and Adjutant had the pleasure of a motor bus ride to Aire for a lecture on artillery cooperation. The next day they visited A1 Subsection to reconnoitre the trenches. On 14 January the battalion moved to billets at Annequin North and at 5:00 p.m. next day relieved 1/Queens' RWS in A1 Subsection, Harley Street. A Company was front left, B Company front right with C Company split between Cuinchy and Braddell Keeps and D Company at Cambrin Support Point. These companies swapped over after four days. Major Douglas was sent to 19th Brigade HQ as the temporary staff captain on 15 January. On 16 January PS/8356 Private John Greenwood of B Company, 18/R Fus was killed by a mortar bomb which landed on the parapet of his trench.[8] Two days later PS/1996 Private Reginald Richardson was also killed.[9]

5   IWM: Donald Price Sound Recording 10168, Reel 7.
6   Hamilton had a BA from Merton College, Oxford, and was gazetted to a commission in 18/R Fus on 6 April 1915.
7   PS/1510 Private Paul Silvester Gardner was buried in Béthune Town Cemetery.
8   John Austin Greenwood, aged 21, from Manchester, was formerly of Bury Grammar School. He is buried in Cambrin Churchyard.
9   Londoner PS/1996 Private Reginald Birkett Richardson, aged 35, is buried in Cambrin Military Cemetery.

Contrary to the idea that many generals stayed in their chateau headquarters, Major General Landon, the commander of 33rd Division, visited the trenches held by 18/R Fus on 18 January. However, Twiss did not relish such visits:

> Met [Major General] Landon & Col Symons at 8 a.m. & take them round the trenches. It is not good enough, having to be up there at 8 a.m. when one does not get to bed till 2 a.m. They simply waste one's time, they are a 'rotten' crowd, how I wish we were in a decent Division. These fellows are all 'eye-wash', & are always writing to us to show to Corps Commander, of whom they seem to be very afraid…[10]

An officer from the divisional staff remembered; 'The G.O.C. is Major General Landon, fairly tall, thin, beaky nosed, a little funny.'[11]

On 20 January No 1 Platoon of the 33rd Divisional Cyclist Company, were attached to 18/R Fus for instruction. The platoon was commanded by Second Lieutenant Herbert West, formerly 21/R Fus, and stayed in the line for two days.[12] On the same day D Company required a reply to German trench mortars and artillery fire was requested. The response was described as 'erratic' with British shells landing partly behind the front line trenches and suggested that artillery liaison was imperfect within 33rd Division. The results were reported to 19th Brigade. Second Lieutenant Dudley Jewell was attached to 180th Tunnelling Company RE. He was killed on 20 January in a mine shaft after trying to rescue men overcome by mine gas. He himself succumbed after going back to recover further men. Colonel Scott wrote to his mother; 'I cannot tell you how much we all feel his death. He had got an extraordinary hold over his men, and they would follow him anywhere.'[13]

The rest of the trench tour was quiet; on 23 January 20/R Fus took over the trenches. Once relieved, 18/R Fus marched to Beuvry North. Sergeant Oswald Blencowe wrote home regarding this stint in the line:

> … we were in for 9 days – 4 days in support + 5 days in the firing line. I wish I could tell you where it was … it was a very … hot corner + I am glad to say I got through this time without having any more casualties in the platoon … In one place the Huns were only about 80 yds away and they were also holding a small crater in front of their line the front edge of which was only about 35 or 40 yds distant … We were fairly safe from shells as the lines were so close together but there was a tremendous lot of sniping and any amount of rifle grenades flying about so one had to be most awfully careful…

---

10 IWM: G.K. Twiss Papers, 17109.
11 IWM: Documents 12059: Papers of G.W.G. Hughes.
12 Herbert Winn West was gazetted to 21/R Fus in June 1915 and later joined the divisional cyclist company. West did not join XI Corps Cyclist Battalion and was transferred to the Intelligence Department at GHQ in April 1916.
13 Second Lieutenant Dudley Mark Hayward Jewell, of 18/R Fus, enlisted in September 1914. He was aged 22, from Selsey, Sussex, and was educated at Felsted School, Essex, where he was in the OTC. He was a farmer in Worcestershire. Jewell was a good all-round athlete who played football and cricket and represented the Worcestershire Gentleman's cricket club. He was buried next day in Guards Cemetery, Windy Corner. See Marquis De Ruvigny, *The Roll of Honour, Vol. 3* (London: Standard Art Book Company, 1916), p.184.

> What impressed me … was the sight of the country round us. Running right across our lines away into the German lines was a road – once a main road with trees each side and full of traffic, and now – I can hardly describe it as one looked along it as dawn was breaking – every tree as far as one could see shot to pieces! Splintered trunks standing out against the sky line. Shell holes – broken stone work – torn barbed wire – equipment old and broken – dead Germans – rifles – absolute despairing ruin everywhere so intense that one could almost touch it… it made a tremendous impression on me…[14]

The day after being relieved was a rest day although fifty men were sent on a fatigue to Annequin. Lieutenant Cyril Skey, 18/R Fus, arranged a concert for the battalion at 6:00 p.m. on the 24th.[15] On 25 January there were several fatigues which required B and C companies to complete one day and A and D the next day. 18/R Fus was at high readiness to go back to the trenches as German activity was expected to mark the Kaiser's birthday on 27 January.

On 28 January 18/R Fus took over B1 Subsection trenches from 5/Scot Rif; A and B Companies were in the front line, C and D were in support at Pont Fixe. Shelling had damaged the communication trenches which required repairs. Snipers troubled B Company on the right and caused slight injuries to three men. The battalion patrolled No man's land in front of their positions on 30 January which suggested that dominating this neutral territory was not a regular activity for 18/R Fus. Next day 18/R Fus was relieved by 21/R Fus and marched to Rue d'Aire. The machine gunners were the last men out and arrived at 4:00 a.m. on 1 February. The battalion was again at readiness to move to the trenches; this time at two-hour's notice. Very early February was taken up with inspections and parades but on 4 February a football match took place against 20/R Fus in a field off Rue de Lille. The result was a one-all draw.

On 6 February 18/R Fus moved to Annequin South but found billets there already occupied by 16/KRRC and other units. The billets were in a bad condition and some had been recently modified by German shellfire. The next day was spent in cleaning billets and providing working parties carrying material to the trenches; these parties were shelled. PS/1556 Private Griffin, PS/3828 Private Keer and PS/6054 Private Hornung (of C Company) were killed and several men were wounded.[16] Colonel Scott wrote letters to the families of these soldiers:

---

14 *1916 brought into perspective – 1986 – Europeana 1914-1918, Europe – CC BY-SA* <https://www.europeana.eu/item/2020601/https_1914_1918_europeana_eu_contributions_5366> (accessed 11 February 2023).
15 Cyril and Charles Skey were the sons of a schoolmaster and from Margate. Both were gazetted to UPS battalions as lieutenants in October 1914; Charles joined 19/R Fus (D Company), Cyril to 18/R Fus (C Company).
16 PS/6054 Private Charles Peter Hornung, aged 20, was born in Melton Mowbray, Leicester, and lived at Horley, Surrey. He attended Marlborough College where he was a capable boxer who was only just beaten in 1912 in a bantam-weight final; in 1914 he won as a featherweight. He was too young for a commission and presumably served in the Public Schools Battalion to kill time. He was killed by a shell going into the trenches. PS/3828 Private Oliver Largent Bell Keer was born in Wickham Market, Suffolk, and lived in Woodbridge. PS/1556 Private William Alfred Griffin, aged 20, was educated at Cranleigh School and was in the first eleven for football. He was wounded and knocked out by shrapnel whilst moving along a communication trench and did not recover consciousness. He had been about to go on leave and a commission application was about to be signed by Colonel Scott. All three are buried in Cambrin Churchyard.

Dear Mr Hornung,
I was so distressed to see in the "Times" that your boy was killed on [the] 7th. I have heard no further particulars, but I feel sure you will not mind me writing to tell you how very deeply I sympathise with you and your family in your terrible loss. He was a splendid boy, and as you probably know. I sent for him on 28th January and promised to recommend him for a commission, which he had thoroughly earned. We had such a pleasant discussion then about him taking up a commission, and his pleasant manners and good sense more than justified my previous opinion of him. Alas! such things are no good now, but I think he was quite happy with us in taking up his share of this gigantic struggle.
With deepest sympathy,
Yours faithfully,
Henry Scott, Colonel. 18th Royal Fus.[17]

All the work done on billets by 18/R Fus paid off on 9 February when the ADMS 33rd Division inspected them; 'Col Daly, ADMS, came round & complimented battn. Had not seen better billets & would report so to the General.'[18]

On 11 February 18/R Fus relieved 2/RWF in Z0 sub sector; A, C and D Companies were in the front line with B Company in Railway Trench, Railway Keep and Factory Trench. The next day there was heavy shelling from 2:15 p.m. to 4:30 p.m.; there were only three casualties but the communication trenches were damaged. Having been instructed in trench warfare, 7/Leinsters was now able to relieve 18/R Fus on St Valentine's Day. The relief was complete by about 11 p.m. and the stragglers from 18/R Fus arrived at the *Ecole des Jeunes Filles* and *Ferme du Roi* at 2:30 a.m. on 15 February.

Following two days in Béthune 18/R Fus relieved the Leinsters again on 17 February, with D, C and B Companies in the front line and A Company in support. According to Kenneth Norman the last six days in the line at Cambrin were marred by the death of Private Oberhoffer, a reserve stretcher bearer, on 17 February. Due to the narrow trenches, it took eight hours to get him to the dressing station. He died of his wounds next day; the day he had been due to go on leave.[19] The next two days were quiet with a little activity from German snipers. On 19 February a company of the Leinsters was attached for three days of instruction. On 22 February 19/R Fus was relieved by 1/Mx. This was the last time 18/R Fus would serve in the front line as a battalion. Out of the trenches a recent draft of 25 men were inspected by the commanding officer having arrived too late to benefit from the experience gained over the winter.

20/R Fus

1916 would be an infamous year for 20/R Fus. Whilst many men vividly remembered the early part of the year not everyone kept accurate records; Herbert Vinden recalled with difficulty; 'My recollections of the first six months are vague. We put out wire, mended trench walls and

---

17   *The Marlburian*, Vol 51, No 758, 2 March 1916, p.19.
18   TNA: WO 95/2423: 18/R Fus War Diary.
19   PS/6149 Private George Henry Joseph Oberhoffer, from Uppingham, was educated at Uppingham School and later became a music master there. He is buried in Béthune Town Cemetery.

patrolled by night in No man's land, but raids on the enemy trenches had not come into fashion or certainly not to the 33rd Division.'[20]

20/R Fus was in billets at Annequin and the start of their year was relatively quiet. There was shelling of the la Bassée Road at 11:00 a.m. on 2 January as part of the wider conflagration involving 19/R Fus but this only cost 20/R Fus one man slightly wounded. 20/R Fus relieved three companies of 18/R Fus in the Z1 Subsector south of the Béthune–La Bassée road (from Boyau 14 to Gun Street); B Company were on the left, C was in the centre and garrisoning Russell's Keep, and A was on the right. 2/RWF were on the right and 16/KRRC were on the left. On the night of 5 January Ernest Brierley recorded sighting a German working party in No man's land whilst he was on sentry duty. He was prevented from firing because a British working party was also out.[21] On the evening of 5 January a man was wounded, presumably whilst out on a wiring party between 5:00 and 8:00 p.m. On 7 January there was a raised level of enemy MG and sniper activity and shrapnel and rifle grenades wounded Second Lieutenant Charles Stuart[22] and eight men.[23] Next day Brierley recalled; 'Our lines heavily shelled during morning. Several wounded but not much damage done.'[24] The War Diary stated that another man received a bullet wound and the Germans bombarded the ground between Cambrin Support Point and the reserve trenches. At 3:00 a.m. on 9 January 21/R Fus relieved 20/R Fus; the latter went to billets in Montmorency Barracks, Béthune.

Training over the next few days consisted of rapid loading (of rifles), grenade throwing and bayonet fighting. Meanwhile, on 7 January a draft of 27 men arrived as reinforcements and stayed with the transport lines in Beuvry until the battalion returned. Donald Price also rejoined on 9 January. On 11 January 1916 PS/4893 Private George Green died of his wounds in Béthune.[25]

On 14 January at 4:30 p.m. 20/R Fus relieved 16/Mx in B1 sub-sector which was north of the canal. 1/Cameronians were on the right across the canal and 7th Battalion Suffolk Regiment (7/Suff R) were on the left at Givenchy. Two companies of 20/R Fus were in the front line and redoubts, another company was in billets at Pont Fixe along with a further company of 5/Scottish Rif. The work mainly consisted of draining the trenches and enemy activity was limited to intermittent shelling of Pont Fixe. On 16 January one man was wounded at 7:00 p.m. whilst on a wiring part. Another two men were wounded next day near billets at Pont Fixe. Another man was wounded in the same place on 19 January. At 5:00 p.m. on 19 January the battalion was relieved by 5/Scot Rif and moved to billets at Beuvry.

On 23 January 20/R Fus relieved 18/R Fus in A1 Subsector with A Company on the left, B on the right, C in support and D Company, 1/Cameronians, in reserve at Cambrin Support Point. 2/RWF were on the left and 100th Brigade on the right. At some stage in January PS/7770 Private Eugene Jeffery, aged 18, was wounded; he died of his wounds on 23 January.[26]

20 IWM: F.H. Vinden Papers, 5565, p.13.
21 Diary of Ernest Brierley, unpublished. With the permission of Angela Jennings. Entry for 5 January 1916.
22 Second Lieutenant Charles Alexander Stuart was gazetted to 20/R Fus in January 1915.
23 One of those wounded was PS/4710 Private Harold Llewellyn Cross of B Company. He was aged 21 and was a 'Gentleman' from Manchester. He enlisted in September 1914 and was appointed as a lance corporal in December 1915. He was wounded right arm, left shoulder and left buttock and returned to hospital in the UK on 16 January 1916. Cross was discharged from the Army in January 1917.
24 Diary of Ernest Brierley, unpublished. With the permission of Angela Jennings.
25 Green was 25 and from Moss Side. He is buried in Béthune Town Cemetery. His date of wounding is unknown.
26 He is buried in Calais Southern Cemetery.

Next day the adjutant, Captain L.D. Cane visited the firing line. Major Twiss recalled that at 9:30 a.m. on 24 January; 'Went up to trenches in the morning, poor Cane, the Adjutant of the 20th Royal Fusiliers was killed, while I was up there, shot through the head by a sniper.'[27] The officers of the UPS were not all disliked by their regular counterparts. There was sadness in the ranks at Cane's demise; Vincent Platt recorded: 'Heard about the sad death of Captain Cane. Jolly hard lines.'[28] Another man was killed and two men were wounded by rifle fire. PS/5079 Private Hope, of A Company, was shot dead by a sniper though his death was recorded as on the 25th.[29] PS/5782 Private Harold Tyson wrote:

> You will be sorry to hear I have lost one of my pals, Freddy Hope was killed in action yesterday (Tuesday) and I am rather cut up at present. Dick and myself were not with him when he was hit and never saw him. He was shot through the head and was unconscious all the time and died before he reached the dressing station. Dick, myself and Crossley were his only friends present when he was laid to rest. He is buried in a little churchyard near the firing line and was buried last night when it was dark at about 6 o'clock. It was very impressive and sad, the guns booming all the time. He died like a brave soldier in face of the enemy and did his "bit".[30]

Donald Price (D Company) remembered his first experience of seeing a dead man on going through the reserve trenches (likely in January 1916); 'I was with a fellow named Walter Cowpe, we were just going into the trenches and saw a stretcher coming along with a dead man on. The first time I'd seen a dead man. They were bringing him out of the trench onto the road … This was the first idea [we had] of what it was all about.'[31]

Over the following days there was a steady drain of casualties. Two men were wounded on 26 January. PS/5308 Lance Corporal Mellodew was one of two men killed on 27 January. His company commander wrote to the grieving family:

> Soon after daybreak this morning the enemy began to strafe us heavily with rifle grenades. One of these burst near your brother, wounding him in the thigh and in the stomach. Everything possible was done for him at the time by the stretcher-bearers and the Medical Officer. He bore his pain bravely, and without a murmur he passed quietly away as he was being carried down the trench. L.–Corpl. Mellodew was a man for whom I had an infinite respect. He was a strong, silent man, in whom one could place the utmost confidence and reliance…[32]

---

27   IWM: G.K. Twiss Papers, 17109.
28   IWM: V. Platt Papers, 17681, p.13.
29   PS/5079 Private Frederick Hope was aged 24 and from Hazelhurst, near Salford. He was buried in Cambrin Churchyard.
30   Letter by Private Harold Tyson, 26 January 1916.
31   IWM: Donald Price Sound Recording 10168, Reel 5.
32   PS/5308 Lance Corporal James Mellodew, aged 28, from Oldham, was educated at Bath College and Rossall School and was learning the business of cotton spinning and manufacturing with his family firm. Mellodew was buried in Cambrin Churchyard. De Ruvigny, *The Roll of Honour*, p.192.

The other fatality was PS/7459 Private Taylor; three others were wounded.[33] Next day there was a heavy bombardment all day and PS/7441 Private Cree, was killed.[34] Around this time PS/7438 Private Dronsfield, a cotton spinner from Oldham, was wounded in the left ankle and was evacuated to the UK on 29 January.[35]

Between 4:30 p.m. and 8:00 p.m. on 29 January 20/R Fus was relieved by 19/R Fus and 20/R Fus marched to billets at *Ecole Michelet* in Béthune. The battalion was there until 7 February. The training on 31 January was not wholly worthwhile; 'Bayonet fighting – physical drill – rapid loading. The latter was perfectly ridiculous simply killing time.'[36] On 4 February 18/R Fus and 20/R Fus played one-another at football and drew 1–1.[37]

The 7 February saw 20/R Fus move to billets at Annequin Fosse where it relieved 2/RWF, though D Company, 20/R Fus, was attached to the Scottish Rifles in the line. On 7 February D Company departed at 2:00 p.m. but suffered from a great deal of confusion after their guide became lost. They took until 9:00 p.m. to find the right trenches; the men were then crammed into dugouts in the support line to sleep.[38] Not everyone had such a luxury, Vincent and Sidney Platt had to then go on fatigue to collect ammunition until 5:30 a.m., during which they had several close escapes from shellfire. During 'Stand To', on the evening of the 7th, Donald Price recorded some irregular behaviour:

> I had a friend called Dick Westmacott ... Dick was very fond of his rum; and Dick says to me; 'What about your rum – do you want it?' ... he gave me sixpence for my rum. And he goes to somebody else and by the time he'd finished he could have had a few. We were standing to ... and the next thing we knew Dick was on the top of the parapet shouting he'd kill the buggers [the Germans]. He shouted his head off dancing about on top of the trench. These bullets came flying around hitting these cans [empty food cans thrown from the trench]; the old Jerries were having a go ... we pulled him down.[39]

An officer arrived to investigate the noise and it was impossible to hide Westmacott's drunken state. He was arrested and was tried by court martial when the company was out of the line. Westmacott was fined £1 and was awarded 21 days of Field Punishment No 1.[40] During the next four days the companies that were in reserve behind Z Sector conducted further training.

---

33 PS/7459 Private Douglas Taylor, aged 21, from Normanby, Yorks, was buried in Cambrin Churchyard.
34 PS/7441 Private Henry Cree was aged 20 and from Cheetham Hill, Manchester. He was buried in Cambrin Churchyard.
35 Dronsfield returned to 20/R Fus on 1 July but was evacuated in early October suffering from lumbago. He was discharged in May 1917.
36 IWM: V. Platt Papers, 17681, p.14.
37 IWM: S. and V. Platt Papers, 17681, p.15.
38 IWM: S. and V. Platt Papers, 17681, pp.16–17.
39 IWM: Donald Price Sound Recording 10168, Reel 5.
40 PS/6709 Private Richard 'Dick' Westmacott was with No 16 Platoon, D Company, 20/R Fus. He was aged 36, was educated at Manchester Grammar School and was a commercial traveller. He was with the South African Constabulary 1902-1903 and enlisted in the UPS in February 1915. After completing his field punishment, he went to the UK in December 1916 and was gazetted to the ASC in June 1917; he relinquished his commission due to ill-health in December 1917.

On 12 February 20/R Fus relieved 5/Scottish Rif in Z2 Subsector with B, C and D Companies in the line and A Company in support. Platoons of 5/Scottish Rif were in Russell's Keep and Lewis Keep. A company of 7/Leinsters was relieved in the line who had been attached to 5/Scottish Rif. 1/Cameronians were on the right with 100th Brigade on the left.

Next day the British trenches were bombarded in the afternoon costing 20/R Fus two men killed and eight wounded (three remained at duty). The two men killed were PS/4902 Corporal Neil Grant and PS/8727 Private James MacGregor of C and D Companies respectively. Both were buried in Cambrin Churchyard.[41] Corporal Grant was 'full of affection and merriness … It was a pleasure to go with him even on an accursed carrying-party up to the trenches when we were 'resting' behind the line.' Hodson was told; '"That steel-plate: loophole. Periscopic rifle. Ian [Neil Grant] was having a shot; rather a lark. All quiet. A bullet crashed through a sandbag. Hit Ian in the eye…".[42] Hodson recalled the atmosphere at the burial service 'his body stitched into a waterproof groundsheet and a sandbag over his bloodied head. Three or four of us were at the newly dug grave, the padre with a surplice on in the moonlight. The guns were shaking and the machine-guns rattling … the Verey lights were shuddering up into the night with their fiery, ghostly, wavering arcs of light.'[43]

On the 14th there was heavy rain and it snowed later. Water filled the trenches and filled men's boots. Those on sentry duty would have found it near-impossible to stay warm. To make matters worse, that night (14–15 February) D Company, 20/R Fus, suffered heavy shelling from 'whizz-bangs' and rifle grenades losing five men in three minutes; 'A rifle grenade let [sic] on the parapet mortally wounding Orr. Whilst attending Orr another came over and killed Gray & Edwards & wounded the other two [Sergeant Holmes and Private Clayden].'[44] Edwards was a close friend of Vincent Platt and Gray was known to Sidney Platt as being a man who made sarcastic remarks.[45] In total on 14 February there were six men killed and nine wounded; which were high for a battalion just holding the line. In addition to the three men of D Company, four other men were recorded killed; PS/4396 Sergeant Ashworth (C Company); PS/5162 Lance Corporal Kirkbride (B Company); PS/4530 Private Brocklehurst and PS/3725 Private Chapman (a stretcher bearer). All seven are buried in Cambrin Churchyard.[46]

---

41 PS/4902 Corporal Neil Grant, aged 23, from Furness Vale, Stockport, and PS/8727 Private James MacGregor, aged 20, from Herne Hill, London.
42 Hodson, *Grey Dawn*, p.198.
43 *The Manchester Evening News* (21 February 1916) stated that a grenade from German trenches exploded and inflicted injuries upon him; he died an hour later. James Hodson, *Return to the Wood* (London: Gollancz, 1955), p.27.
44 The three men of D Company (likely No 13 Platoon) that were killed were; PS/4916 Corporal Robert John Gray, aged 25, from Belfast; PS/8504 Private Charles Robert Edwards, aged 22, from Hale, Cheshire, and PS/5393 Private Burrell Orr, from Belfast. All three were buried in Cambrin Churchyard. IWM: S. Platt Papers, 17681, p.19.
45 IWM: V. Platt Papers, 17681, p.19.
46 PS/3725 Private Alfred Chapman, aged 23, from Newton Heath, Manchester, attended St Wilfrid's School, Newton Heath, and Manchester Grammar School. He was killed whilst acting as a stretcher bearer. PS/4396 Sergeant Joseph Hargreaves Ashworth, aged 28, was from St Annes-on-Sea and served with C Company. PS/5162 Lance Corporal John Berchmans Kirkbride, aged 23, from Manchester, had played football for B Company. PS/4530 Private George Bowden Brocklehurst, aged 27, was from Marple, Cheshire.

15 February was 'quiet' by comparison with two men being wounded. A new officer, Second Lieutenant Algernon Coggin, joined 20/R Fus on 15 February. He was 5ft 10in tall, had a fresh complexion, grey eyes and dark brown hair. Coggin was not an ex-UPS officer having been gazetted from enlisted service with the Household Cavalry.[47] Nonetheless, he would prove himself to be an integral character within 20/R Fus.

On 16 February 5/Scot Rif relieved 20/R Fus who moved to Annequin Fosse having left A Company in reserve to 5/Scottish Rif. On the 16th PS/5527 Lance Corporal Frederick Rice, aged 19, of No 13 Platoon, D Company, received multiple severe bomb wounds and was evacuated to the UK.[48] 20/R Fus was again in reserve and provided working parties and conducted training over the next few unpleasant days as it was cold and rained constantly. A draft of replacements arrived and joined 20/R Fus from 29/R Fus. One of them was Private Dykes who had only attested on 5 November 1915. He marched into Annequin Fosse on 17 February:

PS/9132 Private James Dykes in marching order, likely just before his departure to France. He is an example of the replacements 20/R Fus received in early 1916. Under-age (he was only 18) at this time, he was physically under-developed (he had barely two-months of training), had imperfect eyesight and ill-fitting equipment. (J.N. Dykes Papers)

> The battalion had been relieved on the previous night, after a spell of duty in the trenches. When we arrived at Annequin Foss [sic], we found the rank and file in the process of cleaning up after a very wet and muddy four days there. We were halted in front of battalion headquarters, given the once over by the colonel – very perfunctorily, detailed off to the companies and platoons we were to grace with our presence, and taken to our billets, – roofless, window-less, door-less, ground floor accommodation only, under water, which lapped

---

47 Second Lieutenant Algernon Oswald Coggin, a messenger from Forest Gate, London, had enlisted, aged 22, in November 1914 into the Household Cavalry. He served with them for eight months before being gazetted to a commission in 15/R Fus in July 1915. He embarked for France on 8 February 1916.

48 PS/5527 Lance Corporal Frederick Rice from Lisburn, Country Antrim, was aged 19 and had enlisted in Belfast in September 1914. He was appointed lance corporal in June 1915. He was discharged from the Army in March 1917.

against the duckboards, – our beds –, just for our further comfort, there was a great hole in the wall at the side, through which the wind blew the cold rain on to our "beds".[49]

The men bound for C Company were then inspected by their CSM and allotted to platoons. This was hardly a warm welcome though at least Dykes, McKeown and Sibbit, all joining No 10 Platoon, C Company, did not have to scrape their equipment clean; 'The platoon ... worked assiduously and systematically on the cleaning process, as if they really enjoyed it. Our colonel was very particular about cleanliness and smartness in his battalion, and the troops were anxious to please him. Besides, slovenly soldiers got put "on defaulters"...'[50]

On 19 February 20/R Fus relieved 5/Scot Rif in Z2 subsector, the same as occupied on 12 February. B and D Companies, along with a company of 6th Battalion Royal Munster Fusiliers (6/RMF), held the front and support lines with C Company in Russell's and Lewis Keeps; A Company were in reserve. Again 100th Brigade was on the left and 2/RWF was on the right. To James Dykes this was his first time in the trenches:

> Ten Platoon's destination was "Mytchett Alley", a good distance behind the front line. We were "in support"... My first day's experience of life in the trenches was something in the nature of an anti-climax. It was tedious. I was expecting something – shelling perhaps – I don't know. For all we had heard of the Jerries they may have gone home on leave![51]

On 20 February Lieutenant George Ziegler, of No. 16 Platoon, D Company, was wounded along with two other men but he remained at duty. The next night, at about 2:00 a.m. on 22 February, Lieutenant Clive Fraser was severely wounded near one of the craters. He was shot in the head and received wounds to his face and chin so severe that he later lost his left eye. He was carried out of trenches by Sidney Platt and was lucky to survive.[52] Another man was killed on the night of 21–22 February. This was PS/5082 Private Rowland Hunt, aged 22, from Mitcham.[53]

On 22 February 20/R Fus was relieved by 2/A&SH and returned to billets in Beuvry North. Whilst in reserve 20/R Fus was notionally attached to 100th Brigade (covering A and B1 subsectors) and was to respond in the event of any attack. Here 20/R Fus was engaged in further training. This period was not all work. 20/R Fus held a battalion concert on 25 February 1916.[54] On 28 February there was a football match between 20/R Fus and the Scottish Rifles which the Fusiliers won 7–0.[55] On 29 February 20/R Fus relieved 16/KRRC in B1 subsector with the La Bassée Canal on the right along with 2/Worcs R and the 13/RWF (38th (Welsh) Division) on the left.

49 IWM: J.N. Dykes Papers, 7378, p.8.
50 IWM: J.N. Dykes Papers, 7378, p.8.
51 IWM: J.N. Dykes Papers, 7378, p.9.
52 Lieutenant Clive Stuart Fraser was aged 28 and from Sutton in Surrey. He was born in Batavia and was well-travelled; he had lived in Holland, France, Germany and Malaya and could speak their languages as well as having visited Italy, Canada, Switzerland and Java. He worked as a publisher for Fraser, Eaton & Co. Fraser enlisted in September 1914 and was initially PS/4842 Private Fraser. He was gazetted on 27 October 1914 and was an original 20/R Fus officer; he was promoted lieutenant in June 1915 and served with 20/R Fus in France until wounded. He later served in intelligence and in censorship in Singapore. He was later made an MBE and married in 1935; he died in 1974.
53 He is buried in Cambrin Churchyard.
54 IWM: S. and V. Platt Papers, 17681, p.22.
55 IWM: S. and V. Platt Papers, 17681, p.23.

14

Disbandment of 18/R Fus, 19/R Fus and 21/R Fus

Being turned into an OTC

> *The fate that we have [for] some time contemplated has now overtaken us, i.e. being turned into an OTC…*[1]

> *I soon found that the movement for commissions was stronger than I had expected. To remain seemed as sure a way of losing one's friends as to go.*[2]

For several weeks in early 1916 there had been two main subjects of gossip for the UPS battalions. The first was 'Blighty' leave, which started in February, and the second one concerned commissions. Rumours had increased in late February that applications for commissions might open again and that the battalions might become OTCs. Herbert Vinden, 20/R Fus, surmised that 'the War Office realised that there was a shortage of officer material and [the War Office] thought of the 4,000 educated men that existed in the four battalions of the Royal Fusiliers. It was decided that they should be withdrawn and turned into officer cadet units.'[3]

However, rumours became fact when, on 27 February 1916, the 19/R Fus war diary recorded; 'The fate that we have [for] some time contemplated has now overtaken us, i.e. being turned into an OTC …'[4] The three UPS battalions (18/R Fus, 19/R Fus and 21/R Fus) were to join General Headquarters Troops at St Omer. These battalions were to trickle-feed drafts of potential officers from amongst their manpower back to the UK to attend Officer Cadet Battalions (OCBs). This ended lengthy debates on the future of these units.

Though this news was a shock many men recognised the reasoning for this decision. Nicholas Pease wrote later; 'There was a great shortage of officers, actual and potential, as the casualty rate was very high indeed, so the authorities naturally turned to the U.P.S. Brigade to provide the material they required. Consequently those of us who were considered suitable were offered the prospect of being sent home as cadets to be trained to take up commissions.'[5] Likewise

---

1 TNA: WO95/2427: 19/R Fus War Diary.
2 IWM: H.W. Spurrell 2138.
3 IWM: F.H. Vinden Papers, 5565, pp.13–14.
4 TNA: WO 95/2427: 19/R Fus War Diary.
5 IWM: N.A. Pease Papers, 8230, p.34.

The UPS Brigade had to endure uncertainty as to its destination and future from conception until its time in France. This sketch from April 1916 highlights the further uncertainty facing the men of the three disbanded battalions. UPS men would be sent to most of these destinations during the disbandment. (Sketch by William Norton published in *The Gasper*)

Philip Wright summarised; 'Now after about three months experience of trench warfare, men were being sent home for commissions in various battalions, and as they were 'fresh from the fighting line' were doubly valuable on account of experience gained and by reason of their being admirably suited for the new work.'[6]

According to Spurrell of 21/R Fus the prospect of commissions becoming available was taken in different ways by different groups within the UPS:

> Some had been chafing under the suspension of the grant of commissions in the previous May and were waiting for the first opportunity of getting them. Others were not anxious to take commissions but wanted to get home to England for a time. A third class were unwilling to leave the old battalions but would take commissions if it were broken up. A fourth party would not have commissions at any price.[7]

Leslie Ireland was in the first category having been angling for a commission whilst posted to No 33 Infantry Base Depot. He recorded; 'I am very much afraid my commission is at an entire dead end … probably you would not comprehend the difficulties which beset a 'miserable' private in H.M.'s Army! One reason is that the Commandant refuses to see anyone re Commissions & that more or less settles it.'[8] Ireland later returned to 19/R Fus. Philip Spaull fitted into both the first and second brackets. Spurrell's motives were different:

> I must confess that at first, I belonged to the last party. I had long been attached to the old battalion and had made friends among the men I did not want to leave. But I soon found that the movement for commissions was stronger than I had expected. To remain seemed as sure a way of losing one's friends as to go.[9]

When Spurrell and another die-hard UPS man, PS/2627 Private George Dymond, were told that if they stayed, they would be sent to another infantry battalion. As such, their decision was made for them and they reluctantly applied for commissions. By fortune, the battalion Spurrell might have been gazetted to in 1914, 8/RWF, suffered heavy casualties at Gallipoli and 23/R Fus, to which he would have been posted on disbandment, also lost heavily at Delville Wood. He had made the right choice by accident.

The disbandment of these battalions did not occur overnight and there was still much to be accomplished by these battalions during the Spring of 1916.

20th Battalion Royal Fusiliers 'soldier on'

Why was 20/R Fus omitted from the disbandment list and reprieved from disbandment? There are two themes of varying levels of merit, as to why the battalion was retained over the other three units. First, 20/R Fus may have been considered less suitable for officer training due

---

6   FM: P.N. Wright Diary, RFM.2013.8.1/2, p.24.
7   IWM: H.W. Spurrell 2138, p.66.
8   IWM: L.W.C. Ireland Papers, 14441.
9   IWM: H.W. Spurrell 2138.

to the grammar school and Manchester background of many the men. Meanwhile the other battalions had higher numbers of members from more prominent public schools that might have been considered to make more immediately eligible officer candidates. Second, the battalion may have been considered to have a stronger esprit de corps and more clear value as a fighting battalion. The other three battalions were less cohesive and were therefore more readily able to be sacrificed for officer candidates.

The first reason gained traction in the Manchester press which suggested snobbery from the military hierarchy prevented the battalion having the honour of commissioning many of its men. The following article appeared in the Manchester Evening News regarding the disbandment of the UPS battalions:

> At a concert at Moss Side in aid of 20 RF's Band Fund Captain MacDonald stated that 20 RF had had the compliment paid to it in surviving disbandment whilst the other three battalions would become Officers' Training Corps. However, such a compliment was tempered by the consideration that it's soldiers were less likely to be commissioned. In an article in the Manchester Evening News on 8 March 1916 telling of the disbandment it was clearly stated; presumably quoting the War Office that; 'It is not intended to make any of the battalions an Officers' Training Corps, nor is it intended to confine the granting of commissions to men of the 1, 2, and 4 battalions [of the UPS]. All the battalions will be placed on the same footing in this respect.'[10]

A further question was asked of the Secretary of State for War, 'differential treatment is now being extended to certain of these [UPS] battalions, as an effect of which one battalion, 20th Battalion of the Royal Fusiliers, will be handicapped in respect of its members getting commissions, and whether this battalion is to be kept at the front as an ordinary fighting unit…'[11] Likewise:

> The War Office continue to bungle badly in their treatment of the 20th Royal Fusiliers… Mr Tennant will not go beyond the statement that men of the battalion cannot be considered eligible for commissions unless they are recommended for them by their commanding officer … grave injustice is suffered by men who happen to have a commanding officer who is bent on retaining nearly all his men in the ranks at all costs in flat defiance of the promises which were made to them when they enlisted…[12]

That commanding officer, Lieutenant Colonel Bennett, had an agenda to retain his battalion to enable him to see further action, and seek further advancement, whilst commanding it. He likely refused commissions to avoid depleting the numbers and quality of the men in his battalion and thereby reduce its battle-worthiness. Alternatively his efforts may have been more altruistic in wanting to create the most effective infantry battalion he could. According to one soldier of 21/R Fus the retention of 20/R Fus had much to do with Colonel Bennett's influence and ambition; 'The 20th [R Fus] remained in the line, at what was supposed to be their Colonel's

10 *Manchester Evening News*, 8 March 1916.
11 *Manchester Evening News*, 16 March 1916.
12 *Manchester Evening News*, 30 March 1916.

express wish.'[13] In early June 1916, Erroll Shearn, then a freshly-minted second lieutenant in the Hampshire Regiment, formerly of 20/R Fus, was heading to France and met Colonel Bennett who was returning from leave:

> I don't suppose for a moment he remembered who I was but he was quite charming and incidentally informed me that while the unit he commanded comprised some excellent material it did not compare with the calibre of the crowd under his command up to the time the War Office had weeded out potential officers. We parted on mutual good wishes.[14]

Bennett was clearly concerned about retaining the fighting quality of his battalion at the expense of providing officer candidates, having already lost a lot of high-calibre men. This was likely despite issues of region and class regarding potential officers. 20/R Fus was a more homogeneous battalion and possessed a strong regional association to Manchester. The other battalions were a mixture of regional and school affiliations. This *esprit de corps* in 20/R Fus was also born out of rivalry between the UPS Battalions. A member of 20/R Fus stated the reason for survival was down to a combination of Bennett's attempts to retain it and the battalion being considered to have a strong, cohesive spirit. According to Vinden; 'Our Colonel [Lieutenant Colonel Bennett], a fire eater, protested at the plan as far as the 20th Royal Fusiliers was concerned and convinced authority that he had a shock battalion and averted its withdrawal. I seemed to remember that we rather sniggered at the other three [battalions] rather than being jealous.'[15] A soldier of 20/R Fus was scornful of the abilities of the other UPS battalion when he recalled the Festubert mud, They [rifles] hadn't been much use at Festubert! I doubt any of 'em would have fired… But we did bring 'em out, which was something – the battalion we relieved had left a lot behind and we brought theirs out, too. Then the War Office turned *that* battalion into an officers' training corps, which roused the troops to scornful laughter.'[16] As such the men of 20/R Fus would see their survival as an infantry battalion was down to their own qualities and cohesion alongside the lack of ability amongst their rivals who faced disbandment.

However, the other battalions were far from complimentary about 20/R Fus for remaining behind with 33rd Division whilst they considered themselves better battalions for having been picked for disbandment. Though much was written about UPS men being called 'Chocolate Soldiers' by the RWF, certainly Morgan Williams of 21/R Fus was proud of the good relationship between his battalion and the Regular battalions of 98th Brigade (1/Mx and 2/A&SH).[17] However, he suggested that the poor performance of the Public Schools battalion that remained after February 1916 (i.e. 20/R Fus) should not let down the wider perception of the UPS Brigade and his own battalion. Likewise, some men of 20/R Fus were realistic about the martial abilities of their unit. Vincent Platt stated that; 'We all realise that men such as our battalion is composed of are no use for fighting. In a corner they would probably fight for their lives but they have not the same lust for blood as the rougher classes.'[18] However, this statement, in itself, highlights a

---

13 FM: P.N. Wright Diary, RFM.2013.8.1/2, p.24.
14 Colonel Bennett departed on leave between 25 May and 3 June 1916. IWM: E.D. Shearn Papers, 2033.
15 IWM: F.H. Vinden Papers, 5565, p.14.
16 Hodson, *Return to the Wood*, p.25.
17 Williams, *From Khaki to Cloth*, p.61.
18 IWM: V. Platt Papers, 17681, p.3.

level of snobbery in the battalion towards other Army units comprising men of different social backgrounds. Though there were many merits of keeping 20/R Fus as a fighting battalion at least one member recognised that there was some incongruity with UPS men remaining in the ranks; 'Kitchener hoping to win the war and smash trenches, with shrapnel … as well as with pea shooters … and university-trained men digging latrines … Bob a day for lawyers, men of science, engineers and journalists being blown to blazes and getting lousy, and six bob a day for artisans making munitions…'[19] However, irrespective of surviving as a battalion, 20/R Fus did lose some personnel to commissions which may have depleted quality despite Bennett's best efforts. Herbert Vinden recalled receiving the news of his being assigned to an officer cadet unit:

> I was ordered to parade at the Orderly Room and was marched in by the regimental sergeant major. The Colonel said: "You have been recommended for a commission and will go to the officer cadet battalion at the Chateau of Blendecques near St Omer. March out sergeant major". So I marched out. How or why I should have been recommended, I have no idea.[20]

Vinden was an unusual case as he was one of four men of 20/R Fus allotted for the Blendecques Course suggesting it was an Army or Corps-wide allocation independent of the disbandment of the other three battalions.

GHQ Troop Transition

Though 20/R Fus would be retained the other UPS battalions needed to be withdrawn from the line, replaced by other units, moved to the rear and personnel sorted out for being commissioned. On 22 February the strengths of the three battalions were as follows:

Battalion	Officers	Men	Entraining Location	Entraining Date and Time	Actual Departure Time
18th Royal Fusiliers	31	902	Béthune	11:00 a.m., 26 February 1916	10:19 a.m.
19th Royal Fusiliers	29	939	Béthune	11:00 a.m., 27 February 1916	13:19 p.m.
21st Royal Fusiliers	31	956	Béthune	11:00 a.m., 29 February 1916	13:19 p.m.[21]

The 16/Mx (100th Brigade) also joined GHQ Troops and was billeted at Quiestede, near St Omer, on 26 February.

18/R Fus had enjoyed a longer period in Béthune having left the trenches for the last time on 22 February. According to the War Diary the battalion was inspected by Brigadier General Robertson on 25 February whilst based at *L'Ecole des Jeunes Filles*. Kenneth Norman recalled that General Landon talked to 18/R Fus during a snowstorm before the battalion departed

---

19 Hodson, *Grey Dawn*, p.192.
20 IWM: F.H. Vinden Papers, 5565, p.14.
21 TNA: WO 95/601: I Corps Adjutant and Quartermaster General War Diary.

from Béthune. Landon thanked them for their work, hoped that some would come back to 33rd Division as officers and wished the men luck for the future.[22]

On 26 February 18/R Fus moved to Béthune Station and boarded trains for St Omer. On arrival the battalion detrained and marched to Campagne and took over billets there which required work to get clean next day. Part of the battalion was also billeted in a small village called Renescure which was not a very lively place. Parades were also conducted though the parade ground was waterlogged. However, Norman 'did not mind what happened now as we knew most of us would get to England eventually...'[23] The battalion then paraded by companies and conducted route marches. Generally life in 18/R Fus was fairly quiet with some square-bashing and route marches in the morning and sports in the afternoons. The weather in March was generally cold with snow on the ground.[24]

On 29 February 21/R Fus left Béthune for St Omer and arrived at the latter about 4:00 p.m. Motor buses carried the men to Wardrecques where they arrived at 6:00 p.m. By 9:00 p.m. the men were all billeted. The war diary for 21/R Fus concluded with the departure message given by Brigadier General Strickland of 98th Brigade:

> It is with regret that I am losing your services from the Brigade, but in the interests of the Army generally it is necessary that you should be withdrawn for a time to enable the large number of men who are recommended for commissions to be dealt with. Since you have been in France you have experienced very considerable hardships in the trenches and have been under heavy fire. The former you have borne cheerfully and under the latter you have acquitted yourselves with credit. The conduct of the Battalion has been excellent as was to have been expected of you. The experience thus gained will, I hope, prove of great value to all ranks, especially those who may be raised to commissioned rank. Should you not return to this Brigade you take with you my very best wishes for your future.[25]

Strickland was a sound judge of military affairs and was selected to command a division shortly afterwards.[26]

19/R Fus was the only battalion that left a War Diary covering the period of disbandment. This narrative will act as a core for the last months of all three battalions. On 27 February 19/R Fus entrained at Béthune and departed at about midday. Sturges recorded that the battalion caused quite a stir in St Omer, where they disembarked, as they were all still mud-covered. Even the Commanding Officer was muddy having been in the trenches the previous night seeing that the last men of the battalion got away. The unkempt infantrymen of 21/R Fus were a poor comparison to the smartened men of the ASC in the town.[27] The battalion arrived in Racquinghem at 7:00 p.m. where the men were billeted. Donald Wright wrote home on 28 February 1916:

---

22   LA: K.V. Norman Papers, WW1/GS/1188, p.118.
23   LA: K.V. Norman Papers, WW1/GS/1188, p.118.
24   LA: K.V. Norman Papers, WW1/GS/1188, p.119.
25   TNA: WO 95/2427: 21/R Fus War Diary.
26   One officer of 98th Brigade recorded; 'Strickland, an ambitious fighting-man with any amount of energy, had just left to take over the 1st Division. TNA: CAB 45/132: Letter from Pearson Choate to the Official Historian, 6 April 1936, p.532.
27   Sturges, *On the Remainder of Our Front*, pp.147–148.

... we are all far away from the trenches now. We left the firing line on Saturday & arrived here Sunday night & are in quite good billets & it seems like heaven after a week in the trenches. This is a very large base [St Omer] & the same place where the HAC & Artists Rifles have their OTCs. ... I don't know quite what they are going to do with [us] now we are here but we have all had our names, schools, fathers businesses etc taken for commissions ... I have had to buy a good many things ... for keeping spruce as we now have to keep ourselves as spotless here as when we were in England.[28]

Initially time was spent cleaning up and routine work. Sergeant Medlicott, 18/R Fus, had been detached from his company since 20 February to work in a new Topographical section behind the lines.[29] He got news from his friend Lieutenant Martin back with his battalion; 'A.C.M. was expecting to go on leave any time, but they are moving back & leave is stopped till they get reformed ...'[30] Medlicott opined in a letter on 4 March; 'I haven't heard from anyone yet, but suppose they are gradually disintegrating. I am quite thankful not to be in this process ... Bn won't be broken up I do not think, it will need large drafts & retraining.'[31]

Selection

Having arrived in their long-term billets the process of nominating, selected and prioritising men for commissions began. This varied between units and sub-units. According to Kenneth Norman all those desiring commissions were interviewed by the Colonel. He asked several questions but also offered his thanks for men remaining loyally with his battalion when they could have left earlier. Colonel Scott happily signed Norman's commission paperwork.[32] Medlicott re-joined 18/R Fus and met the commanding officer:

> He said I had better put in a paper for commission at once, after seeing P.E. [Captain Price-Edwards, OC D Company] who said the same, I went to the M.O. & was medically inspected & he filled in a commission paper, which I completed & took back to P.E. for putting before C.O. for his signature. So that's done – it means in course of time a course at Oxford O.T.C. or elsewhere.[33]

On 1 March a list of candidates for commissions in 19/R Fus was drafted for GHQ. Philip Spaull recorded; 'I had my commission papers handed to me which I filled up. They were then handed by me to our Colonel who signed them ... I was medically examined again and found to be fit. I think we are notified by the War Office, the name of regiment and place where stationed, also £50 deposited at Cox's Bank.'[34] He continued: '... we are 30 kilometres from the firing line ... There has been a call on this Brigade for 1,080 men for commission. I put my name

---

28 IWM: S.D. Wright Papers, 17675, letter 28 February 1916.
29 Either 33rd Division Topographical Section or First Army Survey
30 IWM: W.B. Medlicott Papers, 1708, p.183.
31 IWM: W.B. Medlicott Papers, 1708, p.190–191.
32 LA: K.V. Norman Papers, WW1/GS/1188, p.119.
33 IWM: W.B. Medlicott Papers, 1708, p.198.
34 IWM: P.F.F. Spaull Papers, 15453, letter 9 March 1916.

forward and have been recommended by my Company Officers to the Colonel … The rumour here is that when the number has been chosen we shall return to the UK to form a cadet corps.'[35] According to Spurrell of 21/R Fus:

> Soon after we reached Wardrecques, a goodly pile of blue papers were seen to arrive at the orderly room. These were applications for commissions. They were issued to all who intended to apply. After their arrival it was useless to attempt to discuss any other subject. The candidates were taken up before Major Fenn, who was commanding while the Colonel was on leave… If the Major approved of them, they were signed and forwarded.[36]

After some hesitation, Spurrell was persuaded to apply and was accepted. Morgan Williams recalled the procedure for interviews in 21/R Fus:

> We appeared in small groups before Major Fenn, a competent officer with a rather hectic career, who was then in command of the battalion. After scrutinising each record, he barked out the question, "What's your father?"… My reply to his question was "Presbyterian Minister, sir." He jabbed the point of his pencil into a sheet of paper, looked through me, and dismissed me with the others…[37]

At least Fenn signed Williams' papers. Much of Philip Wright's No 8 Platoon, B Company, 21/R Fus, were admitted to hospital in Arques due to scabies and they missed the initial 'rush' on commissions. According to Wright, officer selection relied not on individual ability but also on their platoon commander:

> Those men not yet seen by the C.O. might yet make an appeal, so we determined to take our chance… My platoon officer was present and very kindly said I was 'unsuitable' … it was impossible to say anything for myself… I had never liked the 'silly little ass' and I was sure he didn't like me. … there were more men in No.8 Platoon "turned down" than in any other platoon… At the time I consigned that … officer to the hottest corner of Hades, but now I am not sorry things happened as they did.[38]

Training

Though many men might be selected for commissions, the unit would still need to undergo training to both maintain and develop the ability of its members in readiness for commissions and returning to the line. This was doubly important after months of trench warfare where there had been little time for training. Having joined GHQ Troops in St Omer they came under Brigadier General L.A.M. Stopford who was now commanding the three battalions. He naturally wanted to visit his new units. He inspected 19/R Fus and was clearly pleased with their

---

35 IWM: P.F.F. Spaull Papers, 15453, 9 March 1916.
36 IWM: H.W. Spurrell Papers 2138, p.69.
37 Williams, *From Khaki to Cloth*, p.66.
38 FM: P.N. Wright Diary, RFM.2013.8.1/2, pp.25–26.

appearance in his remarks; 'You could not have a nicer Battn & I would not have missed seeing them. They are very steady in the ranks & extremely well turned out.'³⁹ Though GHQ had stipulated that the three battalions would become Officer Training Corps the training programmes and work routines at the three billeting locations was not aimed at this goal. According to Hugh Spurrell writing about the time 21/R Fus spent at Wardrecques; 'The two months that followed were very happy, though the unpleasant prospect of a speedy separation loomed on the horizon. We settled down in a typical Flemish village on that evening in early March. As we were not to leave it until May we had plenty of time to get comfortably fixed up...'⁴⁰ Harold Wilson recorded on 1 March; 'Reveille 7 o'clock. Breakfast eggs & bacon 8 a.m. morning spent cleaning up kit for inspection in afternoon by Col [Wolrige Gordon]. Inspection 2 o'clock. Col very pleased with turnout. Those recommended by Capt S[hipster]. for commissions (including Cecil & myself), called out & inspected separately. Worked in farmhouse after parade, washing up etc...'⁴¹ The next few days were spent by 19/R Fus with an eight-mile route march to Aire (described as a; 'pleasant march'), kit inspection and company training. Men like Harold Wilson and others found enough time to write letters and read during the afternoons and had enough money to visit an *estaminet* for dinner. A further planned route march was cancelled after it 'snowed like mad' but the battalion conducted a route march to Arques on 7 March. Orders were promulgated that all buttons were henceforth to be shined, a fringe benefit of trench-service now forgotten, and surplus items like gas helmets and rifle ammunition were handed in.

For 19/R Fus a lot of training took place under company arrangements and this will have varied in format depending on the company commanders and platoon officers delivering training. The time out of the line was not very restful; 21/R Fus likewise saw an increase in parades and polishing. According to Spurrell those in authority:

> ... promptly introduce a severe system of training. Fletcher [RSM 21/R Fus] had been rather out of the limelight during the period in the line. War was not his forte and during the days we spent out of the trenches, he had usually enjoyed himself in his own way and not been capable of very much work. But now his hour had come ... So he determined to turn us into soldiers once more.⁴²

One of those who worked hardest to look the smartest was Philip Wright but his exemplary appearance fell afoul of the sergeant major who reprimanded him for using black show polish on his leather webbing equipment. He rebelled and went from being one of the smartest men in the company to one of the most unkempt. He also took over as one of the company cooks after two were sent away for commissions. Of the other two cooks one was an old schoolmaster and the other was a London brewer.⁴³ They made little effort to vary the meals; at times they were busy preparing ingredients, at others they lazed in the sun whilst the cookers did their job. They were dirty and grime-covered throughout their duties.

39 Later Major General Sir Lionel Arthur Montagu Stopford KCVO, CB. TNA: WO95/2427: 19/R Fus War Diary.
40 IWM: H.W. Spurrell 2138, p.68.
41 IWM: H. Wilson Papers, 3606.
42 IWM: H.W. Spurrell 2138, p.69.
43 FM: P.N. Wright Diary, RFM.2013.8.1/2, p.28.

Once back at Racquinhem smartness similarly re-asserted itself for 19/R Fus according to Robert Sturges; boots needed to be shined and buttons polished again having previously been covered in black enamel. Foot and arms drill started again and Sturges felt 'we are slipping back into picture soldiers, which seems rather to go against the grain after the "real thing."'[44] According to Walter Medlicott the training in 18/R Fus was sporadic:

> I came off the guard duty at 10 this morning & did not have any parades this morning, in consequence. This afternoon there was a route march & tomorrow there is a battalion march in the morning & nothing else. They have a difficulty in occupying the time – new NCOs are being instructed to take the places of those who have gone ... physical drill ¾ hr, platoon drill 1 ½ hours, route march a day 2 hours in the usual days' work.[45]

Sending off all the SNCOs with the first batch for commissions was a potentially short-sighted move when considering that battalion efficiency needed to be maintained. After a few parades by 21/R Fus at Wardrecques:

> One morning the R.S.M. appeared, dragging with him, or rather pushing in front, a distinctly unwilling adjutant. The officer passed along the ranks apparently satisfied with the turn-out. Not so Fletcher! He found a dirty button here, a puttee wrongly crossed there. "Disgustin" said he, "disgustin I call it, Sir". Before we knew where we were, we found ourselves marked down for an extra parade in the evening.[46]

Reveille was made two hours earlier and parades and inspections were increased. Spurrell recalled that drill training lasted for three hours a day every day which culminated in the whole battalion being paraded and drilled. Sometimes at platoon and company level this was an opportunity for commission candidates to command their peers at drill though; 'The results, as a rule, were not very effective' according to Spurrell.[47]

Philip Wright recalled that:

> Several fellows occasionally were called out to drill the whole battalion and for the most part acquitted themselves very well. During these 'stunts' the adjutant, the various captains and platoon officers, the regimental and company sergeant majors were all represented by privates of the Batt[alion]. This was very amusing, because they took the opportunity of mimicking any of the officers etc., peculiarities ... and making them look very 'foolish'. One fellow especially was able to imitate Fletcher the B.S.M. exactly, setting off his voice and repeating some of his usual phrases ... that the fellows were so amused as to hinder the 'drilling operations'.[48]

---

44 Sturges, *On the Remainder of Our Front*, p.149.
45 IWM: W.B. Medlicott Papers, 1708, p.201.
46 IWM: H.W. Spurrell 2138, p.69.
47 IWM: H.W. Spurrell 2138, p.69.
48 FM: P.N. Wright Diary, RFM.2013.8.1/2, p.26.

Often, a few hours were also spent on lessons for commission candidates. There were other diversions; on 20 March parties of 60 from 18/R Fus were driven to St Omer by bus for baths.[49] In addition sport events occurred again; 18/R Fus held a sports day on 21 March; D Company won five of the events and the day was generally considered to have been a success.

The first commissions

There was little training accomplished before the first batches of officer candidates were despatched to the UK. The first group of fifty men from 19/R Fus were conducted by Second Lieutenant Charles Skey to No 6 OCB at Balliol College, Oxford, on 14 March.[50] The other two battalions provided fifty men making a contingent of 150 men which presumably formed a company intake for that OCB. The exact selection method is unknown, but they were likely those who were volunteers for commissions, those who had served the longest with the battalion, those who had already proved themselves brave or otherwise capable and those for whom extra training would not be immediately required. Robert Sturges, one of this batch from 19/R Fus, recalled that his battalion spend two weeks doing drill, playing sport and enjoying the *estaminets* before instructions arrived regarding departures for commissions. Sturges was in the first batch to depart and they knew they were 'saying good-bye to a life: a life, it is true, of squalor and dirt, a life of terror and hardship, a life which we hated', however, many had enjoyed that life and were sad to leave it. Colonel Wolrige Gordon spoke to the men before they departed of his pride in them and that he hoped they would apply themselves as officers and make their old battalion proud. As they departed and looked back they could see him stay there, on his horse, watching them march away.[51] Sturges and some of his comrades departed on 13 March 1916 bound for No 6 OCB, and, ultimately for him, a commission in the Motor MGC. His departure ended his narrative as an enlisted man with the battalion. Also amongst this first batch were the Dingley brothers from D Company who were both from Tufnell Park, London. Keith Morgan Dingley, a drapery buyer, was the eldest; the younger brother was William who was a dental student.[52] They had both enlisted on 2 September 1914 aged 23 and 18 respectively. After passing out of No 6 OCB they were gazetted into 13/Suff R in July 1916. Having served together for almost two years they were then sent to different battalions. Also in this draft were the Hunt brothers; Edward Wallis Alleyne Hunt the elder brother was an engineering student; the younger was Charles Henry Hunt. After OCB training the brothers also separated with Charles joining 12/Queen's RWS whilst Edward joined the RFC to be a pilot.[53]

---

49 IWM: W.B. Medlicott Papers, 1708, p.205.
50 Departures mentioned in the 19/R Fus war diary do not always coincide with drafts from other UPS battalions.
51 Sturges, *On the Remainder of Our Front*, pp.152–153.
52 PS/264 Private Keith Morgan Dingley was aged 23 and PS/265 Private William Dingley was aged 21. Both were born at Barnet and resided at Tufnell Park, London.
53 PS/518 Private Charles Henry Hunt was born in Venezuela. PS/519 Private Edward Wallis Alleyne Hunt, aged 24, was born in Melton Mowbray. Both lived in Sidcup, Kent; their father was in the Colonial Police and they spent their early years in La Guaria in the West Indies. They were heading to No 6 OCB. One brother was to die in the war and the other survived.

In a letter home on 14 March, Elton Jones, 21/R Fus, wrote, 'we have been turned into an O.T.C. Our sergeants have already left us and gone home to England to be turned into officers and most likely in a few days or under all of us will come out for the same thing.'[54] All four of the sergeants in Spurrell's platoon in 21/R Fus departed in this first wave. They represent some of the positive and negative qualities of those who would be trained and gazetted first. PS/2813 Sergeant Reginald Talbot Clements Hassell, aged 39 was educated at St Georges College, Harrogate, was a ship-broker and an experienced soldier. He was to be awarded a DCM in June 1916 for patrol work; he departed on 14 March 1916 destined for No 6 OCB and the South Lancashire Regiment (S Lancs). Less popular sergeants also departed. PS/3183 Lance Sergeant Cyril Renouf had attended Yeovil Grammar School and Bruton and was an engineer. PS/3266 Sergeant Hugh Miller Swears was privately tutored in Germany and Switzerland and attended Jesus College, Cambridge; he was to join the Motor MGC and later served with the tanks. PS/2417 Lance Sergeant James Atherton was the company bombing sergeant. He was destined for the Loyal North Lancashire Regiment (LNL). Spurrell's platoon corporal, PS/3016 Corporal Herbert George Merchant, went a week later; this will have gutted Spurrell's platoon and prevented meaningful training concerning either a general military or 'officer training' syllabus.[55] Some of the sergeants, in Spurrell's eyes, were 'duds'; it is probable that they were sent first due to their rank rather than their suitability as officer candidates. Another man from this first batch from 21/R Fus was PS/2874 Sergeant Leslie Norman Jones, of No 3 Company, 21/R Fus. His bravery bringing back the patrol after the Whittington raid earned him a 'Mention' in June 1916; this might have assisted in his selection for the 'first wave'.

Five men of this first batch from 18/R Fus provide a sample of the men 18/R Fus sent. PS/1780 Lance Corporal Keith Freeling Markby of B Coy was a strong candidate having been educated at Marlborough College and employed as a solicitor's articled clerk. He was a writer in The Gasper. PS/1248 Private Edward Alan Fitzherbert Batty, of D Company, was an old boy of Oundle School and was in engineering; he was destined for the Rifle Brigade. PS/6403 Corporal Harold Arthur Langston Pattison, of A Company, formerly of Bedford Grammar School, was bound for the Bedfordshire Regiment. PS/2050 Sergeant Reginald Arthur Shann, of B Company, was educated at Tonbridge School and the Royal Agricultural College Cirencester; he went to the East Lancashire Regiment. The final man was PS/1594 Sergeant Norman Bradford Harris of A Company, 18/R Fus; he too was recognised with a 'Mention' in the June London Gazette. He departed around this time to the RFC.

Though men were departing for commissions there was still considerable uncertainty. Spaull considered it unlikely that the UPS Battalions would be converted to OTCs; 'We do not know what they will do with us. Last Tuesday 50 left for England (Oxford). We were told in orders that we should proceed to England on the 24th inst. We may move earlier so rumour hath it…'[56] Meanwhile other rumours surfaced and Medlicott mentioned in a letter that he: '… won't be able to tell you much of their future because no-one knows it … whether 18th [Battalion] is reformed or joins up with 19th [Battalion] is quite uncertain. WO suggest

---

54 *Letter from Harold Jones to his mother during the First World War 14 March 1916* <https://www.peoplescollection.wales/items/431006> (accessed 20 November 2022).
55 Merchant was destined for the Bedfordshire Regiment. He survived just over a month as an officer.
56 IWM: P.F.F. Spaull papers, 15453, 16 March 1916.

one thing – disbanding – G.H.Q. another – amalgamation.'⁵⁷ Another rumour which reached Philip Wright was that the battalions would be made up to strength with drafts from England and would be sent 'up the line' again.⁵⁸

After further training in 19/R Fus under company arrangements a further batch of 50 men was despatched to the UK for officer training on 19 March. They would go to No 3 OCB at Bristol University. This draft comprised sergeants from the battalion and they were accompanied by Captain Shipster who had presumably recovered from his wound sustained on 2 January. However, this group likely attended No 4 OCB contrary to the war diary. These fifty men comprised 42 men of 19/R Fus, six of 18/R Fus and two of 21/R Fus. Almost all the 19/R Fus soldiers were senior NCOs. PS/818 Sergeant Albert Rissik of C Company, 19/R Fus, was MiD in June 1916. PS/778 Lance Sergeant Arthur Lawrence Piper, of A Company, had been to Whitgift Grammar School and was a Stock Exchange clerk. The Essex Regiment would become his new regimental family. PS/864 Lance Sergeant Harold Saxon, of A Company, was a solicitor who had attended Sherborne School; he was gazetted into the Royal Sussex Regiment.

A third batch from 19/R Fus, which, combined with the other battalions amounted to about one hundred men, left for No 4 OCB at Alfred Street, Oxford, the next day under Lieutenant George Watts. Most of the 19/R Fus men were all corporals or lance corporals. To address this shortfall of NCOs thirty-two unpaid lance corporals were appointed. As drafts of men were despatched the gaps allowed other men to get experience as NCOs. Not every man sent back for OCB training was worthy of gazetting. PS/3741 Private Eric Connold, of A Company, 18/R Fus, was formerly at Brighton Grammar School and was a master draper. However, he was removed from his OCB course as 'unlikely to become an efficient officer'; he departed to 28/R Fus. PS/400 Lance Corporal Stanley Graham of 19/R Fus was also returned to 28/R Fus, presumably also as unsuitable. Meanwhile, some who might not be seemingly eligible turned out to be able to pass the course. PS/2667 Corporal Eric Forbes-Robertson was unlikely to become an effective officer being 48 years old, an artist and the battalion post corporal; 'I shall never forget that figure draped in the longest, ill-fitting greatcoat, hung on drooping shoulders and touching the ground as he walked.'⁵⁹ Nonetheless, as a battalion stalwart, he was sent for a commission and served in the Royal Garrison Artillery. Lance Corporal William Bentley of C Company, 19/R Fus, left with this draft. He recollected the reception at his OCB. His party was met at Keble College by the Dean (an Army captain) and the new-arrivals were allocated to rooms. That night they dined in the Hall as if they were students, in their private's uniforms barely cleaned of French mud, alongside the remaining college students who were mostly from overseas. Parades took place next day in the quadrangle and fresh 'officer-cut' uniforms were issued without rank badges.⁶⁰

According to the 19/R Fus war diary a fourth draft of fifty men from 19/R Fus departed on 21 March to No 4 OCB under Second Lieutenant John Evelyn. However, medal rolls suggest about fifty men of 18/R Fus and the same number of 21/R Fus went to No 3 OCB on 21 March; about fifty men of 19/R Fus went to No 2 OCB at Pembroke College, Cambridge. The draft of 19/R Fus going to Pembroke College was under Second Lieutenant Ivor de la Rue. He was not

---

57   IWM: W.B. Medlicott Papers, 1708, p.209.
58   FM: P.N. Wright Diary, RFM.2013.8.1/2, p.27.
59   LA: WW1/GS/0125: Papers of W.G. Bentley, p.5a.
60   LA: WW1/GS/0125: Papers of W.G. Bentley, p.16.

the best choice for this draft having been educated at Magdalen College Oxford. Harold Wilson departed with this draft though his diary ceased on 4 March. Another two brothers were with this draft; Francis Neville Dyson and William Edwards Dyson were from Kent who had both attended Clifton College, Bristol. They too would be separated with Francis joining the MGC after being gazetted on 4 August. William was only 17 in 1914 and presumably had to await his nineteenth birthday in November 1916 before he could be gazetted and sent overseas with the Devonshire Regiment. Another contributor to The Gasper also went; PS/192 Private Reginald Francis Clements of D Company, 19/R Fus, had been educated at Hereford Cathedral School. In addition, PS/6093 Private Ronald Munro Swyer, formerly of St Paul's School, Cheltenham, and aged 22, was another man who would be 'Mentioned' in June 1916 whilst at their OCB. He too departed from 19/R Fus in March 1916 and ended up with the RFC.

The procedure for seeing these men off, generally in batches of fifty, varied. Kenneth Norman, of 18/R Fus, departed on 22 March with another batch bound for No 2 OCB at Pembroke College, Cambridge. Colonel Scott spoke publicly to each cohort that departed that he was sorry the battalion was being broken up and he wished them luck for the future. Norman's batch left Renescure at 6:00 p.m. and marched to St Omer where they entrained on a third class carriage bound for Boulogne and a ship home.[61]

The marginal improvement in transport highlighted where they were going and that they might become commissioned officers. After arriving in Boulogne at 1:00 a.m. the draft stayed there for the night and waited for the leave boat which departed at 2:30 p.m. next day. The 18/R Fus draft was escorted by Second Lieutenant Norman James Holloway, a well-known Sussex cricketer. By 7:00 p.m. the draft arrived in London Victoria. As they were to report at Cambridge on 24 March, the 18/R Fus party, at least, was ordered to reassemble at Liverpool Street Station at 2:30 p.m. next day enabling the men to enjoy a night out in London.

A sixth batch departed 19/R Fus on 23 March, this time for No 1 OCB at Denham, Uxbridge. Seventy men went there under Captain Reginald Helps and Lieutenant and Quartermaster Frederick Paton. Presumably, as with previous draft conducting officer, they returned to 19/R Fus in France after the men were handed over at their destination. This draft mostly comprised 19/R Fus with a few men from the other two battalions. PS/907 Private Louis Solomon, a signaller with D Company, 19/R Fus, was one of this draft. He was an ex-Dulwich College boy and was a budding war poet. Another man was PS/3233 Private Cecil Sewell, a machine gunner with 21/R Fus, was also with this tranche of officer candidates. He too was a boy of Dulwich College and was an articled clerk. He was later gazetted into the Heavy Branch of the MGC. Two brothers from 21/R Fus departed to No 1 OCB on 24 March; PS/2832 Private Douglas Ian Inglis and PS/2833 Private Charles Allan Inglis were from Drogheda, County Meath, Ireland. They were destined for the Royal Dublin Fusiliers (RDF).

Morgan Williams left for No 2 OCB on 24 March 1916 destined for 10/RWF. This batch from 21/R Fus also included the Start brothers. PS/3230 Lance Corporal Arthur Ernest Start and PS/3231 Private William Start from Loughborough. Arthur was aged 27 and was a council school teacher; William, aged 26, was a clerk in the timber trade. Both were destined for the Leicestershire Regiment after No 2 OCB. 18/R Fus recorded sending seventh and eighth batches to Cambridge and Uxbridge respectively; presumably by this time 19/R Fus had fewer men immediately available for commissions and more training of them was needed.

---

61   LA: K.V. Norman Papers, WW1/GS/1188, p.119 and p.122.

Those UPS men attached to other units also missed out on relatively easy selection for commissions. To Leslie Ireland, languishing at an Infantry Base Depot, there was frustration in a letter to his father on 21 March 1916 after learning that his commissioning application had gone astray: 'Very annoyed to learn the contents of the latter & don't know what to do now. I expect I shall start all over again which will waste at least another 6 weeks. Obviously my papers have been lost between here & the WO [War Office] … The best place the WO could go to is hell!! … Fed up!!…' By 8 April Leslie Ireland was back at 19/R Fus and was 'sent for [by] the Colonel re Commission. My original papers made out last November were in his possession, by what means he obtained them I cannot say. He mentioned that all the cases he had to deal with – & that is anything up to 450 … He also said I could expect to be back in England shortly …'[62]

NCO training under company arrangements filled the time of the remaining men. Being a battalion attached to GHQ they were also required to put on a 'show' for visiting dignitaries. On 26 March the battalion paraded at 9:30 a.m. for an inspection by General Cadorna the Commander-in-Chief of the Italian Army. He expressed himself very pleased with the turn out of the battalion. Medlicott wrote home; 'I am glad as I thought the long line of dripping capes a dismal spectacle.'[63] Cadorna had attended the Allied Commanders' Conference on 12 March 1916 and departed on 26 March.

During this period the battalions were not without losses through illness and injury. On 29 March 1916 PS/3614 Lance Corporal Williams of 19/R Fus, aged 20, from Salisbury, died of spinal fever in hospital at St Omer.[64] In order to maintain the recruiting for the UPS battalions several good men from less conventional backgrounds had been enlisted. Medlicott commented on one batch of Reinforcements:

> A few more cingalees [Singhalese] have come out with the new draft. There are already 5 or 6 in the Bn … They are good athletic fellows & get on well with the others, so they are wrongly called 'the Blacks'; one, Aluwalkiri [sic], was at the base when I was there & is very profuse in his thanks to me … They feel the cold rather…[65]

The supply of immediately ready officer candidates had presumably ceased and those remaining would need further training and preparation before they were suitable to attend OCBs. Training continued for 19/R Fus for the rest of the month with Major Monro, the battalion 2IC, playing a prominent role in lecturing NCOs. Around this time Medlicott suggested that disbandment became a self-fulfilling prophesy as efforts were not made to integrate new drafts of men: '[The]… Colonel thinks the 18th will remain [as] a unit & [will] not be broken up. … there is lots of "spirit" left, enough to go around for the fresh drafts if the officers work hard – but they are very slack just now & I don't wonder. It's hard to carry on these days with any seriousness…'[66] However, the UPS battalions still had opportunities to gain themselves credit:

---

62 IWM: L.W.C. Ireland Papers, 14441.
63 IWM: W.B. Medlicott Papers, 1708, p.212.
64 Brian Moray Williams was buried in Longuenesse (St Omer) Souvenir Cemetery.
65 IWM: W.B. Medlicott Papers, 1708, p.212.
66 IWM: W.B. Medlicott Papers, 1708, p.213.

... the 18th [R Fus] had the good fortune to capture a big Fokker behind the lines on April 10th, 1916. They came on the scene when a private of the Royal Engineers was attempting to convey his delight at meeting a presumed French airman who was trying to restart his machine. The German, finding his hand warmly gripped, tried to look the part; but the 18th Royal Fusiliers instantly recognised the machine, with its Iron Cross, for what it was. They doubled, unslung their rifles, and, thinking the German was trying to pass papers to the other man, opened fire. But their zeal outstripped their performance. The sapper, now thoroughly bewildered, took to his heels; and the 18th took over the machine and the pilot.[67]

The pilot was passed to GHQ to be interrogated and the aircraft was taken to a nearby aerodrome where it was examined by the commandant of the St Omer sector of the Lines of Communication.

Another 'wave' of commissions

It was not until 10 April before the next batch on only twenty-five men from 19/R Fus were ready to depart to No 8 OCB at Whittington Barracks, Lichfield. Another batch of thirty men followed on 13 April destined for No 5 OCB at Trinity College, Cambridge. Routine training continued throughout April. Meanwhile, according to Sergeant Walter Medlicott 18/R Fus sent twenty men to No 5 OCB on 10 April and another twenty on 11 April. The former included Medlicott and Company Sergeant Major Philip Reiss of A Company, 18/R Fus. Medlicott's last letter arrived at his home a few minutes before he did on 10 April.

Whilst the number of soldiers in the battalion was slowly decreasing there were still new officers posted in to replace those who had already posted back to the UK. Second Lieutenants William Folds Cooper, Arthur Cecil Roper, John Lionel Wood, Cecil Wright-Ingle and an unidentified Second Lieutenant Scott, joined on 22 April 1916 and were posted to Second Army at Bailleul for instruction. As part of this training they were attached to battalions of the regular army and did duty with them in the trenches. Wright-Ingle was attached to 2/Leinsters and joined them in the line near Wulverghem for instruction. On 30 April the Germans launched a major gas attack against 24th Division; gas was released at about 12:45 a.m. followed by a heavy bombardment. A German party attempted to enter the British trenches but were prevented by the British wire. During this action Second Lieutenant Wright-Ingle, attached to learn about trench warfare, was killed whilst experiencing it.[68] His front-line career lasted a week.

---

67 H.C. O'Neill, *The Royal Fusiliers in the Great War* (London: William Heinemann, 1922), p.17.
68 Second Lieutenant Cecil Hubert Wright-Ingle, aged 32, was gazetted into 19/R Fus in July 1915. He had attended Cambridge where he had studied for an MA and was a barrister from Eastbourne. He went to France in March 1916 and joined 19/R Fus on 22 April 1916. After instruction at the Second Army School, Wright-Ingle was attached to 2/Leinsters. He is buried at Ration Farm Cemetery.

Infantry battalion postings

Not everyone was lucky enough to get gazetted. The remainders of each UPS battalion were to be posted as privates to other infantry battalions. Communications were not always maintained between the different cohorts. Men were not selected for commissions for numerous reasons; 'Forrest and Bromley were turned down, to everyone's surprise. Before the interview we had all decided that Forrest was the most likely to get taken on. But Mackarness had taken a dislike to him and his advice prevailed. Various strings were pulled in England to try and get this decision reversed, but all to no purpose.'[69] In the initial draft from 21/R Fus not a member of No 3 Company, commanded by Mackarness, was present, suggesting he had not recommended many of his men. This later changed and one draft took over half of Spurrell's platoon. Three of Spurrell's friends; Brandreth, Forrest and Bromley were not accepted for commissions and were sent to 23/R Fus; those who left for OCBs hoped they would get justice and be sent on to an OCB in the future. Brandreth and Forrest were killed before the end of the year; Bromley had to wait until late 1917 for a commission.[70]

Others remained in the ranks for different reasons. Because of suffering from rheumatism Private Claude Briggs was transferred to 26/R Fus rather than being gazetted. One of Morgan Williams' chums, Lionel Bruce 'Tim' Greaves, missed out on selection having been earlier evacuated with trench fever. No 2 Company of 21/R Fus was also delayed from departing due to an outbreak of measles which required quarantine. Medlicott's posting to a survey unit likely delayed his posting to the UK for a commission until 11 April 1916.

To Etaples

Later in March and April the parades for 21/R Fus slackened off. Other than a parade before breakfast, a route march followed by a couple of hours of drill and, other than any guards or camp duties, the rest of the day was left to the men to enjoy. The existence was described by one as '…very lazy and enjoyable…' due to free time and improving weather.[71] On 20 April the following was recorded by the GHQ Adjutant General Branch; 'The following Battalions to be broken up for purposes of drafting to other Battalions of the same Regt: 18th, 19th and 21st Royal Fusiliers.'[72] However, the final fate of these three UPS battalions only became known to their men on about 23 April. Information was received that all three were to be broken up; next day orders arrived for 19/R Fus to proceed to Etaples to accomplish the disband-

---

69   IWM: H.W. Spurrell 2138, p.69.
70   PS/6629 Private James Gerald Forrest was a clerk for a cotton manufacturer from Darwen, Lancashire. He enlisted in April 1915, was blown to pieces outside Delville Wood whilst carrying a message on 27 July 1916, aged 27 and is buried in Longueval Road Cemetery. PS/2492 Private Arthur Killingworth Bourne Brandreth from Birkenhea was educated at Wadham College and enlisted in 1914. He was killed whilst serving with 23/R Fus on 1 November 1916. Brandreth is commemorated on the Thiepval Memorial. PS/4000 Private Frederick Farrar Bromley, from Bolton, later served with the MGC before being gazetted into the MGC in December 1917 and afterwards with the Tank Corps.
71   FM: P.N. Wright Diary, RFM.2013.8.1/2, p.26.
72   TNA: WO 95/26/1: GHQ Adjutant General War Diary.

ment of the battalion. However, in the meantime a draft of 71 men, under Lieutenant Palmer, arrived from Etaples as replacements on 25 April. Next day, Second Lieutenants Clapton and Gillmore, joined the battalion. The humour of this situation was not lost on the war diarist, Lieutenant Philip Ingleson. At 7:15 a.m., on 27 April, 19/R Fus paraded and moved to Etaples via Wardrecques. On 27 April 28 officers and 602 men from 19/R Fus and 17 officer and 501 men from 21/R Fus arrived at Etaples from Wardrecques.[73] 25 officers and 551 men from 18/R Fus arrived at Etaples two days earlier having entrained at Renescure. Over half of the men of each battalion still needed to be disposed of.

The final tasks for the disbandment of the remainder of each of the three battalions would occur at the Infantry Base Depots at Etaples. Though orders to move to Etaples came suddenly, they were greeted positively as being a move nearer to home. According to Hugh Spurrell 21/R Fus reached Etaples on May Day and stayed for three weeks; 'They were strenuous days which we did not enjoy'.[74] RSM Fletcher became even more determined to make the remaining men of 21/R Fus into the best soldiers and wanted to demonstrate their superiority to the other battalions. Spurrell recalled that they outdid the other battalions with ease but made themselves more tired in doing so.[75] Colonel Denison had returned to command the battalion 'although he worked us hard, he soon won the admiration of all. Twice he marched us down to Paris Plage [4 miles from Etaples], gave us lectures on the beach, an hour to ourselves in the town and the took us back again.'[76]

Once at Etaples the three battalions were controlled by the Camp Commandant there. They arrived at 3:30 p.m. and were accommodated in tents at 'U' Camp, Etaples, in the sand dunes. After a time at Etaples 21/R Fus was divided with those who had not applied for, or not received, commissions, being separated from those who had. The battalion transport left first, taking its vehicles and animals, by road, to the depot at Abbeville. Many of the residual enlisted men that were not considered suitable for commissions started to be dispersed to other battalions. Two drafts of men were despatched to 23/R Fus and 24/R Fus respectively. On 28 April a draft of fifty men of 21/R Fus was sent to No 18 Infantry Base Depot destined for 11/R Fus (18th (Eastern) Division). This draft included Philip Wright who wanted to escape cook-house duties:

> My deliverance came in the evening, when fifty of us from B Coy, were ordered to 'pack up', as we were to march to another part of the camp to the 18th Infantry Base Depot, to be transferred to the 11th Battalion Royal Fusiliers. It was to be "Good-bye" to the 21st for good – for which I was sorry. It was a great pity… that the old Batt should have been 'broken up', but as it had been destined for commissions, it could not be helped. We were now wondering what sort of 'crush' the 11th were, and on arrival at the new quarters, tried to find out as much about them as we could…[77]

The 11/R Fus draft left Etaples on 11 May bound for the front. On 28 April 100 men of 19/R Fus were sent to No 12 Infantry Base Depot bound for 9/R Fus (12th Division). After further routine work and cleaning of accommodation a further set of drafts from 19/R Fus were

---

73 TNA: WO 95/4026: Etaples Base Commandant War Diary.
74 IWM: H.W. Spurrell 2138, p.73.
75 IWM: H.W. Spurrell 2138, p.73.
76 IWM: H.W. Spurrell 2138, p.74.
77 FM: P.N. Wright Diary, RFM.2013.8.1/2, p.29.

provided; 40 men went to 12/R Fus (24th Division) under Lieutenant Percy Cottrell; 30 men, under Lieutenant George Watts, went to 17/R Fus (2nd Division) and 30 men under Second Lieutenant Robert Long went to 22/R Fus. Many of the officers of 21/R Fus periodically left; Colonel Denison had allegedly secured suitable employment for each of them. Spurrell's company commander, Betts, went to the Special Brigade, RE; as did Lieutenant Christopher Rathbone. Mackarness went to 22/R Fus and later served as a staff captain.

Training in May 1916 for 19/R Fus consisted of voluntary parades and lectures to train those men applying for commissions. One man described it as a 'course of OTC training'.[78] These lectures were well-attended and the men were generally very keen. The handwriting in the 19/R Fus War Diary changed in early May as Lieutenant Ingleson and Second Lieutenant Charles Barrington-Brown left for staff posts in the UK. Second Lieutenant Bertram Hubbard took over as Adjutant. Next day, an officer of 22/R Fus took 28 men as a draft to 23/R Fus; another dozen men were sent on 7 May. On 6 May the battalion transport, which had travelled by road, finally rejoined 19/R Fus. On 8 May Lieutenant Cottrell also departed to the UK and Second Lieutenant Emerson Stevenson joined 19/R Fus. The battalion had further departures; the 21-strong battalion band went, under the Drum Major, Sergeant Joseph Castree, to join 33rd Division. Also on 10 May the battalion moved to a new camp to join No 33 IBD to make space for No 41 IBD. After a few more days of routine training under company arrangements the disbandment of the battalion accelerated. On 15 May Captain Meares and Second Lieutenants Eathorne and Laidlay, were all transferred to 24/R Fus and joined that battalion on the same day.[79] Meanwhile, on about 11 May 1916, about a dozen men of 18/R Fus were sent to 99th Brigade and were split between 22/R Fus and 23/R Fus; this draft included PS/6673 Private Harry Elliott Harvey, who joined the former, and served as an infantryman for the remainder of the war. He recorded some of his memories as 'Battle-Line Narratives'.

Next day, the 16th, fifty men of 19/R Fus left for No 10 OCB at Gailes in Scotland under Second Lieutenant Stein.[80] At this last moment Leslie Ireland's commission papers came through and he departed with this intake for a commission in the Manchester Regiment. The following day Major Raymond Pelly Houston Monro left for HQ Fourth Army where he was to be given instruction as an assistant provost marshal (APM). The same day Captain J.G.C. Leach of 18/R Fus departed for the RGA, HQ Fourth Army. Three large batches of 19/R Fus personnel departed on 18 May. 97 men went to No 5 OCB at Trinity College, Cambridge, under Captain Helps; 42 men likely went to No 4 OCB in Oxford, under a conducting officer from 18/R Fus, and 135 men went to No 6 OCB, at Balliol College, under an officer from 21/R Fus. On the same day Major Henry Tuite and Second Lieutenant Robert Long went to No 11 OCB in Camberley as instructors.[81]

---

78 IWM: L.W.C. Ireland Papers, 14441.
79 Captain Cecil Stanley Meares had enlisted in 1914 and was gazetted to a commission in 19/R Fus in February 1915. He was promoted captain and commanded C Company, 24/R Fus, which took part in an attack on 30 July 1916, 600 yards east of Waterlot Farm. Meares was killed at the German wire leading his men and was buried in Delville Wood Cemeteryl. Second Lieutenant Francis John Eathorne, aged 24, was killed on 31 July 1916. He is commemorated on the Thiepval Memorial. Nigel Laidlay, aged 23, hailed from Strathpeffen, Ross-shire.
80 Second Lieutenant Oswald Frank Stein was transferred to the Grenadier Guards in June 1916. He had enlisted in 19/R Fus in September 1914 having been educated at Tonbridge School.
81 Formerly PS/1168 Private Henry Mark Tuite; he was educated at Brightlands Prep School, Dulwich;

The RSM of 21/R Fus departed a couple of days before the final drafts left for the UK:

> He carried on his duties to the last minute in the usual way. His face showed no sign that any change was taking place. Then he shouldered his pack and stalked out of the camp alone. Those who saw him raised a cheer; it was meant as a compliment, not as a sign of joy at his departure. Everybody realised that Fletcher, though a holy terror, was really the mainstay of the battalion. His new C.O. got a jolly good man.[82]

On 17 May half of the men of 21/R Fus departed Etaples for the UK under the CO of 18/R Fus, and, according to Spurrell, on 18 May the remainder of 21/R Fus departed under Colonel Denison. They travelled to Boulogne and crossed to Folkestone next day. The Commandant at Etaples recorded on 18 May that a final draft of five officers and 657 men from the three UPS battalions departed Etaples for OCBs in the UK.[83]

The medal rolls also record approximately 115 men of 21/R Fus, mostly of No 4 Company, and including Hugh Spurrell, who departed for the UK on 27 May 1916. This party were all destined for No 10 OCB at Gailes, in Scotland. These men, and the previous draft, are presumed to be a combination of the least credible for commissions; the most 'die hard' advocates of the UPS battalions; those that had needed more training or those that had been previously overlooked by their officers. One example was PS/4113 Lance Corporal Robert Cyril Briscoe Jones, from Whalley Range, Manchester, was MiD in June 1916, presumably due to good work whilst with 21/R Fus. He was sent home with this batch and was later gazetted into the Lancashire Fusiliers (LF). He later became a leading geologist. Conversely, PS/4170 Private Gerald Openshaw, aged 22, from Winton, Manchester, was an old boy of Eccles Grammar School and went to No 6 OCB. He failed to complete the course having contracted an 'avoidable' illness and was instead posted to 5/R Fus.

This final OCB draft left 49 men in 19/R Fus on 19 May; all but three were men classified as 'permanent base' or 'employed'. The former ten men were transferred to 20/R Fus on 21 May, the remainder were presumably recycled by the Infantry Base Depots and joined other units. Lieutenants Watts and Palmer were attached to No 35 and No 20 IBDs on 20 May. The Adjutant of 19/R Fus, Second Lieutenant Hubbard, departed to the UK on 22 May.[84] On 23 May Lieutenant Colonels Denison and Wolrige Gordon went on leave to England; the latter was to report to the War Office for re-assignment on expiry of leave. Certainly 19/R Fus had officially ceased to exist as a battalion. Though records for the other two battalions do not survive with these last moves the formal activities of all three battalions were over. Whilst 2/RWF may have derided the Public Schools battalions for being 'Chocolate Soldiers' the UPS men, and others, did spend a lot of money in their battalion canteens. When the canteen for

---

Dulwich College and RMC Sandhurst. He had previously been a second lieutenant with 2/Border. With this education he had been employed as a schoolmaster. He was aged 28 on enlisting with 19/R Fus in September 1914. He was gazetted on 27 October 1914 and was later the adjutant. Promoted to Major in September 1915, Tuite later commanded No. 3 Company from early March 1916.

82　IWM: H.W. Spurrell 2138, p.74.
83　TNA: WO 95/4026: Etaples Base Commandant War Diary.
84　Captain Bertram John Hubbard, aged 22, an Old Etonian, had only been adjutant since 4 May 1916. He later joined the Grenadier Guards.

21/R Fus was wound up on disbandment there were sufficient profits to buy a gramophone and twenty records for each company to dispose of the funds.

How did this steady decline and departure play out amongst the small teams of men in the UPS battalions? Of eleven men of Hugh Spurrell's section in 21/R Fus, two went with the first batch on 14 March (the section commander, PS/4102 Lance Corporal James Eckersley Hill and PS/3132 Private Hugh Cecil Patterson who was both the platoon sanitary man and an Old Marlburian). They were followed a week later by another three men; PS/4041 Private Herbert Dickie of Manchester Grammar School and Manchester University Medical School, PS/6652 George Alvarez Richardson of the Royal Grammar School and Sheffield Polytechnic and PS/4282 Private Thomas Bagaley Wrack who (according to Spurrell) had an appalling accent and could be garrulous and over-excitable. No more men departed until 27 May when four more men went. PS/2627 Private George Wilfred Dymond had attended Cambridge University and was destined for the Cheshire Regiment; PS/3315 Private Thomas Charles Turton; PS/2824 Norman Stuart Hunt and PS/3261 Private Hugh Spurrell of Llandovery School and Cambridge. The remaining two men were posted to 23/R Fus as privates; PS/6629 Private James Forrest and PS/4000 Private Frederick Bromley (known as 'Brimstone' due to his yellow hair). Only one would not be gazetted. Of these eleven, five would be dead within a year.

Of Walter Medlicott's section in No 13 Platoon, D Company, 18/R Fus the first man to leave was PS/2082 Private Horace Fleming Smith. He had been educated at Whitgift Grammar School; he departed 18/R Fus on 20 March bound for No 4 OCB. Two days later three more men (PS/1930 Corporal Philip John Pither, formerly of Monmouth Grammar School; PS/7818 Private Alfred Harold Pickard, who had been educated at Strand School and Kings College London and PS/7538 Private John Frederick Rogers) departed for the UK to join No 2 OCB. On 23 March PS/3700 Private Percival James Burton departed, likely also for No 2 OCB. One man departed on 10 April (PS/1209 Private Frederick Roberts Arnold); three more left a day later (PS/1811 Sergeant Walter Medlicott previously of Winchester College; PS/7537 Private William Pascoe Rogers, formerly of University College School and PS/1196 Private John Homfray Addenbrooke, formerly of Marlborough College); they all likely went to No 5 OCB. The last man going for officer training was PS/3960 Private Harry Vander Weyden, of West Heath School, Hampstead, who left on 19 May 1916. PS/7329 Private Leonard Bailey of Medlicott's section became Captain Price-Edwards' servant and departed. Three others would remain as enlisted men and were sent to other battalions (PS/7219 Private George Davies, PS/7493 Private Harry Oliver Essex and PS/7222 Private Rowland Peace). By luck this section fared better; only two of the four enlisted men were killed.

Many had been critical of Stuart-Wortley. However, though Lieutenant Colonels Wolrige Gordon and Denison departed to new jobs, their men had a healthy respect for them. The following open letter concerning Lieutenant Colonel Wolrige Gordon was published:

> We are back again in Blighty, but, ... we have one tremendous regret – we have lost our colonel – the best and most wonderful colonel a regiment ever possessed ... We would have gone anywhere and done anything for him ... He was and is our ideal – a great gentleman and a great officer. Wise, tactful, kind – keen, strong and clever, a man of men and a born leader ... We shall remember that far-away look in the eyes – which could change suddenly to a keen, searching glance that missed nothing, and the merry smile which lit his face when he was amused ... I speak not only for myself but for many, nay all of us here in

training, when I say that we shall be finer gentlemen and more efficient officers for having had such a colonel as ours …[85]

Based on his experience, paternalism, and professionalism, Denison appears to have been well-respected.

Meanwhile, 16/Mx was treated differently and returned to the Front Line on 24 April. They joined 29th Division because their place in 100th Brigade had been taken by 1/9 HLI (Glasgow Highlanders). 16/Mx, like 20/R Fus, would also face tribulations over the summer of 1916 that might make many men envy those who had been gazetted from the three disbanded UPS battalions.

---

85 *The Gasper*, 29 April 1916, Vol. 2, No. 12, p.2.

15

Prelude to the Somme – 20/R Fus in the Spring of 1916

The whole thing was such a kolossal adventure

> The first I knew … was the impression I got that the floor of the trench had come up and hit me. I have never experienced an earthquake, but I should imagine it would be like this …[1]

> The whole thing was such a kolossal [sic] adventure … Even the patrols, when one wriggled and floundered in that leprous wilderness, like a landscape in the moon …[2]

20/R Fus in France – March to June 1916

Whilst the remainder of the UPS Brigade departed for a new location and the prospect of a change of scenery this was not the case for 20/R Fus who were to remain in the Béthune area for another four months. This was an important period where the endurance of the battalion was strained but where its fighting spirit was maintained. This further period of routine trench warfare did not seem to deteriorate morale and enthusiasm.

The first day of March found 20/R Fus in the sector north of the La Bassée Canal. There was a steady stream of casualties through 'routine' trench wastage caused by snipers, trench mortars and rifle grenades. PS/4400 Lance Sergeant Frederick Austin of A Company was killed on 1 March and two men were wounded.[3] PS/5181 Private James Lancaster was struck down next day.[4] Three other men were wounded on 2 March; one returned to duty, another did not. During this period of trench duty there were occasional volleys of whizz-bangs but as these were already registered they could have a devastating effect:

> A shelter a short distance away from ours was struck and demolished, burying Bracewell shallowly. He was quickly dug out, but just as he was free, and we were congratulating him

---

1 IWM: J.N. Dykes Papers, 7378, p.24.
2 Gibson, *Postscript to Adventure*, pp.140–141.
3 PS/4400 Lance Sergeant Frederick Austin A Company, 20/R Fus, aged 22, was from Whalley Range, Manchester. He was killed on 1 March 1916 and is buried in Cambrin Churchyard.
4 PS/5181 Private James Lancaster was 24 and from Blackburn. He is buried in Cambrin Churchyard.

on his lucky escape, there was another salvo, and a large splinter penetrated his body. The stretcher bearers carried him away, but we heard that he died shortly afterwards. It really looked as if he was fated to die!⁵

This was PS/4552 Private William Bracewell, aged 22, from Blackpool. He succumbed to his wounds on 2 March 1916 and was buried in Béthune Town Cemetery. One of the men who came to Bracewell's aid was Corporal Ernest Holden who recorded in a letter home, 'I received some shades of glory by going through the casualty lists … It was very little, just a rap on the back of a hand with a piece of shell. I was very lucky as the chap I was helping out of a hole got a piece clean through him. I hear since that he has died.'⁶

On 4 March the war diary recorded that another man died of his wounds along with three other casualties occurring. The fatality was likely PS/7414 Private Francis Baxter who died on 5 March and was buried in Cambrin Churchyard. As if these casualties were not enough on 6 March a German sniper managed to draw a bead on Captain Heinemann, who was commanding A Company, whilst he was with a wiring party.⁷ His wounds proved fatal two hours later. On 4 and 6 March there were heavy falls of snow making trench life even more difficult to endure.

On 8 March 20/R Fus was relieved by 10th Battalion Welsh Regiment (38th Division). The relief was complete by 8:35 p.m. and the battalion marched to le Quesnoy. B Company remained in reserve for 5/Scot Rif at Cambrin Support Point. On 10 March PS/5762 Private Arthur Turner died of wounds in Béthune and was buried there; he had been wounded in the back and legs by a German rifle grenade.⁸ On 12 March 20/R Fus returned to the line in the Cuinchy subsection having relieved 5/Scot Rif. James Dykes and No 10 Platoon were in St Andrew's Trench in support. Here they had a hard time with German trench mortars and rifle grenades and suffered more casualties than during other tours. Sergeant Taunton of No 10 Platoon was wounded and carried off down the line. He was 'lucky' and rejoined the Battalion on recovering. There were lighter sides to the shelling; Private Hardy was one of 10 Platoon's more cheerful members and he was using a latrine when a German trench mortar bomb landed nearby; luckily he suffered no injury but his language was less than courteous to his enemies! To add to the danger there was an outbreak of stomach trouble in 10 Platoon; likely due to drinking contaminated water.

'… some liveliness that was going on…' – A German Bombing Attack

There was no 'lighter side' to be enjoyed on 13 March 1916. At 4:30 a.m. the battalion stood-to in the trenches. The 20/R Fus war diary stated; 'Mine blown by enemy at A21d7.5.2.5 at 6.15

---

5 IWM: J.N. Dykes Papers, 7378, p.11.
6 LIM Archive: Letters of E.A. Holden, courtesy LIM, p.75.
7 Captain John Walter Heinemann was aged 26 and was from Brompton Square, London. He was educated at Marlborough where he was in the OTC and was at Balliol College, Oxford, where he was a rower. He worked for a publisher and enlisted in September 1914 and plated rugby for the UPS. He was one of the original officers of the battalion being selected for a commission in 20/R Fus being gazetted on 27 October 1914. He was buried in Cambrin Churchyard.
8 PS/5762 Private Arthur Norman Townsend Turner, aged 19, was from Pendleton, Manchester. He was educated at Manchester Grammar School and was a keen athlete and swimmer and played lacrosse and cricket. He worked for Williams Deacon Bank. He was buried in Béthune Town Cemetery.

a.m. Crater joined ETNA and GIBBONS: No casualties. Bomb attack by enemy in evening: reported Captain E.T. Wright killed by rifle grenade, 2nd Lieut T.S. Pope wounded, 4 OR killed, 8 OR wounded (one remaining at duty). We hold the near lip of crater.'[9]

Vincent Platt succinctly recorded his impression of the mine explosion; 'Quite a funny sensation. Don't want to be much nearer a mine.'[10] Members of D Company 20/R Fus, under Captain Eric Wright, were detailed to extend No 2 Sap as far as the new crater. At about 10:00 a.m. PS/5261 Lance Corporal John McDowell of D Company was shot in the head by a sniper and killed.[11] When it got dark this party could go into the open, presumably sheltered from enemy view by the near lip of the crater and dig between the sap and the crater. According to Vincent Platt:

> … Sid & I were told off to go and dig in the crater. Crawled to the edge of the crater with machine gunners and saw the Germans working. The Germans began getting into the crater so we went back to sap. Machine gunners & bombers then shifted the Germans from the crater. Germans then bombed us and Sgts Jeffcoat & Williams were wounded. Hunter, Davie Newell, Capt. Wright killed. Awful experience and providential escape.[12]

According to the Manchester Grammar School magazine, PS/5937 Sergeant Alec Williams, 'was just throwing his last bomb when a German bomb fell near and shattered the upper part of his right side. Directly afterwards another bomb fell and shattered the lower part. He had thirty wounds in all, one piece of shrapnel entering his right lung… "Give it 'em, boys! " he yelled, as he lay helpless on the ground.'[13]

The Germans had also taken the opportunity darkness offered to attempt to occupy the crater. According to Sidney Platt about forty Germans had entered the crater. According to Sidney Platt the Germans were, 'engaged by bombers and machine gunners and were wiped out…', but in the process the Germans inflicted several casualties on D Company.[14] At some point in the action a British bombing attack was launched to attempt to capture the far lip of the crater. Godfrey Skelton recorded that:

> … the Germans blew a large mine in No-man's-land on our Company's front (D Company). My platoon, Number 14, was in support trenches at the time. My great friend, Sergeant L.A. Phillips, was ordered to take his bombing section and occupy the lip of the crater formed by the explosion. This was the normal practice as the lip of the crater, being higher than the

---

9   Second Lieutenant Thomas Stephen Pope was educated at Westminster School and had enlisted in 10/R Fus. He was gazetted to 20/R Fus on 30 January 1916 and joined in February. He was wounded on 13 March and later joined 29/R Fus in August 1916. TNA: WO 95/2423: 20/R Fus War Diary.
10 IWM: V. Platt Papers, 17681, p.27.
11 PS/5261 Lance Corporal John McDowell, aged 21, from Shankill, Belfast, had been educated at the Royal Belfast Academical Institution. He had been employed by the Royal Insurance. He enlisted in September 1914 and was with No 13 Platoon, D Company, 20/R Fus.
12 IWM: V. Platt Papers, 17681, pp.27–28.
13 PS/5937 Sergeant Alec Williams had attended Manchester Grammar School. He was treated at the Red Cross Hospital, Daisy Bank Road, Victoria Park. He recovered sufficiently from his injuries to join the Labour Corps and was discharged in March 1919. Anon. Author, *ULULA, The Manchester Grammar School Magazine, Vol 44, June 1916, No 326* (Manchester: Galt & Co, 1916), June 1916, p.115.
14 IWM: S. Platt Papers, 17681, p 27.

ground around it, provided observation over the enemy's trenches. … The enemy had already occupied their lip and were in strength, and in trying to dislodge them by bombs and rifle fire Sergeant Phillips was killed instantly by a German bomb. He was the first of my many close friends to be killed. His body was brought to me so that I could go through his effects, and send them, and a letter, to his sister in Wales.[15]

Captain Wright was the most senior officer on the spot and likely ordered Phillips' attack. Wright was described thus: 'He'd a touch of red in his hair, high cheek bones; a scholar and a civil servant but a born soldier, too, who said he wouldn't have missed the war for anything… And how soon he was killed… He was in a saphead – attracted there by some liveliness that was going on.'[16] Wright had left his part of the line to assist in organising the bombing counterattack.[17] According to Dykes; 'Capt Wright … was killed on the 13th March, a rifle grenade hit him on the head & blew half of it away. We were very sorry to hear of his death, there was not a single man in in the battalion that did not regard him as a brother…'[18] Dykes

PS/5457 Sergeant Louis Augustus Phillips was aged 38 and was from Newport, Monmouthshire. Having represented Wales at Rugby in 1900, he was forced to retire due to a knee injury and became a champion golfer instead. Phillips was commanding No. 16 Platoon, D Company, when he was killed. He is buried in Cambrin Churchyard. (From the papers of Godfrey Skelton, reproduced courtesy of James Skelton)

considered that Wright was one of the most popular officers in the battalion. Up until his death Wright had administered the battalion coffee shop.[19] The Chaplain, Reverent Ernest Mannering, remembered Wright; 'A singularly attractive person … pleasant voice. A man who

---

15   IWM: G. Skelton Papers, 13966, p.42.
16   Hodson, *Return to the Wood*, p.26.
17   Hodson, *Grey Dawn*, p.204.
18   IWM: J.N. Dykes Papers, 7378, p.83.
19   Captain Eric Tracey Wright, aged 22 and from Bayswater had been born in Bombay. He was educated at Wadham College, Oxford and was a member of the OTC. He enlisted in 20/R Fus in September 1914 and was in the first tranche of officers gazetted on 27 October 1914. He was buried in Cambrin Churchyard.

made a success of anything that he touched, who rarely, if ever, made a mistake. A delightful companion in the mess ... with a real fund of humour.'[20]

The action died down with the British and Germans holding their respective crater lips. Vincent and Sidney Platt had the unenviable task of helping remove the dead; 'Carried the dead back to mortuary in Harley St. heavy work. Were then allowed to rest practically all night.'[21] In addition to Captain Wright and Sergeant Phillips the following men were all killed on 13 March 1916; PS/7755 Private John Hunter, aged 20, from Hartlepool; PS/5390 Private David Newell, aged 21, from Belfast; PS/5691 Private George Stevens, aged 27, from Cuckfield, Sussex and PS/5929 Private John Whalley, aged 24, from Chorley, Lancashire,.[22] It is likely that PS/5297 Private Arthur Massey, aged 19, a builder from Alderley Edge, Cheshire, died two days later in Béthune of head wounds which he received on 13 March,.[23]

The work of removing the dead was presumably not able to be completed that night and on the night 14–15 March Sid Platt recalled; 'Fairly quiet night. Went up to sap for my equipment. Helped to carry out Whalley (killed). Very heavy & nasty job. The smell was horrible. All our fellows worn out with work. On guard. Vince on carrying ... helping with Macdowel [sic] (killed). Rifle & machine gun fire active. Germans straffed [sic] us a little at night.'[24]

For Ernest Brierley and other the members of the battalion not in the trenches on 14 March the day was less strenuous but still with some danger; '[Band] Practices morn[ing] and aft[ernoon]. Football. Game interrupted after tea by Allemandes shelling...'[25] On the 16th, PS/5527 Private Frederick Rice, of D Company, was severely wounded by bomb fragments whilst in the trenches. He was carried to the rear by Sid Platt.[26] Other companies may have had an easier tour though when in the line the trenches had to be kept in good order. It was around this time that James Dykes considered this was taken too far when Captain Yorston, OC C Company, had some of his men out in the open one night picking up discarded food tins which had presumably been thrown from the trench.

On 17 March 20/R Fus was relieved by 1/Mx and marched to the *Ecole de Jeunes Filles*. Next day, according to Ernest Brierley, the battalion played 2/RWF at football and thrashed them 9–1.[27] However, Dr Dunn, of 2/RWF, complained that this was because 20/R Fus exempted their team from trench duty and were therefore unbeatable.[28] The 20/R Fus team played 1/Cameronians on the 22nd and beat them five-three.

20 IWM: Reverend E. Mannering Papers, 6756.
21 IWM: V. Platt Papers, 17681, p.28.
22 Four men of D Company were buried in Cambrin Churchyard: PS/7755 Private John Hunter, aged 20, was from Hartlepool. PS/5390 Private David Newell, aged 21, from Belfast, likely served with No 13 Platoon, D Company. PS/5691 Private George Stevens, aged 27, from Cuckfield, Sussex, served with 6 Section, No. 14 Platoon, D Company. He was born in Preston and enlisted in Brighton. PS/5929 Private John Whalley, aged 24, from Chorley, Lancashire, was a colour mixer for a calico paint manufacturer.
23 Arthur Ashton Massey was aged 19, from Alderley Edge, was employed with a building firm. He was buried in Béthune Town Cemetery.
24 IWM: S. Platt Papers, 17681, p.28.
25 Diary of Ernest Brierley, unpublished. With the permission of Angela Jennings. Entry for 14 March 1916.
26 Rice was discharged in March 1917 due to his wounds.
27 Diary of Ernest Brierley, unpublished. With the permission of Angela Jennings, 17 March 1916 entry.
28 Dunn, *War the Infantry Knew*, p.187.

Once out of the line there was a greater than usual effort at smartening up according to Dykes:

> One day, our colonel was pleased to have us turn out for a route march – showing the flag – or at least, showing any natives who might be interested how spic and span the 20th could be! The Colonel suitably mounted and truly resplendent along with his Adjutant, similarly accoutred heading the column; we, the rank and file, in full marching order, also resplendent; the officers and N.C.O.s had seen to that, the inspection when we paraded being most exacting, almost terrifying …[29]

There was a heightened level of spit and polish in 20/R Fus around this time in preparation for an inspection of the battalion by General Gough on 21 March 1916:

> No fatigues were allowed to interfere with the object of presenting as proud and martial a spectacle for the satisfaction, and, we hope, the delight of the inspecting general. The whole of the day was spent in cleaning and furbishing uniforms, buttons etc. The platoon officers and N.C.O.s were on their toes and very critical indeed of our efforts, examining our results closely, and being unsparing in their condemnation of anything below excellence.[30]

Dykes recalled the inspection:

> General Gough, small in stature, but high in rank, arrived in due course. Salutations and impeccable drill followed. The General and his attendants moved along the line of infantry …General Gough stopped occasionally, asked a few questions of a stiffly standing soldier, then recommenced his inspection. I was surprised that I should be a subject for his interest, and in reply to his interrogation, supplied my name, occupation in civilian life, prospects, and did my Army Service perhaps have any effect, derogatory or otherwise, on my future … When the General had taken the salute at the "March Past", he departed, I suspect, with many sighs of relief…[31]

Colonel Bennett was undoubtedly happy with the performance of the battalion and granted the men the rest of the day off. During this rest period Captain Yorston departed ill on 20 March. His position commanding C Company was filled by the Adjutant, Captain Maxwell.

On 25 March 20/R Fus moved to Annequin South and relieved 2/Worcs R. For two days 20/R Fus was in reserve. More new drafts arrived from the Reserve Battalions in the UK as Spring 1916 progressed. One such new arrival was PS/9019 Private Frederick Shield, aged 24, from Leicester. He joined 20/R Fus on 23 March 1916 having been trained by 29/R Fus and travelled via the 'Bull Ring' at Etaples. On arrival he was assigned to a company and sent off for mining and carrying fatigues at Annequin. Shield kept a skeleton diary covering his service in France.

---

29 IWM: J.N. Dykes Papers, 7378, p.12.
30 IWM: J.N. Dykes Papers, 7378, p.13.
31 IWM: J.N. Dykes Papers, 7378, p.13.

On 28 March 20/R Fus moved into the Auchy Right subsector and relieved 5/Scot Rif. This position was further to the right than occupied before and was from Boyau 1 to Boyau 8. 1/Cameronians were on the left and 7/Suffolks (12th Division) were on the right facing the Hohenzollern Redoubt. A, B and D Companies were in the line; Shield was in B Company and this was his first tour of the trenches. C Company was in support; Dykes and his comrades occupied Railway Keep. During this tour of trenches two companies of 16th Battalion Nottinghamshire and Derbyshire Regiment (16/Notts&Derby) were attached for instruction and relieved A and D Companies for much of this tour. Shield remained in the line throughout before being relieved by 16/Notts&Derby on 1 April and returning to the *Ecole de Jeunes Filles* in Béthune. Meanwhile, Dykes recalled his platoon being billeted in Montmorency Barracks.[32] During this tour in the line PS/5754 Private James Thornber was wounded in the arm whilst out on a working party.

On 4 April the battalion marched from Béthune to Annequin South and relieved 5/Scot Rif. Whilst in reserve 20/R Fus supplied working parties. On 7 April 20/R Fus had 17/Notts&Derby attached for instruction in trench warfare and 20/R Fus went into the line with A and D Companies and two companies of the Notts&Derby. The trenches on this occasion were the Auchy Right Subsector. As such B and C Companies, 20/R Fus, did not arrive in the trenches until the evening of 9 April, to relieve two 17/Notts&Derby companies who had completed this training period. Dykes, of C Company, recalled the sector taken over was a 'veritable nest of rifle grenades…'[33] The interim period was not spent in boredom but required more work on mining fatigues. Shield spent this period in the trenches in High Street reserve trench. As the Battalion was relieved by 17/Notts&Derby, between 8:00 p.m. and 10:40 p.m. on 10 April, Shield's company was strafed by Whizz-Bangs.

However, during the period from 4–9 April Dykes' platoon only did one fatigue, on the night 8–9 April, in support of the RE tunnellers near the aptly-named Mine Point. This time they had to empty sand bags filled with mine spoil. After a few hours this work was severely disrupted by German shelling of the vicinity of the mine shaft entrance. The working party cowered in the mine entrance and when it ceased activity commenced speedily and the rest of the night was spent at work. Just as the task was about to finish the Germans detonated a mine against the British mine workings; 'I was in the trench when there was a rumble and oscillations of the ground, this was quickly followed by the sound of an explosion. I was already down on the floor in a prone position. Apart from the fright I suffered no ill-effects, although I was scared and wondered what next would happen…'[34] According to Dykes five miners were buried alive as their shaft collapsed and a nearby RWF man had his neck broken by debris. The mines were blown at 8:00 a.m. on 9 April and these five men were from 251st Tunnelling Company RE.[35] Luckily, the Germans did not follow this explosion with a raid though this was the most eventful night of James Dykes' war to this point. Lieutenant Walker got his platoon away from this hazardous area and back to Annequin. Around this time PS/5062 Sergeant Ernest Holden had a very lucky escape whilst in the trenches. A stray bullet, which could have killed him, struck his ammunition pouch which stopped the projectile. Such luck was not the preserve of every soldier of 20/R Fus.[36]

32   IWM: J.N. Dykes Papers, 7378, p.15.
33   IWM: J.N. Dykes Papers, 7378, p.94.
34   IWM: J.N. Dykes Papers, 7378, p.15.
35   TNA: WO 95/551: 251 Tunnelling Company War Diary.
36   IWM: J.N. Dykes Papers, 7378, p.97.

There followed a period in Béthune from 10 to 17 April for training whilst 19th Brigade was in reserve. On about 15 April Second Lieutenant Walker left No 10 Platoon to become the battalion sniping officer. He was replaced by Second Lieutenant Guy who was less popular according to Dykes, 'we did not all appreciate the change & I am afraid Mr Guy was called some hard names by some members of the platoon.'[37] On 16 April 20/R Fus beat the Divisional Supply Column, ASC, at football in the semi-final of the divisional competition. The score was 1–0 after what was a very fast and equal game. On 29 April the commander of First Army, General Sir Charles Munro GCMG KCB, attended the final of the divisional football league. 20/R Fus played 2/A&SH and beat them 4–0.[38] 20/R Fus had the best football team in 33rd Division until late 1916. Whether the battalion excused its footballers from trench service before these games is unknown; however, a number would be killed or wounded in the months ahead.

During this period those men of 20 R Fus attached to RE Tunnelling Companies remained in the line. German trench mortars claimed the life of PS/5756 Private Thomas Thornber, aged 28, on 11 April 1916. He had been attached to the tunnellers along with his brother, PS/5755 Lance Corporal Percy Thornber. Percy wrote in a letter:

> Tom has been killed. He was working in a sap helping to make a new mine which was just being started when a trench mortar came over and exploded near to him. He was killed outright and suffered no pain at all. As far as I know it was the force of the explosion that killed him, and he was not hit by anything. I was working in another mine at the time, which was about 2 a.m. this morning…[39]

Two other men were wounded by the same mortar bomb. Meanwhile, 20/R Fus was in Béthune until 18 April.

The next period in the trenches (18–21 April) was spent in reserve in the le Quesnoy sector, followed by a stint in the Front line in the Cuinchy left subsector with the left of the battalion on the la Bassée Canal. Beyond the canal was 39th Division and on the right was 2/RWF. On Saint George's Day, 23 April, Shield manned a listening post at the crater near the embankment for five hours. The next night, Monday 24 April, he was on an all-night fatigue repairing a trench that had fallen in. On 24 April two men were wounded and all that Shield wrote that day was; 'Tired out'.[40] On the night of 24–25 April Shield recorded George Wivell was killed.[41] Another man was wounded on 26 April, the day the battalion was relieved by 1/Cameronians and returned to being in reserve at Le Quesnoy. At 6:15 and 8:30 a.m. the battalion 'stood to' in gas helmets due to reports of gas having been discharged further south. Gas was detected having

---

37 IWM: J.N. Dykes Papers, 7378, p.103.
38 TNA: WO 95/2408: AQ 33rd Division War Diary.
39 PS/5756 Private Thomas Thornber, aged 28, was from Burnley and was educated at Burnley Grammar School where he played football and cricket. He was employed by a cotton merchant and was a member of the Mechanics' Institute. He enlisted in October 1914. Thomas Thornber is buried in Vermelles British Cemetery. Express and Advertiser, 19 April 1916.
40 IWM: F.J. Shield Papers, 1385.
41 PS/5949 Private George Banks Wivell, aged 29, from Keswick, was an electrical engineer. He was educated at Hutton School, Preston, and his parents owned Keswick Hotel. He had also been employed in South Africa and arrived back in the UK in 1913. Banks is buried in Cambrin Churchyard. His younger brother, PS/5950 Private Alec Wivell, served with 20/R Fus before joining 23/R Fus; he survived the war.

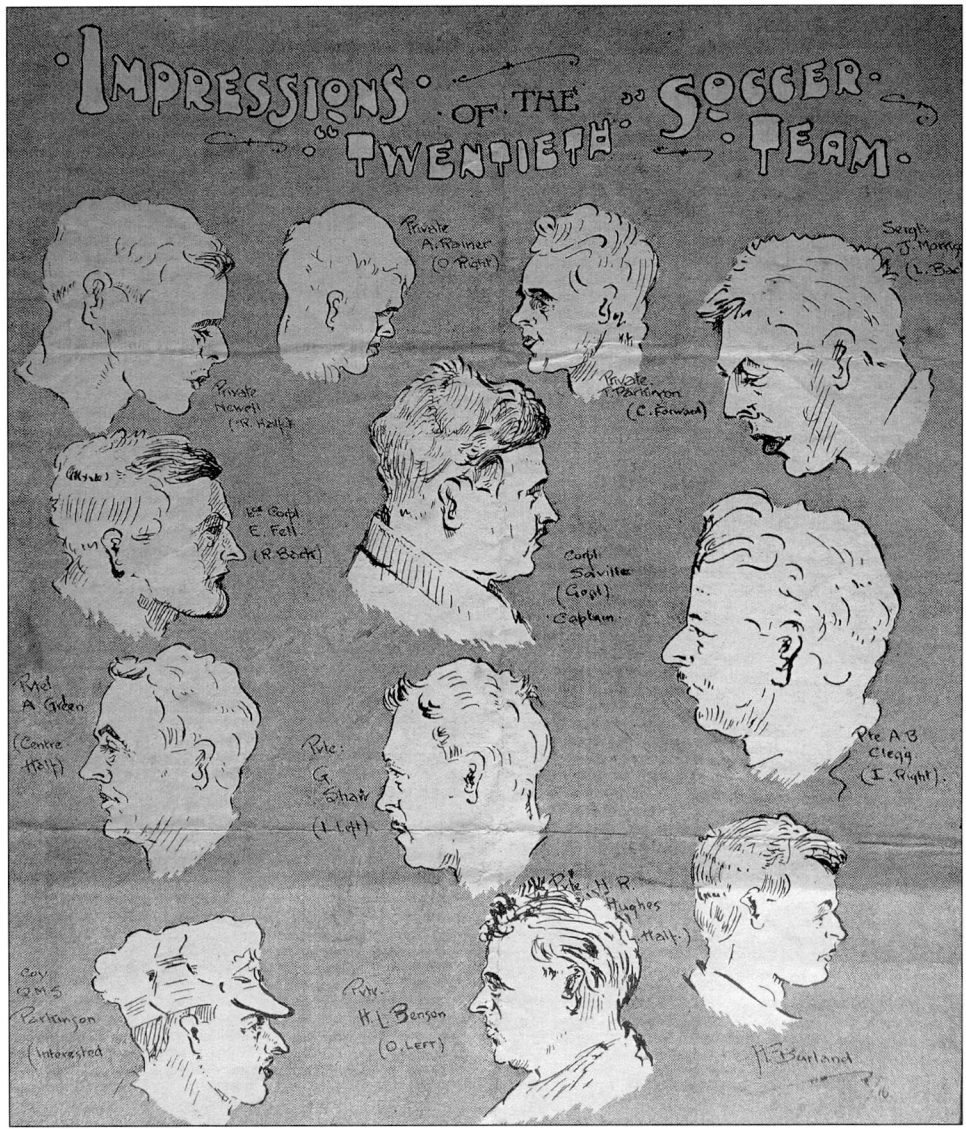

The 20/R Fus football team comprised the following men: In goal was the team captain, PS/5582 Private Arthur Coningsby Saville, aged 24, from Lymm and an old boy of Manchester Grammar School. He was with B Company. GS/47408 Corporal Ellis Fell, of D Company, was the right back; he was a former footballer for Nelson and taught at the Council School at Barrowford. PS/5355 Sergeant James Morrison was the left back. In midfield, PS/5091 Private Herbert Richard Hughes was the left half; he was aged 22, and from Blackley, Manchester. The centre half was PS/4897 Sergeant Alfred Green. The right half was PS/5389 Private Thomas Earls Newel, aged 30, from Belfast. The outside left was either PS/4471 Private Harry Benson or PS/6110 Private Herbert Benson. PS/5616 Lance Corporal George Shaw, aged 26, from Blackpool, a schoolmaster at Baine's Grammar School, was the inside left. The centre forward was PS/6512 Private Thomas Parkinson. At inside right was PS/4638 Corporal Albert Bramhall Clegg, aged 26, from Altrincham, Cheshire. On the right wing was PS/5505 Private Arnold Rainer, aged 21, from Upper Norwood; he served with D Company. The team organiser and coach was PS/5424 CQMS Harold Parkinson.

blown seven miles from Hulluch but had no effects on personnel. Next day Shield recorded doing drill and afterwards being able to swim in the canal. Even whilst in support in Cabbage Patch Redoubt Private Shield still found himself employed as mess orderly.

On 28 and 29 April 20/R Fus was in reserve and suffered two men wounded of which one remained in the trenches. The latter was PS/5366 Private Harold Mumby who had drawn cartoons in editions of The Pow-Wow; he was injured on 29 April but was not evacuated. Next day 20/R Fus relieved 1/Cameronians in the same Cuinchy sector. 11/R Sussex was on the left and 2/RWF were again on the right. On 30 April the war diary for 1/Cameronians stated the following:

> Relief by the 20th Royal Fusiliers commenced about 7:30 p.m. and was completed in an hour and a half. Weather has been fine throughout these 4 days and a great deal of [work] has been done, mostly in building up and revetting damaged fire trenches. Working parties of the 20th Royal Fus came up night and day, at night to help with putting up wire. The men of the 20th R.F. require a great amount of supervision and instruction in this particular, having little notion of wiring. They are a "Public Schools" Battalion and very well educated and excellent in many respects.[42]

The author, likely the Adjutant of 1/Cameronians, does not highlight the respects in which 20/R Fus was excellent but was very specific regarding their inexperience. It is presumed that the UPS Battalion did contribute a great deal to the work done. Likewise, on 23 April the Cameronians diarist grudgingly records that 20/R Fus left their billets in good order on handover.[43]

One 20/R Fus soldier was wounded during the relief. Another man was superficially wounded next day and a further man was wounded on 2 May. James Dykes described the 'Brickstacks' sector in early May:

> 'The Brickfields ... contain a number of stacks, I believe there are 13. We hold 8 or 9 & the Germans hold the rest. Round about the Brickstacks is supposed to be one of the hottest parts of the sector held by 19th Brigade, this sector stretches from the right of Duxbill [sic, Duck's Bill] to *Boyau* 1, a front of perhaps 2 or 3 miles ... There have been one or two V.C.s won round about the stacks. These stacks are huge things anything up to 40 feet in height, they [sic] must be millions of bricks stacked there. The stacks [are] all very solid and have resisted all heavy shells.'[44]

On 3 May PS/4659 Sergeant Frederick Cox, of B Company, was killed.[45] Another NCO wrote to his family; 'Fred was wounded by a shell in the trenches ... He had been engaged for two or three days constructing a new piece of trench. A sudden shower of shells came over, the first of which caught Fred, who was in an exposed position.'[46]

---

42 TNA: WO 95/2422: 1/Cameronians War Diary.
43 TNA: WO 95/2422: 1/Cameronians War Diary.
44 IWM: J.N. Dykes Papers, 7378, p.5.
45 Sergeant Frederick Ratcliffe Cox, aged 25, was a clerk from West Didsbury. He was educated at Hulme Grammar School and was a keen cricketer and footballer. He enlisted on 8 September in Manchester.
46 He was buried in Cambrin. De Ruvigny, *The Roll of Honour, Volume 3*, p.65.

A Minor Operation

On the night 3–4 May 20/R Fus tried its hand at minor operations. In a coordinated operation involving trench mortars 20/R Fus sent two patrols to attack German posts. The assault force comprised bombers from A Company. According to Dykes, a bystander; 'I heard the initial bombardment of the sap by our artillery. This ended as suddenly as it began; a short silence followed, then the exploding of hand grenades – rather more than had been anticipated – and it was some time before a fitful silence again enveloped the area.'[47] According to the War Diary one post was found to be empty; the second was occupied and after a few bombs were thrown the British party withdrew having suffered four men wounded; one of whom was not evacuated wounded. Private Dykes heard more of the raid second-hand from its participants. Apparently the Germans had withdrawn from the sap in question having been tipped-off by the British bombardment. The Germans then returned and threw bombs into their own sap and forced the raiders to withdraw. According to Dykes, A Company were very depressed at the poor result and though there were no fatalities a sergeant was very severely wounded. Sidney Platt, of D Company, merely recorded; 'We had a small attack (bombers) on the German saps. Failure.'[48] On 4 May 20/R Fus suffered another man was slightly wounded but the battalion was relieved by 4/Kings R and moved to billets in *Rue d'Aire*, Béthune. There followed a lengthy period in reserve from 5 to 15 May.

There were some sporting amusements according to Captain Templar, in a letter to the Manchester Grammar School magazine:

> The boxing gloves and footballs came at a most opportune moment. There is to be a Brigade boxing contest on May 10th, and Hodgson-Jones is training the men for it. Last night we were all sitting in the Divisional Coffee Bar, watching two men scrap with a rotten pair of French mitts, and twenty men spoiling for a scrap. Then in came a man with two huge parcels for me, labelled "boxing gloves," and amid huge enthusiasm I pulled out real English gloves—and in less than no time four scraps were filling the air with the shuffle of feet and the smack of gloves. The men are awfully keen on boxing, and the room is packed out at night, as some of the men are quite good fighters and well worth watching. One of them, Farr, was the public school champion. The number of Old Mancunians is surprising; Sergt.-Major Blackstock, Sergt. Jimmie Morrison, Sergt. Holmes, Sergt. A. Holden, Private Roberts, Hodgson-Jones, and heaps more, being Macaulay's old pupils. This afternoon we were able to have two footer matches, one rugger, one soccer, with the balls you sent to us.[49]

On 16 May 20/R Fus relieved 1/Queens RWS in the Auchy left subsection; the relief was complete by 10:15 p.m. 2/RWF was on the right and 1/Mx were on the left. During the following night PS/8706 Private Sydney Heywood was shot in the head, reportedly by an explosive bullet, and lived for half an hour before succumbing to his wounds.[50] On the 17th May,

---

47 IWM: J.N. Dykes Papers, 7378, p.19.
48 IWM: S. Platt Papers, 17681.
49 Anon. Author, *ULULA, Vol 44, June 1916, No 326* (Manchester: Galt, 1916), pp.118–119.
50 PS/8706 Private Sydney Charles Heywood, the son of a grocer, from Ashtead, Surrey. IWM: S. Platt Papers, 17681, p.51.

Frederick Shield recorded in his diary that a 20/R Fus wiring party in No man's land suffered fourteen casualties and that one man, Bannister, with whom he was presumably acquainted, was killed.[51] This likely referred to PS/9348 Private Robert Bannister who died on 18 May and was buried in Béthune.[52] Meanwhile, D Company, 20/R Fus had a bad few days. The war diary merely recorded that 20/R Fus suffered, during the whole tour, a total of three men killed, one died of wounds and 28 men wounded, of whom eleven remained at duty. One of the three private soldiers killed were Private Herbert Standage, who according to Sidney Platt was 'the man who never grumbled'; Standage was killed by a whizz-bang on 17 May 1916.[53] Privates Frederick Ekin and William Patchell were grazed by the same bomb and presumably stayed 'at duty'. However, such stoic actions would indirectly create the circumstances of their deaths in the future. William Haile was shot in the head by a sniper on 18 May whilst in the trenches.[54] The man who was mortally wounded was likely Bannister.

During this tour Shield was employed as a bayonet-man in a team of bombers who were protecting one of the British Saps. In this role, if a bombing attack was required to clear a trench of German raiders, Shield would have one of the most dangerous jobs in the Army. The bayonet men in a bombing party would wait until the bombers had thrown grenades into the next trench bay. When they exploded the bayonet men would charge around the traverse and kill any German survivors there with cold steel. If they timed it badly they could be hit by fragments of British grenades, meet Germans untouched by grenade fragments or be killed by German bombs. Luckily, during Shield's three nights in this role, the Germans did not attack. However, no sooner than he was released from this function on 19 May than he was out all night erecting barbed wire which was far from a sinecure.

After a hard tour 20/R Fus was relieved at 9:30 p.m. on 22 May by 1/Cameronians and after this handover was complete by 11:30 p.m. and the battalion marched to billets in Annequin South where it was in reserve. Shield recorded doing mining and dug-out fatigues.

During this period there was trouble brewing within No 13 Platoon, D Company, due to poor officer-man relations between Captain Toller, their company commander, and his men. Sid Platt recorded an act of indiscipline going unpunished; 'Had to appear at orderly room for having "dirty" rifles. 12 of us 1 hours pack drill. We (4 of us) put air cushions in our packs. CSM caught us but did not report us. He says [Captain] Toller is a rotter. Fancy CSM daring to make such a statement to privates.'[55] This could be a suggestion of poor discipline in 20/R Fus but could also be judged as poor leadership by Toller, poor officer-man relations with his men or a lax attitude from a warrant officer. A further act of indiscipline took place on 26 May when members of D Company refused to turn out to parade at 6:00 a.m. for fatigues when ordered by their company sergeant major. This might appear inflammatory but this was in response to them having returned from a mining fatigue at 3:00 a.m. Instead they breakfasted at 9:00 a.m. and

---

51 IWM: F.J. Shield Papers, 1385.
52 PS/9348 Private Robert Bannister, aged 19, from Preston, was buried in Béthune Town Cemetery.
53 PS/5688 Herbert Raymond Standage, aged 22, from Great Bourton, Banbury. He was educated at Bloxham School (1905–1909) where he was recorded as a quiet, retiring boy. IWM: S. Platt Papers, 17681, p.51.
54 PS/8641 Private William Kenneth Haile, aged 18, was born in Madagascar and lived in Merriott, Somerset. He was educated at the Congregational School, Caterham. Haile is buried in Cambrin Churchyard along with Heywood and Standage.
55 IWM: S. Platt Papers, 17681, p.49.

spent the afternoon playing cards.[56] The Platts' platoon, No 13 Platoon, also refused to obey an order, as a platoon, on 7 June 1916.[57] This suggested an inadequacy of their CSM to command their respect or a lack of understanding by him of what the soldiers of his company were doing.

At 8:30 p.m. on 27 May 20/R Fus commenced taking over the Auchy Left Sub-Sector trenches from the Cameronians with the Suffolk Regiment on the left and the Scottish Rifles on the right. Three men were wounded next day and a further four on the 29th. One of those wounded was PS/7769 Private George Hey, aged 29, a warehouseman from Lower Broughton, who was shot in the arm on 29 May.

At midnight on the night 29–30 May a small British mine was blown to elongate a previous crater and thereby create a greater obstacle against the enemy. On 30 May a company of 2/5 R Warwicks was attached for instruction with two platoons replacing two platoons of 20/R Fus in each of the two companies in the front line. This attachment lasted until 2 June when 20/R Fus was relieved by the Cameronians and the battalion returned to Beuvry. During this tour another four men were wounded with one remaining at duty; one man died of wounds next day.[58]

20/R Fus spent four days in reserve in billets before being relieved by 16/KRRC and marching back to billets in Oblinghem whilst 19th Brigade was in divisional reserve. However, more working parties were found. No 13 Platoon were not the only members of the battalion suffering from lowered morale. By 3 June Shield recorded himself 'fed up' but with the move to Oblinghem, and a 'Bon' billet with straw, his outlook presumably improved. Whilst here the battalion was informed that PS/5969 Private Ronald Woolfenden had been awarded the Military Medal (MM) whilst attached to 251st Tunnelling Company RE. Also around this time a fresh draft of about forty men joined 20/R Fus. These were not UPS recruits from the reserve battalions but 'normal' Royal Fusilier replacements; some had previous service with other RF battalions and had been returned to the UK sick or wounded. However, most of these men were fresh from the UK and a number may have been conscripted under the new Military Service Act.

On 17 June this sojourn was over and 20/R Fus returned to Beuvry and relieved 16/KRRC in billets. Though the battalion remained here until 20 June Shield recorded having to march back to Oblinghem to go through the gas chamber to ensure his gas mask fitted and he knew how to use it. On 20 June 20/R Fus relieved 4/King's in the right subsector at Givenchy; 16/KRRC were on the right and 2/RWF on the left. Sidney Platt recalled that 4/King's were 'very windy'. They had every right to be.

Red Dragon Crater

The following period of trench warfare was not all bad according to Dykes:

> The weather was co-operative, and the Jerries seemingly quiescent. Two nights passed peacefully on a fire-step under the vault of heaven. The third – and all Hell was let loose!!

---

56   IWM: S. and V. Platt Papers, 17681, p.54.
57   IWM: S. and V. Platt Papers, 17681, p.59.
58   This was likely PS/5968 Cyril Wilson, aged 22, from Manchester, who died of his wounds on 2 June 1916 and was buried in Béthune Town Cemetery.

> The first I knew of anything out of the ordinary was the impression I got that the floor of the trench had come up and hit me. I have never experienced an earthquake, but I should imagine it would be like this. I became aware of an indescribable and confused babble of sound, much shouting, and an automatic impulse that I must stir myself and do something. I staggered to my feet and in the now stable trench, groped around and found my rifle and equipment, tin hat, which I balanced precariously on my head, in the approved fashion, and felt that I was ready for action and wondered what form it would take… pandemonium reigned in "Orchard Keep", or to be more exact, forward of it. In fact there was an interlude. There was, audibly, much activity "up front" in the region of the firing line … Suddenly, the demons of hell were let loose in "Orchard Keep", made up of ear-splitting detonations, heralded by the hissing sounds of flying metal and whistling shrapnel. Every conceivable type of explosive appeared to be concentrated on that comparatively small area. It seemed miraculous that we were not all blown to blazes with the first salvo…[59]

Shield recorded; '2 a.m. Huge mine exploded by Germans to left of our battalion. R.W.F catch it heavily. A Company rapid fire. On sentry all through Strafe. Germans driven out. English warplane brought down.'[60]

The mine erupted at 1:55 a.m., opposite the Duck's Bill, under the right company of 2/RWF who were subsequently raided. The German raiders entered the RWF trenches but were ejected almost immediately before a C Company (20/R Fus) bombing party could come to their aid. According to Sidney Platt:

> At 2 a.m. (Thurs) very large mine blown up on left flank of 2nd Battalion R.W.F. lines. C Company caught it badly. Very heavy bombardment. Mine between support and front line of R.W.F. at Pont–Frixe[sic]. Bombardment lasted for 1 ½ hrs. Dutton, machine gunners, did well – 3,000 rounds fired… Dutton changed [the] barrel of his gun during bombardment last night & was the only gun firing. All 4 R.W.F. guns done for. 2 blown up. 1 smashed by bombardment & 1 captured. Huns came over but were driven back.[61]

Dutton had been the only Lewis gun to engage the Germans and it is likely that his fire reduced the ability of the Germans to enter the British trenches. Dykes' diary recorded the state of confusion:

> As is only to be expected, the men without anyone to take command & under such circumstances were panic-stricken, the Welsh and Middlesex I mean & were rushing all over the place causing more confusion than ever. 2nd Lieut Coggin instantly went into the R.W.F. lines & restored order again assisted ably by LCpl Steward & Pte J.B. Young. These 2 men went into the Welsh lines not wearing the[ir] tunics & were obeyed as if they were officers, they inspired confidence into the panic-stricken men & so disposed them as to offer the best defence possible. The third man who was recommended was Sergt Roberts, in charge of No 12 Platoon, he did very good work in the Keep. Lieut Coggin & Stewart & Young were recommended by the Welsh Colonel.[62]

---

59 IWM: J.N. Dykes Papers, 7378, p.24.
60 IWM: F.J. Shield Papers, 1385.
61 IWM: S. Platt Papers, 17681, p.64.
62 IWM: J.N. Dykes Papers, 7378, p.122.

PS/5334 Private James Albert 'Bert' Moore, aged 20, was from Higher Broughton, Manchester and served with No 10 Platoon, C Company, 20/R Fus. He took half an hour to die of his wounds. Moore is buried in Gorre British and Indian Cemetery. His parents later faced a double tragedy. (Reproduced with permission of Sutton Archives)

These were likely PS/5539 Sergeant Francis Roberts, PS/5676 Lance Corporal Eric Roberts and PS/5994 Private James Harold Young. None were seemingly awarded medals. Captain Dunn, of 2/RWF, only mentioned Coggin indirectly: 'Conning [2/RWF] was sent to relieve a 20th R.F. officer [Coggin] whom Sergeant Rush had called in during the early morning when he found his platoon cut off from the Company.'[63] Whether Colonel Crawshay of 2/RWF recommended Coggin, and the others, for awards is unknown, but he recommended his own men.[64]

On 22 June, across 20/R Fus, the war diary recorded four men were killed and Captain Norman Cockell and 22 men wounded by a German artillery barrage prior to a German mine explosion and raid.[65] However, No 10 Platoon, 20/R Fus, in Orchard Keep alone lost nine wounded and four men killed. James Dykes knew all four well.[66] He was:

> ... consequently, well aware of their many sterling qualities, and was proud to have had their friendship. Hill, Kay, Lambert ... and finally, and most tragically of all, Bert Moore,

63 Dunn, *War the Infantry Knew*, p.216.
64 8672 Sergeant C. Rush, 2/RWF, was awarded the Distinguished Conduct Medal (DCM) on 28 July 1916, he 'at once organised his men, attacked and drove them [the Germans] out in disorder.' The London Gazette, 28 July 1916, p.1336.
65 Captain Norman Alexander Lindsay Cockell had arrived in France on 19 May 1916 and served with 20/R Fus for barely a month. He later served with 13/R Fus and 7/R Fus. He was wounded on 5 April 1918. One 20/R Fus man who died on 22 June was killed because of an unrelated mining event. On 21 June the Germans detonated a mine against the 180th Tunnelling Company mine system at The Quarries, near Cité St Elie. PS/7059 Private Joseph Lewis, still attached to that company, was badly wounded in the explosion. He died next day and is buried in Vermelles British Cemetery.
66 PS/5006 Lance Corporal Ronald William Hill, aged 22, from Cheadle Heath, Stockport, had been educated at King's School, Chester; Manchester Grammar School and Manchester School of Music; he was studying for the bar in 1914. He had enlisted in September 1914. Hill was buried in Gorre British and Indian Cemetery. PS/7867 Private Percy Lambert, from Clown, Chesterfield, was buried in Béthune Town Cemetery.

younger brother of George. They had been inseparable, had enlisted together in Manchester in early 1915, had always been together in France, and now the younger was gone. From being jolly and cheerful … George became silent and morose … We were desperately sorry but could only watch hopelessly and helplessly the rapid change in a personality …[67]

Dykes' memories were at odds with his diary which recorded Kay was hit in thirteen places by shrapnel and was presumably evacuated.[68] Lambert had only returned from leave a week before and died about an hour after being wounded. According to Dykes, Hill could be an irresponsible daredevil but was killed outright. In Hill's pocket was found a message stating; 'Do not let your faith in God be shaken, I do not fear death, rather am I proud to be able to lay down my life for my Country'.[69] In No 12 Platoon PS/5617 Private Eric Shaw was killed outright and PS/5709 Private Frank Stott was mortally wounded in the stomach and suffered greatly.[70] Another man, PS/5089 Private Andrew Hughes, died of his wounds bringing C Company casualties to six dead and fifteen wounded.[71] A comrade wrote of Frank Stott's part in the action:

> The limitations of the Censorship do not permit of the full story of his end, thus robbing him in part of his glory, for he died as he lived – courageously, bravely, cheerfully, – a man. The Battalion was met by a situation of grave danger, an event which called upon the last limit of their courage and fortitude – hand-to-hand work with cold steel at the finish. "Stottie" took his share leading and encouraging. After the initial struggle he stayed behind to assist some men who had been buried in a fall of earth thrown up by an explosion. Whilst engaged on this work he was struck by a portion of a 'Minenwerfer' just below the heart, and three hours later he died. He died a soldier and a gentleman.[72]

A few days before the mine was exploded Captain Price-Edwards had joined 2/RWF; he was not a welcome addition. However, his servant, Private Parry, a chef trained in Paris, was greeted with open arms by the 2/RWF mess.[73] Price-Edwards was attached to B Company and it was his company that was decimated by the mine explosion losing two platoons. His body was missing after the battle.[74] PS/7329 Private Leonard Bailey, formerly of No 13 Platoon, D Company, 18/R Fus, was also killed on 22 June 1916 whilst attached to 2/RWF; it is presumed that he was Price-Edwards' servant having transferred with Price-Edwards from their former battalion.[75]

---

67   IWM: J.N. Dykes Papers, 7378, p.26.
68   This was likely PS/5149 Private Ernest Cecil Kay, from Middleton, was educated at Manchester Grammar School; he died of wounds at home on 27 July 1916 and was buried in Middleton St Leonard Churchyard Extension.
69   PS/5006 Lance Corporal R W Hill, CWGC Entry, <https://www.cwgc.org/find-war-dead/casualty/592650/hill,-ronald-william/> (accessed 26 April 2018). IWM: J.N. Dykes Papers, 7378, p.21.
70   PS/5617 Private Eric Shaw was born in Driffield. PS/5709 Private Frank Stott, aged 25, was from Didsbury, Manchester and was educated at Manchester Grammar School. Both were buried in Gorre British and Indian Cemetery.
71   PS/5089 Private Andrew Dunlop Hughes, aged 22, from Altringham, was buried in Béthune Town Cemetery.
72   Anon. Author, *ULULA, Vol 44, June 1916, No 328* (Manchester: Galt, 1916), p.191.
73   Dunn, *War the Infantry Knew*, p.206.
74   Dunn, *War the Infantry Knew*, p.217.
75   PS/7329 Private Leonard Bailey, aged 23, from Acton, was buried in Gorre British and Indian Cemetery.

The 2/RWF called the hole formed by this mine explosion 'Red Dragon Crater'. Though not in the front line directly facing this threat, some men of 20/R Fus had proactively assisted 2/RWF. Others had recovered quickly and had been ready to defend their positions. A fatigue party from 18/Mx (Pioneers) had been working on a nearby communication and helped eject the Germans using spades and picks. Parties from the Scottish Rifles also aided 2/RWF. During the process of rescuing buried men Acting-Sergeant John Erskine, of 5/Scottish Rif, was awarded the Victoria Cross for gallantry. Beneath the ground the German mine had caved-in a gallery of the Shaftesbury Avenue Mine. Trapped within, Sapper William Hackett of the RE helped others escape before remaining behind with an injured comrade only to be entombed underground. Hackett was awarded a posthumous VC. Whilst many UPS men might grumble of fatigues assisting the tunnelling companies, the majority would have willingly avoided this horrible fate.

* * *

Ernest Brierley's skeleton diary gives an impression of hard work and events whilst in the trenches during this tour:

> 20 June – Moved to Brickstacks in evening. Decent dug-out. 21st – Stand to 3 o'clock. Cleaning up trench etc. Rest in aft[ernoon]… Explosion. Big strafe. Ration party. 22nd – Working all morn[ing]. Sleep in aft[ernoon] … working at night. Ration party. 23rd – Carrying party in morn[ing]. Rest in aft[ernoon]. Rain. Sandbagging. Plenty of rain at night. 24th Pumping water out of trenches all day. … Rations at night. 25th – Clearing trenches and sandbagging in morning. Rest in afternoon … Messing about in evening sandbagging etc. 26th – Busy day. Moving ammunition in morn[ing]. Sleep in aft[ernoon]. Moving all sorts of stuff in evening … Moved to Pont Fixe. Beastly and wet.[76]

On 23 June four men were wounded but three were not serious enough to warrant evacuation. The next day Second Lieutenant Francis Brown was wounded along with two other men.[77] On 24 June G/24815 Private Charles Ashwell was killed; he was the first non-UPS other rank to be killed whilst with 20/R Fus.[78] The situation did not quieten down though; on 24 June Shield wrote; 'English strafe, big German retaliation, on sentry through strafe'. Sidney Platt recorded; 'Artillery strafe. Plenty of whizz-bangs over by both sides. Germans getting dangerously near.'[79] Next day saw another British strafe at intervals.

There was an aerial combat which observers recalled between five German Albatross and four British biplanes; five Fokkers and three British machines or five English versus four German aircraft depending on different sources. The result was generally agreed to be a draw with one machine shot down apiece. On 24 June Major Vincent Joseph Kelly of the Royal Munster Fusiliers was posted to 20/R Fus.[80]

---

76 Diary of Ernest Brierley, unpublished. With the permission of Angela Jennings. Entry for 20–26 June 1916.
77 Second Lieutenant Francis Henry Brown had only been in France since 19 March 1916.
78 G/24815 Private Charles George Ashwell, aged 24, a labourer from Upper Edmonton, London, was buried in Béthune Town Cemetery. He may have been the first conscript fatality in 20/R Fus.
79 IWM: S. Platt Papers, 17681, p.65.
80 Kelly was formerly a member of the Royal Irish Rifles who had first been gazetted in 1888. He had

Between 10:00 and 11:30 p.m. on 26 June the battalion was relieved by the Cameronians and moved into support and was disposed with A Company in Givenchy Keep, Moat Farm (Moat House Redoubt), Hilders Redoubt and Herts Redoubt; B Company in Pont Fixe North; C Company in Pont Fixe South, Windy Corner and le Plantin South; D Company was in reserve though Sidney Platt recalled getting some decent accommodation at Windy Corner. Fatigues started at 9:00 a.m. on 27 June with D Company working in Gunner Trench knee deep in water. Vincent Platt also became 'fed up'; his brother probably felt likewise but his morale was likely raised on hearing of a successful raid by 1/9 HLI on the German trenches. On 28 June the fatigues were at night after a day off; the Platt brothers were shifting trench mortar ammunition. The battalion remained there until relieved by the Cameronians on 2 July 1916.

Continual trench warfare and all it entailed, including carrying and fatigue parties, strafing, mine explosions and poor weather, had undoubtedly sapped morale. However, rumours were rife of a major British offensive further south. Little did the battalion know that this offensive would soon become a reality and that reality would have a disastrous effect on them.

On 27 June No 10 Platoon, 20/R Fus received another platoon commander, Lieutenant Wilmshurst, who gave the platoon a lecture on the strategy of the coming offensive 'the coming offensive was on such a scale as to ensure its complete success, the collapse of the German Army, culminating in the final victory of the allies. It was not anticipated that the 19th Brigade controlled area would be involved in the offensive, because it was known the German defences were immensely strong...'[81] Wilmshurst's assessment was correct in many ways but if the fighting would not come to the 19th Brigade front, the brigade would need to move to the fighting.

However, this period of trench warfare had not been all bad according to some. Though gazetted into 21/R Fus, at some point Ashley Gibson, joined 20/R Fus. He summed up the first six months in the trenches:

> I will be honest and admit that I *liked* my first six months of trench warfare, from that second winter of the war till the Somme. There were thrills in plenty, but our battalion remained a unit, our casualties between reliefs, after that sanguinary christening we got, a bare dozen or so every four days in the line. I really believe that barring sudden catastrophes and accidents the most trying feature of trench existence was the smell, characteristic and omnipresent. One smell that was really two smells – corruption and chloride of lime ... yet one got used to it. The whole thing was such a *kolossal* adventure ... Even the patrols, when one wriggled and floundered in that leprous wilderness, like a landscape in the moon, that lay beyond our wire, head and arms in one shell-hole and legs in another, one's hinder-end in parlous jeopardy, while the Boche Verey lights blazed into high heaven and his snipers lost their heads with excitement.[82]

A pitched battle would be a different prospect for 20/R Fus.

---

seen action and been captured during the Boer War in 1899. He retired in 1906 but returned to the colours in 1914 as a major with 9/RMF and he had been in France since December 1915. That battalion was disbanded in May 1916 and Kelly was presumably re-assigned to 20/R Fus.
81   IWM: J.N. Dykes Papers, 7378, p.26.
82   Gibson, *Postscript to Adventure*, pp.140–141.

16

The UPS and the First Day of the Somme

There was nothing for it, but to grit one's teeth and go into it

---

I would have given worlds to have been miles away ... there was nothing for it, but to grit one's teeth and go into it outwardly calm if inwardly shaking...[1]

We went into literally a hail of machine gun fire. I could hear the "smack" of the bullets ...[2]

Though the First day of the Battle of the Somme passed whilst 20/R Fus was in trenches near Béthune, there were many ex-UPS men involved in the British Army's bloodiest day. UPS men were represented, as officers, in differing numbers within almost every division that participated. The different divisions engaged in this battle enjoyed success and failure but sometimes success might still lead to heavy casualties. The ex-UPS officers would be almost entirely those who had left for commissions during the training period in the UK and were predominantly from 18/R Fus, 19/R Fus and 21/R Fus.

On 31 May 1916 Kenelm Dyott, an ex-member of 19/R Fus, observed the Battle of Jutland. A former medical student, Dyott was discharged to complete his studies in the Spring of 1915 and became a naval surgeon on HMS *Tiger*. He recalled the following:

> The enemy's battle cruisers, now plainly visible on our port bow, were also going at full speed and leaving clouds of smoke behind. ... I went to earth like a rabbit. Our first gun went off just as I arrived below – it was 4:53 p.m. For the next few minutes we sat and listened to our own guns. Those in the turret above us made a terrific noise, and the salvoes shook the ship from one end to the other. Soon after we began to get hit, and in quick succession we received several large "prodges" close to our station. The concussion was terrific, and the whole ship trembled ... Then the wounded began to arrive, crowding into the station… The poor men were, for the most part, terribly burnt about the face and hands, quite black from head to foot from charring and a dirty oil deposit ... With hands held limply up, and skin literally dripping from their hands and faces, the men presented the most piteous spectacle.[3]

---

1 FM: P.N. Wright Diary, RFM.2013.8.1/2, p.38.
2 IWM: E.D. Shearn Papers, 2033.
3 The Action of May 31st, The Battle of Jutland, *The London Hospital Gazette*, No 193, January 1917, p.344.

PS/292 Private Kenelm Mitchell Dyott, aged 26, from Tamworth, Staffordshire, enlisted with the UPS on 15 September 1914. He had studied medicine at Oxford University and had formerly served with the City of London Field Ambulance. He was discharged on 26 February 1915 and was gazetted as a surgeon on 24 July 1915. Dyott died, aged 29, on 13 December 1917, when his vessel, HMS *Stephen Furness*, was sunk. (Reproduced with permission of Sutton Archives)

Many infantrymen would not have swapped their trenches or dugouts for such conditions.

There were numerous vignettes displaying the impact ex-UPS men had on the Somme offensive. One example was PS/2037 Private Frank Schumann who had enlisted in C Company, 18/R Fus, alongside his brother Oswald Schumann. They were in their 30s and from Brondesbury, London, and had been educated at Framlingham School. Frank was 6'1", well-developed, and was employed as an export merchant. Both were gazetted to 6/KRRC to serve together and changed their surnames to Walker in August 1915. PS/1975 Corporal Pelham Donovan Ravenscroft, formerly of Harrow and Jesus College, Cambridge, joined 6/KRRC with them.[4] All three went to 2/KRRC in France in November 1915. Oswald was evacuated to England in May 1916 with trench fever. There were numerous raids and minor attacks to distract German attention from the Somme offensive. One operation saw 2/KRRC conduct a diversionary attack on 'The Triangle', a German position near the Double Crassier. One of Frank Walker's soldiers recalled the confused fighting in the German trenches:

> [The trench] … is filled with a medley of men from all our four Companies, and Mr Walker is urging the men to dig fire positions in the parados. On both flanks of the [German] front line the Germans have re-entered the trench and are systematically bombing their way down, crowding us in towards the centre … "Stick it out, men!" encourages Mr. Walker …. I look round at him. His left arm is bandaged and the tunic-sleeve ripped off. His face is streaked with blood, dirt and sweat. He must have had a pretty hectic time by the looks of him! …[5]

The 2/KRRC was forced to withdraw. The writer, a veteran, undoubtedly admired Walker's bravery during this episode. Despite his injuries, Walker remained at duty and set a fine example

---

4   P.D. Ravenscroft, *Unversed in Arms* (London: Crowood Press, 1990).
5   Giles Eyre, *Somme Harvest* (London: Stamp Exchange, 1991), pp.72–73.

of how a UPS man had become a resourceful platoon commander. His battalion would see further action on the Somme.

Second Lieutenant Erroll Shearn (formerly of 20/R Fus) was with 1st Battalion Hampshire Regiment (1/Hamps R)(4th Division). He recorded his thoughts before the main attack on 1 July 1916: 'Our guns loosed off all night & kicked up a hell of a din. We all thought & indeed were told that there would be nothing left alive in the German lines & our attack would be in fact a "walk-over"... We all were pretty confident, though quite a number of my men looked pretty grim.'[6] 4th Division attacked with three battalions in line to capture the first objective. 1/Hamps R formed part of the second wave. Shearn was dressed in officers' tunic with a revolver; he later regretted this attire which made him a target:

> We went into literally a hail of machine gun fire. I could hear the "smack" of the bullets as they hit the ground or sand-bags or whatever. I got hit on my prismatic compass which I carried on the left front of my belt, I felt the impact of the bullet ... I opened my belt and pulled up my shirt to see. There was a small & very neat hole under my left ribs from which a little blood was coming...[7]

Not 'feeling' wounded, Shearn continued forward and across a sunken road before being hit again; 'I felt as though something had hit me in the back & I spun round in a half circle & came down on my bottom with my legs in the air & feeling like pins & needles running down my right leg.' He was badly wounded in the back and hip and crawled back to the sunken road before being evacuated by his batman on a stretcher. 2/Essex (4th Division) had a similarly fruitless day and one of the 22 officers and 400 men lost on 1 July 1916 was Second Lieutenant Gilbert Waterhouse who was a notable ex-UPS poet.[8] One observer reported his death:

> I got over one German trench and was advancing on to a second ... I caught sight of Lt. Waterhouse about 30 yards from me. He had a revolver in his hand and he stood out, a solitary figure. I saw him drop to his knees and began to crawl and I did the same and so did his platoon. I thought at the time that he had seen a M.G. and was avoiding the fire but he may have been hit ...[9]

He was described as being 'a very brave man and he stood out more conspicuously than any one. He seemed so fearless...' The 4th Division attack ended at 2:00 a.m. on 2 July when the last British infantrymen were forced from the Quadrilateral.

Though the three disbanded UPS battalions no longer existed, their spare infantrymen were sent to reinforce other Royal Fusilier battalions ready for the offensive. One 'receiving' battalion was 2/R Fus, 29th Division, which went into action on 1 July 1916 with many ex-UPS men. At 7:20 a.m. the Hawthorne Redoubt, capable of enfilading attacking troops with MG fire, was neutralised by the explosion of a British mine. An assault force consisting of two platoons from 2/R Fus, four MGs and four trench mortars, advanced to secure the crater and support

6   IWM: E.D. Shearn Papers, 2033.
7   IWM: E.D. Shearn Papers, 2033.
8   See Appendix 4.
9   TNA: WO 339/50234: Papers of G. Waterhouse.

the subsequent attacks. This party could only secure the near lip of the crater. According to one company commander; 'The notice, however, given by the Mine was such as to permit the enemy's lines being manned and for the crater to be defended, with the result that the major portion of the first wave of the three Companies of the 2nd R.F. were placed out of action almost immediately, with no one getting into the enemy's trenches.'[10] The British artillery fire was concentrated on the Germans second and third lines leaving the Germans free to resist vigorously in their front line. 2/R Fus was unable to secure the German positions and the elements of D Company eventually retired. The butcher's bill for 2/R Fus was 22 officers and 514 men. One of the officer casualties was Second Lieutenant Ferdinand Reiss, formerly of 18/R Fus, who was wounded but was later awarded the Military Cross.[11] Of the twenty ex-UPS men killed the majority are commemorated on the Thiepval Memorial or were buried in Hawthorne Ridge Cemeteries No 1 and No 2. They were late joiners to the UPS who enlisted in late 1915. Their former employment suggests they were not natural UPS men; jobs included an assistant tea dealer; a student in commercial art, an assistant in a boot shop and the manager of a grocers. They included PS/9298 Private Francis Drieberg, aged 19, and PS/9300 Private Thomas Hodgson; both were born and lived in Ceylon and had travelled back to the UK to enlist.

The Ulster-raised battalions of 36th (Ulster) Division had an initially successful day and penetrated deeply into the German defences but they were ultimately unable to hold their gains. The 9th Battalion Royal Irish Fusiliers (9/RIrFus) lost heavily in crossing No man's land; 'The right centre company appears to have suffered less severely and was seen to penetrate the three German lines and a small body of them was reported to have reached Beaucourt Station.'[12] The battalion suffered 14 officers and 518 men killed wounded and missing. These heavy casualties could be tempered by pride at the way in which the battalion had conducted itself. Lieutenant Geoffrey St George Shillington Cather was the adjutant. He went out during the evening of 1 July to bring in wounded men lying out in No man's land and carried on until midnight having rescued three men. He had already shown extreme bravery but he went out again the next day as his citation for the Victoria Cross stated:

> … Next morning at 8 a.m. he continued his search, brought in another wounded man, and gave water to others, arranging for their rescue later. Finally at 10:30 a.m., he took out water to another man, and was proceeding further when he was himself killed. All this was carried out in full view of the enemy, and under direct machine gun fire and intermittent artillery fire. He set a splendid example of courage and self-sacrifice.[13]

Cather was killed by MG fire and was buried where he fell. His grave was later lost.[14]

Second Lieutenant Henry Walker, 2nd Battalion King's Own Yorkshire Light Infantry (2/KOYLI) (32nd Division) was in support on 1 July. At 7:30 a.m. 2/KOYLI advanced and in

---

10 TNA: CAB 45/138: Official History correspondence.
11 *The London Gazette*, 22 September 1916.
12 TNA: WO 95/2505: 9/RIrF War Diary.
13 *The London Gazette*, 8 September 1916, p.8869.
14 Geoffrey Cather was educated at Rugby School and worked in the tea trade. In 1914 he had enlisted in 19/R Fus and become PS/1269 Private Cather; he served in B Company. He was aged 25, and lived St John's Wood, London. He was gazetted into 9/RIrF on 22 May 1915 and served in France from 5 October 1915.

crossing No man's land the war diary recorded; '2nd Lieut[enant] H.G. Walker, C Coy, who had been previously wounded ... continued to lead his men and jumping over the parapet shouted to his men to come on when he was instantly killed. The hostile machine gun and shell fire were so intense that all efforts to cross the fire swept zone between the opposing lines failed.'[15] A scout, patrolling in No man's land, recovered Walker's identity disc and binoculars from his body which he found just in front of an old and disused trench called 'Dead Man's Lane' about 200 yards from Leipzig Redoubt.[16] One man who deserved greater recognition for bravery on 1–3 July was Lieutenant Harry Catmur.[17] He served with 96th MG Company on Z Day. His section was to follow 16/NF but that battalion was mown down in No man's land. Catmur's company commander recorded:

> On the night of the 1st July he went out into "No man's land" and searched for wounded men of the 16th North[umberland] Fus[iliers] for two hours, under fire, eventually finding and bringing in a severely wounded private. During the day and night he attended to scores of wounded ... On the morning of 3rd July during the unsuccessful attack of the 2nd South Lanc[ashire] Regt he went out to the barbed wire with his servant under intense fire & after disentangling a wounded officer ... brought him in safely. This took between 5 & 10 minutes to accomplish, in broad daylight ... Shortly after ... he got the [machine]gun out of the emplacement ... mounted it in the open on the ridge of the road and swept the German line which was packed with enemy infantry firing at the 2nd South Lancs, who were still attempting to advance & [firing] upon our wounded. After about 5 or 7 minutes he was killed by a shell ...[18]

Catmur's acts were not considered to warrant a Victoria Cross. He was instead 'Mentioned', a poor consolation for his family.[19]

Two brothers, Second Lieutenants Basil Belcher and Wilfred Belcher, formerly of Brighton College, had originally joined 18/R Fus but were, by 1916, officers with 2/R Berks (8th Division). Basil was reported missing on 1 July amongst casualties of twenty officers and 414 men.[20] His body was later recovered from the battlefield. According to a newspaper; 'The battalion made an attack on the morning of July 1st, and came under very heavy machine gun fire, and 2nd Lieut

---

15 PS/3374 Private Henry Gerald Walker, aged 25, from Mirfield, Yorks, was educated at Dollar Academy and Fettes College, Edinburgh, where he was a member of the OTC. He worked for his father's pile cloth manufacturing business. He enlisted, aged 23, on 15 September 1915 and joined C Company, 21/R Fus; he was gazetted in May 1915. Walker went to France in March 1916 and joined 2/KOYLI. He is commemorated on the Thiepval Memorial. TNA: WO95/2402: 2/KOYLI War Diary.
16 TNA: WO339/48402: Papers of H.G. Walker.
17 PS/1346 Private Harry Albert Frederick Valentine Catmur, from Beckenham, Kent, enlisted with 18/R Fus. He was gazetted into the R Sussex in May 1915 and had served in France from March 1916 with 96th MG Company.
18 TNA: WO 95/2398: 96th MG Company War Diary.
19 Catmur is named on the Thiepval Memorial.
20 Basil Henry Belcher was aged 25 and from Newbury. He was educated at Brighton College and studied in Paris. He and his brother enlisted with 18/R Fus and were gazetted to commissions in the 3/R Berks in May 1915. Both went to France in March 1916 and were with 2/R Berks until Basil's death. He is buried in Serre Road Cemetery No 2.

B.H. Belcher was seen to fall just after leading his platoon over the parapet.'²¹ Wilfred Belcher survived the war with an MC to his name and was later ordained and became a bishop. The 8th Division attack had also failed with heavy losses.

7/Yorks R (17th (Northern) Division) was holding the front line facing Fricourt. The battalions of 21st Division and 7th Division on their left and right were to attack at 7:30 a.m. 7/York R was to attack at 2:30 p.m. by which time the village should have been enveloped. According to the war diary; 'Owing to an unfortunate mistake on the part of the officer commanding A Coy, his company assaulted at 7.45 a.m. As soon as they began to climb over our parapet terrific machine gun fire was opened … the company was almost at once wiped out…'²² Worse was to come; at the correct time the battalion attacked, with a fresh company replacing 'A'. They were met by a murderous fire and were shot down. Lieutenant Harold Hillman, 7/Yorks, was an actor in peacetime who wore 'a Harry Tate moustache, which he twitched up and down and side to side like a semaphore'.²³ One of his men recorded; 'Mr. Hillman, command[ing] No. 6 [Platoon], climbed the parapet and ordered his platoon forward. He was shot immediately through the head'.²⁴ Thirteen officers and over 300 men were lost within three minutes.

Private Philip Wright of 11/R Fus (18th Division) described his nervousness waiting for the advance:

> How that night dragged along! … It was possible for some fellows to forget everything and get some rest, perhaps, but not me. It was an absolute impossibility for me to dismiss all thoughts of the coming conflict. I think I am honest enough to say that I would have given worlds to have been miles away. I was positively as nervous as a kitten! Still … there was nothing for it, but to grit one's teeth and go into it outwardly calm if inwardly shaking with the worst type of 'wind-up'.²⁵

Wright's experience mirrored those of thousands of Tommies awaiting the order to attack:

> About six … our bombardment … increased tenfold in its appalling ferocity. A thousand guns of every calibre seemed to be concentrating their utmost efforts to encompass the destruction of any living thing in the foreground. … A little while previously I had been rather perturbed at some Hun machine guns opening fire at the particular spot where I had to go 'over the top'. Great was my relief when after a few minutes of the ensuing 'inferno', no more was heard of them.²⁶

At 7:30 a.m. the time came to advance:

---

21 *Newbury Weekly News*, 13 July 1916.
22 TNA: WO95/2004: 7/Yorks War Diary.
23 Victor Purcell, *The Memoirs of a Malayan Official* (London: Cassell, 1965), p.11.
24 Lieutenant Harold Alexander Moore Hillman, aged 30, from Thornton Heath, Surrey, was formerly PS/490 Private H.A.M. Hillman, 19/R Fus. He was gazetted into 11/Yorks in November 1914 and later joined 7/Yorks. He was buried in Danzig Alley British Cemetery. *The Cuthbertian*, Vol XXI, No 4, December 1916, p.116.
25 FM: P.N. Wright Diary, RFM.2013.8.1/2, p.38.
26 FM: P.N. Wright Diary, RFM.2013.8.1/2, p.38.

After allowing a minute or so for the first two waves to get on their way, we received the signal to get "over the top and the best of luck!" ... Our guns had done their work only too well, there was not the slightest sign of resistance during the first part of the advance... The German wire entanglements had been absolutely blown out of existence by our artillery and the trenches – well it was nearly impossible to find out where they had been...[27]

Lieutenant Colonel Carr, of 11/R Fus, recorded that this successful attack was far from bloodless:

Pommiers Redoubt was practically taken by 11/Fusiliers working round the west side of Redoubt, and was a most bloody fight, no quarter was given or asked for, the bayonet was used here freely. On reaching the final objective the left flank of the 54th Brigade was in the air as the troops on the left had been hung up early in the attack, the machine guns of the 11/Fusiliers were used to play on Germans seen retiring on this sector towards Mametz Wood.[28]

Privates Charles Coupe, Norman Cleveland and John Armstrong of 21/R Fus had been part of the fifty-man draft from B Company, 21/R Fus, who, like Philip Wright, joined 11/R Fus. All three went into the attack 'like bloodhounds let loose from the leash.'[29] Coupe and Cleveland were listed as missing along with Frederick Hunter and Tom Lever who had both served with 19/R Fus.[30] Wright survived this attack but the Somme was not finished with 11/R Fus.

Second Lieutenant Tudor Evans (formerly of 21/R Fus), of B Company, 8/E Surrey, was involved in one of the best-known escapades of the battle.[31] He was one of the four officers of Captain Nevill's company; that company 'kicked-off' the attack with two footballs.[32] The company 'had to face a very heavy rifle & machine gun fire...'. Another officer recalled; 'Many of the officers in my battalion were struck down the moment they emerged into view.'[33] Evans was one of many men killed.

1 July 1916 was a bloody day for the British Army and many ex-UPS men were killed or wounded during the fighting. More UPS-men would die in the coming weeks and months, especially once 33rd Division, and 20/R Fus, was committed to this offensive.

27 FM: P.N. Wright Diary, RFM.2013.8.1/2, p.38.
28 TNA: CAB 45/132: Letter from Lieutenant Colonel Carr, 23 March 1930.
29 O'Neill, *The Royal*, p.113.
30 PS/4022 Private Charles Coupe, aged 24; PS/7077 Private Norman Cleveland, aged 21; PS/8508 Private Frederick Hunter, aged 20; and PS/8566 Private Tom Lever, are commemorated on the Thiepval Memorial. PS/7112 Private John James Armstrong, aged 20, is buried in Dantzig Alley British Cemetery.
31 PS/2637 Sergeant Tudor Eglwysbach Evans, aged 24, from Penarth, Glamorgan, had been university educated. He had been gazetted to a commission in May 1915 and had gone to France in January 1916. He was buried in Carnoy Military Cemetery.
32 Ruth Harris, *Billie, The Nevill Letters 1914–1916* (London: Julia MacRae, 1991), p.198.
33 Harris, *Billie*, p.199. J.R. Ackerley, *My Father and Myself* (London: Penguin, 1968), p.50.

17

20/R Fus on the Somme

This is murder, not war

None of us was afraid – we had laughed at death too often.[1]

This is murder, not war. Many casualties. Dead & wounded English and Germans everywhere.[2]

The 20th Battalion Royal Fusiliers was about to undergo its first, and most notorious, battle. According to Godfrey Skelton during the early part of the year, up to this point: 'We did little really but occupy trenches and try to avoid being killed by the ceaseless shelling, trench mortars, fixed rifles and machine gun fire, and rifle grenades, sometimes fired from batteries of rifles … It was hard suffering all these casualties day by day, one's friends of fifteen months or so were either killed or wounded …'[3] After a relatively sedentary first half of 1916 from the start of July onwards everything changed rapidly. 20/R Fus would transition from trench holding to playing a part in the biggest British offensive of the war to that point. Just getting to the Somme and the start point of the great attack, which would affect the battalion and every member of it, is of interest in understanding what happened next and the legacy of that action.

The start of July 1916 saw the battalion situated in support north of the Canal at Givenchy between that village and Pont Fixe. A, B and C Companies occupied a series of 'keeps' in the support line and D Company was in reserve behind them. On the 2nd the battalion relieved the Cameronians in the Givenchy Right Sub Sector; on the right were 16/KRRC, 100th Brigade, and on the left were 5/Scot Rif. Shield recalled small events that made life less mundane:

2. [July] Up to firing line at Cuinchy, left Brickfields [sector], trenches badly damaged.
3. [July] English plane struck by shrapnel bursts into flames – falls near le Plantin. Pilot and observer badly shaken. Germans cheering. On patrol at night to German sap.[4]

---

1 Hodson, *The Soul of a Soldier*, p.87.
2 IWM: F.J. Shield Papers, 1385.
3 IWM: G. Skelton Papers, 13966, p.46.
4 IWM: F.J. Shield Papers, 1385.

Shield was also out on patrol on 5 July. On 3 July alone one man was killed and three wounded of D Company by one aerial torpedo.[5] There were other hazards according to Dykes:

> German snipers were very active and had a great nuisance value. They were excellent marksmen and patient in the extreme, as we frequently found to our cost. It had been a general instruction, for some time, that sentries should never look over the parapet twice together at the same spot. This order was sometimes forgotten or carelessly disdained, and, the marksmanship was so deadly, that we suffered casualties in consequence … Brearley was shot through the head and killed instantly.[6]

According to his diary Brearley was wounded severely whilst on sentry and died two days later.[7] PS/5265 Private Thomas George McKinney, aged 23, was likely a sniper with D Company having been attached to a telescopic rifle course, was also wounded on 3 July 1916 by fragments from an Aerial Torpedo. He later succumbed to gangrene at St Omer.[8] Another man was wounded by shell fire; '[PS/4934 Private John] … Hardy, a man who was a general favourite with the rest of the platoon. He was wounded in the right arm & was smiling when he went out of the trenches, he had received a very nice 'Blighty' wound and was quite happy in consequence.'[9] His wound was a fractured humerus which would later result in his discharge.[10] Between the 3rd and the 8th whilst the battalion was in the line five men were killed and fifteen were wounded; three of the latter remained at duty however.[11]

On 5 July 2/RWF had its revenge for Red Dragon Crater by raiding the German trenches. The event was a great success with several dug outs bombed or destroyed and prisoners taken; all for minimal casualties. The only RWF officer killed was Second Lieutenant Raymond Hollingbery, who had originally enlisted with 20/R Fus, along with ten 2/RWF other ranks.

Seemingly to mark the end of the 20/R Fus stay in this area there was a large mine initiated on 8 July at the Ducks Bill; a final act of revenge; 'The explosion was of gigantic proportions, and in broad daylight appeared even more terrifying than when the ears alone recorded the impressions.'[12]

---

5   The man killed was PS/7763 Private Noel Peter Coupe, aged 17, from Chorlton-cum-Hardy, Manchester. He was buried in Gorre British and Indian Cemetery. One of those wounded was PS/5486 Private Reginald Power but he later returned to the firing line.
6   IWM: J.N. Dykes Papers, 7378, p.26.
7   This was PS/4563 Private Fred Crossland Brearley, aged 21, from Bolton, who was killed on the 3rd; he presumably died of his wounds and is buried in Béthune Town Cemetery. He was in No. 5 Section, No 10 Platoon, C Company. IWM: J.N. Dykes Papers, 7378, p.125.
8   McKinney was born in Carnmoney and lived in Belfast. He attended the Royal Belfast Academical Institution and Agricultural College in Glengormley. He was also wounded in March 1916. McKinney is buried in Longuenesse (St Omer) Souvenir Cemetery.
9   IWM: J.N. Dykes Papers, 7378, pp.125–126.
10 PS/4934 Private John Hardy, aged 33, from Hale, Cheshire, was educated at Bath College and was a merchant by trade. He returned to the UK on 26 July 1916 and was discharged in September 1917.
11 On 5 July PS/5605 Private Cecil Sefton, aged 25, from Newark-on-Trent and PS/8254 Private Alfred Bertram Krauss, of D Company, aged 19, from South Norwood, London (educated at Selhurst Gramar School, Croydon and worked in the accountant's department of the LBSC Railway) were both buried at Gorre British and Indian Cemetery. PS/8899 Private Ernest Patrick Lynn, aged 33, from St Annes-on the Sea, Lancashire, died of his wounds and was buried in Béthune Town Cemetery.
12 IWM: J.N. Dykes Papers, 7378, p.26.

Raymond Archibald Robert Hollingbery was aged 22 and from Kensington. Gazetted into the RWF in June 1915, he arrived in France on 6 May 1916. Hollingberry is buried in Gorre British and Indian Cemetery. (Reproduced with permission of Sutton Archives)

Between 9:15 p.m. and 11:15 p.m. the battalion was relieved by the 1st Battalion of the Hertfordshire Regiment and marched to Annezin where it remained on the 9th tidying itself up. On 8 July Dykes had heard rumours that the battalion would go out of the line and after a rest move to another part of the front. Next day he recorded that their platoon commander, now Mr Wylie, briefed them to lighten their loads as they had much marching ahead but did not know where they were going. Shield scribbled in his diary for 9 July, 'Informed we are to take part in the advance. To leave late tonight. Destination unknown.'[13] These were the orders the UPS men had been waiting for; orders for 20/R Fus, and 33rd Division, to take part in the Battle of the Somme. James Hodson summed up the feelings of his comrades in his dramatized account; '…the Somme, after Cambrin, would be like starting to fight all over again; they were old soldiers, and yet they knew little of real warfare, of assault, of hand-to-hand conflict … they were exalted, like schoolboys off to the footer match or the principle cricket duel of the season. Novelty robbed the prospect of its horror.'[14]

Journey South

The stay in Annezin would be brief. The battalion boarded a train at 3:00 a.m. on the 10th for an unknown destination but the realisation that it was heading south towards the Somme quashed all previous rumours. The train arrived at Longueau near Amiens at 11:00 a.m. The onward march to billets in Poulainville was initially pleasant but as the sun grew hot the dust and flies caused irritation. James Hodson remembered the emotions at the time, 'when the march [southwards] began … one could ponder a good deal. Already we had been eight months in trenches … we felt like recruits again. We felt that we knew nothing of it all, this fierce hand-to-hand

---

13   IWM: F.J. Shield Papers, 1385.
14   Hodson, *Grey Dawn*, p.240.

fighting, this *real* warfare where anything might happen at any time. We were awed a little, like new boys at school.'[15] He also summed up the desire to get on with the job at hand; '…here at last was some work … that would shorten the war, and no temporary agony was too great and nothing too terrible to achieve that end. No wish was in our hearts to postpone the day. We knew its inevitability. None of us was afraid – we had laughed at death too often. Yet none of us was gay or happy.'[16] On the other hand there were numerous physical hardships to be overcome. James Dykes wrote of one hard stint of marching:

> Our next four hours of existence were only remarkable for the scorching sun, the thick dust raised by the hundreds of army boots, the ever rising temperature, the faces lined with the dust … dust which had stuck to the perspiration which ran down and dropped off the chins of those marching men, and, worst of all, the persistence of millions of flies which hovered over those damp and dusty faces, causing intense irritation and frustration.[17]

On the 11th another early start at 6:00 a.m. saw the battalion depart for Vecquemont where it was billeted. Morale was still high despite the early start, 'the air was fresh and pure, the countryside charming; good to be alive. Marching songs, grave and gay, the vast majority exceedingly rude, I fear, enliven the proceedings. Badinage was exchanged with A.S.C. units encamped on the roadside…'[18] According to James Hodson; 'On our march down to the Somme, we had trudged along sunlit roads, enjoying it; I remember a night when a regimental band played Gilbert and Sullivan – the air of a fete was abroad. A fresh billet in a different village every night, lots of fresh eggs, a good deal of wine, and larks in the haylofts and barns…'[19] The men were still relatively carefree; after arriving in billets in a French village; 'Within a short time half the battalion were strolling about the village, men burdened with huge round French loaves, tinned fruit, eggs, chickens, chocolate, milk. Others were washing feet in pails, or shaving, or scraping puttees, or cleaning rifles. By eight or nine o'clock most of them were in bed, not sleeping, but … smoking or reading or writing…'[20]

However, Shield recalled that; 'Everything (what could be got) frightfully dear'.[21] Not every man of the battalion would continue towards the Somme. Donald Price recalled one event on the march:

> When we got round about the Somme it was a summers day and we could hear these guns at some distance and we were all paraded on the road, we'd finished the march, and the Regimental Sergeant Major who was with us, he was an old fellow, a very lovely chap, I forget his name now, and he said goodbye to us on the road. He wasn't coming with us from there on and somebody else took over. He was the RSM, an old regular, and a nice chap, but he was getting on; I remember his grey hair quite well.[22]

15 Hodson, *Soul of a Soldier*, p.86.
16 Hodson, *Soul of a Soldier*, p.87.
17 IWM: J.N. Dykes Papers, 7378, p.28.
18 IWM: J.N. Dykes Papers, 7378, p.29.
19 Hodson, *Return to the Wood*, p.37.
20 Hodson, *Grey Dawn*, p.242
21 IWM: F.J. Shield Papers, 1385.
22 IWM: Donald Price Sound Recording 10168, Reel 8.

It is possible that Price was confused and described his CSM being replaced. He may have been replaced temporarily by Robert Armour, a Regular, who had transferred across from the Cameronians. Armour was considered, 'as fine a soldier as ever I have seen'.[23] PS/8860 Warrant Officer Class 1 Leonard Raven, aged 40, had been the RSM since September 1915 and served as RSM again in 1917. James Dykes and his good friend John McKeown were also separated after an eight-month friendship as the latter, a batman to Lieutenant Walker, had to leave Dykes' section.[24]

These moves were all part of a steady move southwards prior to taking part in the Battle of the Somme. At Vecquemont all packs and spare kit were left behind for storage. At 2:00 p.m. on the 12th the battalion marched to Buire-sur-L'Ancre arriving at 6:00 p.m. On 13 July the battalion waited at Buire to be committed to the offensive; the day was spent awaiting orders for an onward march towards the battle. This pause gave the 20/R Fus men the chance to take stock of what was around them. The first German prisoners were seen:

> … a fairly considerable party of them marching through the village away from the fighting area. They were dishevelled, dirty, looked tired and very apprehensive. Without exception they looked as if they had had a very rough time. They seemed listless and depressed, and not particularly concerned with what might lie ahead of them.[25]

There were feelings that the Battle of the Somme was going well and that the Germans were on the back foot as the battalion marched south but Skelton noted, 'there seemed to be few prisoners in the prisoner-of-war cages but many of our wounded coming down the line.'[26] Whilst the battalion was at Buire James Hodson remembered: 'That late afternoon the cavalry – thousands upon thousands of them – came at a walk-march through – British cavalry, Indian cavalry, the British staring stolidly at you, the Indians grinning. The last word in fitness and efficiency, with their sabres, their rifles, their lances, their machine-guns, how they stirred our hearts! It was good to be a soldier…'[27] Sid Platt succinctly reported; 'Indian cavalry ride through [village] (about a division). Fine lot … Indian & our cavalry passed through – also mounted machine guns and R.H.A. Very impressive site. Never seen so many mounted troops before. Passing through all the day…'[28] Others were more realistic; PS/5782 Private Harold Tyson, of A Company, chose to write his will.

On 14 July the men of the battalion were equally unmolested until about 11:00 a.m. Around that time they marched onwards to Méaulte and took over billets. No sooner had the men finished their dinner than they were turfed-out again and got into formation. The battalion moved further forward to a bivouac site not far from the side of the road between Méaulte and Bécordel-Bécourt arriving at 6:00 p.m. From this high ground the men could see the town of Albert and the trenches beyond. On this day some cavalry of the 7th Dragoon Guard and

---

23   L/17086 WO1 Robert Armour was formerly a soldier with 1/Cameronians and went to France in August 1914 as a corporal. See IWM: Donald Price Sound Recording 10168, Reel 8 and Terry Norman, *The Hell they Called High Wood* (Wellingborough: Patrick Stephens, 1984), pp.141–142.
24   IWM: J.N. Dykes Papers, 7378, p.31.
25   IWM: J.N. Dykes Papers, 7378, p.29.
26   IWM: G. Skelton Papers, 13966, p.47.
27   Hodson, *The Soul of a Soldier*, p.88.
28   IWM: S. and V. Platt Papers, 17681, p.72.

Brigadier General Charles Mayne was aged 41 when he took command of 19th Brigade in July 1916. Commissioned into the HLI in 1895, he served in the Ashanti, Nigerian and the Blue Nile campaigns and was awarded the DSO in 1902. Mayne went to France in August 1914 as a captain where he was wounded. In October 1915 he took over command of 20/LF. Clearly an able and experienced officer who was a strong candidate for brigade command, he was described warmly by an officer of 33rd Division: '... a familiar figure in his great height and rugged strength, striding over the battlefield, possessed of that quality which he so happily shared with his General, unwearying patience and good cheer.'*

* G.S. Hutchison, *The Thirty-Third Division in France and Flanders 1915–1919* (Uckfield: N&MP, originally 1920), pp.174–175.

the 20th Deccan Horse, possibly those horsemen seen a few days before, attacked south of High Wood and were repulsed with heavy casualties. According to Skelton who saw the battlefield later; 'Alas, the attack was a shambles against uncut barbed wire and the ground covered with shell holes, and machine gun fire…Casualties among men and horses were dreadful and no more cavalry attacks were made…'[29]

Though probably unknown to most of the men, the Brigade Commander, Brigadier General Robertson, departed to command a division; his place was taken by Brigadier General C.R.G. Mayne DSO. Mayne was likely selected above Lieutenant Colonel Chaplin of 1/Cameronians whose disappointment was doubled by being left out of the fighting as a 'reserve brigade commander'.[30] Mayne was taking over at an awkward time on the eve of a battle and having had no opportunities to understand the qualities and weaknesses of his subordinates and the component units of his brigade.

Arrival on the Somme

The battalion was now nearing the Somme battle zone; the next move commenced at 4:00 a.m. on the 15th and took the form of a long march up to an encampment site near Mametz Wood which had only recently been wrested from the enemy. The 5/Scot Rif war diary noted:

29 IWM: G. Skelton Papers, 13966, p.48.
30 Andrew Davidson, *The Invisible Cross* (London: Heron Books, 2016), p.213.

> The Brigade moved up at 4:30 a.m. Order of March: The Cameronians, 5th Scottish Rifles, 20th Royal Fusiliers, 2nd Royal Welsh Fusiliers. The morning was very misty and the air thick with gas from lachrymatory shells. The old front lines, British and German, were soon passed and after marching about 2 miles over the system of German trenches, the Brigade formed up behind Mametz Wood.'[31]

The solitude for many on the march left them with time to mull over what they saw. Dykes described some German trenches on the march forward to Mametz:

> The condition of the German trenches had to be seen to be believed. They were shocking indeed. How could any living thing have possibly existed under the intensive bombardment which had been directed against these earthworks? Debris and equipment lay about in uneven piles, especially where efforts had been made to clear the way for transport, which was vital, and for the passage of fighting men. It was difficult enough to negotiate in daylight, but in darkness well night impossible... As we progressed with difficulty, more and more devastation, silent and motionless forms covered by ground sheets, so that only the lower parts of bodies were visible, empty sandbags covering heads and faces ... Not much glamour and glory apparent![32]

Hodson described getting gassed whilst marching forward to Mametz; 'Eyes began to smart and tears ran down their cheeks. Small shells were dropping near, exploding with a quiet pop, and fizzing. Tear-gas! They pulled out gas masks and enveloped their heads in flannel soaked in chemicals and stared out through round circular windows and breathed out through a revolting tin tube. Silence.'[33] As the battalion moved uphill the mist thickened; eventually the gas dispersed and the men could remove their masks. Dykes also described this gas:

> We ... immediately had to don gas masks and goggles to combat tear gas from shells dropping in the vicinity. The acrid fumes caused intense irritation to the throat and eyes, the effect being similar to the onslaught of a severe cold, causing severe coughing, streaming eyes which itched abominably so that you felt you must rub them. The goggles did not give complete protection, and breathing was difficult with the mask over the face. Fortunately the density of the gas was not such that long inconvenience and long unpleasantness was suffered.[34]

Marching through the mist and gas was an eerie experience; 'Crash! Went a gun on our left ... Ah! A figure sprawled awkwardly on the bank with a haversack dyed red ... And then it was gone in the fog. Next, a horse, its head bloody and smashed ... The mist made things come and go ... a wounded man on a stretcher, his arms and hands hanging limp over the sides...'[35] These seemingly passing images left Hodson feeling 'hazy and uncertain. I felt sick and fearful.'

31 TNA: WO 95/2422: 5/Scot Rif War Diary.
32 IWM: J.N. Dykes Papers, 7378, pp.31-32.
33 Hodson, *Grey Dawn*, p.246.
34 IWM: J.N. Dykes Papers, 7378, p.31.
35 Hodson, *Soul of a Soldier*, pp.89–90.

The battalion fell out similarly to how they had rested after field days in England and made bivouacs, 'they were sitting in an undulating valley … spread higgledy-piggledy over a large area … not in lines, nor were they digging in. … A few minutes later he heard a report, saw smoke from an explosion, and pieces of a man leaping into the air twenty feet high. … He was angry at the lack of care…'[36] Dykes remembered that; 'On instructions, we immediately commenced making shelters for ourselves, by means of entrenching tools. This task was simplified by the presence of numerous shell craters which pitted the ground, and we soon had adequate cover.'[37] Frederick Shield recorded; 'Early morning moved up through English & Boche lines to past Mametz. Ground in awful condition. Equipment, the dead, horses dead and wounded, English and Germans, everywhere. Bivouac Mametz Wood. Bodies still there. [We were to] Take part at 9 a.m. in attack on Martinpuich – two divisions. Some strafe. Retaliation [by] our guns …'[38] According to Dykes the following happened that day (15 July):

> Around 7 a.m., our platoon officer [Lieutenant Wylie] came along to brief us regarding the programme for the day. The whole of 33rd Division was concerned and involved. The 98th and 100th Brigades were to attack at 8 a.m. The task of confronting the attackers was the comparatively simple one of "mopping up" a sector of German trenches. It appeared that the resistance of the defenders could only be a token opposition by reason of the terrific beating to which they had already been subjected. When the attacking brigades had attained their object … the 19th. Brigade would advance through the gap thus created and capture the village of Martinpuich … What followed might be rather more difficult, namely, the containing of the enemy counter attacks …[39]

Though the short notice of this attack might suggest poor planning the approval that Dykes gave to this plan suggested he had confidence in the ability of the officers of his battalion; 'Every eventuality had been guarded against, and provision made to combat all the wiles of the Germans.'[40] Dykes' platoon, and the rest of the battalion rested in their shelters awaiting the call to action.

As the day dragged on the optimism dissipated and news filtered in that the attack had been a failure. Rumours were of heavy casualties to the attacking battalions which would only have sowed doubt and affected morale. Morale might have benefited from the fact that whilst British shelling of German positions could be heard to be heavy and continuous the German response was meagre. The 15th was quiet and according to Dykes the only casualty was Corporal South of No 9 Platoon who was wounded but evacuated for treatment.[41]

There were gruesome sights in Mametz Wood:

> We were within easy reach of the revolting sights in the wood, and many men went in to see for themselves what hand-to-hand fighting meant in terms of horrifics! Any who

---

36 Hodson, *Grey Dawn*, pp.247–248.
37 IWM: J.N. Dykes Papers, 7378, p.32.
38 IWM: F.J. Shield Papers, 1385.
39 IWM: J.N. Dykes Papers, 7378, p.32.
40 IWM: J.N. Dykes Papers, 7378, p.32.
41 This may have been PS/5650 Private Edward Baxter Southern, aged 24 and from Southport. He enlisted in September 1914 having been educated at Charterhouse and Oriel College, Oxford. He died of wounds on 21 July and was buried in Mericourt-L'Abbe Communal Cemetery Extension.

were perhaps too squeamish to look at fearful 'tableaux', were given food for thought by the vivid accounts of the returning witnesses, of bodies piled high in some places where it was plain to see that a stand had been made in that spot, of fierce hand-to-hand combat, of broken and bloodstained weapons, of an eerie silence which pervaded the wood of complete desolation.[42]

A less gruesome sight, but one which might have concerned the more adroit soldiers of 20/R Fus were some German dugouts which were on the opposite side of the hill to the bivouac area; these were deep and well-constructed. Imagining that the Germans on the receiving end of the British artillery might be protected by similar constructions would have cause concerns. There were also horrors by the roadside, 'horses, mules and men slaughtered by shell fire and just cleared away into the roadside ditches to make way for more troops and guns, etc… In the hot sun the smell was awful and the air full of flies.'[43] Because Lieutenant Wylie was made company 2IC, he was sent back to the battle reserve. Dykes' platoon was taken over by Second Lieutenant Blaauw ('Beelaw' according to Dykes).

The battalion spent the day of 15 July in the wood and the subsequent night. At 3:00 a.m. on 16 July the battalion moved to relieve the 16/KRRC in support on the Contalmaison–Longueval Road from the junction north of Bazentin-le-Grand for 400 yards to the east, and to a point on the track running towards High Wood. 16/KRRC had attacked High Wood from this location and had suffered horrible casualties. Dykes recalled; 'Early morning of the 16th July, the battalion formed up on the road, and marched in fours in the direction of the fighting area, behind the two Scottish battalions. Marching in such close order seemed to some of us to be tempting providence.'[44]

The Cameronians and Scottish Rifles had also been called forwards. At 1:00 a.m. on 16 July the former had been called forward to High Wood to take over from the Glasgow Highlanders. As an HLI guide lost his way, a Cameronian company lost 49 men to German MG fire. The battalion was withdrawn from near High Wood during the afternoon. Frederick Shield's diary recorded the following; 'Still digging – many casualties. Cams from our front forced to retire leaving us to hold the line.'[45]

Having spent so long in deep breastwork trenches at La Bassée several members of 20/R Fus recalled feeling more exposed whilst serving in the rear areas on the Somme; 'I do not think Captain Toller [OC D Company] felt any fear at all and seemed to think that we were equally brave. He paraded us several times in closed ranks even when enemy shells were falling quite near. This seemed quite senseless to risk casualties in this way.'[46] Such attitudes later had fatal consequences; 'In moving to this position in darkness we were heavily shelled and our Number 13 Platoon, mostly from the Belfast area, had many killed and wounded by a 5.9[inch] shell which landed in their ranks; we just had to march on leaving the dead and dying and wounded to the stretcher bearers…'[47] However, Skelton's account mistakenly recorded this event as occurring on 19 July. Dykes, further forward in the column, recalled:

42 IWM: J.N. Dykes Papers, 7378, p.33.
43 IWM: G. Skelton Papers, 13966, p.47.
44 IWM: J.N. Dykes Papers, 7378, p.33.
45 IWM: F.J. Shield Papers, 1385.
46 IWM: G. Skelton Papers, 13966, p.48.
47 IWM: G. Skelton Papers, 13966, p.49.

… we had only marched a short distance from our late bivouac, when we heard an ominous rushing sound over our heads, followed almost immediately by a deafening explosion behind us. The high explosive shell, for such it was, exploded plumb in the centre of No. 13 Platoon in "D" Company. It created terrible havoc, killing seven men outright, and wounding fifteen others, many of them seriously.[48]

Vincent Platt, of No 13 Platoon, recorded this horrific fact in such a matter-of-fact way; 'Unfortunately half our platoon was wiped out by shell that dropped right into us. Marvellous escape for both Sid & I.'[49] Sidney Platt could likely not even bring himself to write about this event. Three of those killed on 16 July 1916 are known to be members of D Company; PS/5435 Private William Neville Patchell, aged 25, from Belfast; PS/5486 Private Reginald Colin Power, aged 25, from Atherstone, Warwick; and PS5676 Corporal Eric Hilton Frederick Stewart, aged 21, from Belfast. Power was educated at Haileybury School and was the son of a surgeon. His platoon commander, likely Lieutenant William Mair, wrote; 'If there was a call for volunteers he was one of the first to answer, especially if there was danger in it.' An NCO wrote 'He was a great man and a fine soldier.' He was regarded as being a grouser by Vincent Platt.[50] He, and likely the others, were buried by the roadside by Bazentin Wood but his body was never recovered. All three are commemorated on the Thiepval Memorial. This one event cost 20/R Fus over twenty casualties.[51]

Despite this incident, the battalion moved along the Contalmaison–Longeuval Road and took over from the KRRC. Dykes recalled that when the battalion reached a cross roads the Battalion HQ was set up and the companies continued onwards before digging in along a bank along the side of the road. After an hour of digging they had established themselves but were order to move to a new location and repeat this activity; presumably an officer had made an error with the location. During the day 20/R Fus suffered casualties from shellfire and according to Dykes; 'The stretcher Bearers were kept busy indeed.' Whilst some platoons were hit badly, Dykes' No 10 Platoon lost only two men wounded. However, shelling was not so heavy as to preclude the inquisitive and the souvenir hunters from examining the area: '… occupied German lines vacated the day previous. Found plenty of souvenirs, helmets, greatcoats, rifles, blacking, scent, field glasses, black bread, cigars etc… Also found violin strings & blankets which we used.'[52] Those examining the German dead found four 'corpses' to be still alive which provided extra work for the 20/R Fus Stretcher Bearers. During the afternoon there was drizzly rain.

At the close of the 16th 20/R Fus was withdrawn back to Mametz Wood. The rain made conditions under foot difficult, and it was easy to slip over on the now slick road surface. At

48 IWM: J.N. Dykes Papers, 7378, p.33.
49 IWM: S. and V. Platt Papers, 17681, p.73.
50 IWM: S. and V. Platt Papers, 17681.
51 PS/9684 Private Edward Andrews, aged 21, from Oxford, was likely a member of No 13 Platoon. According to his obituary his platoon was hit by a shell moving to support positions which caused twenty-one casualties. Andrews was recorded as being buried by the roadside in Bazentin-le-Grand. Other fatalities on 16 July which are presumed to have been killed with No 13 Platoon were; SR/162 Private George Bowman, from Brixton; G/24620 Private Stanley Basil Ewins, aged 23, from Twickenham; and PS/7848 Private George Edward Peffer, aged 21, from Barnsbury, Islington. Andrews, Ewins and Peffer were presumably buried by the roadside like Power and are named on the Thiepval Memorial. Bowman was buried in Mericourt-L'Abbe Communal Cemetery Extension.
52 IWM: S. and V. Platt Papers, 17681, pp.73–74.

10:00 p.m., 20/R Fus returned to bivouacs in Mametz Wood having lost 42 men, including the 22 men from No 13 Platoon.[53] Morale was likely lower after another day in the battle zone, more casualties, and without accomplishing a significant military task. The weather left the majority soaked-through, and weapons and equipment clogged or covered in mud. It was not until midnight that Dykes' comrades, and likely others, were able to get some rest. No sooner than they had the Germans sent over a salvo of shells into the bivouac area adding to their woes. Dykes was so drained he and a comrade slept very heavily. Frederick Shield recorded simply in his diary; 'Coop blown to pieces'. This referred to PS/4661 Private Robert Coop, aged 21, from Oldham.[54] Shield added; 'This is murder, not war. Many casualties. Dead & wounded English and Germans everywhere.'[55] His sense of inevitability was mirrored by Dykes' earlier comments on there not being 'much glamour and glory apparent!'

The situation could have been significantly worse. 100th Brigade suffered heavily in attacking High Wood and lost about 41 officers and 1,440 men.[56]

The battalion woke up having had a very wet night; Dykes had slept in a puddle. The day saw a return to dry weather and a chance to clean up and dry equipment. Many men were able to rest all day and the subsequent night. Some members of the battalion took part in a fatigue to carry ammunition up to the forward positions. As an NCO Godfrey Skelton found allocating these hazardous duties difficult; 'This was a very dangerous job, moving through shell barrages and machine-gun fire… One would have liked to have detailed single men only but this was not possible, one just had to work down the list regardless of personal circumstances.'[57] James Dykes acted as a mess orderly on 17 July and thereby missed this fatigue. Even in the forward battle area, with action expected at any time, the normal battalion routines continued. Likewise, though the battalion did not indulge in smartness in the forward areas it did ensure that key personal equipment was clean and serviceable; rifles, gas helmets, iron rations and goggles were inspected during the day. Shield again noted the following; 'Raining hard. Wet through. Shells everywhere. Got under limber to shelter. Limber blown to bits by gas shell. Unhurt. Made a dugout during day. Hear we move up to dig ourselves in tonight…'[58] With that, his diary ended. Despite what was ultimately a quiet day on 17 July the battalion lost a further 13 men wounded and three killed including PS/4686 Lance Corporal Basil Coupe whose brother, Noel, had been killed earlier in the month.[59] The battalion had now lost five percent of its manpower without engaging the enemy.

---

53 Another fatality on 16 July 1916 was PS/7528 Private Stephen Castle, aged 19, from Dudley. He had enlisted with 18/R Fus and was attached to the HQ of 19th Brigade in an unknown capacity. He died of his wounds and is buried in Daours Communal Cemetery Extension.
54 PS/4661 Private Robert Coop, aged 21, was born in Oldham and lived at Werneth. He was the assistant secretary of the Moss Lane Mill in Heyside. He was killed on 15 July 1916 and was buried in Dantzig Alley British Cemetery, Mametz.
55 IWM: F.J. Shield Papers, 1385.
56 TNA: WO95/2408: 33rd Division Admin War Diary.
57 IWM: G. Skelton Papers, 13966, p.48.
58 IWM: F.J. Shield Papers, 1385.
59 PS/4686 Lance Corporal Basil Coupe, aged 23, was from Whalley Range, Manchester was a clerk with an insurance company. He was buried in Heilly Station Cemetery Méricourt–L'Abbé. PS/8826 Private Percy John Crowe, of D Company, was aged 18 and from Ipswich. He is commemorated on the Thiepval Memorial. PS/7419 Private Alfred Knowles was from Leigh, Lancashire. Knowles was a member of Leigh Harriers running club and well-known long-distance runner. He died of his

The next day was also quiet and other than inspections personnel were allowed to explore the surrounding area. Rather than face the horrific sights of Mametz Wood again, Dykes opted for a fatigue party carrying ammunition like the one the previous day, under Second Lieutenant Everard of No 9 Platoon. The Germans traversed the Contalmaison–Longeuval Road with artillery fire and caught the party. Second Lieutenant Frank Everard was wounded in his left thigh and hip on 18 July.[60] Half of the party became casualties before they withdrew having left their loads by the side of the road to carry the wounded. The party returned, commanded by a sergeant, and delivered much of the ammunition to its destination at the foot of the windmill near Bazentin-le-Petit. No sooner than they had arrived back than the battalion was paraded for the trenches.

A 'forlorn hope'

The battalion went forward again on the 18th, at 5:00 p.m., relieving 5/Scot Rif in the front line between the Quarry and the Windmill. This was not a safe position even though it was out of view of the enemy. The Scottish Rifles had lost eight officers and 146 men over two days; predominantly through shell fire. Though much of the battalion was positioned on a reverse slope, and not observed by the enemy, a piquet needed to be placed on the top of the ridge to observe for any signs of any enemy attack. This task fell to half of Dykes' No 10 Platoon, C Company, which was supposedly hand-chosen by Colonel Bennett. In the event of an attack this party would send up a green flare to request reinforcements and would sell their lives dearly to delay the enemy. The actions of this party provide an interesting tactical vignette of action on the Somme and deserves especial examination.

The party, a 'forlorn hope' according to Dykes, was led forward by Captain Maxwell, OC C Company, and consisted of Second Lieutenant Blaauw, a sergeant, a corporal, a lance corporal and fifteen men.[61] They advanced past dead bodies from both sides and the detritus of the previous few days of battle. Dykes described the scene in greater detail:

> Arriving at the original front line, we came upon what had apparently been a strong point, – a barricade of all sorts of odds and ends, sandbags mainly, with a space, a loop hole for a rifle of machine gun. More dead bodies lying in amazing postures and attitudes were grouped around this point. We had to climb over this barrier, and continue to advance along the track, at the side of which had been dug a shallow trench, not even breast high,

wounds at 63rd Field Ambulance and was buried in Dartmoor Cemetery, Bécordel–Bécourt.

60 Second Lieutenant Frank Leslie Everard was from Kettering, Northants. He was educated at Bridlington Grammar School, where he was a member of the OTC. He was an engineer and was employed by Tangye's in Birmingham. He enlisted, aged 19, in September 1914 with 19/R Fus (PS/326) and was promoted to sergeant. He was gazetted to a commission in 19/R Fus in March 1915. He later served with 28/R Fus and went to France on 20 May 1916 joining 20/R Fus three days later and was allocated to C Company, No. 9 Platoon. He was wounded in the left thigh and hip on 18 July according to Dykes whilst on a working party. Once recovered, he served with 29th Training Reserve Battalion and 52nd Graduated Battalion of the Queen's Regiment. He applied for the RFC in October 1916. He still had a bullet in him in July 1917.

61 This party included: PS/5739 Sergeant Basil Taunton, PS/9132 Private James Dykes, PS/4599 Private Fred Carter, Private Whittaker and Private Ellis, amongst others.

and obviously done in a hurry. More dead bodies in the open and then we reached our destination. It was a short length of trench, about 25 yards or so in length – quite shallow – a breastwork – nothing more! Whilst a Scottish officer and N.C.O. got on with the business of "handing over", we had a good look over our responsibility for at least the next 24 hours; we could not find any cause for satisfaction of jubilation. We were just short of the top of the ridge, there was no interference with the field of vision to left and right of our position, or behind, but from about 50 yards in front, there was only skyline, the ground dropping away after that distance. The German advanced post, we were informed, was some 40 yards on the further side of the ridge. Away to our right was High Wood, in German hands. We could see no signs of desecration in that area. It looked very green and lovely, untouched by man's destroying influence.[62]

Presumably Captain Maxwell departed having seen the party to their destination. The Scots stated many of the casualties had been caused by premature British shells. During the handing over of this forward position on 18 July 1916 a shell landed at the feet of Second Lieutenant Blaauw and damaged his right leg which was later amputated. He was wounded, by what Dykes believed was a premature shell, whilst alongside a Scottish officer and sergeant.[63] The sergeant took charge. Second Lieutenant William Moult was sent forward to command the party vice Second Lieutenant Blaauw.[64] In the meantime PS/5769 Sergeant Basil Taunton gave the party their orders.[65] They were to hold the post at all costs for their 24-hour stint there. Five flares would be sent skyward in the event of an attack to warn the main line of an attack. A single sentry was to be on duty in the breastwork by day, two at night, with another sentry further forward as near the skyline as they could safely reach. Other than a few of the men crawling out to German bodies in front in search for souvenirs or loot, little happened.

---

62 IWM: J.N. Dykes Papers, 7378, p.35.
63 Second Lieutenant Henry Thomas Gillman Blaauw, aged 43, was born in 1874 and was from Wivelsfield, Haywards Heath. He and was educated at Eton College until 1893 and Oxford (where he was a member of the University Volunteers). He attended the Royal Agricultural College, Cirencester, and was a farmer. He was quite old to be a subaltern having enlisted, aged 40, in the Sportsman's Battalion. He was discharged to a commission on 16 January 1915 and went to France on 9 March 1916. He originally served with 21/R Fus, later joined 20/R Fus, and subsequently took over No. 10 Platoon, C Company. After losing his leg he served with 30/R Fus in the UK until March 1917 when he resigned his commission to return to farming.
64 William Moult, aged 32, was from Sale, Cheshire; his trade was described as both manufacturer and employer. He enlisted with the UPS in September 1914 and became a sergeant (PS/5344) in A Company, 20/R Fus. He was gazetted in April 1915 and joined 29/R Fus. Moult went to France on March 1916 and attended a two-week instructional course.
65 PS/5769 Sergeant Basil Taunton, aged 31, was from Tunbridge Wells. He attended Tonbridge School in 1899 and was a clerk for the Post Office. He had originally been commissioned in the Kent Volunteers (Cyclists). Taunton enlisted in Exeter in September 1914 and served with C Company, No 10 Platoon, as Platoon Sergeant. He married Amy Whiting in Epsom on 17 April 1915. He was wounded in March 1916 but returned to 20/R Fus.

A distressing incident

The night was quiet and early the next morning, 19 July, an artillery officer and his orderly arrived, presumably in response to complaints of British shells bursting prematurely but possibly to set up and OP for bombarding German territory. Against advice, the gunner officer crawled out towards the ridge top. Private Whittaker was still on sentry in the forward position.[66] He tried to stop the artillerymen by waving to warn them of an active German sniper. The artillery officer and his bombardier continued to crawl until the officer was hit; 'Suddenly there was the sound of a rifle shot, we saw the officer make a convulsive movement and then lie still.' The bombardier went to his aid as did Whittaker and 20/R Fus stretcher bearers were called forward. The wounded officer was recovered but had to remain at the breastwork until dusk when he could be safely evacuated. Meanwhile, Whittaker was hit by the sniper as he crawled back to his sentry post. After lying still he began to crawl towards the German lines. PS/4599 Private Fred Carter, Whittaker's close friend, broke cover and ran to warn Whittaker, but was shot dead, likely by the same sniper.[67]

Sergeant Taunton then tried to locate and distract the sniper:

> He put his shrapnel helmet on his rifle barrel, and raised it slowly in the air, the idea being to draw the attention of the sniper away from Whittaker. We in the trench were watching the proceeding with agonising intensity. The sound of the shot rang out, we saw the tin helmet shudder on the rifle, and heard the metallic ping of the bullet; the next minute the sergeant had turned and was crawling back towards us, a wounded man. The only possible explanation could be that the bullet had penetrated the helmet, had been diverted in its flight, had then shaved his hand, and finally made a slight flesh wound down the side of his face. The wounds bled fairly freely, but he was not incapacitated, the bullet, in it's erratic flight had also encountered the sergeant's wrist watch, and put that out of action…[68]

Whittaker turned back and, after crawling slowly, eventually made the British trench; he had been wounded in the buttocks but was able to be evacuated.[69] All this happened between 8:30 and 9:30 a.m. on 19 July. After a quiet day Private McMillan, the platoon runner, arrived and passed an order to evacuate the position. The troops quickly collected their kit and withdrew. On arriving back at the main line a British bombardment commenced of the German forward posts; shells would have landed dangerously close to the British post if it had still been occupied.

Luckily this was not a long stay. At about 9:00 p.m. on the 19th, 20/R Fus was relieved by 4/King's. According to James Dykes this was at about 8:30 p.m. when the battalion was relieved and the tired fusiliers returned towards Mametz; the Battalion was halted before they arrived and dug in and made bivouacs in Bazentin-le-Petit Wood. Though many might have expected

---

66 Likely to be PS/5921 Private Donald Gordon Whittaker, from Cranford, New Moston, who had been educated at Manchester Grammar School.
67 PS/4599 Private Fred Carter was just short of his 19th birthday and was from Marple Bridge, Derbyshire. He was reportedly killed rescuing a wounded officer; the other man with him was awarded the MM. Carter is commemorated on the Thiepval Memorial.
68 IWM: J.N. Dykes Papers, 7378, p.37.
69 Private Whittaker was evacuated back to the UK in late July 1916 having been severely wounded, twice. He was later gazetted to a commission in the KOYLI in April 1917.

a rest was due they were, in fact, to attack. This was to execute plans for a renewed attack by 33rd Division on High Wood. According to Vincent Platt: 'Stayed in support all day, and were relieved at night under the impression that we were going back. However, we had to "about turn" and went over the top to take the wood…'. Other men in the battalion had a quiet day in bivouacs. Private Don Price was playing with German bombs with his friend Sid Warhurst during the afternoon of 19 July. Sid did not let go of a bomb in time and it burst above them; a fragment lodged in Sid's eye and he had to be evacuated.'[70]

This stint in the line cost the battalion three officers wounded (Ziegler, Everard and Blaauw) and a further 28 casualties from amongst the men. At some point Lieutenant George Ziegler had received multiple wounds; he was hit in his left thigh, right knee and left elbow.[71] On 18 July 1916 three men from 20/R Fus were killed or died.[72] Next day, 19 July, a further three men were recorded as killed.[73] By the end of 19 July 20/R Fus had already lost twelve men killed, seventy-four wounded and one missing. This suggests that the battalion was already depleted, tired and in need of a rest. It was not in the best of conditions in which to conduct offensive operations.

However, there was no time to question these orders as the men had to prepare themselves to battle. Rations and ammunition needed to be issued; rifles cleaned; water bottles filled and the platoon officers needed to explain to their men what little of the plan they knew. James Dykes was briefed by his platoon commander, Second Lieutenant Moult. According to Private Donald Price (of No. 16 Platoon, D Company, 20/R Fus); 'I had no idea, and I don't think any of my pals had any idea what we were supposed to do, where we were going, how we were going to get there, who we were going to attack. Nothing at all as far as I was concerned.'[74]

It was not until Price went to collect a shovel and ammunition and an officer came around with rum that he knew they were going to storm the wood. According to Sergeant Godfrey Skelton (No 14 Platoon, D Company) the NCOs of 20/R Fus were not briefed prior to High

---

70   PS/6511 Private Sydney Warhurst, aged 25, was from Stretford, Lancashire, and worked for a corn milling firm. He enlisted in January 1915 and joined No. 16 Platoon D Company, 20/R Fus. Warhurst was wounded accidentally on 19 July 1916 by a bomb fragment. He later served as a corporal with the Royal Wiltshire Yeomanry (No 321067) and was discharged in February 1919. See IWM: Donald Price Sound Recording 10168, Reel 8.

71   Lieutenant George Geoffrey Ziegler, aged 19, was from Birkenhead. He was educated at Haileybury and Richmond Schools and at Cambridge (Gonville and Caius College 1913) where he joined the OTC; he was a medical student. He was well-known as a wing three-quarter, and his reputation preceded him from Haileybury. He did not get a blue at Cambridge as he was tall and easy to tackle. Ziegler enlisted in 20/R Fus in September 1914, aged 19, and was allocated PS/6001, after which he was gazetted to 20/R Fus on 27 October 1914. Ziegler was promoted to lieutenant in February 1916 and was IC No 16 Platoon, D Company and was wounded on 19 July 1916. See TNA WO 339/67877: Lieutenant G.G. Ziegler Officer File.

72   PS/5326 Private Frank Milnes, aged 23, was born in Manchester and lived in Rochdale. He attended Manchester Grammar School and was employed as an insurance Clerk. He was buried in Bazentin-le-Petit Communal Cemetery Extension. PS/5667 Private Arthur Smith was born and lived in Chorlton-on-Medlock, Manchester. He died of his wounds and was buried in Flatiron Copse Cemetery, Mametz. PS/6833 Private Frederick James Dittmer, aged 26, was born in Stepney and lived in London. He also served with D Company and is commemorated on the Thiepval Memorial.

73   PS/4499 Private George Cecil Blomeley, aged 23, was born in Bury and lived in Altringham. He was employed by the Union Bank of Manchester. PS/4554 Private Thomas Bradbury, aged 21, was born and lived in Stalybridge. Both are commemorated on the Thiepval Memorial.

74   IWM: Donald Price Sound Recording 10168, Reel 8.

Wood on the exact plans or enemy defences. When their officers later became casualties Skelton and his fellow NCOs were inadequately briefed to take over command.[75] Private James Dykes remembered being briefed in their bivouacs by his platoon commander but only remembered two salient points. Firstly, casualties were to have only first aid applied and then be left for the stretcher bearers; 'The second instruction, strongly stressed; we must not take prisoners. We had no time to be concerned with them. The Platoon Officer [Second Lieutenant Coggin] informed us that the Germans were not taking prisoners, to us the implication was that there was to be no quarter asked or given. This was, simply, Death or Glory!'[76] Another account published in a newspaper stated; 'We asked no quarter and we gave none…'.[77] These accounts were contrary to divisional orders whereby a divisional POW collection point was to be established at Fricourt.[78]

20/R Fus was not at full strength as it went into battle. For example, Dykes' No 10 Platoon (C Company) was now commanded by Second Lieutenant Coggin, vice the wounded Blaauw. No 10 Platoon consisted of Sergeant Basil Taunton; Corporal Phillip Steer; three lance corporals and twenty-seven men; one officer and thirty-two enlisted men instead of a platoon establishment of one officer and fifty-three men.[79] Likewise, No 13 Platoon, D Company, which had consisted of UPS men from Belfast, cannot have recovered from its previous decimation on 16 July.

Though many may have displayed outward nonchalance, many men harboured inner fears:

> We were dismissed and sent back to our funk holes to get as much rest as possible in view of the strenuous day ahead of us. There was no sleep for me; my mind being full of the ordeal which lay ahead of us. I felt tensed up and could think of nothing but the coming encounter. What sort of exhibition would I put up? I don't think I was altogether frightened, rather more worried about my reactions. I hoped I would not "let the side down".[80]

Dykes' platoon paraded at 12:45 a.m. and after all personnel were found to be present the platoon marched off towards High Wood.

On the eve of the coming battle the state of the battalion should be noted. The battalion had suffered several casualties in officers and men. This had both depleted the numbers but had also eroded the efficiency of the unit as commanders had to adjust to new roles and new subordinates. The men were tired from uncomfortable conditions, physical effort and a lack of sleep. To some men this may have numbed them to fear and further hardship in others it may have eroded morale. However, the previously high morale of the battalion was still present with the desire to; 'Speed on the time for our turn to begin. Speed on the decision!'[81] In the words of Private Dykes; 'We were soon on the move, in the direction of our destiny!'[82]

---

75   IWM: G. Skelton Papers, 13966, p.49.
76   IWM: J.N. Dykes Papers, 7378, p.39.
77   *Burnley Express*, 29 July 1916.
78   TNA: WO95/2408: 33rd Division Administrative War Diary.
79   TNA: WO24/910: Infantry Battalion War Establishment, p.3.
80   IWM: J.N. Dykes Papers, 7378, p.37.
81   Hodson, *The Soul of a Soldier*, p.93.
82   IWM: J.N. Dykes Papers, 7378, p.38.

18

The Fight for High Wood 20th July 1916

Either dear 'Blighty' or death

For most of us it would be a decision, either dear 'Blighty' or death, and we were glad to make it.[1]

The time had come for the attack on High Wood. 20 July 1916 was the first major offensive operation by 20th Battalion Royal Fusiliers (20/R Fus) but would be the costliest day of fighting for the battalion. Before examining this battle, the background and context of the fighting must be examined to enable the significance of these operations to be understood. With the British and French advances on the southern end of the Somme battlefield on 1 July, and subsequent days, the conditions were met for the next phase of the Somme Offensive. This took place on 14 July and aimed to capture Bazentin-le-Petit and Delville Wood using four British divisions. This attack was essentially a success, and the line was advanced. The Germans were seen, in places, to be in full retreat and there was no opposition to reaching High Wood; a potential breakthrough was foreseen but the exploitation of this opportunity failed leaving British divisions facing High Wood and Longueval with Delville Wood beyond the latter. A further set-piece attack would be needed to push troops onto the top of the ridge beyond High Wood; possession of this dominating feature would dislocate the German defences. Despite 33rd Division already having tried to take High Wood about a week beforehand they would need to try again.

An assault on High Wood, planned for 19 July, was cancelled due to the state of the weather.[2] During the morning of 19 July HQ 33rd Division received orders for the conduct of various reliefs with a view to the division attacking on 22 July. Orders for these reliefs were despatched but at 3:45 p.m. revised instructions arrived; the attack was to be brought forward to 20 July. This would relieve pressure on XIII Corps. The objective was to secure High Wood and the Switch Trench with 5th and 7th Divisions attacking on the right of 33rd Division. HQ 33rd Division despatched revised orders at 5:00 p.m. though the instructions for the artillery bombardment was not received until 6:00 p.m. 19th Brigade had also been unable to establish any posts or strongpoints from which it could observe the Switch Trench directly and Corps HQ was also pushing to get all these posts joined up, as well as preparing for the attack. HQ

---

1   Hodson, *The Soul of a Soldier*, p.92.
2   TNA: WO 95/2408: 33rd Division A&Q War Diary.

33rd Division sent out two further missives at 7:15 p.m. explaining the artillery plan and the location of a German MG position at the western tip of High Wood which would enfilade any attack. Foremost in this communication to 19th Brigade was the following from the divisional commander; 'I also informed the G.O.C. 19th Infantry Brigade that the C. in C. attached the greatest importance to the capture and holding of High Wood.'[3] Though capturing the Switch Trench running through the wood and out either side was important, High Wood was given prominence as the objective.

The Plan

According to the 19th Brigade war diary the order to attack was received at 5:30 p.m. on 19 July. In the time available 20/R Fus and 2/RWF needed to be relieved by battalions of 98th Brigade and other preparations for the attack made. At 2:00 a.m. the two assaulting battalions were to be formed up outside the barbed wire of the front-line positions facing northeast towards High Wood. 1/Cameronians (under the 2IC, Major Hyde-Smith) and 5/Scot Rif were to attack on the left and right respectively with 20/R Fus in support and 2/RWF in reserve near Bazentin-le-Grand Wood. In addition, 2/Worcs R (100th Brigade) were to push forward posts on the left flank to protect the advance of the Cameronians but this would be their only involvement in the fighting. 7th Division would attack on the right of 19th Brigade.

The bombardment of the wood would become intense at 2:55 a.m. and at 3:25 a.m. the bombardment on the forward edge of the wood would lift and the assaulting battalions would storm the wood. At 3:35 a.m. the barrage would lift completely from High Wood. The earlier order stated that the infantry would assault at 3:35 a.m. though this would allow the German defenders on the wood edge time to recover. The assaulting battalions were to 'press right on in rear of the barrage even at the risk of losing a few men from our own fire.'[4] On entering the wood the Scottish Rifles would clear to the eastern corner and establish a strongpoint there before holding the northeastern and southeastern faces. The brigade diary recorded; 'The Cameronians, leaving one Company to deal with the enemy's position at the Northwestern Corner of the WOOD, were to push on to the Northern apex and consolidate that portion of the WOOD when gained.'[5]

20/R Fus was in support and would advance with their right on the Bazentin-le-Grand–High Wood Road and enter the wood from the south fifteen minutes after the Cameronians and Scottish Rifles assaulted. The Royal Fusiliers would sweep through the wood and clear it of Germans who might have been missed; they would also presumably destroy any German dugouts. According to Captain Dunn (MO, 2/RWF); 'The 20th Royal Fusiliers were to support the attack, follow the 5th S.R. and, pivoting on their own left, mop up the Wood from east to west'.[6] They would be followed by three sections of 11th Field Company RE and two companies of 18/Mx (Pioneers). Four MGs of 19th MG Company were to accompany 1/5 Scot Rif. Once the wood was clear the engineers would build strongpoints at the north, east and west corners

---

3   TNA: WO 95/2405: 33rd Division GS War Diary.
4   TNA: WO 95/2405: 33rd Division GS War Diary.
5   TNA: WO 95/2420: 19th Brigade War Diary.
6   Dunn, *War the Infantry Knew*, p.230.

and a 'keep'. The pioneers would start a communication trench to the rear. The Switch Trench, having been captured, would be blocked either side of the wood.

Unknown to the British, the Germans were only holding the wood with three companies. *7. Kompanie, Infanterie Regiment 165.* (7/165), and *5. Kompanie* (5/165) were holding the southwest face, whilst *10. Kompanie, Infanterie Regiment 72.*, was holding the south corner and southeast edge. On their left flank the Germans had further companies of *72/IR* along the eastern of the two roads between High Wood and Longueval. On the German right flank *Infanterie Regiment 93. (93/IR)* was holding posts from the western corner out across the open ground.

The written orders drafted by 20/R Fus for the attack on High Wood have not survived though Private James Hodson suggested that an un-named colleague may have kept a copy.[7] Though it was not known to 20/R Fus, who were following behind, 1/Cameronians had a plan to try and reach the wood with minimal casualties when the barrage lifted. The Cameronians would initially advance in column, 'shake out' into four columns, each of four platoons with B Company leading. When that company reached a point 40 yards from the right of High Wood the other three companies would form up to their left. These manoeuvres left the forward platoons of the Cameronians extremely close to the wood and when the barrage lifted from the wood they would charge with the forward platoons in extended order and with supporting platoons in file behind. This proximity to the wood might enable maximum surprise to be achieved. If it worked, it would reduce unnecessary casualties from any German counter-barrage which would fall behind the attackers. If the battalion did not surge forward then, come daylight, the men would be exposed in open ground opposite the wood and could be picked off by German riflemen and machine gunners.

If these tactics worked for the Cameronians and Scottish Rifles, they endangered 20/R Fus. The latter would need to follow these battalions in support a safe distance behind and would have to lie down and wait further out from the wood in a more likely site for the German counter-bombardment. They would also need to advance through that barrage to clear the wood. Both eventualities would see the supporting battalion exposed to more significant casualties from artillery fire despite the assaulting battalions suffering less.

The Advance

These operations commenced after dusk on 19 July when the 2/Worcs R pushed forward their posts from the crossroads northeast of Bazentin-le-Petit towards the western corner of High Wood. At 2:00 a.m. the rest of 19th Brigade was positioned in attack formation with the two assaulting battalions positioned outside of the British barbed wire obstacles. In simple terms these battalions would advance across the open, under cover of the bombardment, and wait a safe distance from High Wood before assaulting it when the barrage lifted.

The complex arrangements required to get the Cameronians and Scottish Rifles to their assault positions would be difficult on a parade-ground but were carried out successfully in the darkness. According to Private Dykes; 'From the commencement of our advance, up to our arrival at our positions outside High Wood, there was no reaction from the Germans, consequently no interference with the plan of campaign. Not being much of a military expert myself,

---

7   Hodson, *Return to the Wood*, p.35.

The Fight for High Wood 20th July 1916 271

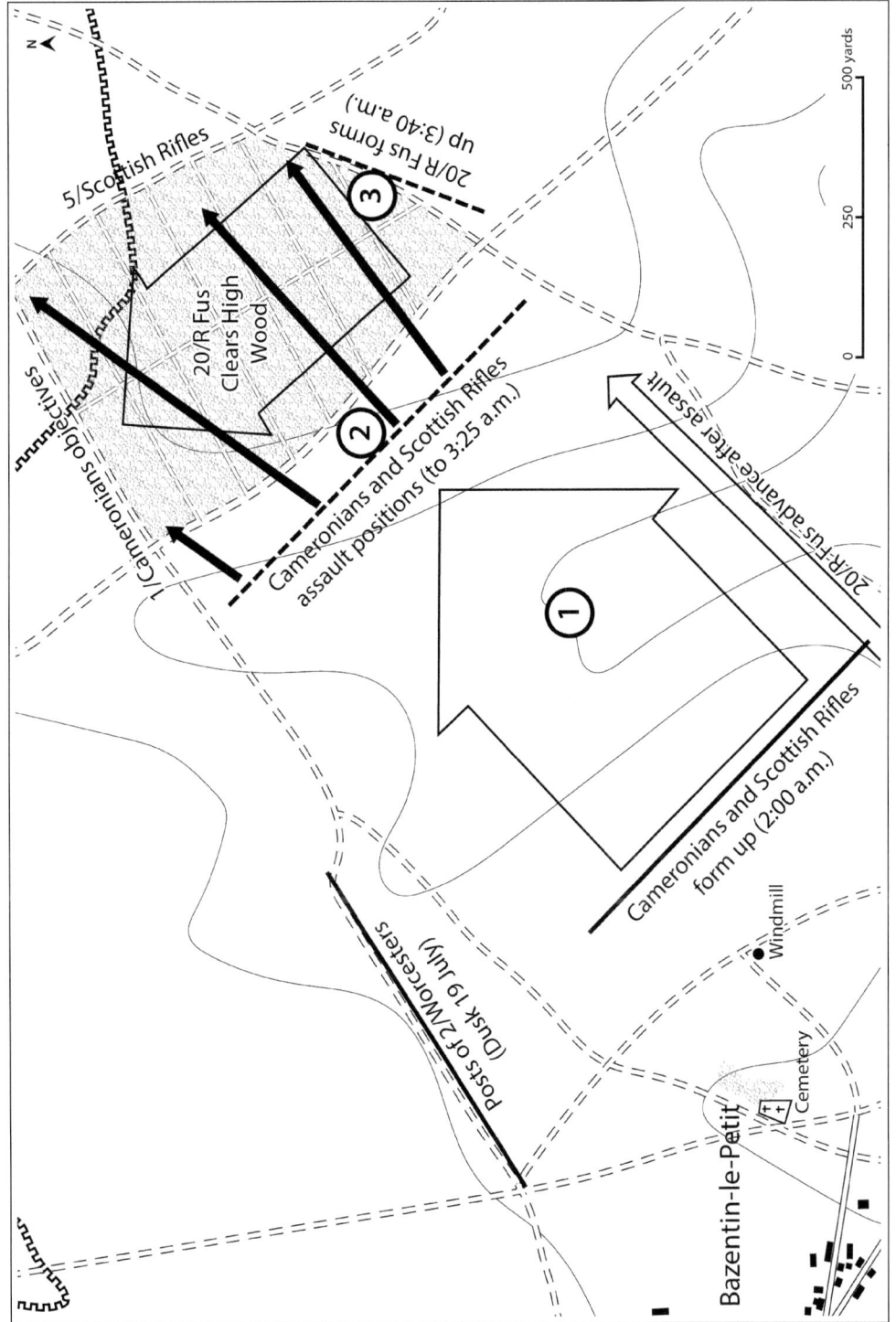

Map 5 High Wood, 20 July 1916: Plan of attack.

the whole exercise was most confusing and complicated to me. No doubt it was very cleverly carried out.'[8] Preceded by scouts, the two leading battalions advanced to their positions just short of the wood.

According to Dunn the Germans dropped shells on the middle of the advancing columns and most of the casualties were inflicted on 20/R Fus; this caused forty of them to withdraw to cover in the roadside ditch along with some of the REs.

The Bombardment

According to Donald Price, after advancing over 1,000 yards up a slope towards High Wood, they lay down in the dark about 200 yards short of the edge of the wood. When Dykes' company reached its jump-off point they were ordered to lie down and load ten rounds in their rifles, presumably the order being whispered from man-to-man along the lines. According to Dykes:

> Suddenly, all Hell was let loose! From quiet except for an occasional shot, there was pandemonium when our artillery blazed into activity. Seemingly, every gun in the area must have been concentrated on High Wood alone. The effect was dumbfounding! An exciting and horrifying sight! Incessant noise assailed our ears accompanied by a lurid glare which seemed to envelope the whole horizon in front. Curiously enough, the scene and uproar brought to my mind greatly enjoyed and keenly anticipated visits to the fireworks displays at Manchester Belle View when I was much younger. This imitation was many times magnified, and I did not find it in the least enjoyable. Red flares were climbing up into the sky from the wood, and the German guns became active. Fortunately for us, the fire was not well directed, and the shells passed well over our heads and burst a long way away.[9]

The bombardment became intense at 2:55 a.m. The German defenders, realising this was a precursor to an assault, initiated a counter-barrage to the southwest of High Wood. This caused casualties within the assaulting battalions. Lieutenant Colonel Kennedy of 1/5 Scot Rif was wounded in the face and back and was evacuated leaving that battalion without a commander at a sensitive time. Because of their proximity to the wood this German shelling did not affect the Scottish battalions as seriously as it might have done. Whilst the German barrage did not significantly affect Dykes' company it did force supporting troops from 20/R Fus, and their accompanying engineers and pioneers, to creep forward closer to the leading Scottish battalions to escape it. This led to intermingling between different battalions just prior to the attack. Conversely, according to Skelton, many men of 20/R Fus were killed by the German counter-barrage; this included many stretcher bearers. Second Lieutenant Ashley Gibson (20/R Fus) heard of events at High Wood second-hand from his company commander:

> There was an awful muck-up from the start. Either there was a mistake in the timetable, or we had the wrong bearing given us when we went over. Because there was the whole of two companies, halted out there in extended order, with our own shells dropping all

8   IWM: J.N. Dykes Papers, 7378, p.38.
9   IWM: J.N. Dykes Papers, 7378, p.38.

over the shop. The barrage ought to have lifted by then, but it didn't, not for some time. Everything was as black as pitch, but we could hear heaps of our chaps getting knocked out. Two platoon commanders, young H. and P., blown to bits where they lay by direct hits. Then the barrage did lift, and we went on and, of course, ran into Fritz's. That wasn't so bad, somehow…[10]

The two officers were likely Second Lieutenants Thomas Hine and John Price; both were reported missing.

Whilst the bombardment was underway, the order was passed along the lines of the men of 20/R Fus to fix bayonets and for every man to open and close the bolt of their rifles to chamber a round of .303 ball ammunition. This had presumably only been done now to remove the risk of a soldier tripping over and discharging his weapon on the approach march and thereby alerting the German defenders.

Second Lieutenant Stewart (1/Cameronians) was with the left-hand company of the Cameronians, and remembered just before Zero Hour that the forward face of the wood was just a series of exploding shells in which he thought nothing could survive. He need not have looked at his watch to tell when to charge as the barrage lifting at Zero Hour would be the signal.[11]

The Attack

As the barrage lifted, the leading battalions attacked. According to Harold Tyson (likely A Company, 20/R Fus); '…our boys were magnificent. They advanced without flinching into a hellish barrage of heavy and shrapnel stuff as coolly as they used to do their training. An old regular in the battalion said he had never seen such a splendid movement from new troops.'[12] There was heavy MG fire from the Switch Trench when the assaulting infantry entered the wood. On the 1/Cameronian front, due to Stewart's company commander being killed there was a delay to the left-hand Cameronian company assaulting the wood when the barrage moved. When they were ready to advance the German MG and rifle fire had grown too intense. Stewart's company was forced to try and dig in a short distance from the edge of High Wood. The German strongpoint just northwest of the wood was consequently left untouched. The rest of the Cameronians and 1/5 Scot Rif pushed into the wood when the barrage ceased. The Fusiliers were to advance into the wood from the southern end fifteen minutes after it had been swept through by the leading battalions and clear it of any remaining enemy. However, as they waited to enter after the leading battalions the German retaliatory barrage inflicted casualties on the RF and the RE Company.[13] The casualties and confusion encouraged the RF to bunch up and become mixed with the leading troops and was a cause of the confusion in the wood.[14] Dykes recalled then being ordered to advance in single file parallel to the front of the wood

10 Gibson, *Postscript to Adventure*, p.154.
11 Alexander Stewart, *A Very Unimportant Officer* (London: Hodder, 2009), p.101.
12 Norman, *The Hell they Called High Wood*, pp.150–151.
13 Colonel H.H. Story, *History of the Cameronians (Scottish Rifles) 1910–1933* (Aylesbury: Hazell Watson & Viney, 1961), p.121.
14 Martin (ed), *Fifth Battalion, Cameronians*, p.74.

PS/5782 Private Harold Holmes Tyson, aged 22, attended Manchester Grammar School and was employed as a bank clerk. He enlisted in September 1914 and was a bugler who also played the side drums in the 20/R Fus band. Tyson was appointed lance corporal in February 1917. In September he attended No 15 OCB at Romford and was gazetted to 52/NF in December 1917. He was attached to the RAF in July 1918. He was demobilised in January 1919. Tyson is wearing the uniform of a Northumberland Fusiliers subaltern dating this photo to early 1918. (Reproduced with the permission of Paul Tyson)

and then at a given point each man in the line turned right so Dykes' C Company approached the wood in line abreast. In their eagerness to get into the wood Dykes and some of his comrades got mixed up with the Scots. As this mixed party moved through the trees they heard German MGs and experienced German shells being fired into the wood. An anonymous writer from 20/R Fus wrote:

> ... we lay in lines while our artillery blew the wood to pieces, and the Boche shrapnelled us... Then the rockets went up and forward we went with bayonets fixed. We went through the wood like a tooth-comb, bayonetting, bombing and shooting. It was gorgeous. I've never enjoyed myself so much in my life. Then we got to the far side of the wood and dug in. It was grand to see that the wood was ours and we all went about shaking hands.[15]

According to Dykes; 'The wood seemed to be peopled by small bands of roaming "heroes" looking for someone to fight.'[16] Groups met and exchanged personnel as men found other members of their sections or platoons; Dykes met and accompanied PS/8983 Private Gordon Wesley Powell, a 20/R Fus man from the same draft as him, but they split when Powell found members of his own platoon.[17] Dykes was left with a man from his platoon, PS/5333 Private George Moore, who had lost his younger brother a month before. Between them they carried a box of rifle ammunition.

15   Manchester Evening News, 5 August 1916.
16   IWM: J.N. Dykes Papers, 7378, p.39.
17   Powell was later recorded as killed on 20 July 1916.

The Fight for High Wood 20th July 1916 275

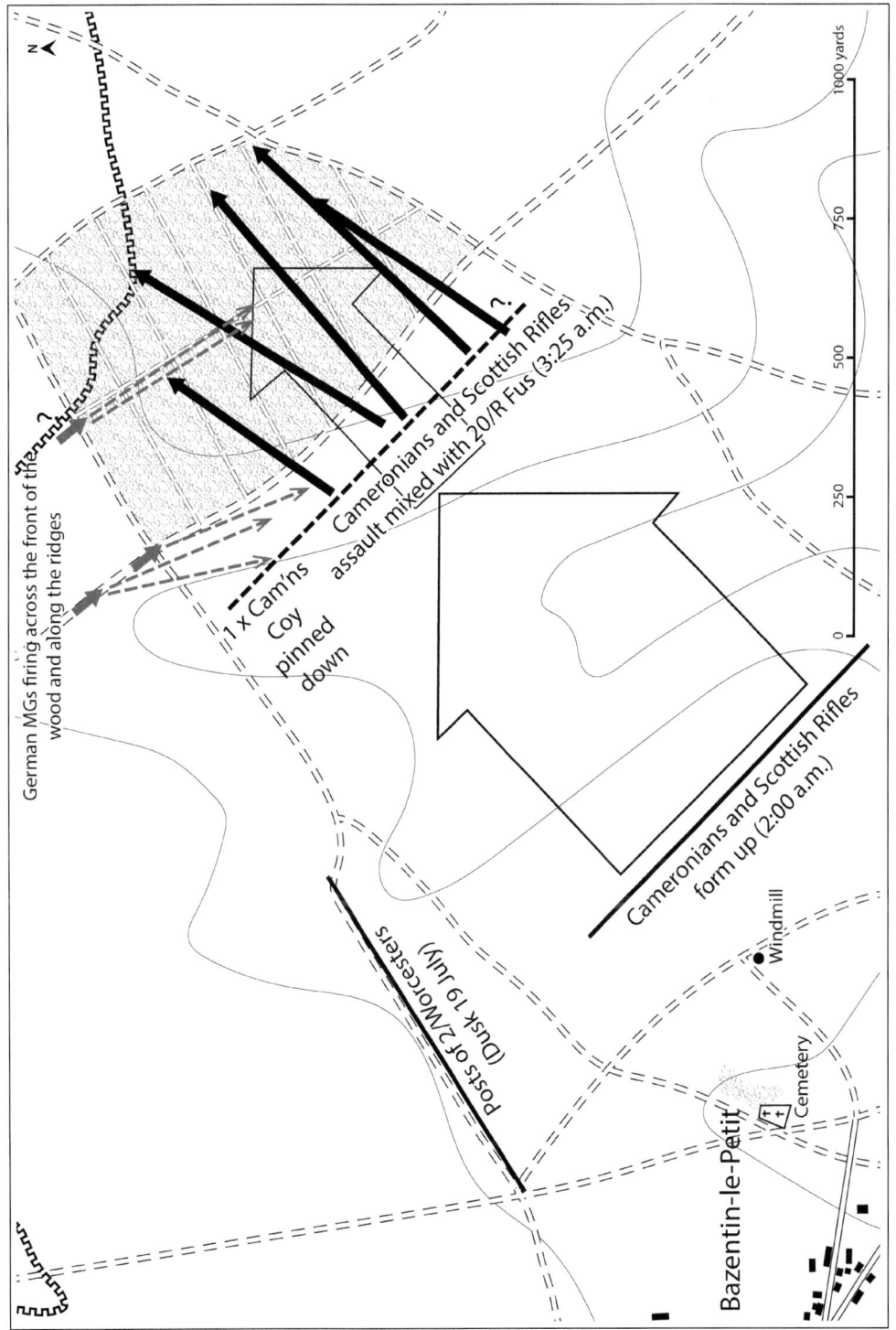

Map 6 High Wood 20 July 1916: 19th Brigade assault.

Having roamed the wood Dykes' party reached their objective:

> Eventually, we came to what must surely be the edge of the wood. There was open ground beyond the trees. Surely this must be where we had to dig in, and prepare an unpleasant reception for any who dared to dispute our right to be there. Anyhow, we began to make the area just inside the trees lining the wood as strong a point as our limited resources would allow. During this operation we were joined by a further small body, mostly Scots, I think, including an officer. We were fairly well spread out, perhaps 100 men, mostly in improvised defences scooped out in shell holes. … there had been some interference with our efforts to protect ourselves by rifle and machine gun fire from what appeared to be a trench some distance away from the edge of the wood, perhaps a distance of 50 yards. We suffered a few casualties, but nothing was done to put a stop to the nuisance. Obviously there was a lack of coordination, and we just had to put up with it![18]

This trench was presumably the 'Switch Line' outside the wood. This would place Dykes' party on the north-eastern side of the wood with the German-held Switch Line entering the wood a short distance to his left.

The Clearance

Dykes' party was separated from their company and were carried to their objective mixed with the Scottish assaulting battalions. The right half of the Cameronians reached the far side of the wood with relatively light losses, but the left half was held up by fire from the strongpoint on their left. They required assistance from trench mortars to loosen the German grip and with cooperation from 20/R Fus drove the Germans out of the strongpoint. However, heavy German fire made it impossible to occupy this empty position. The Scottish battalions had reached most of their objectives. Meanwhile, 20/R Fus needed to sweep through the wood and 'mop up' any remaining German defenders who had been overlooked in the Scottish advance. The clearance of the wood was made difficult due to the casualties sustained by 20/R Fus, its mixing with other battalions and the darkness. Gibson's second-hand account continued:

> [20/R Fus] … got into that infernal wood, full of Boches of course, and we were all mixed up with the Cameronians, but between us we took Fritz's Switch [Trench] and most of his strong points, not all of them. It was so damned dark, though, except for the crumps, and nobody knew which chaps were whose … Our chaps were bayonetting mostly; it was all they could do.[19]

According to Gibson his company commander in the UK had behaved like a 'bloodthirsty ruffian' who encouraged aggressiveness when in charge of bayonet training. However, when a German soldier who had appeared in front of him in High Wood he had not been able

---

18   IWM: J.N. Dykes Papers, 7378, p.39.
19   Gibson, *Postscript to Adventure*, p.154.

to kill them.[20] Vincent Platt (No 13 Platoon, D Company) recorded little in his diary; on 19 July: 'Our artillery gave them jip, then we took the wood, taking several (30) German prisoners. I spied the first Hun & went for him with my bayonet whereupon he cried like a child.'[21] Whether Platt spared or killed this man is unknown. Between 6:00 a.m. and 8:00 a.m. three German officer PoWs and thirty-one other rank PoWs were received by Brigade HQ. The Scottish attack overwhelmed 5/165 and the company commander, *Leutnant der Reserve* Schneider, suffering from gas poisoning, was captured along with six of his men. They escaped during the confusion of a later attack.[22] 10/72 lost its commander and was driven back; this exposed the remains of 7/165 to attack from behind as well as the front. The latter company fought back and sent for assistance.

James Hodson wrote of a Lewis gun team during the wood clearance; '"If this is taking a wood," I thought, "it is easy." We made for the left, along the top of a trench half-filled with broken trees and soil. Germans dead, Germans wounded, Germans feigning death lay in the trench. How thorough the artillery work had been! Equipment, rifles, and men poked out of the debris…'[23] The same men picked their way through the wood 'mopping-up'; '…running bayonets into the undergrowth, firing the machine-gun down dug-outs. The man on my right fell forward and remained kneeling, a hole in his head, dead… We heard the ta-ta-ta-ta-ta of a machine-gun … and we got into the trench behind a fallen tree. I crawled a few yards and watched. Grey helmets bobbed among the foliage twenty yards away with astonishing effrontery.' Hodson and his comrades engaged them with their rifles and Hodson thought he shot two Germans.

An anonymous account, from a Manchester newspaper, described the clearance; 'Away we went, with eyes alert and bayonets fixed, groping our way through the network of shrubs and shell holes, and carefully searching the bushes lest perchance Fritz might be concealed, until we arrived at the other side. Here they dug themselves in and were heavily shelled.'[24] Bayoneting was the order of the day and the use of grenades according to Skelton; 'Our battalion then attacked in support, my company bombing all dugouts on the way killing the occupants. We got as far as the far edge of the wood and occupied an old German trench but it was not the real enemy position. This was cleverly sited in dead ground about a quarter of a mile away.'[25] Skelton was likely to the right of Dykes' party and was also looking at the German Switch Trench. An observer from 11th Field Company RE considered that:

> … the assault went absolutely as expected the 5th S.R. and the Cameronians carrying their objectives. The clearing of the wood [by 20/R Fus], however, was too hurried, evidently not being taken systematically, the clearing parties moving in compact bodies up the open spaces, leaving enemy snipers and not clearing the dug-outs. There is no doubt that the

---

20 Gibson, *Postscript to Adventure*, p.154.
21 IWM: S. and V. Platt Papers, 17681, p.74.
22 Albrecht von Stosch (ed), *Schlachten des Weltkrieges 1914–1918, Somme-Nord, II Teil* (Oldenburg: Berlin, 1927), p.105.
23 Hodson, *The Soul of a Soldier*, pp.97–98.
24 Manchester Evening News, 12 August 1916.
25 IWM: G. Skelton Papers, 13966, p.50.

enemy retired into his dug-outs, and when the parties had passed, came out, mounted his M.G.'s and shot down & sniped the parties which had passed through.[26]

This unsystematic clearance is understandable when considering the complex terrain, darkness and the casualties suffered, which likely encouraged this 'bunching' and further increased casualties. Nonetheless, the author of the war diary for 11th Field Company blamed 20/R Fus for inadequate 'mopping-up', despite not all the wood having been captured and held.

The wood had been taken and partially consolidated except for the north and northwest corners. These were dangerous omissions and the enemy possessed a foothold in the wood from which to base any likely counterattack. According to the 19th Brigade report, 'Owing to the casualties amongst Senior Officers and the confusion and intermingling of Units, inevitable in wood-fighting, things now became rather involved.'[27] 20/R Fus even gathered some of the engineers that accompanied them and employed them as infantry to hold sections of the wood.[28]

'I found myself completely alone…'

Donald Price also went into the attack with 20/R Fus and recalled, years later, the confused fighting:

> We came under an enormous barrage. … it was appalling, screams of dead, screams from all over the place for stretcher bearers and by the time we got to the wood we were nearly annihilated. Before we'd got up to the top of the wood at all. I remember there were four of five of my pals, we were still usually together you see, and we got up, and it was getting light now and we were getting to the wood and there was an enormous bombardment of shells and bullets and god knows what. I remember the five of us got through this hedge onto what they called a ride, a big pathway, like a roadway… One fellow was talking away there, Dick Nolan he was, and we started going up there [the ride], I had a bomb in my hand, how the hell I got it I don't know, I threw this bomb where I thought there was movement and we kept to this ride. The shelling was coming down and bullets were flying all over the place. And this bombardment was terrific. And we made a big mistake – we kept to this roadway to go forward, the undergrowth was so thick we couldn't go anywhere so we kept to this road. The unfortunate part of that was that's where he [the Germans] had got his machine guns trained on. Before we'd finished I found myself completely alone. Some of the lads had got knocked out … some 100 yards into the wood which was only a quarter of a mile long.[29]

Price, exhausted from a working party the night before, found a dugout, where three seriously wounded Germans were lying, and slept. The rides throughout the wood were covered

---

26 TNA: WO95/2414: 11th Field Company RE War Diary.
27 TNA: WO95/2420: 19th Brigade War Diary.
28 TNA: WO95/2414: 11th Field Company RE War Diary.
29 'Dick' Nolan was possibly PS/7057 Corporal Leonard Nolan who survived High Wood. IWM: Donald Price Sound Recording 10168, Reel 8.

by German MGs. Due to the undergrowth the UPS men followed these rides through the wood and presented better targets. The rides became killing grounds as it grew lighter and any movement attracted German fire. Numerous bodies were already there; many were from the previous attack.

Captain Templar, commanding A Company, wrote of his Company Sergeant Major, Robert Blackstock:

> During the preliminary bombardment we lay side by side, and he chatted away quite unconcernedly under a perfect canopy of shrapnel. Then at the word 'Go' we went forward together, and he was invaluable in keeping the line, rallying stragglers and generally organising his immediate portion of the attack. It was fine to hear his voice ringing out and giving orders just as if we were on parade. Then we got to our objective and dug in, and he was among the first to go – a shell hit him and he died at once.[30]

Though casualties amongst officers and NCOs were heavy those that remained helped to lead their troops to their objectives. Holding these positions would be the next problem. James Hodson recalled what his party did next; 'We moved under orders from an officer to a small circular emplacement abutting on the end of an old trench. This emplacement ... guarded the whole left flank of the British position and proved to be the succouring spot for our own soldiers ... for many hours. It represented 'home' to many a wounded man ...'[31] Hodson and his comrades held this position and cared for British and German wounded that crawled in. He did not record them repelling any counterattacks.

Consolidation

After the initial struggle there was a temporary lull. After it became daylight, it was generally quiet in the wood whilst both sides sought to understand their respective predicaments. Dykes also caught up on some sleep until:

> I was wakened by the sound of an aeroplane, passing fairly low overhead. It was a glorious day now ... The next thing I became aware of was the sound of rifle shots and confused shouting. Looking out hastily, and grabbing my rifle, I saw some German-uniformed men coming towards us with their arms raised above their heads. I saw no weapons in their hands and got the impression they were coming over to give themselves up. The shots were coming from our men! The Germans gave up what ideas they had had about surrendering, turned, and ran back ... leaving a few of their number lying on the ground.[32]

Quietness reigned again after this unpleasant incident. The aeroplane was likely a machine of No 3 Squadron RFC which was attempting to ascertain the extent of the 19th Brigade advance

---

30   ULULA, Volume XLIV, No 328, October 1916, p.186.
31   Hodson, *The Soul of a Soldier*, pp.98–99.
32   IWM: J.N. Dykes Papers, 7378, p.39.

in High Wood. Part of this period of consolidation involved evacuating the wounded; Vincent Platt wrote of his short time in High Wood:

> We tried to clear the wood but could not in the left-hand corner where a small party with a machine gun held out. Sid was wounded with [a] bullet in [the] thigh. We absolutely went through it later and near the end I got hit on the head with a shell. Hurrah. Received many kindnesses from RAMC including drink of hot tea, being the first for a week.[33]

Godfrey Skelton failed to describe a comprehensive account of events within High Wood but recalled:

> It is impossible to write of all the incidents of the twenty-four hours in that wood; they would fill a book themselves. The falling and burning trees, the dead and dying, aircraft bombing us, the deeds of bravery and examples of the loss of nerve, going forward to rescue my second sergeant, by name Harris, with a stretcher uplifted towards the enemy trenches and the Germans respecting [it] by not firing at us. Most of our officers and senior N.C.O.s were now casualties and junior N.C.O.s assumed command of small parties.[34]

PS/4921 Sergeant Frank Harris, aged 30, from Rusholme, Manchester, did not leave High Wood alive.[35]

The German account recorded 'Scottish' efforts to capture the southeastern part of the wood which were repelled by *12. Kompanie, Infanterie Regiment 72* and *4. Kompanie, Reserve Infanterie Regiment 99*. The official account records; 'Scottish bodies lay in the eastern part of the forest like seeds.'[36] Many of these bodies would also have belonged to men from 20/R Fus. However, portions of 20/R Fus had reached their objectives in High Wood despite having suffered many casualties, including many officers. In the words of one un-named contributor to Dunn's account; 'From quite early the unfortunate R.F. men seemed to be getting killed all over the place to no purpose. The whole operation had been conducted in confusion almost from the start, and for want of superior direction it became a shambles.'[37] Lieutenant Colonel Bennett, commanding 20/R Fus, had not shied away from leading his men. According to one writer; 'Our Colonel led the men right into the wood, and then went down early in the engagement with a bullet through the shoulders.'[38] After being wounded, Lieutenant Colonel Bennett withdrew from the wood during the lull. He was later met by officers of 2/RWF about 10:00 a.m.:

> After summarising the situation he added, in a torrent of ejaculations, that nothing was being done; everything was chaotic; no one was in command; not a message, not an order had been received from Brigade. When the CO [2/RWF] recovered his breath, he

---

33 IWM: S. and V. Platt Papers, 17681, p.75.
34 IWM: G. Skelton Papers, 13966, p.50.
35 He is commemorated on the Thiepval Memorial.
36 von Stosch, *Schlachten des Weltkrieges*, p.104.
37 Dunn, *War the Infantry Knew*, p.232.
38 *Burnley Express*, 29 July 1916

suggested to Colonel Bennett to call at Brigade [HQ] before going to the dressing station, but this he did not do.[39]

According to James Hodson of 20/R Fus: 'It's said he called at Brigade H.Q. on his way down to the dressing station and played hell with the staff for getting us slaughtered uselessly. We accepted that he was capable of abusing the staff.'[40] Dunn highlighted the issue of command and control and suggested that the HQ of 19th Brigade was too far to the rear and could have occupied a closer, and better protected, position from which to direct the battle. The Divisional HQ, at Fricourt Chateau dugouts, was also equally remote. Communications were almost impossible due to telephone wires being cut by shellfire, smoke and dust obscuring vision for signallers and runners being unable to reach the wood and return. Dunn's comment that; 'The cost in all the lower ranks of preserving some Generals of brigade and division, and some members of their Staffs, is beyond reckoning, but must be stupendous.'[41]

Counterattack

The situation facing 19th Brigade at about 9:00 a.m. was summarised in the Brigade diary:

> Owing to the casualties amongst Senior Officers and the confusion and intermingling of Units, inevitable in wood-fighting, things became rather involved, but at 9 a.m. a report was received from the Officer Commanding 1st Cameronians to the effect that we were holding the Southern half of the Wood, meeting with strong resistance from the enemy in the Northern half, that casualties amongst officers had been heavy, and that hostile reinforcements were being brought up.

Those German reinforcements would soon be employed. Dykes was again awoken by firing on his left; his party stood ready and waited:

> Shortly, we saw figures moving amongst the trees. They were khaki clad, and were obviously falling back towards us, rather precipitately, next we saw grey clothed men, moving from tree to tree, plainly making good use of the cover afforded by the trees, and firing steadily at our troops. George Moore and I fired at the advancing Germans … but it was a difficult business, because there was always the danger of hitting our own men. The position was not improved when the retreating men, whose numbers were being augmented as they passed over funk holes, by the occupants of the holes, thus increasing the confusion and disorder which appeared to be getting out of hand with considerable rapidity. Obviously we were very soon going to be left behind … Moore and I looked at each other and with one accord got out of our holes and made off in the direction of the retreat. Without doubt, the panic which had affected the retreating men had been communicated to us.[42]

39 Norman, *The Hell they Called High Wood*, p.143
40 Hodson, *Return to the Wood*, p.44.
41 Dunn, *War the Infantry Knew*, p.233.
42 IWM: J.N. Dykes Papers, 7378.

Leadership might have quelled this crisis but many officers and SNCOs were out of action. Dykes' panic was increased when Moore fell. Thinking he had stumbled Dykes 'exhorted George to get a move on. He made no movement. I … tried to lift him. I could not move him at all. When I took a closer look, the reason was clear. Moore had gone to join his brother!' Dykes ran on and joined a calm party of men entrenched in a clearing where a Scottish officer was in charge. Here the Germans were engaged and kept at arms' length by rifle fire. As the German infantry made no effort to advance further Dykes felt relatively secure. However, a German party cut them off from the rear and the Germans closed in. From the German side, the commander of 9/93, *Leutnant der Infanterie* Viola, found the British on his right flank and rear. One of his platoon commanders, *Vizefeldwebel* Lubrich, threw his platoon into a counterattack against a superior enemy and both inflicted heavy casualties, captured fifty Scottish soldiers and recaptured a German MG. This may have been the force that outflanked and captured Dykes' unlucky party. The brave Lubrich was badly wounded in the stomach for his efforts.

Some of the men around Dykes may have withdrawn out of the wood. The machine-gunners that accompanied 1/5 Scot Rif had to abandon their guns when the infantry withdrew and may have added to those retreating. Further to the right of Dykes' group were Godfrey Skelton and his comrades. Skelton recalled that 'At about ten o'clock in the morning the Germans attacked in strength, massing in the dead ground in front of the position we held. Our company, by rapid rifle fire and one machine-gun section beat off the attack on our immediate front but were enfiladed on the flanks and had to retire, taking our wounded with us.'[43] This enfilading was likely because of the withdrawal of the mixed party on his left which comprised Dykes and some of the Scottish Rifles. However, Dykes' situation was desperate. He had his rifle damaged by a German bullet:

> I looked around me in a demented sort of way for a replacement … I could not see one, even though there must have been many lying about. I was very frightened, and was wondering what it felt like to be shot out of hand … When it seemed likely that the Germans would rush us and finish the unequal struggle, a badly wounded sergeant shouted "surrender", a white article of some description was put on the point of a bayonet and raised in the air' … we began to clamber out of our improvised shelters, those of us who were able to do so. Half a dozen Germans holding their rifles at the ready, came towards us, whilst the remainder, still amongst the trees, kept us covered, so that there should be no funny business … About a score of us moved forward, the others being in no condition even to leave the holes. I was scared and wondered how long it should be before being shot![44]

Dykes survived the battle as a prisoner of war.

At about 11:00 a.m. the situation in the wood changed; the German counterattack recaptured the western strongpoint but otherwise supposedly did not re-enter the wood. Officers had become fewer, casualties mounted, control and direction of troops decreased. Some British troops panicked and fell back to the southern side and some withdrew out of the wood into the open. They were rallied by officers and NCOs. One of those who checked this withdrawal was the 2IC of the Scottish Rifles who had been sent to the wood to take over his battalion. Some

---

43 IWM: G. Skelton Papers, 13966, p.50.
44 IWM: J.N. Dykes Papers, 7378, p.40.

men had also undoubtedly withdrawn to the safer west side of the wood. A new line was formed about 200 yards into the wood. According to Dunn, Father McShane, a Roman Catholic Chaplain, had rallied some men of 20/R Fus who had withdrawn from the wood and placed them at Crucifix Corner on the Bazentin-Longueval Road. However, there were plenty of UPS men still fighting and dead and wounded RF men littering the wood. Major Abbay, presumably the 2IC of 20/R Fus, also arrived from the 'battle surplus' around this time to command the remnants.[45] With few officers left, RSM Armour rallied men and brought some semblance of order in part of High Wood. He arranged for a trench to be dug across the wood.[46] Skelton held this new trench and recalled; 'The heat of the July day was intense and water bottles were soon empty and we suffered much from thirst.' Getting fresh supplies of water and ammunition were a problem; 'The carrying party came up over open ground, bringing water in petrol tins that had not been cleaned properly. The water was undrinkable … This party lost several men through shell fire.'[47] The sun was hot, the smell was terrible and there were flies everywhere.[48]

The British hold over High Wood was incomplete, and the lodgement was vulnerable to counterattack. Fresh British troops would be needed to re-launch an attack on the northern half of High Wood. 2/RWF would be deployed from brigade reserve to conduct this attack.

The 2/RWF Advance

At 11:00 a.m. orders reached 2/RWF to advance to High Wood and complete its capture; these originated from the earlier message from the Cameronians that the north of the wood had not been secured. The German counterattack had altered the situation; the RWF would now need to capture much more of the wood. As many of their men were still on fatigues the battalion was significantly under strength. The fact that 2/RWF was partially committed to supporting tasks meant that it was not an uncommitted reserve and was not able to exploit success effectively. At about noon 2/RWF advanced to the wood. They first had to reach the wood over open, shell-swept, ground. During the advance by 2/RWF one of the main critics of 20/R Fus, Lieutenant Robert Graves, was wounded in the chest by a shell that burst three yards behind him; he did not reach High Wood and had to rely on second-hand information on the conduct of the battle.[49]

---

45   Major Bryan Norman Abbay, aged 35, was born in Plomesgate and lived in Earl Soham. He was gazetted into the Essex Regiment in February 1900 and as a lieutenant, in 1908, he transferred into the Indian Army where he joined the 27th Light Cavalry and later served with the 18th Cavalry at Risalpur (6th Cavalry Brigade). He went to France between January and June 1915 and in November 1915 was promoted to major. Abbey joined 20/R Fus as 2IC in January 1916 and temporarily commanded it in February 1916 when Colonel Bennett took leave. He later served with 16th Light Cavalry and the 18th Cavalry. Married in 1913, Abbay was a lieutenant colonel by 1922. He died on a fishing expedition in Kenya in 1947.
46   IWM: G. Skelton Papers, 13966, p.50.
47   Norman, *The Hell they Called High Wood*, p.142.
48   IWM: G. Skelton Papers, 13966, p.47.
49   Graves received a fragment of shell which entered his back at the lower angle of his right scapula and exited two inches above his right nipple. He was able to return to France in late 1916 but returned to the UK in February 1917 with bronchitis. After working as an instructor at an OCB he later served in Egypt with a Garrison Battalion. Graves, *Goodbye to All That*, pp.181–182. TNA: WO339/23299:

According to Dunn the only formed unit in the wood when 2/RWF arrived at High Wood was a weak company of 1/5 Scot Rif on the eastern corner. The 2/RWF formed itself up in a deliberate manner on the southeastern face of the wood, gathered in its carrying parties and some of its officers conducted a reconnaissance. Colonel Crawshay discussed the situation with the surviving representatives of the units within High Wood; a major from the Cameronians and a lieutenant from the Scottish Rifles; 20/R Fus was represented by a major, likely Abbay.

At about 2:00 p.m. 2/RWF was ready to advance through the wood. Lieutenant Stewart observed members of the RWF attack through the wood across his front to attack the MG that had held Stewart's men up. The RWF advanced using tactics of fire and movement advancing in platoon 'rushes' controlled by officers with their whistles; the sign of a well-trained unit. He described them as 'coming on in splendid style ... a pretty and fine sight to watch...'[50] The RWF suffered casualties as they passed Stewart's position. Another officer of the Scottish Rifles described the RWF; 'This small, controlled force was a most effective contrast to the large loose mass that had been herded into the wood that morning, when one attacking battalion and one in support, not mixed with it, would have been ample ... They were obviously out for blood, and were most heartening...'[51] The RWF, with limited assistance from the existing survivors, cleared the wood after heavy fighting but suffered many casualties. 19th Brigade sent forward two trench mortars to assist in the defence of the wood and at a critical juncture twenty bombs from these mortars silenced a difficult German MG.[52] With the wood reported recaptured there was another lull whilst both sides took stock of the situation.

Soon after 2/RWF retook High Wood, Crawshay was informed that he bore the title 'O.C. High Wood' and the remnants of the Cameronians and Scottish Rifles were under his command. Dunn alleged that the 20/R Fus reinforcements at Crucifix Corner were prevented from returning to the wood. The surviving major of 20/R Fus, likely Major Abbay; 'entrenched himself in the Army List. "I am sainior to Major Cra'shay", he reiterated, and sat tight; and he let his men at Crucifix Corner sit tight.'[53] This was clearly not a time to be discussing seniority.

The 20/R Fus personnel at Crucifix Corner were presumably about forty men who had withdrawn, leaderless, to a roadside ditch after their columns had come under heavy German artillery fire in the open on the approach to High Wood. They had witnessed heavy casualties to their peers and were likely shocked and temporarily demoralised. They were later described as ineffective or idle ('fainéants') when Father McShane took charge of them and led them from Crucifix Corner up to High Wood. Whether they had remained there due to a lack of orders from Abbay; were battle-shocked, tired and leaderless; or were shirking, is open to conjecture. Under stout leadership, they returned to do their duty. Major Crawshay had few men to command from 20/R Fus; they were disorganised, had lost two-thirds of their number and most of their officers. A Sapper officer, Captain Pressey, found the commanding officer of 20/R Fus unhelpful; that officer stated that his men in the Wood were demoralised and tired.[54]

Crawshay recognised that none of the battalions in the wood were strong enough to hold it against a fresh German attack. The news of the capture was relayed to HQ 19th Brigade, but

    Officer File R. Graves.
50   Stewart, *A Very Unimportant Officer*, pp.103–104.
51   Norman, *The Hell they Called High Wood*, p.147
52   19th Light Trench Mortar Battery also furnished guards to escort German POWs.
53   Dunn, *War the Infantry Knew*, p.237.
54   TNA: WO 95/2414: 11 Field Company RE War Diary.

Crawshay urgently requested reinforcements, or his force would be unable to hold the wood. At about 5:00 p.m. he sent the following:

> Only 200 men left in 19th Brigade. These are too spent for further action. High Wood clear of enemy except N.E. corner from Switch – Edge of Wood, both inclusive. Machine gun in Switch about S.3.b.8.8. very active; also snipers in Switch. Cannot hold position the night. Fresh battalion should relieve us … Very heavy casualties to the Brigade from our own artillery…[55]

This message arrived at 19th Brigade at 7:00 p.m. and likely accelerated the preparations for relief. Whilst reconnoitring the northern part of the wood, at about 6:00 p.m., an officer of 11th Field Company came under MG fire from the north corner, where there were many Germans. He reported that this sector could not be held, nor a strongpoint dug.[56] Whether the northern corner, and the Switch Trench, had not been captured by 2/RWF, or had been relinquished having been rendered untenable by German fire, is difficult to determine.

At about 5:00 p.m. orders arrived for the consolidation of the wood by digging a NE–SW trench across the wood to face the Switch Trench; this would however give up the western strongpoint to the enemy. Front and support lines were to be constructed and consolidated with a strong point built at the eastern tip which was in the support line. This order was confirmed by the GSO2, 33rd Division, Major Pakenham DSO, at about 6:00 p.m. All was relatively quiet in High Wood; Sergeant Major Armour of 20/R Fus was able to walk unhindered along the Flers road despite being warned that he would draw fire.

The line that would be consolidated consisted of two RWF companies on the right with two RF 'Companies' on the left, likely comprising the meagre survivors. There was a Vickers MG in the centre and one on each flank. The other two weak RWF companies were in support. The Cameronians and Scottish Rifles were mingled along the southwest edge of the wood with a company of the latter at the eastern corner. A sapper officer inspecting the new line considered that the RF had not dug a good trench and had, in places barely dug six inches because; 'Manual labour was distasteful to them, and was not seriously enforced, for they were officered from their own ranks.'[57] This comment on the quality of UPS platoon and company officers is somewhat unfair. 20/R Fus had few officers left, and equally few NCOs, having been in action since early morning, unlike the comparatively fresh 2/RWF. Exhaustion, rather than idleness, was a more probable reason for any shortfall in trench digging.

Donald Price awoke late that afternoon to find all three wounded Germans in his dugout had died. Presumably now thinking clearly, he went out to rejoin his battalion and found his sergeant major, likely Armour, and three other men and held a position on the edge of High Wood. According to Price; 'We got mixed up with another regiment … there was no case of authority; nobody could tell you what to do. … Most of them [officers and NCOs] had been killed by that time; even the commanding officer … had been wounded…'[58] Skelton's and Amour's party held a small area of the wood. Some men of 20/R Fus may not have reached the

---

55   TNA: WO 95/2405: 33rd Division GS War Diary.
56   TNA: WO 95/2414: 11 Field Company RE War Diary.
57   Dunn, *War the Infantry Knew*.
58   IWM: Donald Price Sound Recording 10168, Reel 9.

wood and a few may have withdrawn from the wood, along with members of other battalions, but this party, along with other groups, held firm throughout the day.

Most of the soldiers in High Wood had experienced sixteen hours of heavy fighting. At dusk a heavy German bombardment forced the thinned-out number of defenders back to the southern half of the wood. Hodson recalled this shelling at about 8:00 p.m.; 'We watched the first one fall, away on our left … and then they crept nearer, each one twenty-five yards closer … They were shelling with devilish system and relentless accuracy. Assuredly one would drop on us. We heard it coming and flattened ourselves against the trench-side … Soil and smoke covered us. "Anybody hit?" we all cried …'[59] The shelling became more rapid and the survivors did not believe they would escape the wood.

Holding the Wood

Later the Germans started a creeping barrage on the wood. When it lifted a mixed party of about twenty men from the Royal Fusiliers, Cameronians and Scottish Rifles drifted back from the centre shouting; 'He's coming over.' These men were from posts further forward that had been overlooked. Dunn recorded that they were officer-less rather than demoralised and at this point they were rallied relatively easily by a Royal Fusilier officer who took them back but was shot moments after doing so. His prompt action, recognised by Dunn, likely avoided the spread of any panic. Private Harold Tyson, a stretcher bearer, recalled that Lieutenant Bert Wallwork was the last 20/R Fus officer casualty who; '…fought bravely and died a hero.'[60] Tyson wrote to Wallwork's mother:

> I was in a support trench in the wood near the road which divides the wood roughly into two parts. About seven o'clock at night we were subjected to a barrage fire and counter-attack. Bert (I knew him as Bert at home) had a rifle and gathered about twenty men or so who had fallen back on the support line and set off to regain the old position. He advanced up the road [presumably 'ride'] and had got as far as a clearing when he was hit by a sniper or machine gun, and a man came back trying to find a stretcher. He was not killed outright but was fatally hit. I know his wounds were dressed and he was given morphia and was being evacuated. He died just outside the wood very shortly afterwards …[61]

Wallwork's actions helped preserve the 20/R Fus line.[62] The Chaplain, Ernest Mannering, wrote to Wallwork's family that; 'He was hit just before the battalion was relieved after a very trying 30 [sic] hours or so of fighting for an important position. He did not die quite at once, but

---

59 Hodson, *Soul of a Soldier*, pp.101–102.
60 Norman, *The Hell they Called High Wood*, p.141.
61 TNA: WO 339/19828: Wallwork Officer File.
62 Lieutenant Herbert Wallwork, aged 22, was born in Walkden, Lancashire. He attended Eccles School and Rossall School and was a member of the OTC. Wallwork was a Mill Manager by trade. Enlisting in September 1914, he was gazetted into the 20/R Fus on 27 October. Promoted lieutenant in July 1915, he served as a platoon commander in C Company and battalion bombing officer. Reported missing on 20 July 1916, his body was buried by another regiment, but the grave was subsequently lost. Wallwork is commemorated on the Thiepval Memorial.

Lieutenant Herbert Wallwork was likely one of the last 20/R Fus platoon commanders left standing in High Wood. (Reproduced with permission of Sutton Archives)

those who saw him tell me that he was unconscious and not apparently in much pain, we have hopes that his passing was peaceful.'[63]

The defending German troops were treacherous: 'I was in the Wood when a party from 2nd R.W.F. fired on a group of Germans who came to give themselves up and then started to sling stick bombs when within range.'[64] A wounded man from 2/RWF also recounted this story to a 20/R Fus stretcher bearer.[65]

Concurrently, the RWF experienced the difficulties of fighting in the wood first-hand as their outpost line was rushed by the Germans and some men were forced to withdraw to the edge of the wood. The RF allegedly withdrew when it believed relief from 18/Mx had arrived. The latter battalion had withdrawn when it learnt of the situation in the wood without the RF having gone back to their positions. It is possible that after Wallwork restored the situation, and was killed, that a further withdrawal by some personnel occurred. The 18/Mx war diary records a little of what occurred:

> Major Best [OC 18/Mx Companies] gave orders accordingly and reported to OC 2nd Bn R Welsh Fusiliers at S.10.b.1.8. While he was there an orderly brought a written message from the R.W.F. in the wood to say that the 20th R.F. had evacuated the front line, and on the heels of the orderly a major of the R.W.F. [Major Crawshay] ... came confirming the message and saying the Germans were advancing.[66]

The two 18/Mx companies were ordered by Crawshay to advance through the wood and meet the enemy with the bayonet. Crawshay guided two Middlesex platoons; one platoon went to the east side of the wood and another platoon moved to support 20/R Fus men holding the support

63 TNA: WO339/19828: Wallwork Officer File.
64 Robert Graves, *But Still it Goes On* (London: Jonathan Cape, 1930), p.37.
65 Dunn, *War the Infantry Knew*, p.241.
66 TNA: WO 95/2417: 18/Mx War Diary.

line in the centre. Elements of 20/R Fus in the front line may have been forced back but those in support had held firm.

An officer of 11th Field Company reconnoitred the wood at 9:00 p.m. and reported that the Germans recaptured the northern corner at about this time and commenced to try and clear through the wood. Around this time an unknown Captain on 20/R Fus reported that his front line had been penetrated by an unknown number of the enemy and that his support line was about to be broken.[67] This was reported to Major Crawshay but many of the troops in the wood were demoralised having spent the day under heavy fire. A heavy German bombardment, including 8-inch guns, commenced at dusk according to the Scottish Rifles; 'Huge shells crashed into the wood and its occupants – the confined space making it an easy mark for the enemy's gunners. Those who had withstood attacks throughout that devastating day were inclined to lose heart – such, indeed, as did not lose their lives…'[68] The German shelling, the darkness and the state of disorganisation amongst British troops made a counterattack impossible. The RE recorded that this barrage lasted from 10:30 p.m. to midnight and was aided by at least one British heavy gun.

Relief and Withdrawal

After severe fighting, the wood had been captured but not completely secured. 19th Brigade had been entirely committed with no reserve available. It needed to be relieved by fresh troops if the wood was to be held; let alone taken in entirety. 100th Brigade was instructed to take over the ground gained; this was ordered at 7:00 p.m. According to 19th Brigade, in the time it took for these troops to arrive the Germans had shelled the wood so heavily, with 8-inch shells, and units had become so depleted, that the northern half of the wood had to be evacuated and the southern half held.

The 1/Queen's RWS and 16/KRRC were provided from 100th Brigade to hold the remaining gains. According to Dunn the handover with the arrival of two companies of the Queen's and one of the KRRC did not last very long. According to the 20/R Fus war diary, the battalion held on until relieved by 100th Brigade at midnight. The survivors were withdrawn to their former camp at Mametz Wood. Price was told to make his own way back when they were relieved that night. Before leaving Skelton took a party of 16/KRRC to the strongpoint at the north-eastern corner of the wood; this had been held by Sergeant Bowers of 20/R Fus; all that Skelton found were obliterated defences and bodies. PS/4508 Sergeant Frederic William Bower was lucky to survive High Wood.[69]

Private Harold Tyson summarised the battle:

> We got the wood and then had to consolidate under heavy fire from three sides and ordered to hang on at all costs 'til relieved. Machine guns swept us, heavies bashed us, shrapnel lashed us, but we held on. It looked as if we should never get out. Snipers in the trees took

---

67 TNA: WO 95/2414: 11th Field Company RE War Diary.
68 Martin, *Fifth Battalion, Cameronians*, p.77.
69 Bower was from Timperley, Cheshire, having been educated at Manchester Grammar School. He was an engineer when he enlisted in September 1914. A former Timperley Hockey player, he was subsequently gazetted in December 1916, afterwards remained with the 20/R Fus.

a heavy toll as also did the machine guns. The Hun counter-attacked first on the left and then on the right but was turned back. Then they tried shelling us out and they shelled behind us to stop reinforcements from coming on. The bosche is a treacherous foe: he fires at stretchers and at helpless men with indifference. Some came with hands uplifted and shouting "kamerad" to a battalion on our left. They did not fire but as soon as they drew near started throwing bombs they had hidden. This lot of bosches were stopped for all time. After about 24 hrs we handed over the wood to the relief and the survivors filed out still defiant though tired and shaken by the strenuous work. Men who have been at Mons said that this last fight was worse than anything they had faced before.[70]

Tyson maintained a hatred of the Germans long after the end of the war because of such treachery.

The German official account stated that British attacks were unsuccessful and do not mention the whole wood being captured at any stage; nor the Switch Line being penetrated. The German account recorded that; 'For the most part, the forest was clear of the enemy on the 20th of July.'[71] 35 prisoners and a few MGs had been captured when I/62 had cleared the forest early in the morning down to the top of the southernmost tip. Though this statement downplayed the British foothold that remained, 19th Brigade had not captured and secured the whole wood. That foothold would pose a problem to the German defenders who had also needed to bring up reinforcements to strengthen their grip on the wood.

Losses had been shocking both quantitatively and qualitatively. According to Skelton thirty officers and 650 men of 20/R Fus went into the wood; four officers, twelve NCOs and 150 men returned.[72] Sergeant Ernest Holden, of C Company, had taken a platoon of forty-five men into High Wood; he and two others came out. He recalled his feelings immediately after the battle:

> … we came down [the line], again at midnight, worn out in mind and body if not in spirit, we have lived through such a life as few who live it are destined to survive … I have asked myself time and time again why I have been singled out for God's especial mercy and protection when comrades, better men than myself, have been struck down by my side. … when one was stunned and blinded and half suffocated by the inferno which raged at times, when fellows one had known all one's life … were killed before one's eyes … I asked myself whether, after all, life amidst all this fiendishness and misery was one of any worth and I almost envied the dead, for at any rate they had done with these horrible things and had made their peace with God.[73]

This was one day in a long war for 20/R Fus but this one day would have a lasting impact on the fortunes of the battalion. Before examining the contribution of the battalion to the remainder of the Battle of the Somme the dramatic impact of High Wood on the battalion must be examined. This will highlight the difficulty of returning the battalion to anything like its previous form.

---

70 Norman, *The Hell they Called High Wood*, pp.150–151.
71 von Stosch, *Schlachten des Weltkrieges*, p.153.
72 IWM: G. Skelton Papers, 13966, p.50. Official documents do not state the strength of 20/R Fus immediately before High Wood.
73 LIM Archive: Letters of E.A. Holden, courtesy LIM, p.87.

19

High Wood, the Aftermath

Have you seen my brother?

England ought, and will, be proud of them when the tale is told.[1]

There was no one like 'em, horse or foot![2]

Whilst the tactical operations of 20/R Fus in High Wood are difficult to trace with accuracy the impact on the battalion is much easier to quantify but not to qualify. Losses were intolerably heavy and the extent of these losses in terms of personalities and numbers must be examined. Likewise, post-High Wood, the extent to which 20/R Fus could recover from these losses would significantly affect how the battalion might perform in the future. Meanwhile, the lasting controversy of the fighting for High Wood must also be explored in relation to allegations of cowardice and unreliability levelled against men of 20/R Fus. This will assist in assessing the lasting reputation of 20/R Fus.

According to Hodson, the survivors of 20/R Fus consisted of 'little knots of men sitting about, some already cleaning their rifles, some washing and shaving in water gathered from shell-holes, some like myself just lying and thinking. So this was the end of our glorious battalion. Well, we had done magnificently … to some of us, with our tears, came added strength of purpose.'[3] Likewise, Harold Tyson recalled the aftermath:

> I think the most pathetic incident of all was the roll call the morning after coming out of action. The survivors answered and then had to give information of comrades who were missing. Ernest Newton of Monton (school) was wounded as also was councillor Gardner and son. Chris Nelson was wounded, Mardock [Murdoch] wounded, Sergeant Horrocks (who was recently on leave) killed.[4]

---

1 *The Burnley Express* and *The Burnley News*, 29 July 1916.
2 Hodson, *The Soul of a Soldier*, p.106.
3 Hodson, *The Soul of a Soldier*, p.106.
4 Personal account by Harold Tyson, unpublished.

PS/4671 Private John Eadon Cook, aged 24, was a designer for a worsted cloth manufacturer from Denby Dale. He was killed in High Wood and is commemorated on the Thiepval Memorial. It is possible that his brother, Frank Eaden Cook, was the man searching for his sibling after the attack. (Reproduced with permission of Sutton Archives)

PS/5626 Sergeant Ernest Shorrock was a teacher. His company commander, Captain Templar, wrote: 'Tell Shorrocks' parents that he died fighting splendidly, and did wonderful work reorganising the men'. PS/5363 Private Frederick Alexander Murdoch, an elementary school teacher, died of his wounds. Tyson also added; 'The fighting was bad enough, but far worse was the agony afterwards, in billets back at Mametz. I'll never forget the heart-rending inquiry repeated over and over: "Have you seen my brother?"'[5]

It took a day or two to comprehend the enormity of the losses to the battalion. Casualties, on paper, changed several times over subsequent days. Initially 19 officers and 450 men were estimated with another tally stating that four officers were killed, three were missing believed killed, one missing and eight wounded (including Lieutenant Colonel Bennett). About 375 men were killed, wounded or missing. A later consolidated report stated 62 killed, 256 wounded and 161 missing or 479 men lost.[6] Of the missing relatively few were captured. Before going into action on 20 July 20/R Fus had lost already suffered twelve killed, 74 wounded and one man missing. Proportions of casualties are difficult to assess as the exact numbers immediately before High Wood were not recorded. Skelton stated casualties amounted to 75% of those going into battle; the most conservative figures state 60 percent with the most likely figures being 70–73 percent[7] Officer casualties were percent. Of the four survivors, three minor carried wounds.

5   IWM: E. Stoneley Papers, 7716, Unknown newspaper cutting, 'The Final Agony'.
6   TNA: WO 95/2408: 33rd Division A&Q War Diary.
7   Skelton states 488 casualties out of 650 men. Realistic estimate of either 392 casualties from 563 men or 479 out of 650 men; conservative estimate of 392 casualties out of 650 men.

19th Brigade Casualty figures

19th Brigade July 1916	Before High Wood				High Wood				15–22 July 1916
Unit	Killed	Wounded	Missing	Total	Killed	Wounded	Missing	Total	Grand Total
20/R Fus	12	74	1	87	50	182	160	392	479
2/RWF	9	48	0	57	26	166	35	227	284
1/Cameronians	19	110	4	133	46	156	163	365	498
5/Scot Rif	7	137	2	146	32	129	282	443	589
19th MG Company	0	4	0	4	1	8	1	10	14
19th TM Battery	0	2	1	3	0	0	0	0	3
Total	47	375	8	430	155	641	641	1,437	1,867[8]

The losses by 20/R Fus should also be seen in context by comparison to other 19th Brigade units (see Table 4). The reduced losses suffered by 2/RWF were likely explained by it going into action under-manned due to its detached working parties. 19th Brigade suffered an estimated eighty officers and 1,867 men during July 1916.[9] Likewise, 98th Brigade and 100th Brigade suffered 1,127 and 1,617 Casualties respectively by comparison. It was not until 26 July that 33rd Division could accurately assess the casualties it had suffered from 15 to 22 July; these amounted to 262 officers and 4,964 men. The words of the divisional commander, Major General Landon, must have sounded hollow when he stated in his report; 'I trust that these casualties will be accepted as proof of the gallantry and determination displayed by all ranks.'[10] Numbers of casualties might indicate the spirit of a battalion to keep fighting until success was achieved. However, they were a poor means of trying to alleviate scrutiny from Corps and Army Commanders regarding partial success or failure at High Wood. Beyond the quantity of casualties, their quality, and qualities, also require examination.

A battalion might be made up of 1,000 men but a battalion was more than just a collection of soldiers. Losses amongst other ranks would impact on the ethos and character of the battalion. Likewise, the officer casualties would severely impact on the leadership and administration of the unit. The effect of these losses must be appreciated.

In numbers of officers Captain E. Toller; Lieutenants H. Wallwork, S. Rawson, J.H. Palmer and J.B. Evans and Second Lieutenants J.T. Price and E. Coventry were all killed. Second Lieutenant T.C. Hine was missing and was later presumed killed. The following were wounded;

---

8   The Cameronians also lost two men missing who had arrived with a draft having served with the 33rd Divisional Cyclist Company; both were formerly from the UPS. 25541 Private Ian Edward Gilchrist and 25587 Private Harry Bertram Sunnocks were formerly PS/6540 and PS/6854 respectively; both are commemorated on the Thiepval Memorial. Gilchrist, from Ipswich, had planned to be a Presbyterian Minister. TNA: WO 95/2408: 33rd Division A&Q War Diary.
9   TNA: WO 95/2408: 33rd Division A&Q War Diary.
10 TNA: WO 95/2408: 33rd Division GS War Diary.

High Wood, the Aftermath 293

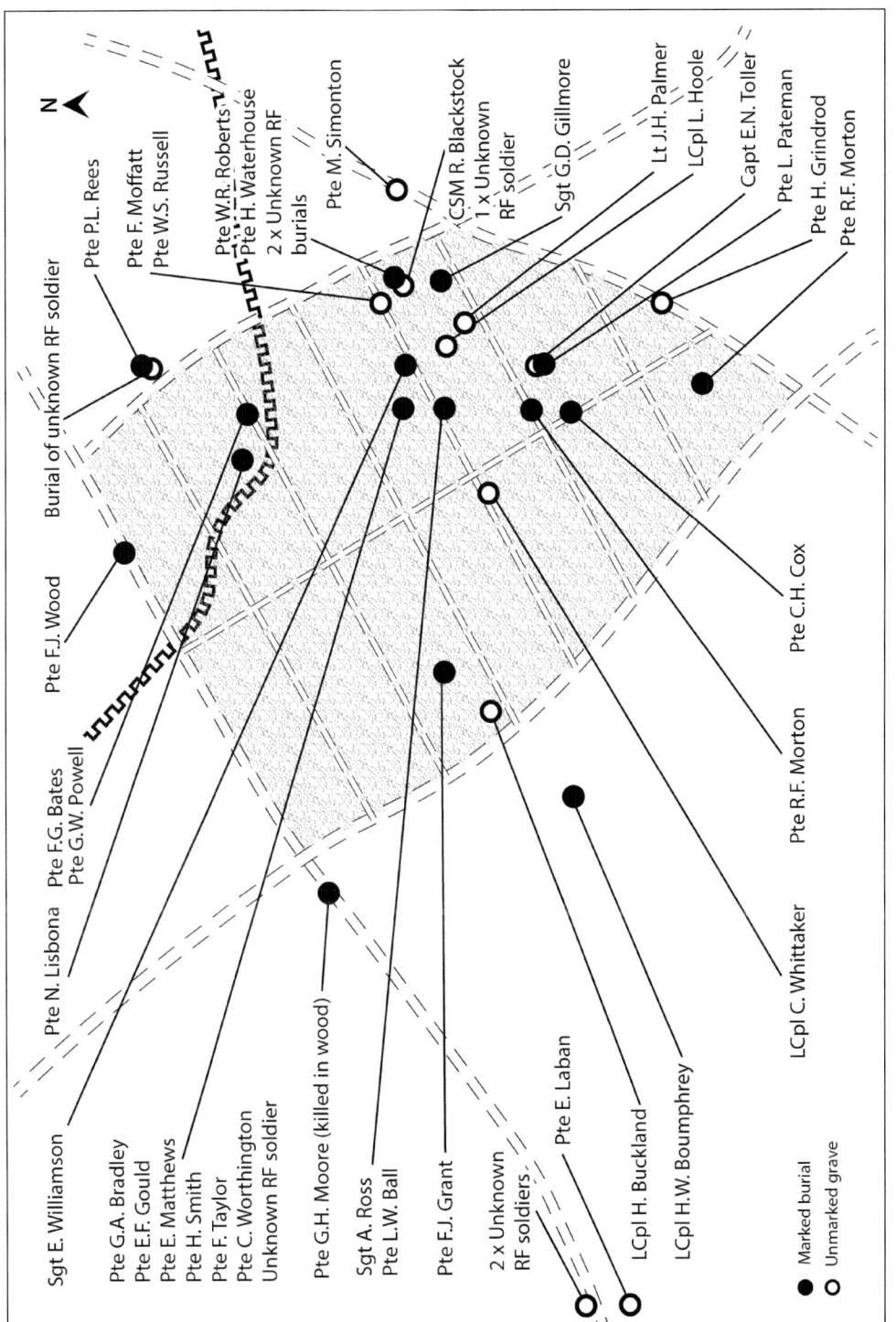

Map 7 High Wood, 20/R Fus burial map.

Lieutenant Colonel C.H. Bennett DSO; Captain D.W. Hollingsworth; Second Lieutenants S.M. Bell, F.T.R. Ives, W.R. Brooke, L.J. Cooke, G. Herbert and N.F. Fabricius. Captain A. Maxwell, Lieutenant A.V. Coggin and Second Lieutenant R. Hume were wounded but remained at duty. The performance of these men in battle and their fates also highlight the loss to 20/R Fus. Lieutenants G.G. Ziegler, H.T.G. Blaauw and Second Lieutenant F.L. Everard were wounded just before High Wood.

The bravery by Bert Wallwork in helping to hold the line has already been described. Some 20/R Fus men thought Captain Toller a bad company commander, but he was nonetheless killed in battle leading his men. However, accounts of his death, and those of other officers, demonstrate the confusion that reigned in High Wood.[11] According to PS/5236 Corporal Ernest Lock; 'In the clearing of High Wood, Capt. Toller went too far forward. The Germans attacked from the extreme end of the wood. Capt. Toller was reported killed…'[12] Another man saw Toller lying dead having been shot through the head and killed outright.[13] He and all the other dead were left behind in the wood. According to the adjutant:

> The Battalion has lost an Officer whom it would be indeed difficult to replace. No one could have been more devoted to or adored by his men. No one could have shown a finer example of courage in the face of the enemy than did Northcote. He was at all times cool and collected, and at all times caring for his men, and he met a fate which any true soldier could have wished for. He was loved and admired by all ranks…[14]

Lieutenant Stuart Rawson also did much to lead his men in High Wood.[15] Rawson's company commander stated in a letter:

> Nothing has affected me so much as losing dear old Stuart, who had been with me so long. You do not need me to tell you of his unfailing cheeriness and generosity, but you cannot realise what a splendid officer he was. He was always so thoroughly reliable and efficient … His men loved him as we officers did and would do anything for him. Before the attack Rawson was the cheeriest of the lot and went into action … full of jests. In the wood he did wonders in rallying the men and consolidating the position. About mid-day he came to my dug-out to tell me that he was holding the line with a handful of men and that the enemy

---

11 Captain Edward Northcote Toller, aged 31, was born and lived in Kettering, Northamptonshire. He was a solicitor by trade. He enlisted in September 1914 and was gazetted to a commission in January 1915 after having been a company sergeant major. Initially he commanded No. 13 Platoon, D Company, 20/R Fus and later took over D Company being promoted captain in March 1916. He was missing, believed killed, after 20 July 1916; his body was found and buried in Caterpillar Valley Cemetery in 1917.
12 TNA: WO 339/56294: Officer File for E.N. Toller; Statement by Corporal Lock.
13 TNA: WO 339/56294: Officer File for E.N. Toller; Statement by Private Sidaway.
14 *Rugby School Archives* <https://rugbyschoolarchives.co.uk/RollofHonour.aspx?RecID=390&TableName=ta_rollofhonour&BrowseID=14> (accessed 8 August 2021).
15 Lieutenant Stuart Milner Rawson, aged 24, was born in Ealing and lived at Coleherne Court, Kensington. He attended Tonbridge School, where he was a member of the OTC (1908–1909), and the University of London and was an engineer by trade. He enlisted in Westminster in September 1914. A sergeant by February 1915, Rawson was gazetted in March 1915 and promoted to lieutenant in January 1916. He is commemorated on the Thiepval Memorial.

looked as though they were going to attack. Typical of the man that, where others would have rushed to me with an excited and exaggerated tale, Rawson treated the whole affair lightly and yet gave me an exact account of how matters stood. I promised him reinforcements and asked him to stay down in my dug-out until the shelling had slackened down but in his usual cheery way he said, "No thanks, I must get along back to the men. We'll pull through all right. Cheery-o, old man." Then, just as he got to the top of the steps he was hit by an exploding shell.[16]

An anonymous letter writer may also have recorded Rawson's moment of death; '…While another officer was standing at the top of the steps of a Boche dug-out a shell burst. The officer cried, "My God! I'm done," and fell down the steps. He died a quarter of an hour later.'[17] Another officer stated; 'No one set his men a finer example of courage in the face of the enemy.'[18]

Second Lieutenant James Evans was the 2IC of D Company at High Wood. He was believed killed on 20 July having been severely wounded in the head and stomach. He had his wounds dressed in shell hole but was later reported missing.[19] According to Gibson both Hine and Price were killed by shellfire before entering High Wood.[20] Lieutenant John Palmer was reported killed; his body was later found in the wood near the eastern corner.[21] What became of Second Lieutenant Coventry is unknown but he too died.[22] Whilst in High Wood Second

16 Anon. Author, *Tonbridge School and the Great War of 1914 to 1919* (London: Whitefriars Press, 1923), p.272.
17 *Manchester Evening News*, 5 August 1916.
18 Anon. Author, *Tonbridge School*, p.272.
19 Second Lieutenant James Bansall Evans, aged 31, was born in Garston, Liverpool, and lived in Heaton Moor, Stockport. He attended Wyggeston Grammar School in Leicester and was employed as a commercial traveller. He was apprenticed with an architect and surveyors' firm in 1904. He enlisted in September 1914 and served in A Company. Promoted to corporal in December 1914, he was commissioned in March 1915. Promoted to lieutenant in March 1916, Evans is commemorated on the Thiepval Memorial.
20 Second Lieutenant Thomas Charles Hine, aged 31, was born in South Cave, Hull, and lived in Farnham, Surrey. He attended Worksop College and became a schoolmaster at Grenham House Preparatory School. He enlisted in Westminster in September 1914. Promoted to sergeant in November 1914, he was gazetted in May 1915 and arrived in France on 23 March 1916 having served with 29/R Fus. He commanded No. 16 Platoon, D Company. He too was missing after High Wood. Second Lieutenant John Thomas Price, aged 25, was born in Llanbadarn, Radnorshire and lived at Kingsland, Herefordshire. He was a bank clerk but also served in the Brecknock Battalion of the Territorial Force (1908–1912). He enlisted in September 1914 with 18/R Fus, was a corporal by October 1914 and a sergeant by December. He was gazetted in July 1915 and posted to 28/R Fus. He went to France on 20 May 1916 and joined 20/R Fus two days later. Hine and Price are commemorated on the Thiepval Memorial.
21 Lieutenant John Henry Palmer, aged 29, was born in Orleans, France, and lived in Timperley, Cheshire. He was an electrical engineer by trade. He enlisted in 19/R Fus in September 1914 and was formerly a CQMS with that battalion from October 1914. He was gazetted in January 1915. He went to France and joined 20/R Fus on 20 May 1916. He was reburied in Serre Road Cemetery No 2.
22 Second Lieutenant Eric Coventry (formerly Zeederberg), aged 28, was born in Cape Town and lived in Upper Norwood. He was a broker and worked for Walker Brothers, London. He enlisted in the Scots Guards as 'Zeeder' in May 1915 but was discharged in August 1915 to a commission having changed his name to 'Coventry'. He was gazetted into 22/R Fus and served with 27/R Fus. He arrived in France on 20 May 1916 and joined 20/R Fus and is commemorated on the Thiepval Memorial.

PS/5267 Sergeant Thomas McCreath was born in Liverpool and lived in Wapping. He died of his wounds on 24 July 1916 and is buried in Heilly Station Cemetery Mericourt-L'Abbe. (Reproduced with permission of Sutton Archives)

Lieutenant Fabricius was hit in the middle of his outer right thigh by a fragment of shell. He needed assistance to walk back to the dressing station but made it safely; he had a 5-inch by 2-inch wound in his leg. He later needed a stick to walk but returned to the front.[23] Second Lieutenant Frank Everard was hit in his left thigh and hip on 18 July during a fatigue party. Lewis Cooke also suffered a bullet wound to the outer muscles of his left thigh on 19 July. Walter Brooke suffered gunshot wounds to his left knee and right leg at High Wood. Second Lieutenant Sydney Bell was wounded in the jaw and tongue and never returned to France.

These wounded officers also left gaps. Some returned after convalescing, but many never rejoined. There was no obvious replacement for Colonel Bennett except for Major Vincent Kelly who took temporary command having been with the battalion less than a month. The ten other officers who were wounded and departed were also a loss. For example, Lieutenant George Ziegler was respected by Donald Price as 'a grand lad' for his kindness, sporting prowess and informality 'other officers would ignore you …'[24] Price would obey other officers but he would do anything for Ziegler. Losing these officers who had known their men for almost two years would catastrophically impact the efficiency and effectiveness of any formation.

However, a core of officers remained. Some came out of High Wood with their reputation enhanced. After High Wood Second Lieutenant Algernon Coggin carried a 'purple weal on his temple, badge of victory in a stand-up duel with a Boche officer in High Wood…'[25] This was the nature of his wound at High Wood.

---

23 Second Lieutenant Niels Frederic Fabricius, of B Company, was born in Burma in 1896 and was educated at Wolverhampton School. He was an articled clerk to an auctioneer and later described himself as an agriculturalist. He enlisted with 21/R Fus in September 1914 in Worcester. By July 1915 he was a sergeant with 29/R Fus. He was appointed to a commission with 29/R Fus in September 1915 and went to France on 20 May 1916. After recovering from his wounds from High Wood he was again wounded during a patrol action whilst serving with 7/R Fus near Ribécourt; he was hit by a bomb causing multiple small wounds to his lower limbs. He died in 1960.
24 IWM: Donald Price Sound Recording 10168, Reel 7.
25 Gibson, *Postscript to Adventure*, p.152.

PUBLIC SCHOOLS' BRIGADE, TWENTIETH BATTALION ROYAL FUSILIERS.
Top Row.—E. Swire (Colne), G. Heap (Colne).
Second Row.—Quartermaster Sergeant Parkinson (Burnley), E. C. Lambert (Nelson), P. Swire (Colne), H. Smith (Colne), C. Curl (Colne).
Third Row.—R. Scaife (Colne), T. Fryer (Foulridge), G. E. Wilson (Colne), Corporal E. Cass (Colne), A. Green (Colne), J. S. Brown (Nelson), F. G. Renwick (Nelson).

Group photograph from 1915 showing men in 20/R Fus from Colne and Nelson in Lancashire. Of these men Sergeant Ernest Swire, Lance Sergeant Edward Cass and Private Harold Smith were killed in High Wood. Private Frank Renwick was also wounded there and Private Percy Swire was lucky to survive. Lance Corporal Edmund Lambert survived High Wood and was promoted to corporal a few days later. In addition Quartermaster Sergeant Harold Parkinson managed the successful 20/R Fus football team and Private Alfred Green was one of the players. (Colne and Nelson Times War Album).

The casualties amongst other ranks were also disastrous with 50 killed and 160 missing (many of whom were also killed).[26] As early as 4 August 1916 a wreath was placed at the foot of the Boer War memorial in Manchester in the memory of those killed whilst serving with 20/R Fus on 20 July 1916. One case study can be understood by looking at a small cohort of men from Burnley. A letter from PS/4558 Sergeant Albert Bradshaw published in the Burnley Express, and The Burnley News, on 29 July 1916 records his view of the battle and a small number of the many casualties:

> Our division was told off to take a wood, but the cost was terrible … At three a.m. our artillery commenced to bombard, and the Huns replied. It was all hell let loose. At 3.25 the word was given to our men, who went forward on what will be one of the biggest adventures we shall ever see. We took the wood and held out against the devils all day and until into the next night, when relief came to the few who were left. Our Colonel led the men right into the wood, and then went down early in the engagement with a bullet through the shoulders. Frank Bracewell was very slightly wounded and got well out of it early. I cleared him off to hospital. It was a Godsend to him as most of his company was wiped out later. Ellis Fell was wounded and cleared. Dick Lawson fought like a hero and was then wounded in the back and put in under cover a shell hole. Afterwards he was hit again and killed. Sergt. Swire, Sergt. Cass and Harold Smith, all of Colne, were killed. 'Jerry' Hartley was badly wounded and cleared to hospital. Frank Renwick was also wounded. I am full of sorrow. Our men fought like tigers but were outnumbered and outflanked. Every man was a hero, and England ought, and will, be proud of them when the tale is told. You will understand how unutterably sad we all feel, but we have the satisfaction of knowing that we did our best. We asked no quarter and we gave none…[27]

However, Ernest Holdon, one of the survivors, wrote in a letter; 'The Burnley paper had a very full account [of High Wood]. Sgt Bradshaw's account was rather overdone, but of course he only heard of things from the back as his duties did not take him into the [censored].'[28] PS/9113 Private Richard Thomas Harold Stuttard was wounded in the right shoulder; PS/4981 Private Alfred Victor Hedges suffered a flesh wound to the back and PS/4476 Private Alan Bellingham, who reported severe hand-to-hand fighting, was shot through one hand and the other arm by an MG bullet.[29] The men mentioned above were from the Burnley, Colne and Nelson area; others, like Corporal Alan Proctor, Private Percy Swire and Private James Medcalf, came through

---

26 See Appendix II for casualty list.
27 *The Burnley Express* and *The Burnley News*, 29 July 1916.
28 LIM Archive: Letters of E.A. Holden, courtesy LIM, p.90.
29 Stuttard was aged 24 and was from Burnley, Lancs. He worked for his family business which controlled two mills. He enlisted in November 1915 and arrived in France after four months of training. On recovery he transferred to the MGC and was discharged in December 1918. Hedges was 21 years old and was from Burnley. He enlisted in September 1914 and was educated at Giggleswick School and was the assistant agent to Lord Shuttleworth. On recovery he was gazetted to the 9/Manch R. Promoted and twice-wounded, he was awarded the MC in March 1918. Bellingham, aged 23, was the son of a Burnley registrar; he enlisted in September 1914. He returned to the UK and following recovery, was gazetted in July 1917 and killed on 26 October 1917.

uninjured along with three others of 20–30 who were formerly with 20/R Fus.[30] Swire wrote to Harold Smith's family recording his death:

> Our company had to leave the front line to dig a second line and had to pass open ground swept by machine-gun fire. Harold and I were together and were just getting under cover when he was hit. I went back to him but could not do much. He was hit in the stomach, and died almost immediately, without much pain. I gave him some morphia, and stayed with him until he went off.[31]

Amongst the dead was a pair of brothers from Altrincham. PS/5971 Corporal Cecil Worthington and PS/5972 Private Arthur Worthington, were aged 33 and 35 respectively; they were both born in Altrincham, Cheshire, and lived in Manchester. Cecil was an insurance inspector and Arthur was a bank clerk.

All these individual losses had a marked effect on the morale and enthusiasm of the survivors. One of those killed was PS/5920 Lance Corporal Cedric Whittaker; he was likely a close friend of Hodson who found out about his death after the battle; '"Heard of Cedric?" a friend asked. I shook my head. "Died in my arms," he said, his eyes growing wet as he spoke. I turned my head and walked away, and a sob burst from my lips. We had been as brothers, Cedric and I, our wives almost as sisters. I wept like a child.'[32] Whittaker had been married to his wife Dora on 19 December 1914 at the parish church in Epsom.

The losses amongst soldiers from Manchester and the surrounding area were also very significant.

The Wounded

The wounded also had much to endure. Some were more fortunate than others. Sidney Platt was treated at a Casualty Clearing Station at Daours and left at 8:00 p.m. on a hospital train. He arrived in hospital at Rouen at 8:30 a.m. next morning. His brother Vincent, having been wounded in the head, spent the night at Corbie and got away on a Red Cross train at 11:00 a.m.; he arrived in Rouen at 9:30 p.m. After a bath and cocoa he was tucked up in white linen sheets in a hospital bed. He went to le Havre on 23 July and was back in the UK next day. Sidney Platt was discharged on the 24th, presumably having only been superficially wounded, and at 6:00 p.m. arrived at No 5 Infantry Based Depot. He took advantage of the facilities and relative freedom and went to the cinema that evening. Many of those wounded were never to return to the battalion. For example, PS/5148 Private John Franklin Kay, aged 19, from Bury, had been educated at Bury Grammar School where he was a member of the OTC and was employed as a builder. He suffered a severe arm wound at High Wood that led to his discharge in June 1917.

---

30 GS/48298 Corporal Walter Proctor, aged 24, from Burnley, enlisted in the UPS in September 1914. He survived High Wood but was wounded on 27 October 1916 in the head and left leg. He was gazetted in the E Lancs in August 1917. He died in a motor accident in June 1919. PS/6514 Private James Medcalf, from Burnley, survived High Wood and later transferred to the Royal Army Pay Corps.
31 *The Burnley News*, 29 July 1916.
32 Hodson, *The Soul of a Soldier*, p.106.

PS/5981 Corporal Basset Wright, aged 26, was born and lived in Derby. He attended Marlborough College 1906, Kings College Canterbury and Oxford where he was awarded a BA from Oriel College in 1909. Lost sight of at High Wood he was reported missing believed killed. Wright's name is inscribed on the Thiepval Memorial (Reproduced with permission of Sutton Archives)

Many wounded still in High Wood after 20/R Fus departed would not enjoy such luxuries. Many wounded could not be got into safety for treatment let alone be safely evacuated. According to a letter written to his family by Ernest Mannering, the Chaplain, PS/5204 Private Joseph Henry Leather, of B Company, was wounded in the legs in High Wood. Sadly several men were killed trying to bring him in and shortly afterwards he was killed outright by a shell. His body could not be recovered.[33]

Those Captured

Several men of 19th Brigade were captured at High Wood. Corporal Lawson was later found to be a POW and wrote home from Germany:

> I shall never forget the 24 hours I spent in the wood we attacked and my many providential escapes. I was hit in the evening in a corner which was shortly afterwards evacuated by our people and was consequently picked up by the enemy next morning. After dressing my wound they sent me to a collecting station and then by train to our last place…[34]

Sadly, Lawson later died whilst in captivity.

Privates James Dykes and Charles Hutchinson were captured together; the latter wounded; '…I could not help thinking that it would be the easiest thing in the world to dispose of us there

---

33   IWM: J.H. Leather Papers, 2783, Letter 20 August 1916.
34   *The Burnley News*, 2 September 1916.

PS/6516 Lance Corporal Richard Neville Lawson, aged 36, was born and lived in Nelson. Over six foot tall, he was educated at St Paul's Church of England School and the Municipal Secondary School and was a deputy town clerk. Lawson died a PoW in hospital at Osnabruck in September 1917 and is buried in Cologne Southern Cemetery. (Reproduced with permission of Sutton Archives)

and then. I was very frightened and must admit that I was very co-operative; but then, so were we all. Life is very sweet, even as a prisoner of war, with the prospect of rough, even brutal treatment ahead.' The prisoners had their pockets looted by German infantrymen and had all their weapons and equipment removed from them. Those who could walk were ushered to the rear; Dykes hoped that those who were too seriously wounded were treated by the Germans. PS/8048 Private George Sergeant of No 1 Platoon, A Company, recorded the immediate moments after entering captivity:

> Fifteen or seventeen of us were wounded and captured … except for a German striking each of us with a leather belt as we passed by, I did not see any instances of deliberate cruelty. We had to march 4 or 5 miles before we reached the Dressing Station, which I think was near Bapaume. Here we were well treated; we had our wounds dressed and were given an injection against tetanus. They gave us a little coffee and bread. The same evening we marched on to the railway…[35]

The prisoners in Dykes' party had to contend with British shelling of the road which was intended to hinder German reinforcements. They were ushered to a German dressing station where some were treated and thence to a brigade HQ and later to a divisional HQ where they were individually interrogated by a senior officer. Of the men in Dykes' party there was only one lance corporal and four men from 20/R Fus. The POWs were then loaded onto a train bound for Cambrai. As they got further from the battlefield their fear of execution diminished but likewise, they were increasingly treated as lowly prisoners by their guards. Though they faced a relatively safe future it would be one of hardship and monotony which few men would envy.

35 TNA: FO 383/392: POW Treatment of Private G.H. Sergeant.

Morale

The battle had been traumatic, and the UPS casualties would have a lasting effect on morale. Initially numbness and fatigue set in and only afterwards would the extent of the casualties be appreciated. James Hodson recalled after the battle that 'I was obsessed by the idea of brewing tea … We sat in the hole drinking the boiling liquid in little sips. I felt as safe as I did … in England … When we awoke the sun was high, and the cookers were up with the letters and parcels. We had slept as peacefully and utterly as children – or the dead …'[36] Despite a good night's sleep the realisation that Hodson's platoon had only eight men left was sobering. The fatalities in High Wood were unfathomable and arbitrary; 'The best and the worst men had been killed together. Men I knew for 'rotters' were alive and well; men I knew for pure and noble lay dead and stinking. Life seemed ruled by chance and nothing by order of merit or worth…'[37] A feeling of bitterness at the casualties was clearly in Hodson's mind as he wrote. Many men, such as Harold Tyson, had pride in their achievements; 'I am proud of my battalion; every officer and man was a hero and fought as keenly as ever did on his playing field at school.'[38] Tyson could also rejoice at having had several extremely narrow escapes from death; 'I was very lucky. I found out afterwards that either a bullet or piece of shell had gone through my gas helmet satchel … and out again, without touching me and a splinter of shell had gone through my haversack.'[39] Harold Tyson, a relatively short man, credited his height for his survival; 'Being a little chap saved my life that day. I could keep my head down. All the six-footers were knocked out.'[40]

Meantime, Donald Price's morale was not too badly affected by High Wood despite him barely escaping the battle alive. Once he was safe his morale improved. He stated that the band came out to meet the returning battalion and he felt both proud and that he had properly participated in the war. The survivors were treated like heroes and they had the bitter-sweet perk of sharing out the food parcels from those who were known to be dead or wounded. There was ambivalence to the battle and the survivors had to balance the sorrow of the losses with the pride and confidence of the success they had achieved. However, those who survived grew closer and created a potentially more cohesive cadre for the battalion. James Hodson provided a few vignettes of this; '"Hallo, Jim!" said a kindly, friendly voice … "Awfully glad to see you," continued the quartermaster sergeant, holding out his hand. He had been no particular friend of mine till now. … Everyone I saw held out his hand and gripped mine. We were like a band of brothers long sundered.'[41]

Replacements

Around that core of survivors a new battalion would need to be forged from newly arrived replacements. The numbers of casualties did not reflect the quality of the officers and men lost and the extent to which they had incorporated the UPS ethos. The ability to replace these men

---

36　Hodson, *Return to the Wood*, p.44.
37　Hodson, *The Soul of a Soldier*, p.107.
38　Norman, *The Hell they Called High Wood*, pp.150–151.
39　Harold Tyson unpublished personal account.
40　IWM: E. Stoneley Papers, 7716, unknown newspaper cutting – 'The Final Agony'.
41　Hodson, *The Soul of a Soldier*, pp.105–106.

with further UPS men would also determine the extent to which the character of the battalion might remain after High Wood.

On 21 July the battalion rested until 4:00 p.m. when it moved to Buire-sur-l'Ancre. It spent the next few days licking its wounds there. Godfrey Skelton recalled; 'There [at Méaulte] we were reinforced by new drafts, many of them older men from the Derby Scheme, then in operation at home. New officers arrived and training started to bring the companies back to strength.'[42] These officers included Second Lieutenants Stewart Humphreys and Oscar Thomas who arrived with a draft.[43] The casualties were further evidenced by the need for large numbers of replacements to be absorbed during the remainder of July; 19th Brigade records 470 were received (15–31 July); the 20/R Fus War Diary states 701 (24–31 July).[44] As such, it is unlikely that many of the missing men rejoined later. Reinforcements, across 33rd Division, were very varied in terms of quality and background:

> A large majority of reinforcements received had very little training, in many cases amounting to only 3 or 4 months or less. Steel helmets were not supplied with any drafts. There were very few NCOs with any of the drafts, which greatly hampered the training of reinforcements. A suggestion was put forward and approved by the army that cavalry NCOs should be supplied. Very few specialists – Lewis gunners, signallers etc. – were sent up with drafts. The system of indiscriminate reinforcing of battalions is likely to seriously injure their efficiency. In many cases it could be avoided as drafts of a battalion in one division were sent to a unit having no connection with it, whilst men of the latter were sent to the former.[45]

According to the Divisional HQ war diary; 'In nearly all cases reinforcements have been sent from the base in a most indiscriminate manner ... every effort was made to exchange these drafts and the matter was taken up by the Corps and 4th Army.'[46] Major General Landon concurred with a request submitted by 100th Brigade on behalf of 1/9 HLI to 'swap' some of their reinforcements. They were a kilted TF battalion but had received large drafts of men who were from non-kilted regiments; many of the reinforcements were not medically fit enough for front line service.[47]

In Hodson's opinion, 'The battalion's ranks were replenished – largely with Cockneys who were, in the eyes of Hardcastle and his friends, conscripts. At all events, they had just come out to France. Their language and manners left much to be desired ... [The originals] ... felt like veterans, as if responsible for them; they stiffened them, prided themselves on doing it ...'[48] This

---

42   IWM: G. Skelton Papers, 13966, p.51.
43   Second Lieutenant Stewart Francis Humphrys, aged 24, was a surveyor from Croydon; he was born in Ottawa. He enlisted in the King Edward's Horse in 1913 and was already a sergeant when the unit arrived in France in early 1915. Returning to the UK for a commission in December 1915, he was originally gazetted into 14/R Fus and arrived in France on 18 July before joining 20/R Fus on 21 July. Humphrys attended a Lewis Gun Course on 23 July and rejoined on 3 August. Second Lieutenant Oscar Thomas was a draughtsman from Wandsworth Common. He enlisted with 8/R Sussex and served in France from July 1915. He was gazetted to the Royal Fusiliers in November 1915.
44   TNA: WO 95/2423: 20/R Fus War Diary. TNA: WO 95/2408: 33rd Division A&Q War Diary.
45   TNA: WO 95/2408: 33rd Division Administrative War Diary.
46   TNA: WO 95/2408: 33rd Division Administrative War Diary.
47   TNA: WO 95/2405: 33rd Division GS War Diary.
48   Hodson, *Grey Dawn*, p.263.

Cecil Isom (pictured in 1919 with the 26/R Fus football team) was an example of the replacements that 20/R Fus received following High Wood. Though not a UPS man, he had experienced the Gallipoli campaign with the London Regiment and was a dedicated member of his new battalion. (Reproduced with the permission of Pat Isom)

was a generalisation; some had extensive active service experience. One of the replacements was GS/63351 Private Cecil Harry Isom, from Peckham, who joined 20/R Fus on his nineteenth birthday – 25 July 1916. Isom was not a Derby Man, nor a conscript, but a Gallipoli veteran from the London Regiment. A whole reinforcement draft of men like Isom also joined 20/R Fus. At least these men were demographically in keeping with the Royal Fusiliers despite them not being UPS men nor having a connection to Manchester. Reverent Mannering wrote of his batman who came from this draft and who had served at Gallipoli with the Londons; 'B ___ a big rawboned country bred man from S. Wales who had risen to be a buyer with a well-known London firm … he was amazingly quiet & deft, in many ways an ideal batman … he had something of that slow passive strength, which comes from the soil …'[49] In addition, Hodson considered that during the autumn of 1916 the officers that join 20/R Fus were in many cases 'neither brave nor gentlemen.'[50] There would be some clear exceptions to this judgement. The failure to replace High Wood losses with returning UPS men, or even those men trained by UPS Reserve Battalions, severely impacted on the future ethos and performance of the battalion.

Despite the dilution of their numbers, and their increasing disillusionment, many UPS men stuck with the battalion. Others, like Donald Price, could not wait to escape. Price was found to be under-age (he was only eighteen) and was sent to work at a headquarters at Auxi-le-Chateau behind the line until he came of age.

---

49 IWM: Reverend E. Mannering Papers, 6756.
50 Hodson, *Grey Dawn*, p.266.

A Lack of Medals?

Despite the hard fighting and significant casualties suffered by 20/R Fus its men were not rewarded with many decorations. The battle did bring some medals to the battalion; five men were awarded immediate MMs after High Wood. They were PS/7053 Private Thomas Ashworth; PS/4635 Private Stanley Clayton; PS/5268 Private Charles MacIntyre; PS/5598 Private Francis Gordon Scott and PS/8735 Private Sidney Snow. Snow was recognised for bravery whilst despatch riding; he was carrying a message by bicycle when his bag was blown from his back. Clayton was rewarded for being one of two men who rescued a wounded officer from no man's land. The other of these men was PS/4599 Private Fred Carter who was sadly killed in the same brave act but received no award. Notice of Scott's award in the newspapers referred to 'gallantry on the battlefield' but the Manchester Grammar School magazine stated he 'rendered conspicuous service in keeping open the communications between him and the Officers in command of the Companies. This had to be done all day under very heavy shell fire, machine gun and rifle fire.'[51] McIntyre was awarded the MM for carrying messages under heavy fire.

Casualties amongst officers likely stopped many brave acts being recorded in this way. Likewise, confusion and darkness mixed units up and made it difficult to identify the names of those who may have acted bravely. It was also difficult to reward bravery for those who were killed. Posthumous acts of valour could only be rewarded with the VC, which might not have been justified, or with a Mention in Despatches; there was no 'middle ground'. One of many men who were considered to have behaved like heroes, but whose gallantry went unrewarded, was PS/5609 Private Thomas Sedgley. His company commander wrote of him; 'We know he was wounded, and whilst wounded went and magnificently worked a machine gun. This gun was captured, and we think your son was captured with it. On behalf of the remaining officers and myself I would like to tell you how very highly he was honoured in the regiment for his courage.'[52] His sergeant major also wrote to this effect. The Lewis gun crew in question had all been killed or wounded.[53] PS/5967 Lance Corporal Frank Wilkinson, an old boy of Hurstpierpoint College, was a collector of birds' eggs. He was killed, aged forty, having chosen to enlist despite his age. According to his school magazine; 'He met his death trying to save another man who was lying out in a crater wounded and helpless. He had dressed his man's wounds and was killed in trying to regain the trench.'[54] Such examples may have been efforts to try and soften the loss to families of their loved ones. However, in the immediate aftermath 20/R Fus was awarded proportionally less immediate awards for Military Medals compared to some other battalions in 19th Brigade.[55] Significant officer casualties meant that the UPS officers were largely not recognised and they too were unable to provide citations for their men.

---

51 *ULULA Magazine*, December 1916, p.257.
52 Manchester Evening News, 18 August 1916.
53 PS/5609 Lance Corporal Thomas Wilson Sedgley, aged 23, was born in Southport and lived in Withington; he attended Manchester University and was a labour exchange clerk. He served with C Company and was a friend of Sidney Platt. He is commemorated on the Thiepval Memorial.
54 Hurst Johnian, November 1916, p.333.
55 There were seventeen immediate MM awards to 1/Cameronians; four to 1/5 Scot Rif; fifteen to 2/RWF and five to 20/R Fus. The low number to 1/5 Scot Rif was likely due to high officer casualties.

Writers and sources

Whilst Graves' popular publication *Goodbye to All That* gained prominence post-war, there were few writers left within 20/R Fus whose accounts were more prominent and could match his work. The fight for High Wood removed a number of those who had chronicled 20/R Fus from within the battalion. Private Frederick Shield, of B Company, was wounded in the left arm and was evacuated to hospital. Private James Dykes, of C Company, was captured and spent the remainder of the war a prisoner. Both Sidney and Vincent Platt, of D Company, were wounded and departed the battalion. None of these men returned to the battalion. Numerous other soldiers who might have left accounts of their service perished in battle or died of their wounds. Donald Price, another survivor, whose oral history assists in understanding the battalion, also departed shortly after High Wood.[56] This 'attrition' amongst authors was such that the number of accounts available to give detailed information on the battalion after High Wood were significantly reduced. Some authors became more remote from the battalion. Whilst at Méaulte, Sergeant Skelton was selected to command the Brigade Sapping Platoon and he picked surviving members of D Company as NCOs.[57] The morale of those men who remained, and the authors who recorded events, James Hodson and Godfrey Skelton, was also degraded.

This reduction in sources requires greater reliance to be placed on the battalion war diary, which was compiled by consecutive adjutants, and from sources external to the battalion. If examining events at High Wood alone the War Diary was often insufficiently detailed to be an authoritative source for events. For example, the battalion war diary recorded the assault on High Wood as follows:

> 19th Brigade attack at 3.25 a.m. on HIGH WOOD: attack by 1st CAMERONIANS and 5th SCOTTISH RIFLES. 20th Royal FUSILIERS in support, 2nd ROYAL WELSH FUSILIERS in reserve. North corner & North west corner of wood not taken, rest taken & consolidated. 2nd ROYAL WELSH FUSILIERS came up about 12 noon. A front & support line consolidated across wood from East to West along with a strong point about S.4.d.2.8. in support line. Held on to the position until relieved by 100th Brigade at midnight then withdrew to old bivouacking around MAMETZ WOOD.[58]

With such a lack of detail other official sources (war diaries for 19th Brigade and for other units) and unofficial sources (personal accounts and unit histories of other battalions) must be relied upon.

High Wood was a watershed moment for the battalion and henceforth officers and men were divided between those who had experienced the maelstrom and those how had not. However, Second Lieutenant Edward Chapman arrived around this time and his letters home, and Cecil

---

56 PS/5473 Private Donald Price was with D Company, 20/R Fus. In September 1916 he was withdrawn from front line service due to his age and went to the Third Army School awaiting his 19th birthday. He rejoined his battalion in March 1917. IWM: Donald Price Sound Recording 10168.
57 The role of the platoon was to dig and wire trenches quickly to enable separate positions to be joined rapidly together to form a complete front line. This was expected to be done under shellfire and flares by night.
58 TNA: WO 95/2423: 20/R Fus War Diary.

Isom's diary, provide an insight into the battalion after High Wood. Likewise, Ashley Gibson also joined the battalion shortly after High Wood giving a further perspective.

This momentous engagement also reduced the number of men who could contribute of The Gasper. PS/5366 Private Harold Mumby was wounded in the head and right shoulder by a grenade explosion and was evacuated to the UK. The last issue was produced on 30 September 1916, entitled 'The (Last) Gasper'; 'The Gasper goes the way of all flesh. Not for lack of support, but because the members of the Brigade whose organ it was, have become so scattered that the paper has really ceased to be the organ of anything in particular …'[59] An attempt to form a roll of honour for the Brigade also failed.

However, the Somme campaign was not yet over. An instruction issued by HQ 33rd Division on 24 July stated; 'All ranks must remember that they are now taking part in open warfare and that … the trench warfare to which they have become accustomed in the past must be dismissed from their minds …'[60] Despite such optimism there was plenty of heavy fighting still to be experienced and 20/R Fus would soon take to the field again.

---

59 *The Gasper*, 30 September 1916, p.6.
60 TNA: WO 95/2408: 33rd Division A&Q War Diary.

20

UPS men in the later Somme fighting, July to August 1916

Hell with the lid off

> *The horrors of war are indeed horrors; those at home cannot possibly conceive what intensities of suffering were undergone in that hell.*[1]
>
> *From now on we lived in hell with the lid off. Everywhere one sees dead men, dead horses and smashed up guns and vehicles …*[2]

The first day of the Somme had seen many ex-UPS officers and men fight their first major battle. Many had not survived and few had survived unscathed. Most officers had departed the UPS before the Brigade left for France and had undergone apprenticeships in trench warfare with their newly adopted regiments. As the fighting progressed there were increasing numbers of ex-UPS men who had been gazetted after serving in France with the UPS battalions and after training by the OCB system. By the end of the offensive, it was this breed of ex-UPS officer who were commanding platoons and even companies. Throughout the campaign, ex-UPS officers and men contributed by their actions, by their leadership and sometimes by their death or wounding, to helping erode and force back the German Army. This section will use case studies to describe the experiences of these officers and men but it cannot do justice to all of the participants.

8/R Fus on 7 July

The men sent from 18/R Fus to join 8/R Fus, 12th Division, were next to see action. This division attacked Ovillers and 8/R Fus was to capture the enemy front and support lines and the church. The advance received heavy German MG fire from the right flank and the attacking waves suffered heavy casualties, especially amongst officers. The battalion captured and held the German positions but lost twenty-three officers and 640 men. Lieutenant Alban Arnold,

---

1   Fusiliers Museum: RFM.2013.8.1/2: Diary of P.N. Wright, p.46.
2   TNA: CAB 45/132: Letter by The Hon Bede Clifford to the Official Historian, 30 May 1934, pp.485–486.

formerly 18/R Fus, was killed on 7 July 1916 in a heavy barrage of artillery and MG fire.³ He was a keen footballer and had been a county cricketer. One of his men remembered him: 'Mr Arnold always keeps goal for the 18th – he is the son of the vicar at Fareham – an awfully nice fellow & a first rate goalkeeper…'⁴ Another UPS officer with 8/R Fus was Lieutenant Henry Franklin formally of 21/R Fus. He was aged 32, from Twickenham, and was a civil servant. He had married in January 1916, arrived in France in April, and was killed at Ovillers. His widow gave birth to a daughter four days after his death. At least sixteen former UPS-men were killed in this action. 9/R Fus also embarked on this attack. Lieutenant Frank Street, aged 46, formerly of A Company, 18/R Fus, was killed in the process along with two other ex-UPS men.⁵ According to his school, 'When the war broke out he was a master at Uppingham, and lately married. Despite his age, he felt it his duty to go to the war, and the loss of so excellent a man will be widely lamented.'⁶

Early in the morning of 14 July 11/R Fus was placed on the alert to go forward and support the attack on Trones Wood. Private Philip Wright recalled:

> … a party – myself included – of us were detailed to carry up bombs to the Middlesex in Trones Wood. We … entered the sunken road in front of our trench, went along so far until we reached a spot nearest to and directly behind our objective. … From this spot we intended making a dash with our bombs through the Hun barrage to Trones Wood. Dodging shells, it may be mentioned, is not nearly so amusing as – say – dodging snowballs. Pieces of shrapnel were literally raining down and shells fell unpleasantly close. We were much relieved to reach the wood with our bombs. And there, what terrible sights met our eyes. The wood was like a ghost of its former self. Trees stripped bare of bark and leaves with huge limbs torn off … And in the trenches – helpless wounded men lay mingled with the dead crying out for succour … The horrors of war are indeed horrors; those at home cannot possibly conceive what intensities of suffering were undergone in that hell.⁷

Another talented and literary UPS man, who had contributed to The Pow-Wow, was Raymond Walter ('Raya'). He was a second lieutenant with 10th Battalion Cheshire Regiment (10/Ches R) having been posted to a cadet school at St Omer in April 1916 and was gazetted in June. His career as an officer was brief. On the night of 14–15 July 1916 the battalion made an attack on trenches south of Ovillers. An observer on the brigade staff recorded what occurred 'at 2 a.m. we had another try; this time they were waiting for us and got their machine guns on to our fellows as soon as they started. The poor 10th Cheshires tried hard, and I wish they could have brought

---

3 Alban Charles Philias Arnold, aged 23, attended Twyford School near Winchester, Malvern College and Magdalene College, Cambridge where he was awarded honours in Modern and Medieval Languages in 1914. He was the wicket keeper for his school and a cricket blue at Cambridge. Arnold also played for Hampshire and enlisted with 18/R Fus in September 1914.
4 IWM: Documents 1708, 1985-11-05: Papers of W.B. Medlicott, p.219.
5 PS/5838 Private Frank Green, aged 32, formerly of 18/R Fus, was killed with 9/R Fus and is commemorated on the Thiepval Memorial. PS/8609 Private Reginald Elson, formerly of 19/R Fus is buried in Ovillers Military Cemetery.
6 Street was educated at Westminster School and Uppingham School, in 1892 he was captain of the OUAFC and was later a master at Uppingham.
7 Fusiliers Museum: RFM.2013.8.1/2: Diary of P.N. Wright, p.46.

it off as they want something to buck them up a lot …'[8] Raymond Walter was killed in action on 15 July when the battalion lost 14 officers and 386 men.[9]

After losing 282 men attacking la Boisselle on 3 July, 8th Battalion Gloucestershire Regiment sorely needed reinforcements. On 11 July they received a draft of 67 men from the Royal Fusiliers which included members of 20/R Fus. No sooner had they joined than 8/Glosters were needed for another phase of the offensive. On 22 July the battalion prepared to attack the German switch line north of High Wood. This night attack failed, and five ex-UPS men were killed. A survivor wrote to PS/4453 Private William Balmforth's family: 'Your son was in my section. I last saw him about 10 a.m. He had been wounded with a bullet through the upper part of his left leg. We bandaged him up and then told him to get out to the medical officer the best way he could, leaving him … in a shell-hole …'[10] Balmforth was recorded as 'missing'.[11] The battalion attacked again on 30 July which cost the lives of another six ex-UPS men.

On the night of 23 July 2/KRRC was involved in an attack near Pozieres at 12:30 a.m.. The battalion took the first line of enemy trenches and held them for several hours. Pelham Ravenscroft, the Lewis gun officer, recalled an 'appalling time'. Sat in a trench the whole night waiting for the next shell … The trenches were blocked with other Regiments who had lost their way and wounded all the time the show was on. I spent most of my time binding up people … Very bad show indeed. Much too hurriedly prepared…'[12] Frank Walker's luck ran out. His last moments were observed:

> [We] … took the German front line which was separated from our own by about 600 yards. We held it for three hours and then were forced to withdraw. Witness saw Capt. Sherlock and Mr Walker lying within about 6 yards of each other, close to a shell hole between the opposing lines. … Mr Walker, who was … seriously wounded, refused to be moved until the others, particularly the Colonel, were all taken in. Witness did not finish bringing in the other men before daylight, when it was impossible to reach the two officers …[13]

Walker is commemorated on the Thiepval Memorial. One of his men, Giles Eyre, already a veteran, constantly mentioned Walker's paternal presence and bravery.[14] Eyre recorded one of his comrades stating; 'There ain't no one in the Batt. like Mr Walker, and you can swank as much as yer likes. We knows 'im and wouldn't swop 'im for nuffink. Any'ow he wouldn't own such a bloody measly lot of coves as yer platoon is!'[15]

8   Edwin Astill, *The Great War Diary of Brigadier Alexander Johnston 1914–1917* (Barnsley: Pen and Sword, 2007), p.175.
9   PS/5888 Private Raymond Walter enlisted, aged 23, at Manchester in September 1914. A salesman from Chorlton-cum-Hardy, he served with C Company, 20/R Fus and was friendly with James Hodson who described him as 'an abiding joy' and that he was one of those 'who are sleeping but who will never die.' Walter is commemorated on the Thiepval Memorial.
10 *Manchester Evening News*, 16 August 1916.
11 PS/4453 Private William Balmforth, aged 20, was from Rusholme, Manchester. He was educated at Manchester Grammar School and worked for a cotton spinning company. He was the son of the editor of the *Manchester Evening News*.
12 Ravenscroft, *Unversed in Arms*, p.45.
13 TNA: WO 339/39097: F.B. Walker Officer File.
14 Eyre, *Somme Harvest*, p.233, p.249.
15 Eyre, *Somme Harvest*, p.168.

'... hell with the lid off ...'

If 20/R Fus had a difficult baptism of fire on 20 July, 22/R Fus, and its sister battalion, 23/R Fus, had an atrocious time at Delville Wood. Amongst the ex-UPS men posted to 22/R Fus (2nd Division) was Harry Harvey of 18/R Fus who chronicled his experiences:[16]

> "Stand to! ... [Germans] Coming over on the left!" The shrapnel barrage overhead slackened; grey patches moved through the shrubs fifty yards off; stabbing bursts from Lewis guns; rifle-fire crackling like enormous rattles; cries and curses; drifting smoke – then – for a space – a quietness! ... Nightfall, with its perpetual barrage, and the brilliant light of falling Verey flares, ghostlike between the havoc of the wood. Again the rosy flickering flames of burning Longueval through the lattice of shattered tree trunks. Wounded crawled away. Dead, some few of them, hurriedly buried. Dawn! ... Aching yes, wearied bodies, minds distraught! ... Dead on all sides, stark, ashen, horrible![17]

An officer wrote in his diary; 'From now on we lived in hell with the lid off. Everywhere one sees dead men, dead horses and smashed up guns and vehicles. Our gunners are splendid ...The roar and din of guns is terrific.'[18] 23/R Fus suffered heavy casualties at Delville Wood on 27 July of which eleven ex-UPS men were killed and two more died of their wounds; most were from 21/R Fus. Amongst them was PS/266 Company Sergeant Major Otto Ditzen, aged 32, from Ealing. He had been MiD in June.[19] PS/2474 Private John Gilchrist Burnie, aged 24, was educated at Shrewsbury School and St John's College Oxford. Though he was strongly recommended for a commission he was rejected due to his eyesight.

August 1916 – 'Over you go my lucky lads...'

A draft of 29/R Fus, which included Private George Cooke, was posted to 21/NF. In a further change this draft was attached to 13/DLI (23rd Division). Cooke went into an attack with 13/DLI on 4 August 1916 near Contalmaison:

> When we went over [the top] we met stiff opposition and had to retire. As we got back we had to take up positions on the fire step and were given the order to give rapid fire. ... later we had the order to attack again, as we were proceeding I stopped by the largest shell hole I had seen ... I had my right leg forward, then I felt a terrible knock on my thigh, I put my hand on my knee and my hand was covered with blood. I managed to get back to the trenches and call for help, an RAMC man was there in no time...[20]

Cooke was one of the 'lucky' ones; four other ex-UPS men were killed with the same battalion.

---

16   22/R Fus received about thirteen men of 19/R Fus, about eight from 19/R Fus and six from 21/R Fus;
17   Harvey, *Battle-Line Narratives*, p.105.
18   TNA: CAB 45/132: Letter by The Hon Bede Clifford to the Official Historian, 30 May 1934, pp.485–486.
19   Ditzen was buried in Sucrerie Military Cemetery, Colincamps.
20   Liddle Archives: WW1/WF/01/C11: Papers of G.K. Cooke, p.6.

PS/1338 Private Alfred Ridsdale Carr enlisted with 18/R Fus, aged 20, on 3 September 1914. Educated at Highgate School and a student of cloth manufacturing, he was gazetted to the RSF in May 1915 and was captured on 12 August 1916 with 6/7 RSF. Carr was repatriated on 21 January 1919 and was demobilised shortly afterwards. (Reproduced with permission of Sutton Archives)

On 3 August, 8/R Fus attacked Fourth Avenue, northwest of Pozieres. PS/10149 Private Albert Clayton had been posted to 8/R Fus from 29/R Fus and recalled:

> Sergeant Downes, the father of his platoon, chanted "Over you go my lucky lads," as if he were running a crown and anchor board, and we climbed out. I wasn't sorry either to get out on top, for the trench had become a death-trap during the last few minutes, when the German barrage fell on it. The night was black … The darkness of the veil of night was intensified rather than relieved by the stabbing flashes of bursting shells, and their explosions crashed so close around me that I was continually astonished to remain alive. I saw figures in dimly in the enveloping gloom, staggering and stumbling about. Whether they… were hit, or because they had tripped over the entangling wire, I didn't know…[21]

The battalion lost 150 casualties including eleven UPS men killed. Next day 9/R Fus attacked Ration Trench along with bombing attacks by 8/R Fus. The two battalions lost another sixteen UPS replacements.

On the night of 12–13 August Lieutenant Alfred Carr of 6/7th Battalion Royal Scots Fusiliers (6/7 RSF) attacked at 10:30 p.m. between Pozieres and Martinpuich. The company took its objective but, according to Carr:

> … the platoon of which I was in command was detailed to move forward 50 yards as a covering party during the consolidation of the objective. This move was duly carried out. On visiting the right flank I found the flank in the air and touch [had been] lost with the right companies of the battalion. I set out to try and establish touch with the right and had proceeded about 100 yards from my right flank when I was surprised by a party of five of the enemy and was taken prisoner.[22]

21 Albert Clayton, *Long Before Daybreak* (Unknown: M.J. Duckworth, 2020), p.46.
22 TNA: WO 339/47984: A.R. Carr Officer File.

PS/1258 Lance Corporal Robert Harold Beckh of Haileybury and Jesus College, Cambridge, enlisted in 18/R Fus on 2 September 1914, aged 20, along with other university friends. He had intended to become a clergyman.

A UPS War Poet Killed

It was not only on the Somme that UPS men were killed during the summer of 1916. On 15 August 1916 the war diary of 12th Battalion East Yorkshire Regiment (12/E Yorks) recorded; 'A patrol consisting of 2/Lieut R.H. Beckh and four O.R. left our lines to inspect the Ferme Cour Lavoué [sic, 'd'Avoué' near Festubert] … Finding the place strongly held, and being fired on, they retired, but again struck the German lines. A machine gun opened fire on them, and 2/Lieut Beckh and one OR were killed…'[23] In the weeks before Beckh had written the following poem:

> *Nine-Thirty o'clock? Then over the top,*
> *And mind to keep down when you see the flare*
> *Of Véry pistol searching the air.*
> *Now, over you get; look out for the wire*
> *In the borrow pit, and the empty tins,*
> *They are meant for the Hun to bark his shins.*
> *So keep well down and reserve your fire –*
> *All over? Right: there's a gap just here*
> *In the corkscrew wire, so just follow me:*
> *If you keep well down there's nothing to fear.*[24]

---

23 TNA: WO 95/2357: 12/E Yorks War Diary.
24 Robert Beckh, *Swallows in Storm and Sunlight* (London: Westminster Press, 1917), p.101.

7/KRRC prepared to attack Orchard Trench on 18 August. Charles Whitley commanded a leading company and Edmund Page led the company in support (both formerly 21/R Fus). Page recalled; 'The bombardment and retaliation were terrific. Charles led his company with two pistols and a stick with wonderful gallantry. A piece of shrapnel the size of the top of a fountain pen went straight through his left forearm before the attack started, but we could not persuade him to go back. His company did extraordinarily.'[25] Whitley's bravery was an excellent example of the leadership by ex-UPS men serving as regimental officers during the Battle of the Somme.

The same day, 3/RB, including Andrew Buxton, attacked at Guillemont:

> The trench we were in was far from pleasant as it was only partly dug, and our men were digging it deeper all they could … We got over well and took a certain number, about 16, prisoners, even though the other Company had been some time in the Bosch line. We then dug hard to make the German trench good to hold. It had largely been levelled by our shells, but still had several deep dug-outs. A lot of Bosch bolted out to meet our first line as they got to the trench and doubled back towards us all on their own! I think we only just got the trench, as there were a lot of Bosch there. Our shelling was intense, especially the ten minutes before we advanced. Impossible to describe what the noise of possibly thousands of guns all firing as hard as they could on to the Bosch line, both H.E. and shrapnel… the time of the shelling and to keep absolutely up to it are essential, and to know when the shelling 'lifts' on further. The Bosch counter-shelling was very heavy, but nothing like ours. I could not make anyone hear a word except by shouting in their ears at the top of my voice … my whistle only reached a few yards![26]

He confided in his diary that; 'Frequently our own shells were doing in our own men.'[27] Robert Vernede of the same battalion was left behind with the transport whilst two of his brother officers were sent into action; both were killed. When officer casualties required Vernede to go forward he had to borrow a revolver as he had lent his to one of the dead officers. On arriving in the captured positions Vernede recollected:

> I found C Coy at last. H.Q. in a 30ft. deep Boche dug-out, choked with dead Germans and bluebottles, and there we had our meals till we started back at 4 a.m. this morning (five days). In between that time I certainly spent some of the most unpleasant hours of my life. It seems that the Batt. had done extraordinarily well and gained the first of two objectives. The second was to be won that night and next day we were to be relieved. Unfortunately a Batt. on our right had been held up and we had to wait for them in a trench choked with our dead and Boche wounded and dying for two days and then do another attack … I had never seen them [his men] so glum. … The push itself is done in hot blood: but the rest is horrible, digging in when you are tired to death, short rations, no water to speak of, hardly any sleep and men being killed by shell-fire most of the time.[28]

The Battle of the Somme pushed officers like Vernede and Buxton to breaking point.

25   IWM: Documents 16924: Papers of E. Page.
26   Woods, *Andrew Buxton*, pp.211–212.
27   Woods, *Andrew Buxton*, p.206.
28   Robert Vernede, *Letters to his Wife* (London: Collins, 1917), pp.166–167.

Air War over the Somme

Many UPS men were commissioned into the RFC in the summer of 1916 either as pilots or as observers. The activities of two brothers (William and Kelvin Crawford) formerly B Company, 18/R Fus, provide a case study of the UPS involvement in the skies above the Somme. In early October William was posted to France and to No 24 Squadron RFC commanded by the famous Lanoe Hawker VC. Kelvin followed on 16 October. Ten days later, at 8:00 a.m., Kelvin was engaged in an aerial combat and was slightly wounded early in the action with some of his controls damaged. He continued to patrol and managed to fire a half-drum of ammunition into a German aircraft from close range causing it to drop 2,000 feet out of sight.[29] He was credited with having shot down a *Halberstadt* near Bapaume. On the morning of 17 November the two Crawford brothers took off in a group of six DH2 scouts, from No 24 Squadron RFC, escorting other aircraft near Bapaume and Achiet. The patrol was attacked by six enemy aircraft and one was sent gliding home by the British flight commander. There was a further offensive patrol in the afternoon involving the Crawfords and three others, again to Bapaume. A hostile machine was spotted over Grevillers Wood at 2:25 p.m. and the patrol attacked. Kelvin Crawford and two others fired on the aircraft and drove it off. Later two of the patrol attacked a further enemy aircraft and forced it to dive emitting smoke. William Crawford was flying a DH2 during both forays and was reported missing after the second patrol. Later, the Red Cross reported that William was killed during an air combat.[30] He was likely the sixth victim of *Leutnant* Otto Hohne of *Jasta 2* around Bapaume–Warlencourt.[31]

September 1916 – 'he immediately proceeded to attack Germany all on his own …'

The offensive planned for 15 September 1916 was to advance the line using tanks to support the infantry. To command the first tanks the Heavy Branch of the MGC (HB MGC) needed technically minded and intelligent men. In April 1916 several ex-UPS men were gazetted to commissions in the Motor MGC (MMGC); of the forty-nine tank commanders on 15 September, seven were ex-UPS men. Second Lieutenants Alec Leslie Arnaud, Leonard John Bates, Thomas Frederick Murphy and Jethro Edward Tull were with C Company, HB MGC, and commanded tanks supporting the Guards Division; all four had served with 18/R Fus. Along with five other tanks they advanced in three groups. Tull's tank had to withdraw early on having suffered a smashed set of driving wheels. Arnaud's tank also ditched, near Ginchy, whilst driving through the darkness. Murphy's tank broke down and by the time he had got a replacement he was unable to advance with the Guards Division. These issues were teething problems associated with bringing a new weapon onto the battlefield and cannot reflect badly on the tank commanders involved. Meanwhile, Bates' tank reached the second objective, engaged several MG posts and had helped to capture numerous prisoners. Compared to the rest of the ex-UPS tank commanders of his section, Bates had had a successful day. Francis James Arnold,

---

29 TNA: AIR 1/911/204/5/836: Air Combat Reports 1916.
30 William Charlton Crawford, aged 23, is commemorated on the Arras Flying Services Memorial.
31 *Kelvin Crawford* <http://www.theaerodrome.com/aces/england/crawford.php> (accessed 18 February 2018).

formerly of 19/R Fus, was also with C Company. His tank, Tank C14, No 509, supported 56th (1st London) Division. On reaching the German front line trench his tank did great execution on the garrison. However, due to a last-minute change in his orders, which he likely misinterpreted, Arnold pressed on beyond this trench, 'he [Arnold] did not wait for the infantry to consolidate the point that he was dominating ... Having shot up the Huns on their front line he immediately proceeded to attack Germany all on his own.'[32] Finding no infantry were following him Arnold's tank returned later to find them. His tank also assisted in breaking up a German bombing attack but became stuck fast near Beef trench and was irretrievably lost. It was not Second Lieutenant Storey's day either; his tank of D Company was ditched near Delville Wood before being able to get into action.[33] He provided no assistance to 14th (Light) Division but proved himself whilst commanding a tank in a raid at Guedecourt a few weeks later according to his DSO citation; 'For conspicuous gallantry and initiative when in command of a Tank. He took his car up and down the enemy trenches, working until all his petrol was exhausted and only two of the crew were un-wounded. He is reported as having been responsible for the taking of between 200 and 300 prisoners.'[34]

20/R Fus had fought hard for High Wood earlier in the Battle of the Somme. Second Lieutenant Andrew Henderson (formerly 18/R Fus) commanded Tank C23, alongside three other tanks, in support of the 47th Division attack on High Wood. Henderson's tank entered the wood but became hopelessly stuck having gone barely 50 yards. The crew fought on with their 6-pounder guns and MGs to good effect. Despite wasting the effectiveness of their tank support, 47th Division still managed to secure High Wood. Once secure, the task of clearing the masses of dead within its borders, including those of 20/R Fus, could begin. Meanwhile, 20/R Fus had moved beyond High Wood and would be engaged elsewhere on the Somme. Furthermore, the ex-UPS men still had many other engagements to fight during autumn of 1916.

---

32   Trevor Pigeon, *The Tanks at Flers* (Cobham: Fairmile Books, 1995), p.67.
33   Educated at Grocers' Company School and Cranleigh School, Storey was employed at a railway stores facility in Argentina. He had formerly served with 18/R Fus.
34   *The London Gazette*, 11 May 1917, p.4588.

21

The Rest of the Somme

Festubert compared with this was a garden party, and High Wood a night out

---

High Wood might be considered the 'high water mark' for the effectiveness of 20/R Fus but it was not the end of the Battle of the Somme for the battalion. The subsequent phases that concerned 20/R Fus must also be described. This was a period when 20/R Fus had to face the strain of further heavy fighting, albeit interspersed with periods of rest. In the background, the losses at High Wood meant that both the Mancunian and Public Schools aspects of the battalion were in decline. As the battalion became more heterogeneous so the ties holding the remaining 'gentleman rankers' to the battalion were loosened and those that departed increased. 20/R Fus increasingly became a Royal Fusiliers service battalion with increasing links to London within its ranks. Godfrey Skelton recalled:

> I had applied for a commission several times before this but with no result, as the Commanding Officer, an old Regular soldier, Colonel Bennett, refused to allow any men to go, especially the N.C.O.s, pleading that it would affect the Regiment's efficiency, but my application was still on the files at Brigade Headquarters.[1]

Commission applications for original members of the battalion were dusted-off and re-submitted as some chose to re-apply. Some 'originals' left during the Autumn for OCBs in the UK. However, with the Battle of the Somme raging, there was a risk for the men involved that further fighting might consume them before their commissions came through. These months represented, for some, a race. This was a race between the grinding progress of the Battle of the Somme, and its voracious appetite for casualties, and the slowness of Army bureaucracy. Some would be winners and others would lose in the coming months.

This period saw a new wave of 20/R Fus personnel learning, developing and coming of age as soldiers in and amongst the Somme fighting. New platoon commanders would find their feet and newly joined personnel, be they returning veterans, Derby men or conscripts, would also develop as soldiers to carry forward the name of the battalion as the 'originals' became

---

1 IWM: G. Skelton Papers, 13966, p.51.

Lieutenant Colonel William Brooksbank Garnett became the next Commanding Officer of 20/R Fus. He was aged 41, had attended Charterhouse, and worked for the Inland Revenue until 1899. He served part-time with the 24th Middlesex Rifle Volunteers and went to South Africa in 1899 with the City Imperial Volunteers. Garnett acquired a taste for soldiering and joined the RWF in 1900. By 1913 he was a captain and the Adjutant of 7/RWF having served in China, India and Ireland. He went to France with 1/RWF in November 1914, married in July 1915, and was MiD in January 1917. He was awarded the DSO in June 1917 and became a brigade commander in May 1918.

scarcer. In this way, the remainder of the Battle of the Somme saw the character of the battalion morph and change. The number of men present to chronicle the battalion was also in decline.

Battalion training continued until 6 August when 20/R Fus marched to a bivouac near Bécordel and Méaulte to act as divisional reserve. According to Cecil Isom his company bivouacked on a hill near Fricourt where they had to make little shelters using ammunition boxes; despite this seemingly Spartan billet Isom considered himself 'quite comfortable'. At 1:00 a.m. next day Isom spent his first occasion under fire as he, acting as a stretcher bearer, accompanied a digging party up to Mametz Wood where plenty of German shells were coming over. 19th Brigade remained in reserve from 7 to 12 August.

On 10 August, Major W.B. Garnett arrived from 1/RWF to take command of 20/R Fus. In the absence of Colonel Bennett a replacement was needed, and Garnett was clearly an eligible officer. His arrival coincided with the first rainy day since late July. He had attended Charterhouse School, had served in South Africa and was an experienced infantry officer. However, he was unpopular in 2/RWF when he later served with that battalion. Garnett was popular with the 'originals' of 20/R Fus. According to Hodson; 'They had a new colonel, who, supposing they were an original Public Schools battalion man … would sign you a chit that empowered you to buy a bottle of whiskey. He had a soft spot for them, and they admired him enormously. He wore a monocle, and they referred to him as "Gus." "Decent old stick, Gus," they said…'[2] Bennett's decision not to despatch more men for officer training before High Wood had likely diminished his reputation amongst the original UPS men.

Back into Action

On the night of 12–13 August 20/R Fus handed in their packs at Fricourt and marched towards the trenches in fighting order. The battalion returned to Mametz Wood where it relieved 2/A&SH in support. The relief was complete by 4:00 a.m. on 13 August and the battalion was in dug outs. The next few days were uneventful with the battalion providing working parties. On 18 August 20/R Fus was attached to 98th Brigade to support their attack and moved up to Crucifix Corner and awaited orders at 3:00 a.m. The plan was for 98th Brigade to attack Wood Lane at 2:45 p.m. At 4:20 p.m. orders were received to send up two companies to support 1/4 Suff R and 4/King's. An hour later the other two companies of the battalion were ordered forward and relieved 4/King's alongside one of those companies sent up earlier. The fourth 20/R Fus company attached to 1/4 Suffolks was delayed. Luckily Cecil Isom stayed at the Regimental Aid Post which was sited at Crucifix Corner.

After a night in the front line the battalion was relieved by the Cameronians at 4:00 a.m. on 19 August. The Fusiliers, once free, returned to Crucifix Corner where they remained all day. That evening 20/R Fus returned to Mametz Wood and rejoined 19th Brigade. On the night of 20 August Cecil Isom was with a working party in High Wood. Whilst he was away the German artillery bombarded Mametz Wood for four hours with gas shells which caused several gas casualties. On this occasion Isom's party may well have not bemoaned their fatigue party.

At 10:00 p.m. on 21 August 20/R Fus again moved two companies forward in preparation for helping an attack by 100th Brigade, if necessary. They took over the right of the line from 5/Scot Rif but were not used in this action as the 100th Brigade attack failed. However, Isom recorded that they received a 'warm reception', presumably from the German guns rather than the Jocks. At 8:00 a.m. on 22 August the remaining two companies arrived and took over the rest of the Scottish Rifles positions. The battalion was holding from the right edge of High Wood to 700m to the southeast; halfway between High and Delville Woods. During the day Lieutenant Eric Alfred Walker was killed.[3] An observer at divisional headquarters recorded some of the problems encountered with these attacks:

> Aug 22. We had a small attack last night & failed again. These repeated failures are really depressing. There may be many reasons for it but one is that we are constantly having to take over new ground & attack from there instead of attacking from ground we have got to know. The artillery complain of the same sort of thing. Another reason is that I am afraid

---

2 Hodson, *Grey Dawn*, p.268.
3 Lieutenant Eric Alfred English Walker, aged 24, from Teddington, is buried in Thistle Dump Cemetery, High Wood, Longueval, having been killed on 22 August. He was born at Buenos Aires and was educated at Bedford School where he was a member of the OTC for four years. Walker was an engineer by trade and was a member of Twickenham Rowing Club. He enlisted in September 1914 and served as a corporal with D Company. He was gazetted to a commission in 20/R Fus in April 1915. Walker was promoted to lieutenant in March 1916 and attended a telescopic rifle course suggesting he was employed overseeing battalion snipers.

the morale, of our division at any rate, is not what it was. Living for days & weeks under appalling shell-fire, unsuccessful attacks, heavy losses especially in officers, all these sap the determination of the men. We came down here one of the finest divisions in the Army & now we don't seem able to accomplish anything. … The 33rd Div has lost its reputation. It is rather sad.'[4]

On 24 August C and D Companies of 20/R Fus were relieved by two companies of 5/Scottish Rif in order that they were freed up to support an attack by 1/Queen's RWS of 100th Brigade. The Queens were attacking New German Trench at 5:45 a.m. Two platoons were sent forward to occupy part of the trenches taken over by the Queen's and they took over a bombing post at the corner of Wood Lane and New German Trench S11a33. They also began to dig a trench to connect this newly captured point with the British 'Sap A' at S10d95. This new communication trench was completed during the night of 25–26 August and the CO of 20/R Fus reported:

> … that the new trench was completed from S10b9.4 to S11a3.3 – Its total length is 300 yards – Its depth varies from 4ft to 6ft with the exception of 25 yards which consists of a chain of holes 2ft deep – In the deeper part of the trench a few sentry posts have been made. The parties working were hampered considerably by M.G. fire & Rifle fire & intermittent bursts of shrapnel & two of the officers in charge were unfortunately killed – The trench begun the previous evening did not give as much assistance as was hoped and it was found that the nearest points of the two trenches already begun were 210 yards apart. The work will be vigorously pushed on tonight from SAP A cut.[5]

Lieutenant Humphrys was deeply involved in running this enterprise but was killed in the process.[6]

Around this time in the trenches Hodson recorded being in a sap about 20 yards from the German positions on the right of High Wood; 'It was a hot place, but he ['Hardcastle'] was … sitting crouched in the bottom of the trench. Yedborough shouted, "Wake up, soldier. Green lights!" Green lights were the enemy's S.O.S. appealing for artillery support. Down came the barrage. In the middle of it Hardcastle felt his left ear stinging … But the wound was too small even to be dressed.'[7] Hodson's disappointment at not receiving a 'Blighty' wound on this occasion was palpable.

At 5:00 a.m. on 26 August 20/R Fus was relieved by the Scottish Rifles and returned to Mametz Wood; relief was complete by 7:00 a.m. After resting the remainder of the day at 7:30 a.m. next morning 20/R Fus moved to Montauban Alley where they, and the Cameronians, came under 98th Brigade. The day after Ashley Gibson got back from convalescence he went to billets in Montauban Alley; 'That night it rained, and there were no dugouts in our part of Montauban Alley, only pits in the ground, in which the surface flood gathered steadily. The

---

4   IWM: G.W.G. Hughes Papers, 12059, pp.41–42.
5   TNA: WO 95/2420: 19th Brigade War Diary.
6   Second Lieutenant Stewart Francis Humphrys, aged 24, from Croydon, was killed on 26 August 1916. He is buried in Thistle Dump Cemetery, High Wood, Longueval. His bravery was recognised with a Mention in Despatches in January 1917, four months after his death.
7   Hodson, *Grey Dawn*, p.269.

Boche seemed to know the spinney on our front had something behind it, and signified his anxiety in the usual manner.'[8]

28 August was spent on working parties. According to Gibson not all the men of the new drafts were up to the task. One carrying party was:

> ... Too exciting ... for some of the new draft. There was a gentleman who, misliking the turn Fritz was giving to things in the middle of one of these adventures, threw down the bundle of wiring material he was carrying and invited unmentionable outrage if he bore it a yard further. There was another alternative, which the sergeant-major saw, and took. Others were wavering, and the moment was critical. He drew his revolver and saved one mutineer a court-martial.[9]

Such an account may be exaggerated. It has not been possible to identify the name of the executed man, nor the sergeant major in question.

On 29 August 20/R Fus was relieved by 2/RWF and moved to Fricourt Wood. On 31 August 33rd Division was relieved by 24th Division and 19th Brigade moved back to Ribemont where it was billeted. During this tour in the line in August 1916 20/R Fus lost two officers killed, two wounded at duty, 23 men killed, 137 wounded, including 19 who remained at duty, and eleven men missing.[10] These were light casualties by comparison to other battalions in the maelstrom

---

8 Gibson, *Postscript to Adventure*, p.144.
9 Gibson, *Postscript to Adventure*, p.144.
10 The following men of 20/R Fus were killed or died during August 1916 (unless otherwise stated they are named on the Thiepval Memorial): PS/4950 Corporal James Sainsbury Harwood, aged 27, from Whalley Range had been born in Rawalpindi, India, and was the son of a bandmaster and was a member of the Halle Choir, Manchester. He died of wounds on 3 August and is buried at St Sever Cemetery Extension. PS/5214 Corporal George Litherland, aged 21, was from Sale, Cheshire. He died on 10 August 1916; his body lies in Caudry Old Communal Cemetery. SPTS/4803 Private Clarence Reid was killed on 16 August. PS/5491 Private William Harold Quarry, aged 25, Belfast, Ireland, was killed on 17 August and was interred in Flatiron Copse Cemetery. GS/6178 Private Samuel Greenland was killed on 17 August and is buried at Millencourt Communal Cemetery. GS/63505 Private Harold Fairchild was killed on 17 August. PS/4795 Private Frederick William Ekin, aged 23, from Belfast, was educated at the Royal Belfast Academic Institution and had undergone a year at Queens University where he had joined the OTC. He served with D Company, No 13 Platoon. He was previously grazed by a whizz-bang on 17 May 1916, suffered a wound to his arm on 20 July and died on 18 August in Edinburgh War Hospital, Bangor; he is buried in Belfast City Cemetery. PS/5969 Private Ronald Woolfenden, aged 21, from Reddish, Stockport, attended Stockport Grammar School and Victoria University, Manchester. He enlisted in September 1914. He was awarded the MM in June 1916 for bravery whilst attached to 251st Tunnelling Company (gazetted 10 August 1916), likely for investigating a tunnel in a recent crater. He was killed on 18 August and is buried in Daours Communal Cemetery. GS/21568 Lance Corporal Charles Squires was killed on 18 August; he is buried in Flatiron Copse Cemetery. PS/7068 Private Ernest Taylor was killed on 19 August. PS/2520 Private Robert Phillip Carr, from Derby, had enlisted with 21/R Fus. He was killed on 20 August and lies in Flatiron Copse Cemetery. PS/2803 Lance Corporal Ivan James Howell, aged 20, Walsingham, Norfolk, and 2765 Private George Gibbons, both died on 21 August. Both men were buried in Millencourt Communal Cemetery. PS/4696 Sergeant Harold Edwin Lancelot Colville, aged 23, from Stockport and PS/5091 Private Herbert Richard Hughes, aged 22, from Blackley, Manchester, were killed on 23 August. Colville died of his wounds and Hughes was killed in action. PS/8044 Private Bryan Heath, aged 19, from Worcester, had attended Worcester Royal Grammar School (1903–1913). He worked at the Ledbury and Evesham

of the Somme. However, they were significant for a battalion that had not been engaged in any major attacks and which was trying to rebuild itself after heavy casualties in July.

Gommecourt – Cushy Trenches

On 1 September 20/R Fus moved to billets in Molliens-au-Bois via a relatively relaxed march. The men's packs were carried on the transport wagons and the battalion stopped for lunch in a field at 1:00 p.m. for an hour. Despite marching 10 miles during the day the battalion arrived at Molliens at 3:30 p.m. Their destination next day was Bernaville, and they rested there on 3 September. Next morning the brigade moved onwards to Bonnieres and on 5 September moved further to Oeuf with Croisette being the destination on a subsequent day. These marches steadily built up from carrying fighting order to start with to carrying full pack by the end of these moves. At Croisette the battalion had another day's rest. Over the next few days 20/R Fus moved via Sibiville and Le Souich before arriving in Bienvillers-au-Bois and Pommiers on 10 September. Isom recorded on 10 September; 'Marched to Bienvillers, 18 mls [miles] F.P. [full pack] 1,200 yds to line. Billeted in barns. Very quiet, no shelling, in Reserve. No rest for the wicked.'[11] After a further day in Pommiers the battalion prepared for the trenches. On the night of 12–13 September 20/R Fus relieved 8/S Staffs (17th Division) in Z Section at Hannescamps. This was complete by 10 p.m. 20/R Fus held from E11b06 to E17c14 – trenches Z62 and Z75. Three companies were in the line and one was in reserve. The Cameronians were on the right and 46th (North Midland) Division was on the left. According to Ashley Gibson:

> When they pushed us in again, it was at Gommecourt, become by some chance of battle a "cushy" refuge … Our first line was no more than a breastwork in the chalk, behind it the troops could stroll about on the grass and pick buttercups. Even in front No Man's Land was 700 yards across, a country of undulating meadow, dotted with bramble patches and an exciting osier bed where we were sure the Boche snipers lurked. Sometimes our patrols tried to put this matter to the test but had no luck. The best nocturnal fun I had was snipping samples from the enemy's wire. They were trying a new brand, which bothered our artillery.[12]

---

branches of the Capital and Counties Bank. He died on 23 August from wounds and is buried in Puchevillers British Cemetery. PS/8736 Private Alfred Henry Guy Sadd, aged 20, from Bishop's Stortford, attended the University of London. He was a Civil Servant with the Inland Revenue Office. He served with B Company, 20/R Fus, and died on 24 August from wounds and is buried at Heilly Station Cemetery. PS/4899 Private Percy Gray, aged 23, from Sale, Cheshire, and was educated at Manchester Grammar School. He died on 25 August from wounds received at High Wood on 20 July and is buried in Ashton-upon-Mersey. PS/7840 Private Robert Bannister, aged 20, from Moston; PS/7863 Lance Corporal John Thomas Reed; PS/10194 Private Francis James Wood (formerly 29/R Fus); and SPTS/4899 Private Thomas Clark were killed on 25 August. GS/2248 Private Henry Filer died of his wounds on 25 August at Heilly Station Cemetery. PS/7061 Private Harold Robinson, attended Bloxham School (1905–1909). He died at Chatham Military Hospital on 3 August from wounds received at High Wood. He is buried at Chitterne.

11 IWM: C.H. Isom Papers, 16336.
12 Gibson, *Postscript to Adventure*, pp.144–145.

Sergeant Skelton agreed that Gommecourt was 'comfortable' after two experiences near High Wood. This was another area where 'live and let live' reigned. Isom even recorded that he was returned temporarily to his platoon because he was not needed as a stretcher bearer in this sector because it was so quiet.[13] However, one man, GS/63432 Private Sherriff, was killed on 13 September 1916.[14] No man's land was so wide that patrolling was common but patrols were not pushed to near the German parapet. On 15 September Isom went out on patrol for three hours and went to within 150 yards of the German trenches before returning. His party was complimented by Captain Hodgson-Jones, D Company, as no other patrol had been closer to the German lines.[15] According to Hodson; 'The summer of 1916 waned and autumn came … Sometimes the trenches were "cushy," as at Pommiers, where their front line dug-out had wire beds, and where the line was so quiet that each afternoon one of the party slipped down to the village for eggs and oats, and fresh milk and fruit.'[16]

On 16 September 2/RWF took over the sector and 20/R Fus moved back to Bienvillers-au-Bois leaving two platoons in Fonquevillers Redoubt. Edward Chapman, with C Company, also found this a pleasant time:

> Our [company officers'] mess is very jolly. The Padre messes with us. All four of us were at Oxford, and I enjoy our meals immensely. To be among such congenial companions is delightful. One of them, named Moult, took his degree in chemistry at Oxford; In addition, he is an artist and a musician. The other, Gibson, is a journalist. Both are older than I am by several years, and I am called Chapboy in consequence![17]

Working parties kept the battalion busy until 21 September. These included carrying timber up to the line and working on communication trenches. On 21 September 20/R Fus relieved 2/RWF at Hannescamps again. The Cameronians were again on the right and 4/Leics R (46th Division) were on the left. Some patrols were sent out; Isom was out on patrol from 1:00 a.m. to 3:00 a.m. on 23 September and on the 25th he accompanied his company sergeant major to search a disused dugout in No man's land believed to have been used by a German sniper. They found nothing but two nights later Isom, his CSM, a corporal, and another man, waited near the dugout in case the German sniper returned. They stayed out until 5:45 a.m. when it was starting to get light. All they had seen and heard was a German wiring party that had been heard singing and whistling. Interspersed with these night patrols Isom and his chum were allowed to walk out of the trenches into Bienvillers-au-Bois to get food for friends.[18]

This stint in the line ended on 27 September with 2/RWF again taking over and 20/R Fus going again to Bienvillers. This quiet pair of trench tours cost 20/R Fus one man killed, two officers and eleven men wounded. The officers wounded were Second Lieutenants Frank Cannell and Frederick Stocker.

---

13   IWM: C.H. Isom Papers, 16336.
14   GS/63432 Private William Sherriff, formerly 1/Londons, joined 20/R Fus on 25 July 1916. He is buried Hannescamps New Military Cemetery.
15   IWM: C.H. Isom Papers, 16336.
16   Hodson, *Grey Dawn*, p.266.
17   IWM: Documents 1799, Catalogue Date 1992-02: Papers of Captain E.F. Chapman, p.24.
18   IWM: C.H. Isom Papers, 16336.

A Divisional 'Diversion' – Rest and Training at Doullens

On 28 September 20/R Fus moved from Bienvillers to Humbercamps where it was billeted. Ashley Gibson summed up the move on 30 September:

> To Doullens we came, marching, that last day, twenty-five kilometres to get there. Grinding up the long hill between the poplars, we realised we weren't so tough as we had been, but from the crest we saw this was a land of promise and picked up the step like heroes. Billets were all they should have been, and the Four Sons of Aymon a tavern without peer and without reproach. The shops stocked every luxury. Deserting their own canteens, the troops ran riot among the patisseries, and in the intervals of kit inspection spat freely and promiscuously the seeds of fresh pomegranates.[19]

At Doullens 20/R Fus enjoyed a further ten days away from the line during which time the battalion trained. This included time spent on the ranges and a revolver range took place in the moat of the citadel on 8 October. According to Chapman; 'Yesterday we had revolver practice for officers and certain N.C.O.s. One … N.C.O. let off his weapon too soon and shot a B Coy officer through the calf! The bullet then grazed my puttee, without tearing it. It was only a very slight wound; but it is a serious thing to lose an officer at the present moment.'[20] That officer was Lieutenant Algernon Coggin. According to Ashley Gibson Coggin had a 'hole in his calf that he got sitting next to me in the old moat at Doullens, watching some rather hectic revolver practice.'[21] The range was being run by Lieutenant Wylie, the Lewis Gun Officer (LGO) assisted by Robert Armour, the Regimental Sergeant Major. The unfortunate soldier that shot Lieutenant Coggin was G/24670 Private George F. Coates who was a No 2 in one of D Company's Lewis Gun teams. This was the first time Coates had fired a revolver and because he had fired several rounds successfully he was left alone by Lieutenant Wylie to shoot. Coates stated his hands were cold and he could not re-cock the revolver so he brought it in closer to his body to do so. One of the officers behind him shouted for him to keep his weapon pointing down-range. However, instead of pointing his weapon in a safer direction Coates span around to see who had addressed him leaving the weapon pointed at Lieutenant Coggin. His hand slipped off the hammer which accidentally fired the revolver.[22] Coates was tried at Field General Court Martial on 10 October for wounding Lieutenant Coggin and was awarded two weeks of Field Punishment No 1.[23]

According to Hodson; 'They enjoyed Doullens, where they took dinner at hotels like gentlemen, these disreputably clad privates. "Brass hats" and battalion officers stared at them, but they summoned the waiter and drank their champagne as though they were brigadiers.'[24]

---

19 Gibson, *Postscript to Adventure*, pp.145–146.
20 IWM: E.F. Chapman Papers, 1799, p.28.
21 Gibson, *Postscript to Adventure*, p.152.
22 TNA: WO339/33386: Coggin Officer File.
23 G/24670 Private George Frederick Coates enlisted on 11 December 1915. He likely went to France with a draft and joined 20/R Fus on 16 June. He later transferred to 363rd Company, Labour Corps and was discharged in May 1918 due to wounds.
24 Hodson, *Grey Dawn*, p.266.

Whilst in Doullens, 20/R Fus rehearsed a planned operation against the target which were four lines of trenches near Rossignol Wood. Gibson's C Company was to be involved in the attack:

> On the rolling downs above Doullens we dug, with the photographs and Fritz's maps to help us, exact facsimiles of Fat, Fun, Fair and Fancy [Trenches], and took them about three times a day. We knew, to an inch and a second, where and when each supporting barrage would drop and lift, when and where to touch off the aeroplane flares, who was coming though us, when to expect our lunch, and what to do with the prisoners. Every man of every section knew his duty, and over and over again perfected himself in the same.[25]

Such were the constant rehearsals that the local French population soon knew what the plans were. According to Gibson General Rawlinson visited battalion HQ and briefed the officers of 20/R Fus that the coming attack would not involve them but that their training had been a deception plan which had drawn German troops and guns to near Rossignol Wood and not near the planned site for the battle.[26]

Whilst at Doullens Second Lieutenant J.C. Smyth joined the battalion on 1 October. Having recovered from his head wound in March Lieutenant Thomas Moyle rejoined on 11 October having served as a company commander in 29/R Fus. Also whilst at Doullens Cecil Isom chose to join the battalion band as a bugler. The band was generally billeted with battalion HQ and often remained with B Echelon out of the line.

On 11 October the battalion moved to Lucheux where it relieved the Cameronians. Next day Lieutenant Coggin returned from hospital. No sooner was he back than Second Lieutenant Oscar Thomas was wounded accidentally on 16 October. On the 18th the battalion transport departed for Ville-sur-l'Ancre. Next day, 20/R Fus departed Lucheux for the same destination travelling largely by motor bus. The battalion was in billets and whilst at Ville-sur-l'Ancre Captain Allen Maxwell, likely commanding one of the companies, departed to 26/R Fus. On 21 October 20/R Fus marched to the Citadel, between Fricourt and Bray, where it encamped in tents and huts having taken over from 1st Battalion West Yorkshire Regiment (1/W Yorks). Next day the battalion marched to Bernafay Wood and took over bivouacs from a battalion of the King's Own at 6:00 p.m. Meanwhile the battalion transport and B Echelon remained in bivouacs near Carnoy, commanded by Major Modera.

Lesboeufs – 'it wasn't ground, but porridge'

On 23 October the battalion took over Straight Trench, north of Ginchy, where it was in reserve to 11th Brigade, 4th Division. Hodson recalled they were in 'support trenches'; 'Their trench was a foot deep in water; no dug-outs save for the sergeant-major.' Hodson remembered that he slept on a shelf of clay in the trench wall, 'Every hour or so he awoke with the cold, shivered, scratched himself, and pressed himself that half inch farther on. With the dawn he slid off and stood in the water. In the nebulous light he could make out other men's

---

25 Gibson, *Postscript to Adventure*, p.147.
26 Gibson, *Postscript to Adventure*, pp.146–148.

forms sitting on narrow shelves … some sleeping, some just sitting, dully, vacantly, waiting … Waiting for what?'[27]

Though in reserve, they were merely awaiting their own turn in the line and there were casualties, likely from shellfire.[28] 24 October saw 19th Brigade relieve 11th Brigade. The 20/R Fus war diary recorded that next day, at 4:00 a.m., 20/R Fus relieved 1/R Warwicks and 2nd Battalion Royal Dublin Fusiliers (2/RDF) in the left sub-section of the right brigade. Two companies were in the front line (A Company right, C left) and two (B and D) were in support in Ox Trench, Shamrock Trench and Foggy Trench. 2/RWF were on the right, 1/Mx were to the left. Hodson's company had moved off for the front line at 8:00 p.m. (24 October) in pitch darkness; they had blundered around much of the night; 'They toiled and slithered and fell and rose and fell again; the mud was slimy, slippery as glass or ice…'. After eight hours Hodson's Lewis gunners eventually arrived in the front line trench which was garrisoned by a dying sergeant of the Warwickshire Regiment. According to Skelton:

> The Sapping Platoon occupied Ox Trench on the line occupied by the Royal Field Artillery and at nights we went forward through a sea of mud and shell holes to dig in where required after various attacks had failed to join up. There were no duckboards and we stumbled our way carrying fighting order, i.e. no packs, but with picks and spades, sandbags, sometimes stakes and coils of barbed wire. At the scene of the previous attacks the dead lay where they were killed, lying in the mud and the debris of war. There were many wounded who had been lying in the open for days…[29]

According to Hodson there were a few casualties from sporadic shelling during the night on the way up to the trenches.[30] Some of those reported killed on 25 October were likely killed in this way. Others may have reached the trenches but were struck down during the day.[31] One of those killed was PS/8002 Sergeant James 'Ponny' Ponsford, aged 19, who had been educated at Shrewsbury School. PS/5639 Corporal Robert Simon wrote to Ponsford's family that their son,

---

27 Hodson, *Grey Dawn*, pp.270–271.
28 GS/13067 Lance Corporal Philip Foster and GS/63595 Private Alfred Crane were killed on 24 October and are commemorated on the Thiepval Memorial. Foster had formerly served with 2/R Fus at Gallipoli
29 IWM: G. Skelton Papers, 13966, pp.52–53.
30 Shellfire likely accounted for the men killed on 25 October; many of whom were missing. The following men are commemorated on the Thiepval Memorial: PS/8713 Private Alfred Arthur John Watts, aged 20, from Barking, Essex, had served with 20/R Fus throughout the campaign from November 1915. GS/63385 Private Ernest Vincent, aged 20; GS/63479 Private Henry Cook; GS/63482 Private Horace Clover, aged 21, had served at Gallipoli; GS/63546 Private William May; GS/63588 Private John Colyer, aged 22, and GS/63651 Private William Cornwall.
31 GS/63506 Private Henry Fry, aged 19, from Homerton, London, had joined 20/R Fus on 12 July and had formerly served with 3/10 Londons. He presumably arrived in the trenches near Lesboeufs and was likely killed by a shell, along with another man, and they were buried together. Their grave was found in 1937 and Fry was identified by an ID disc in his pocket bearing his former 3/10 Londons regimental number and a 3/10 London's stamp on pieces of his boot. Both men were buried in London Cemetery and Extension, Longueval. The unidentified man carried a knife stamped 'W.H.'; this suggests the unidentified body might be either May's or Cornwall's.

'was a great favourite with every man in the platoon.'[32] Gibson gave his general impressions of this period:

> We were on a new line now, farther south and east. Our front was fluid, and had no proper landmarks, but its limit each end was a low mound of broken sherds only discernible in the wilderness of mud when one got right up to them. Their names on the map were Morval and Lesboeufs. Le Transloy and the enemy were in front, and near, but we couldn't see them. The Boche had, though, some advantages of position, and it appeared could see us. There were occasional and by no means continuous furrows that gave us partial shelter. But dodging from one to the next was difficult when everything squelched under your feet, and sometimes you slipped in up to your waist, or further, and all the time you were being sniped with whizz-bangs – stretcher bearers and everybody. To us this was a new phase, stationary warfare in ground where one couldn't dig in, because it wasn't ground, but porridge.[33]

The 19th Brigade War Diary recorded on 25 October that:

> The trenches were found to be very disjointed and in a bad state, which was to be expected after the rain of the past two days and the recent heavy fighting in this area. Lesboeufs was continuously shelled by the enemy and communication in consequence rendered difficult owing to the lack of communication trenches. The enemy shelled the back area with considerable intensity but seemed afraid to fire much on the front line owing to the difficulty of seeing exactly where the trenches were.[34]

Chapman described the immediate situation on taking over:

> The trenches we occupied were very shallow and rough and marked the furthest point reached in an attack. We did not know exactly where the Boche line was. You were liable to be sniped at from almost any direction, as the Boche still hung on in shell holes. We could see a sniper only 20–30 yards in front of our trench. He killed one man and wounded another. We had a lot of shots at him … but without any luck! … Ultimately, I think he was disposed of by our shells. He must have been a plucky devil. *Requiescat in pace*. … We had only been in these trenches about 2 hours, when four Germans came over and gave themselves up. The poor devils were shaking with cold and fright.[35]

---

32 PS/8002 Sergeant James William Prior Ponsford, aged 19, had been educated at Greenhill School, Moseley, Lickey Hills School and Shrewsbury College, where he was an OTC member. He was formerly an Accountant with Charlton and Long, Birmingham, from October 1914 before enlisting in the UPS on 30 June 1915. He had likely served as an instructor with 29/R Fus before he departed for France on 3 August 1916 and joined 20/R Fus. He had served in the trenches for two weeks prior to his death. He was initially buried in Grass Lane Cemetery in 1920, unidentified and with no cross, but when this cemetery was concentrated and moved to AIF Burial Ground, Flers, his body was identified by his ID disc. De Ruvigny, *The Roll of Honour*, Vol. 2, p.252.
33 Gibson, *Postscript to Adventure*, p.152.
34 TNA: WO 95/2420: 19th Brigade War Diary.
35 IWM: E.F. Chapman Papers, 1799, p.33.

The next night (25–26 October) Chapman's men spent trying to improve their shallow trenches and lost two men to sniping and shelling. The next morning 'there was the most beautiful sunrise, one of the most peaceful and lovely I have ever seen.' A quiet morning followed.

Lieutenant Edwin Wilmshurst had been appointed as an acting captain, commanding A Company, on 25 October. Next day, C Company was shelled heavily and he was severely wounded in the head. Wilmshurst's family were told that he was wounded whilst being shelled by both enemy and friendly guns.[36] He was evacuated to No 34 Casualty Clearing Station and later went to a hospital at Rouen where his situation worsened. He died of blood poisoning and acute meningitis on 1 December 1916.[37]

Hodson also experienced 'friendly' shelling on 26 October; 'The following day our own artillery shelled them. High explosive from behind began to drop in the trench away on their right. Their signallers stood on the trench top at infinite risk to wag their flags and stop the artillery from firing, but before they succeeded there were many casualties…'[38] He was potentially referring to 2/RWF on the right of 20/R Fus; only one man was recorded killed in 20/R Fus on 26 October.[39] The next morning they were bombarded heavily according to Hodson, 'the second Lewis-gun, twenty-five yards away, was hit and most of the crew killed. Hardcastle was ordered to re-establish the gun. It was a horrid business. One man lay in the trench, his stomach shot away … Hardcastle ignored him … pulled out the gun and began to clean it.' He proceeded with this task as the bombardment continued relentlessly and the mortally wounded man groaned and later died.

Also on 27 October Lieutenant Algernon Coggin (C Company) and Second Lieutenant Frederick Henley (A Company) were killed.[40] Chapman wrote home in a letter on 28 October 1916:

> Yesterday was saddened by the death of a fellow subaltern [Henley], who was killed by a shell. One moment I was talking to him; I went away, and the next moment the shell came and killed him. Poor little chap. These things are very sad… Meanwhile another C Company Friend [Coggin] had been killed, and also a sergeant [Roberts], a perfect gentleman, whom I almost loved. So it has been a sad time.[41]

Henley's family were informed by telegram on 1 November 1916. According to a letter from Lieutenant Colonel Garnett, Henley was killed by shell fire whilst commanding his company in an advanced trench.[42] According to PS/9074 Private James Barlow, Henley was mortally

---

36 IWM: Documents 11046, Papers of F. Henley.
37 Acting Captain Edwin Roy Wilmshurst is buried in St Sever Cemetery Extension.
38 Hodson, *Grey Dawn*, p.273.
39 This was GS/63598 Private Alfred Leary, from Deptford, who had joined on 22 July having previously served with 4/4 Londons.
40 Coggin is buried in Guards' Cemetery, Lesboeufs.
41 Chapman was presumably referring to Henley (who was 5ft 5in) rather than Coggin (5ft 10in). IWM: E.F. Chapman Papers, 1799, p.33.
42 Lieutenant Frederick Henley, aged 21, and from Ilford, Essex, was an old boy from Merchant Taylors' School and a member of the school OTC. He was a stockbrokers' clerk and he had enlisted with the Inns of Court in August 1915. He was gazetted in December 1915 to 14/R Fus after a training course at Cambridge University OTC. Henley went to France in early August 1916 and was on a course 14–24 August 1916. He commanded a platoon in D Company but was temporarily commanding A Company when killed.

wounded in the chest and stomach by the shell and was killed instantly.⁴³ Henley, nicknamed 'Baby' by his brother officers, had temporarily taken over A Company after the wounding of Captain Wilmshurst the day before. His company commander, Hodgson-Jones, of D Company, stated in a letter to Henley's family that he had put new life into the men he took over 'organised splendidly and had no fear…'.⁴⁴ Not only did Colonel Garnett write but also the 2IC, Major Modera, who evoked a very 'public school' attitude to the death of an officer; 'He [Henley] is one of those many brave fellows who has lost his life playing the game for his side, and our hearts thrill with pride when we realise the great sacrifice he made.'⁴⁵ There was potentially an element of untruth in these letters to lessen the blow of Henley's death to his family as Chapman wrote home that he also had been in charge of A Company since Wilmshurst's wounding.

Coggin is buried in a small cemetery behind the trenches south west of the village of Lesboeufs on the afternoon of 28 October along with GS/63473 Private William Banks who is buried nearby. Their graves were later centralised at Guards' Cemetery, Lesboeufs.⁴⁶ Lieutenant Colonel Garnett had hoped to bring Henley's body in and bury him next to Coggin, but this was presumably not possible; the location of Henley's grave is unknown.

After dark on 27 October the Scottish Rifles relieved 20/R Fus who were able to withdraw to Hogsback Trench to be in brigade reserve. This relief was complete at 4:00 a.m. on 28 October. On 27 October Major Herbert Hickley returned to 20/R Fus and took command of A Company. This company needed an enduring leader having had several permanent and temporary company commanders in the last few weeks.

It was planned for 19th Brigade to attack Hazy and Boritska Trenches in conjunction with 152nd French Division on 28 October but the latter were not ready, so the attack was postponed a day. On 29 October the Scottish Rifles and Cameronians attacked Hazy and Boritska Trenches. 20/R Fus provided two 100-man carrying parties to assist the two assaulting battalions on 19th Brigade and strengthen the defences. The sapping platoon of 20/R Fus was to report to the Scottish Rifles; presumably to assist in digging communication trenches. 20/R Fus was also under orders to send forward further reinforcements. In the process 20/R Fus lost another eight men killed on 28–29 October.⁴⁷

---

43 IWM: F. Henley Papers, 11046.
44 IWM: F. Henley Papers, 11046.
45 IWM: F. Henley Papers, 11046.
46 IWM: F. Henley Papers, 11046.
47 On 28 October 20/R Fus lost the following four men killed who are commemorated on the Thiepval Memorial: PS/5243 Lance Corporal James Llewellyn George Mann, aged 20, from Fulham, had enlisted in 20/R Fus in October 1914 and served in C Company; GS/63342 Private Charles Parker, had served with 1/Londons at Gallipoli; GS/63439 Private Charles Chalk, aged 38; and GS/63653 Private Henry Wilde. Also commemorated on the Thiepval Memorial and killed on 29 October: PS/5015 Corporal Cresswell Campbell Higgs, aged 24; G/26945 Private William Wingrove; GS/63414 Private George Giles; and GS/63450 Private Matthew Morgan. Another three men died of wounds over the next couple of days and were buried at Grove Town Cemetery, Méaulte: PS/5205 Private William Rowland Leach, aged 26, from Rochdale, was educated at Manchester Grammar School and was employed as an assistant clerk to the Rochdale Board of Guardians. He died on 31 October. PS/5357 Lance Corporal John Albert Morris, aged 23, from Blackpool had been wounded in both legs. SPTS/4895 Private Harold Smith was also wounded in both legs and, like Morris, died on 1 November.

At 2:00 p.m. on 30 October 20/R Fus was relieved by 16/KRRC and marched away from the front line. At some stage during the day Lieutenant William Fraser Mair was wounded in his right thigh by a shrapnel ball.[48] They marched to a hutted camp halfway between Montauban and Carnoy. Edward Chapman gives an idea of how quickly comfort could be restored:

> We got here about 8 p.m. last night, rather wet, as the roads were often ankle deep in water. … you can imagine our delight at finding dry huts, with our blankets and valises all ready, and some boiling hot stew all ready for us! We turned in very merry and are none the worse today. We were all drenched from the waist downwards; even my great trench boots were full of water. It was delightful to be able to get dry things on at once. To add to my delight, I found two letters from home waiting for me.[49]

Warmth, food, rest, cleanness, dryness and contact from home could all restore fatigue and flagging spirits endured by the appalling conditions in the line. Between 22 and 30 October 20/R Fus suffered three officers killed (Coggin, Henley and Roberts) and two wounded (Wilmshurst and Mair) along with eighteen men killed, four missing, 44 wounded and four slightly wounded.[50] Ashley Gibson was on leave and missed this stint in the line. On his return he was informed of the casualties:

> They told me Coggin had gone, our best subaltern; Roberts, our most admirable sergeant – buried, with half the platoon, and both dead when they were dug out. Coggin hadn't been long with us but had proved his worth. One had expected him, somehow, to come through now, what with the purple weal on his temple, badge of victory in a stand-up duel with a Boche officer in High Wood…[51]

Sergeant Roberts was described as; '…an intelligent well-educated man. His father is Canon of Manchester. He is smart and well set up and should make a good officer.'[52] His death caused a curious situation to come to light:

> Roberts was one of our "old contemptibles," the only one of all the Epsom old guard who hadn't got a commission, and he and all of us wondered rather wistfully why this had

---

48 William Fraser Mair was from Manchester; he attended Rossall School (1901–1904) and joined the school OTC. He was employed as a surveyor. Mair enlisted in September 1914, aged 26, and was quickly appointed as a company sergeant major in October 1914 with A Company. He was gazetted on 20 March 1915 but did not go to France until March 1916 having presumably served with 29/R Fus. As an officer, he commanded No 13 Platoon, D Company but also attended gas and signalling courses. After being wounded on 30 October he returned to the UK and served with 5/R Fus. He did not see further front-line service and commanded 273rd Prisoner of War Company and served with 51st Graduated Battalion, Royal Fusiliers.
49 IWM: E.F. Chapman Papers, 1799, p.34.
50 20/R Fus suffered 22 men killed or died of wounds during the month or straight afterwards. PS/7836 Private David Smith, aged 19, from Prestatyn, Flintshire, also died on 8 October 1916 as a POW and is buried in Niederzwehren Cemetery, Kassel.
51 Gibson, *Postscript to Adventure*, p.152.
52 TNA: WO 339/77171: Second Lieutenant Roberts Officer File.

not been so. We discovered the reason when we searched B.H.Q. dispatch boxes for his records. His papers had come through months before and owing to headquarters casualties and muddles arising therefrom had got mislaid. So we buried Second Lieutenant Roberts, which was all we could do.[53]

That Sergeant Roberts is buried as a second lieutenant demonstrates a high turnover in officers in battalion HQ through casualties. It also demonstrates a lack of coherence of administration by the Adjutant, and the orderly room, though this would have been under difficult campaigning circumstances. 20/R Fus also had to process more than the average battalion's worth of commissioning paperwork. The incident also highlights unreliability in the battalion war diary.[54] It must be wondered whether other men killed and wounded after High Wood might have been serving with 20/R Fus 'in error'.

This administrative problem may have preserved Hodson's life; he was recommended to work in the battalion HQ orderly room. Whilst there, Hodson was also able to get Colonel Garnett to sign off his commission paperwork before, 'he saw the Brigadier, who bestowed upon him a comprehensive stare and a kindly word and signed his papers.' Whilst in this role, Hodson provided an impression of battle procedure within 19th Brigade and HQ 20/R Fus whilst not in the line:

> He slept in a Nissen hut. Every night towards midnight or in the small hours movement orders came from brigade. The despatch-rider would stumble in and flash the torch questioningly. … Hardcastle would stir … the rider would … flash his lamp while Hardcastle read. "Operation orders. Shift again in the morning. Seven-thirty. Always do move us in the middle of the night…" he threw off his blankets, pulled on boots, tunic and greatcoat, and sallied of to find the adjutant [Captain Modera], who would sit up in his blankets, made a few rapid calculations in his head, and dictate battalion orders to Hardcastle. When Hardcastle had written them out he sallied forth once more to rouse the various company commanders … and the sergeant-majors … Moving office every day… arranging transport,

---

53 PS/5539 Sergeant Francis Roberts, aged 21, was from Fallowfield, Manchester. He attended Rossall School and was in the OTC and was an 'agriculturist'. He enlisted in September 1914 and was promoted to sergeant in May 1915. He travelled to France with the advance party on 12 November 1915. Roberts had commanded No 12 Platoon, C Company. He was killed on 27 October 1916 and is buried in Guards' Cemetery, Lesboeufs. He was recorded as being appointed to a commission on 23 October 1916 and that his commissioning papers had come through but gone astray. A request was made after his death for him to be gazetted posthumously; 'Was in command of a platoon at date of death and appeared in Part II Regimental Orders as appointed to a commission. Promotion was not communicated owing to conditions of active service.' See also, Gibson, *Postscript to Adventure*, p.152.
54 The War Diary should have been kept up to date daily and despatched up the chain of command to GHQ at the end of each month. If Gibson's account is correct, then the war diary entry stating Roberts was discharged to a commission on 23 October was doctored at the end of the month by Colonel Garnett (who initialled the changes) or the Adjutant. This strongly suggests that the war diary was written after 27 October, and likely at the end of the month, rather than day-by-day. This questions the reliability of the diary as a source as the author could be selective in compiling the diary at the end of the month rather than recording everything that happened daily, as it happened. In Roberts' case the diary may have been 'doctored' to not only honour a popular casualty but also to cover up poor administration.

working between mouthfuls of tea and speaking with mouthfuls of food … decidedly not dull…[55]

This despatch rider had orders that would take 20/R Fus back to the front line.

Lesboeufs Again

The 20/R Fus enjoyed a short rest in the hutted camp near Carnoy. On 2 November Lieutenant Thomas Moyle was recorded as being wounded whilst visiting the trenches; he was hit by a rifle bullet in the thigh which got him sent back to 'Blighty'.[56] This was presumably part of a reconnaissance by the officers of the battalion prior to taking over the trenches next day.

On 3 November 20/R Fus went into the trenches for the last time during the Battle of the Somme. The battalion relieved 4/Queen's RWK on the night of 3–4 November. Two companies were in the front line, one company was in support at Thistle, Windy and Union Street Trenches and the fourth company was in reserve at Cow Trench. Battalion HQ was located in a sunken road southeast of Lesboeufs. On the right was 2/RWF and on the left 7/E Yorks (17th Division). At Lesboeufs, according to Ashley Gibson:

> The mud got no solider. It was dashed cold but wouldn't freeze. We went in again, back at Lesboeufs, and came out with my company seventy greatcoats short. There was an awful row, but how can a man drag half a hundredweight of dripping, slime-sodden felt on top of his arms, ammunition, and equipment, through three miles of morass under shell-fire? They saved their gas masks, which was something.[57]

Likewise, Hodson recalled Winter on the Somme; 'Rain … and mud. Snow … and mud. Frost… and mud. Broken trees and landscapes bare of all save rusty barbed wire and pools of water and ruined brickwork … and mud. Deserts of mud. Mud in which mules got embedded to their bellies and were shot … This was life, this unending agony of filth and cold and exhaustion…'[58]

The enemy defensive scheme posed a problem in this sector in that the Germans did not hold an obvious defensive line. Their front was held by posts sited in shell holes or short sections of trench with many containing a machine gun. These posts were difficult to identify and German MG fire could halt advances from unexpected angles. At some stage during this period in the line Sergeant Skelton had a narrow brush with death:

> One night the trench we were in was hit by a large shell and blown in, we lost several men buried and I is buried up to my shoulders but was dug out by the others unharmed.

---

55   Hodson, *Grey Dawn*, pp.281–282.
56   Thomas Moyle recovered from his physical wounds but also suffered from neurasthenia. He requested permission to resign his commission due to ill health to allow him to renew his medical training at Cambridge. He was discharged unfit in June 1917.
57   Gibson, *Postscript to Adventure*, p.155.
58   Hodson, *Grey Dawn*, p.270.

> Throughout the Somme battle I seemed to bear a charmed life; men were killed beside me, in front of me and behind me, but somehow I was not hit.[59]

On 4 November the sunken road housing battalion HQ was shelled heavily throughout the day. At some point Captain Yorston was wounded but remained at duty. Chapman was warned that his company (C Company) would be taking part in an attack next day:

> On the evening of Nov. 4th we came under the heaviest shell fire I have ever experienced. This was simply a sign of nervousness on the part of the Boche…. we only had 2 killed and a few slightly wounded. But that evening, with the prospect of the attack, my chances of ever coming home again seemed very small indeed! I am not afraid of being killed, but all the same I am very glad to be still alive![60]

The next day, on 5 November, there was to be an attack by 17th Division on the left, 100th Brigade on the right and the French beyond them. The part set for 20/R Fus in this operation was stated in 19th Brigade operational orders as; 'At Zero + 5 the 20th Royal Fusiliers will launch a strong Bombing attack up the Sunken Road and clear it of the enemy, afterwards establishing posts in it to join up with 17th Division.'[61] This post was to be where Summer Trench intersected the Sunken Road northeast of Lesboeufs. This position was just short of the crest of a small rise and was presumably intended as a reverse slope position from which to provide early warning and a delay to any future German advance. According to Ashley Gibson:

> D Company were detailed for a small show on their own. The line, such as it was, wanted straightening, up beyond Dewdrop and Summer trenches, so-called optimistically in the new maps. The lie of the ground was again unfavourable. In supports we heard Fritz start up with his machineguns and knew exactly what was happening. One officer brought them back. My excellent batman and I crawled forward in the middle of the operation to see if the supports could do anything… Wylie's grey and ravaged countenance (our oldest, he was forty-three, and now looked sixty) peered over the wet sandbags, and his dull eyes, opaque like those of a dead man, moved stiffly in their sockets. "No thanks," he said. "But have you got your brandy flask? Mine's empty."[62]

The 17th Division assault failed and the 20/R Fus party was assailed by fire from its exposed left flank. Lieutenant Butchard of 20/R Fus was killed at the start of the advance.[63] Though they made several attempts they were only able to advance 100 yards. Meanwhile, on the right,

---

59 IWM: G. Skelton Papers, 13966, p.53.
60 IWM: E.F. Chapman Papers, 1799, p.36.
61 TNA: WO 95/2420: 19th Brigade War Diary.
62 Gibson, *Postscript to Adventure*, p.156.
63 Lieutenant Robert Archibald Butchard, aged 23, was born in Gravesend, Kent, and lived in North Fleet, Kent. He was educated at Dartford School, New College, Herne Bay and worked as an analytical chemist for a cement company. Though not an ex-UPS man he had enlisted with the Honourable Artillery Company in August 1914 and served with the battalion in France before being gazetted to a commission in 10/Queen's RWK and undergoing training with 31/R Fus. He joined 20/R Fus on 16 September 1916. Butchard is buried in Guards' Cemetery, Lesboeufs.

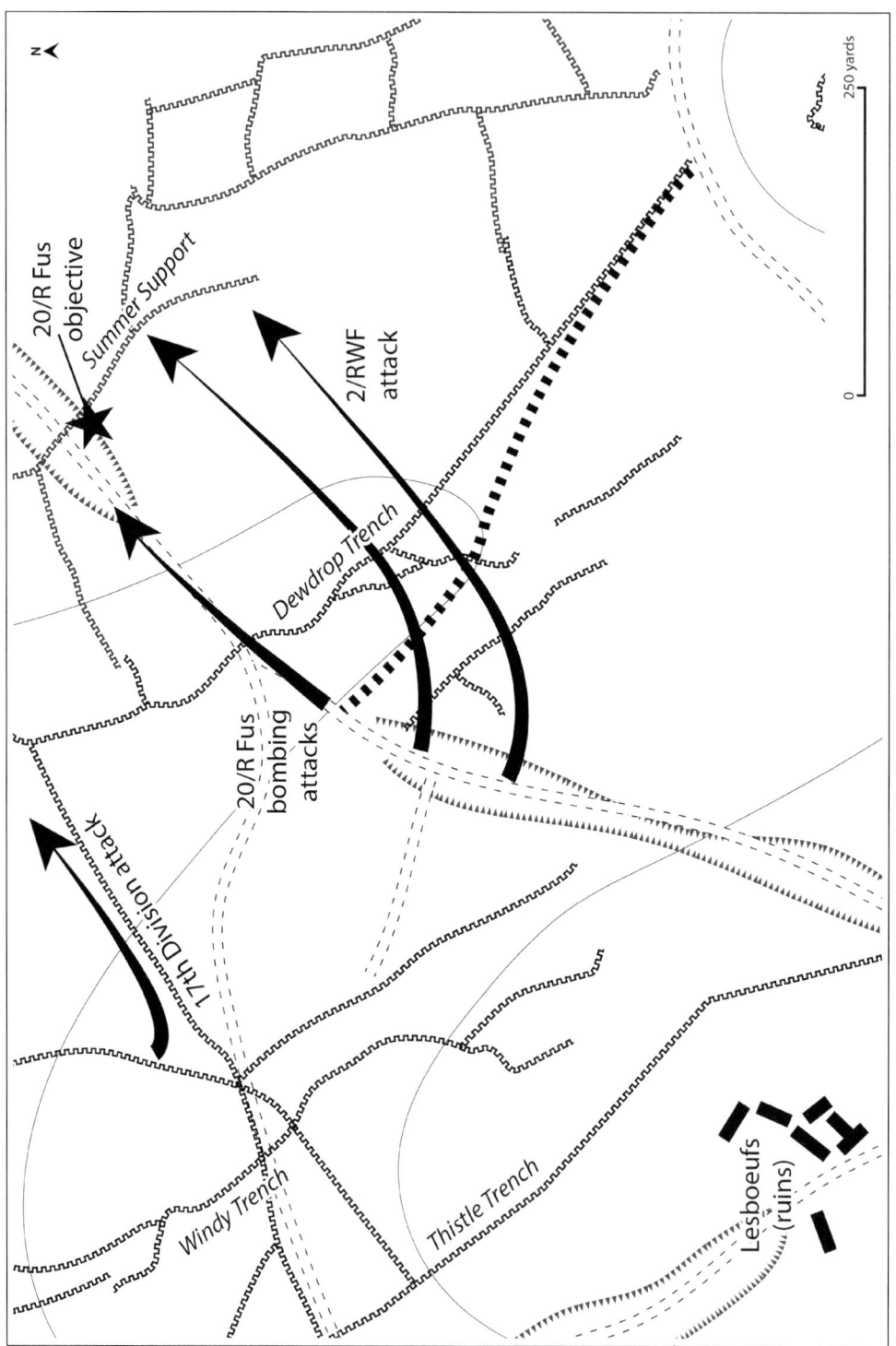

Map 8 Lesboeufs 5-6 November 1916.

PS/4693 Lance Sergeant John Frederick Cobb, aged 26, was educated at Central Higher-Grade School and Manchester Grammar School. He enlisted in September 1914 and was wounded in March 1916. Cobb was killed on 6 November and is buried in A.I.F. Burial Ground. (Reproduced with permission of Sutton Archives)

2/RWF conformed to the 100th Brigade attack and were able to establish a new outpost line which was improved into the new Front line. Captain Hollingsworth wrote to the family of PS/5973 Lance Sergeant George Hulme Worthington:

> George was killed in action on November 5th, 1916. ... he died gloriously and painlessly ... one of my officers led a party of bombers in an attack; the officer was killed, and your son took command, but a few yards further he was hit by a bullet and died instantaneously. The same night he and the officer (Lieut. Butchard) were buried side by side. ... he was one of the few men who had been with us from September 1914 and he was always cheerful, keen, and most reliable.[64]

Though the 19th Brigade War Diary recorded heavy artillery fire on 6 November it stated there was no infantry action that day. However, 20/R Fus launched a third, and ultimately successful, attempt to establish the required bombing post. Lieutenant Henry Bulbeck was shot through the head and killed whilst conducting a bombing raid near the Sunken Road.[65] He is buried

---

64 PS/5973 Lance Sergeant George Hulme Worthington, aged 21, from Wigan, was with B Company. He had attended Wigan Grammar School, where he played cricket and football; he was also a member of Wigan rowing club and was a capable cross-country runner. He was articled to a chartered accountants firm. George Worthington is buried in A.I.F. Burial Ground, Flers. His brother, Thomas Augustus Worthington, was killed whilst serving with the Cameronians on 29 October 1916. *The Wigan Observer*, 2 December 1916, p.8.

65 Lieutenant Henry Edmund Bulbeck, aged 22, from Clapham, was educated at Battersea Polytechnic where he received a BSc. After serving at Gallipoli he returned to 16/R Fus in the UK before he was attached as a replacement to 20/R Fus in August 1916. On 6 November 1916 he was shot through the head and killed. He is commemorated on the Thiepval Memorial.

where he fell but his grave was later lost. Though Butchard and Bulbeck were not originally UPS officers they had served in the ranks or with other units and had certainly not lacked bravery.

Survivors

On the night of 7–8 November 20/R Fus was finally relieved by the Middlesex Regiment and withdrew to camp at the Briqueterie near Bernafay Wood. Cecil Isom had been left out of the line and was employed to guide the battalion into camp; 'Moved to Bernafay Corner. Met Batt coming out of line. They came in in twos and threes. All done up and covered in mud. Up all night guiding them to their tents.'[66] During this tour of the trenches 20/R Fus had lost two officers killed and two wounded (one of whom remained at duty) and 22 men killed, 68 wounded (with eight still at duty) and fourteen men missing.[67] 20/R Fus would soldier no more during the Battle of the Somme. If the losses sustained at High Wood had not been appalling

---

66   IWM: C.H. Isom Papers, 16336.
67   These included the following men commemorated on the Thiepval Memorial unless otherwise stated: GS/63412 Private William Bowles and GS/63534 Private Ernest Powell, aged 32, were killed on 4 November 1916. The following were killed on 5 November 1916: PS/5030 Lance Corporal Robert Holt, aged 23, from Rochdale, was a member of B Company, 20/R Fus. PS/7855 Private John Lawrence Dowling, aged 29, from Nottingham, is buried in Guards' Cemetery, Lesboeufs. GS/63035 Private Walter Chick is buried in Thiepval Anglo-French Cemetery. GS/63354 Private Leonard Bowtell; GS/63400 Private Percy Ellis; GS/63382 Private Harry Bovey; GS/63424 Private Joseph Charles Moore, aged 25; GS/63448 Private John Kelly; GS/63592 Albert Alfred Inwood, aged 21; GS/63628 Private Arthur Kimpton; GS/63634 Private William John Nicholls; GS/63703 Private George Dawes and GS/63645 Private Henry Hawes had almost all joined 20/R Fus in late July 1916 and served for about three months before being killed and commemorated at Thiepval. Killed on 6 November: PS/4693 Lance Sergeant John Frederick Cobb, is buried in A.I.F. Burial Ground. PS/8897 Lance Corporal James Leach, aged 20, from Rochdale, was also buried in A.I.F. Burial Ground as were PS/5881 Lance Corporal Henry Edward Wainwright, aged 26, from Manchester; GS/63455 Private Frederick Burlie, aged 30; GS/1638 Private Joseph Simpson, aged 22; GS/63447 Private John Calnan, aged 28, of C Company; GS/63441 Private Thomas Howe and GS/63406 Private Edmund Duggin, of B Company. PS/10112 Private Leslie Bowman Hind, aged 19, and GS/10541 Private John Hutchinson Tomlinson, aged 18, were both buried in Grove Town Cemetery. Tomlinson died on his wounds at No 34 CCS. GS/63438 Private Thomas James Pooley was with No 11 Platoon, C Company. His death was presumed on 6 November. Several men were killed on 7 November: G/35372 Jesse Bull, aged 23; GS/63661 Private Alfred Dwight, of No. 12 Platoon; GS/63698 Private Stephen Morgan; GS/63430 Private George Perkins and SPTS/4657 Private Leonard Ward are commemorated at Thiepval. GS/63312 Private John Ferguson, aged 24, served with 1/Londons at Gallipoli; he is buried at A.I.F. Burial Ground, as were SPTS/4890 Lance Corporal John Fox and GS/63604 Private Arthur Walker. G/35376 Private Bertie Lyes, aged 26, and GS/15374 Private Frank Fisher, aged 27, both reside in Grove Town Cemetery. Several of those wounded later died: GS/66396 Private Claude Ernest Easte died of wounds on 10 November and is interred in Grove Town Cemetery. GS/63418 Private Frederick Wootten died on 16 November and is buried in St Sever Cemetery Extension. PS/4709 Corporal Walter Lucas Croft, aged 23, from Derby, had already been wounded once in March 1916. He died of his wounds on 17 November and is buried at Etaples Military Cemetery. PS/8972 Private Luther Bailey, aged 24, from Trafford Park, Manchester, had served with 28/R Fus and later joined 20/R Fus. He died of his wounds in the UK on 18 November and is buried in Eridge Green Churchyard. GS/63419 Private Joseph Lynch died of his wounds on 19 November and is buried in St Sever Cemetery Extension.

enough, the contribution of 20/R Fus to the remainder of this offensive involved the loss of a further seven officers killed and nine wounded and 64 men killed, 264 wounded and 29 missing.[68] There were not only physical wounds. After High Wood the strain of further operations on the Somme affected some officers and men significantly. According to Gibson; 'My company commander was cracking up. Day by day his limp became more pronounced; inguinal hernia it was, which he called sciatica. Myself, at times companion of his spasmodic and twitching slumbers, had some knowledge of what stuff his dreams were made of.'[69] Though Gibson names him 'Murgatroyd' his identity is unknown.[70] Gibson had not been the only one to detect this deterioration in his superior; Captain W. McConnell, RAMC, the battalion MO, took the officer in question to one side and sent him home to the UK for a rest.[71]

The difficult conditions that the battalion had endured, rather than the casualties, made these two periods in action hard to bear. According to Hodson, regarding Lesboeufs; 'Festubert compared with this was a garden party, and High Wood a night out.'[72] The battalion needed further rest and a chance to refit but it also had a winter in the trenches to endure. Hodson, Skelton and others might yet win their races to be commissioned in time. Others had not been so lucky; especially Sergeant Roberts, who, though in life had been denied a commission, was awarded one in death. However, many ex-UPS officers and men endured the second half of the of the Somme campaign.

68   Casualty records on the CWGC suggest that 97 officers and men were killed or died of wounds.
69   Gibson, *Postscript to Adventure*, p.153.
70   This is not Second Lieutenant Arthur Murgatroyd, 20/R Fus, who was not gazetted until 10 December 1916.
71   McConnel remained with the battalion as MO until September 1917.
72   Hodson, *Grey Dawn*, p.272.

22

UPS men later in the Somme fighting – September to November 1916

An excellent chance of getting back to Blighty for Xmas

> *Things are beginning to move out here & we may be in the thick of it any time.*[1]
>
> *… an excellent chance of getting back to Blighty for Xmas …*[2]

In early July, early August and late September three batches of ex-UPS officers were gazetted after officer training by the OCB system. This influx of officers reinforced depleted units who had suffered casualties earlier in the fighting.

Second Lieutenants Douglas Candy and Wilfrid Clarke were sent to 1/Bedfords in mid-September 1916; two of twenty-nine officers who joined that month alone. Clarke had been the Company Sergeant Major of C Company, 18/R Fus. Clarke and Candy were in an attack on 24–25 September when the battalion successfully captured its objective, a sunken road west of Morval. Clarke recorded the following personal account:

> It was the same old sights and sounds which you must know so well. The line of spouting earth along the Bosch trench at zero time. The waves of infantry moving slowly forward against the leaden death; above you the air is so thick with shells that they seem almost to roof out the sky; all round the dead and wounded, German and our own; behind and in front the splitting crash of guns. The prisoners come over in hundreds, and you light the red flares, while the contact aeroplanes go over, and the second line passes through to take the village. I remember stepping gingerly over the broken German wire and feeling a numb red hot feeling in my arm. My rifle dropped and lay stupidly in front of me, "Go on, boys" I heard myself shout, as the whole scene whirled round, and I crept bleeding to a shell hole, all the happy careless smile of confidence gone, all the wild exultation of success struck from you, and every fiendish terror, clutching at you cold heart, as the barrage grows more intense and the air seems thick with bullets meant for you. No more gay happiness of leading on, a keen eye on the weak spot in the line of bursting shells, and a brain working

---

1   FML: RFM.ARC.2482.47: Documents of H.D. Etheridge, letter 27 September 1916.
2   Liddle Archives: WW1/GS/1188: Papers of K.V. Norman, p.240.

quietly and quickly in the tremendous din. Just a maimed body of flesh and blood, lying among the crashing steel…³

Clarke's period of active service lasted a week. Candy was mortally wounded.⁴

Thiepval Stormed

On 26 September 11/R Fus was destined to capture Thiepval. This fortified village had already blunted several attacks since 1 July and PS/6924 Private Philip Wright of the battalion, felt a sense of trepidation. 11/R Fus would follow 12/Mx and the attack would be supported by tanks and an indirect-fire barrage from MGs; the advance would commence at midday instead of dawn. Whilst waiting for 'Zero' three men of the MG Section were blown to pieces by a shell; they were formerly of 21/R Fus, and close friends of Wright,.⁵ Wright just had to carry on:

> Shortly before noon we lined up in a communication trench ready for "over the top"… For the last few minutes before the time, our artillery were sending some huge 15" shells right into the heart of Thiepval to blow in their dugouts… And then – we went over. Besides my equipment & rifle I was carrying a bag of four drums filled with ammunition for the Lewis Gun. The air was full of bullets singing about like bees. Fritz was sending over his usual 5.9's, the place was a perfect hell ten times worse than the offensive of July 1st. First a lump of shrapnel hit me a 'wallop' on the head (thanks to my tin hat) and knocked me silly for a minute. … I found a small hole in my own left thigh … I also had a slight flesh wound in the left arm just below the shoulder… I got my first view of a 'tank'. Crawling along on our left came the monster in the direction of Thiepval. … From a neighbouring trench trooped about twenty Huns, all unarmed and holding up their arms. Someone had to take them back … I took charge of the party … My leg by this time was getting rather stiff and painful and we had to dodge shells galore on the way back…⁶

Wright survived this battle, but eleven ex-UPS men did not.⁷

In late September 1916 Captain George Maurice Reginald Gillett of 6/Leics R (formerly 18/R Fus) returned from leave in the UK. According to his friend, David Kelly, Gillett knew

---

3   IWM: Documents 18911: Papers of W.K. Clarke.
4   Candy is buried at Bronfay Farm Military Cemetery, Bray-sur-Somme.
5   These were likely PS/2524 Private Francis Heriot Gregory Clayton, aged 23, and PS/6929 Private James Edwin Budge, aged 19; both are commemorated on the Thiepval Memorial. Clayton had been educated at Marlborough and Hertford College, Oxford.
6   Diary of P.N. Wright, Fusiliers Museum, RFM.2013.8.1/2, p.55.
7   In addition to Budge and Clayton of 21/R Fus the following were killed: Formerly from 19/R Fus – PS/7471 Corporal Arthur William Pritchard and PS/8532 Private Dennis Hall, aged 19. From 20/R Fus – PS/7124 Sergeant Harry Ryals Burgess; PS/8747 Private William Heywood, aged 35, and PS/9968 Private Sydney Pugh, aged 20. From 29/R Fus – PS/10066 Private James Frederick Travis, aged 19; PS/10675 Private Howard William Myers; PS/10690 Private Charles Herbert Roberts, aged 19, was an undergraduate from King's College, Cambridge and PS/10999 Private Frank Horace Smith, aged 24. They are commemorated on the Thiepval Memorial except for Hall and Roberts who were buried in Lonsdale Cemetery and Bray Vale British Cemetery.

it was his last leave having presumably foreseen his death.[8] He was killed, aged 33, on 26 September at Guedecourt having fulfilled his premonition.[9]

October 1916 – Sudden Death at Bayonet Trench

20/R Fus did not fight a battle on 7 October 1916 but it was one of the worst days for UPS casualties. 8/R Fus and 9/R Fus had both received drafts of ex-UPS men on the disbandment of the UPS Brigade and absorbed reinforcements from the UPS Reserve Battalions to replace their casualties. Robert Tudor remarked on 3 September:

> I can't think why it could not have been arranged that the men from the 28th could not have been sent all to the same Batt. out here. The fellows in this Batt. are first class fighters and good fellows, all the same the few men who came out to join this Batt from the 28th have not much in common with the original men of the 9th.[10]

On 7 October the 12th and 41st Divisions attacked side by side northwest of Guedecourt and three Royal Fusilier battalions would attack concurrently; 32/R Fus (41st Division) on the left; 8/R Fus in the centre and 9/R Fus on the right (12th Division). The latter two battalions were to capture Bayonet Trench. The attack was at 1:45 p.m. and coincided with the German relief; the enemy trenches were double-manned and German MG and artillery fire caused such heavy casualties that the attack failed. According to Private Albert Clayton:

> … somebody blew a whistle, and away we went… I stumbled along, trying to keep in line as we advanced in open order; not too close to the next man, not too far away. I could see a long undulating line of advancing troops stretching away down south … There was, mixed in with the general uproar, a good deal of "rat-tat-tat-ing" from somewhere. I couldn't tell quite where, as I was mainly occupied in … trying to avoid the many shell holes. About halfway across I looked up and about, and to my intense astonishment, found I was walking entirely alone. The whole of the advancing troops in my near vicinity had melted into the ground…[11]

The few survivors of 8/R Fus withdrew to their start line. They lost 35 men killed, 113 wounded and 96 men missing. The front three companies of 9/R Fus were decimated by MG fire. The only men to reach the objective were a few wounded men that the Germans dragged into their trench. 9/R Fus lost nine officers and 313 men. Second Lieutenant Reginald Still (formerly 19/R Fus) was killed with 8/R Fus and Captain David Cuthbert (formerly 20/R Fus) was lost by 9/R Fus.[12] Forty-four men of 8/R Fus were ex-UPS men; mostly from 28/R Fus and 29/R Fus. A further thirty-three fatalities from 9/R Fus had served with UPS battal-

---

8   Kelly, *Ruling Few*, p.92.
9   Gillett's name is inscribed on the Thiepval Memorial.
10 Letters by R Tudor, Royal Sussex Regiment Museum, letter 3 September 1916.
11 Clayton, *Long Before Daybreak*, p.83.
12 Still and Cuthbert are commemorated on the Thiepval Memorial.

ions. Amongst those killed were several well-educated men who had joined the UPS late, but could have helped perpetuate the ethos of 20/R Fus, had they been sent there.[13] Arguably too old to be an effective infantryman, Robert Tudor had a dangerous job as a stretcher bearer; he was killed in action on 7 October. His body was not recovered, and his name appears on the Thiepval Memorial.[14] PS/10544 Private Raymond Ashby was a close friend of Basil Peacock's from 29/R Fus. Peacock was under-age and had been left behind; he soon received a letter from 'Raymond' stating; 'Remain where you are as long as possible, this is no place for you. It is hell on earth and our casualties are heavy…'[15] Peacock never heard from him again. Ashby served with A Company, 8/R Fus. He had been wounded at Pozieres on 3 August 1916 and awarded the DCM for gallant conduct for capturing 20 to 30 German prisoners despite being wounded in four places:

> Raymond Ashby was again in a most terrible advance on October 7th, 1916, near Le Transloy and Flers; it was against fearful odds, and his Battalion was almost completely annihilated; he was leading his platoon in the centre of attack and was hit, when not far from our lines, and was seen to fall wounded by two of his comrades, one of whom tried in vain to find him … The War Office reported him as "Missing" …[16]

PS/10721 Private Geoffrey Kenyon was killed with 8/R Fus whilst trying to save a wounded comrade.[17] Likewise, PS/10224 Sydney Hutton was slain by a piece of shrapnel whilst helping an injured soldier.[18] The bodies of most of these men remained on the battlefield.[19]

The adjutant of 32/R Fus wrote later; 'The German MG positions in Bayonet Trench were consequently untouched [by shell fire] and at zero [hour] very heavy MG fire was opened on our front line. No one succeeded in covering more than 20 yards of ground and very heavy loss was

---

13 PS/10092 Private Thomas Sydney Pinder, aged 26, attended Manchester University and was awarded an MA in French. Pinder was an assistant master at Welshpool Country School. PS/10497 Private William Forster, aged 23, was educated at Downside and Trinity College, Cambridge; he was employed as an articled clerk. PS/10788 Lance Corporal Robert Dudley, aged 35, was educated at Warwick School and Rossall School, Fleetwood; he was employed as a corn merchant. PS/10077 Private Douglas Bradley-Smith, aged 35, had lived in Paris and had attended University College London; he was employed as an insurance secretary. PS/825 Private Rhys Counsell Roberts, aged 50, had attended Marlborough College; he became a solicitor in 1890. PS/9916 Private Thomas Arthur Thomas, aged 25, attended University College, Cardiff, and was a mining engineer. PS/10408 Private Richard Bodley Furley, aged 35, was educated at Haileybury and attended Keble College, Oxford.
14 PS/10351 Private Robert Edward Tudor, aged 41, had worked in Argentina for several years and had only returned to England just before the war to establish a poultry farm at Plaistow. He enlisted in April 1916.
15 Peacock, *Tinkers' Mufti*, p.36.
16 *The Leightonian*, December 1917, pp.173–174.
17 PS/10721 Private Geoffrey Kenyon, aged 30, was from Thorne, Yorkshire, and had been educated at Charterhouse. He is commemorated on the Thiepval Memorial.
18 PS/10224 Sydney Frederick Hutton, aged 32, was from Ashton under Lyne, Lancashire and was educated at Sedbergh School (1898–1901) and in Germany. He attended New College Oxford and was awarded a Theological BA in 1910. He worked for a shipping company before being ordained in September 1913. Hutton enlisted in January 1916 and had applied for a chaplaincy before he went to France in August.
19 They are commemorated on the Thiepval Memorial except for Furley who is buried in Beaulencourt British Cemetery, and Thomas who is buried in Grevillers British Cemetery.

sustained. The battalion, relieved the next night, was 2 officers and about 80 [men] strong.'[20] 32/RF had not received many ex-UPS men reinforcements but lost Second Lieutenant Eric Dunfee Meredith and 26/R Fus lost Captain Christopher Simpson; both had formerly served with 19/R Fus.[21]

12/RB attacked Rainbow Trench on 7 October. They had five officers in the first wave of the attack. Three were ex-UPS officers; Second Lieutenants Phillimore Dashwood and Walter Simmonds had served together at 19/R Fus and No 4 OCB. Three days before the battle Second Lieutenant Oswald Blencowe had been attached having served with 18/R Fus. One of Dashwood's soldiers wrote to the officer's family:

> He was a good soldier; well-liked by all his men of his platoon, who were always willing to do anything for him. Our battalion made an attack on October 7th and I'm proud to say Lieut. Dashwood led his men on magnificently. He was wounded by a sniper, in the lungs, but suffered very little pain. … I will always be proud to say I was a servant to an officer who will always be so remembered by all his men of his platoon …[22]

Dashwood was dragged to a dugout, where he died. Blencowe and Simmonds were also killed and half of their men became casualties.[23]

1/1st Battalion London Regiment was involved in a divisional attack by 56th (London) Division near le Transloy on 7 October. Amongst their number were two brothers, Second Lieutenants Charles and Reginald Arden, formerly scouts from 18/R Fus. The first objective was Spectrum Trench which the left-hand company secured. The right-hand company was stopped by severe MG fire and suffered about 80 percent casualties. In this operation Reginald was killed; Charles was wounded in the right leg by a shell but survived. Second Lieutenant Willoughby Bernard (also ex-18/R Fus) was also in this attack; he too was wounded.[24] Second Lieutenant John Coucke, also of 18/R Fus, was one of the lucky ones having been sent on leave just beforehand; he only rejoined after the battalion had suffered these heavy losses.

On 12 October 2/LF attacked Zenith Trench near Lesboeufs. Second Lieutenant William Clifford Bolton, a High Wood survivor, had held a commission with 2/LF for sixteen days. Bolton later recalled:

---

20 TNA, CAB45/133.
21 Meredith and Simpson are commemorated on the Thiepval Memorial.
22 TNA: WO 339/58844: Dashwood Officer File.
23 PS/249 Ronald Phillimore Dashwood, aged 22, attended St Lawrence College, Ramsgate, and Birmingham University where he studied engineering. He enlisted with A Company, 19/R Fus. He was gazetted to 15/RB in July 1916. He is buried in Bancourt British Cemetery. PS/887 Sergeant Walter Sangster Simmonds, aged 33, was an old boy of Eastbourne College, who attended the University of London and was a schoolmaster. He served with C Company, 19/R Fus. He too was gazetted to 15/RB. PS/1283 Sergeant Oswald Charles Blencowe, aged 26, had attended Dragon School, Oxford (1902–1908) and St Edward's School, Oxford, and was a schoolmaster at a prep school. He was a platoon sergeant in D Company, 18/R Fus. Blencowe was gazetted to 9/OBLI and joined 12/RB three days before his death. Simmonds and Blencowe are named on the Thiepval Memorial.
24 Willoughby Newton Bernard, formerly PS/1270, lance sergeant, 18/R Fus, was gazetted in August 1916.

Second Lieutenants Charles Edward Arden, and Reginald Douglas Arden, were from South Norwood and were educated at Exeter School, Devon. They enlisted with 18/R Fus in September 1914; both were later appointed as lance corporals and became scouts in C Company. Reginald had previously served with the 4th Devon Volunteers and had gone to Canada in 1908 and was a Bank Clerk. Charles was a rice planter in India and had previously been with the Bihar Light Horse Volunteers. Both brothers were sent for commissions from 28/R Fus in March 1916 and were gazetted to 1/Londons in August 1916. Reginald was killed and his body was lost. He is commemorated on the Thiepval Memorial. Charles was badly wounded. (Reproduced with permission of Sutton Archives)

> Owing to the nature of ground [the] position of German trenches could not be seen ... & [I had] no general idea of situation of objective to be gained. When [the] barrage moved forward I was soon of opinion that first wave, with which there was no officer, was not advancing fast enough & as a result I became merged into this wave & endeavoured to urge men on faster. In company with 12 men or so reached first line & number of prisoners sent back ... Then I pushed on with as many men as I could muster, only a few, direction having been lost & gaps caused in my original line through enemy M.G. fire. [The] Bit of trench proved eventually in my opinion to be part of the enemy communications trench reached. [I] Realised we were partially cut off, as could see in rear & to the right that attack was being held up. Trench only v[ery] short & blown up in parts. Managed to kill sniper & other man with revolver from trench. At dusk shadows of body of men seen disappearing into gloom. Followed same but soon lost touch. Eventually landed up behind German strong post which I thought was part of our own front line ... Eventually surrounded by Germans from trench [and,] having finished all ammunition resistance [was] useless. ...[25]

25   TNA: WO 339/72989: W.C. Bolton Officer File.

Right: PS/4510 Lance Corporal William Clifford Bolton, aged 19, was an articled clerk from near Stockport, Cheshire. He enlisted with 20/R Fus on 5 September 1914. After surviving High Wood he was sent to the Cadet School at St Omer on 9 August 1916 and was gazetted on 26 September. He was repatriated on 19 December 1918 and was demobilised in March 1919. (Reproduced with permission of Sutton Archives)

Bottom right: Second Lieutenant Ernest Airlie Holden, formerly 20/R Fus, had already had two lucky escapes; one when a bullet hit his ammunition pouch in March 1916, and at High Wood when he was one of three survivors of a party of forty-five. His luck ran out at Stuff Redoubt. (Reproduced with permission of Sutton Archives)

The British were not the only ones attacking. 8th Battalion Loyal North Lancashire Regiment (8/LNL) faced German storming parties that attempted to recapture Stuff Redoubt. After some hard fighting the Germans were repulsed but Second Lieutenant Ernest Holden, formerly C Company, 20/R Fus, and a graduate of the St Omer Cadet School, was severely wounded. Holden died five days later having had both legs amputated.[26] He had been an officer for just two weeks.

On 14 September 1916 Terence Doherty had returned to France as a second lieutenant; he joined 11/Essex. He wrote home on 11 October from Trones Wood having just come out of the trenches; 'I am very well and enjoying myself as best I can under the circumstances. We get good food here, of course we buy a lot of extras ourselves…'[27] On 15 October his battalion attacked near Guedecourt; though it captured its two

26 Holden is buried in Contay British Cemetery.
27 IWM: Documents.12961, Catalogue date 2003-12-22: Papers of T.O'C. Doherty, p.75.

Terence Doherty prior to his disfiguring wounds sustained in October 1916. He would continue his military career despite losing an eye. (Reproduced with the permission of Gillian Häkli)

objectives the German bombing counterattack re-took the first objective cutting off the leading British troops. Doherty was one of four officers wounded. He wrote home a week later:

> I was hit on Sunday morning during an attack. I am sorry to say my right eye was so badly smashed about that it had to be taken out yesterday afternoon. I can see fairly well with the left eye, so I have good reasons to be thankful. I was also hit in the left jaw just below the ear, but that is not worrying me much, except that I can't eat anything hard…'[28]

1/Cameronians absorbed a draft of men from the 33rd Divisional Cyclist Company in early May 1916. The company had been partly created from men from the UPS battalions and several 20/R Fus men found their way back to 19th Brigade. Three were killed or mortally wounded on 28 October. 25576 Private Bertram Hopewell and 25543 Private Thomas Hopewell were cousins from Lancashire and were apprentices in the cotton trade. The third man, 25590 Private Thomas Augustus Worthington, from Wigan, was educated at Wigan Grammar School and was a commercial traveller.[29] His brother, George Hulme Worthington, was killed a week later with 20/R Fus.

November 1916

Second Lieutenant Kenneth Norman, of 1/Queen's RWS, had seen the 'Battle of the Somme' film whilst in the UK; he was now experiencing the battle first-hand. On the night of 2–3 November he and his platoon set out for the trenches. After a hellish night march with lost

---

28 IWM: Documents.12961, Catalogue date 2003-12-22: Papers of T.O'C. Doherty, p.79.
29 Bertram Hopewell, formerly PS/5046, 18/R Fus, is buried in Carnoy Military Cemetery and Thomas (PS/5047, 18/R Fus) resides in Grove Town Cemetery. Worthington (formerly PS/5970, 20/R Fus) is commemorated on the Thiepval Memorial.

guides, casualties from shellfire, and across thick mud and shell holes, they arrived at Antelope Trench. Their rations had been lost or soiled during the march, they were soaked through and were in a trench with mud so thick they could barely move. Sentries were posted and officers went back to battalion HQ to report the position. Despite their condition and situation, their orders were to attack Bariska Trench; 'No one felt in the mood for fighting under these conditions although, as someone said, it was an excellent chance of getting back to Blighty for Xmas… this was about the only consolation one had.'[30] Norman continued:

> At 3 p.m. a terrific bombardment opened on all sides, someone said it seemed as if hell opened upon us. This lasted until 4 p.m. during which time the men got a bit nervy particularly the married ones. … I fortunately had a large tin of cigarettes in my haversack which I passed round to the men in my platoon. This bucked them up considerably. … At a few minutes to four the Huns who seemed to have some idea of an attack opened a terrific machine gun fire which swept a few inches above our parapet … At four o'clock over we went, I leading my platoon by a few yards. Machine guns were now playing havoc amongst our men. At 4.5 [4:05 p.m.] I was hit in the chest by a machine gun bullet … I fell like a stone but was quite conscious …'[31]

Norman was helped to the Dressing Station by one of his men. Norman's close friend Leonard Perkins was killed by a sniper; Reginald Lloyd and William Gross (both formerly 21/R Fus) were also killed.[32] Kenneth Norman's recollections of conditions during the latter stages of the Battle of the Somme were typical of many other newly gazetted ex-UPS subalterns.

Battle of the Ancre, 13–18 November 1916

Walter Medlicott joined 24/R Fus as an officer on 1 October 1916. On 13 November the battalion advanced at 5:45 a.m. Medlicott remembered the moments before zero hour:

> … the guns started to rain every conceivable explosive on the Hun – six minutes before zero – the noise is appalling, nothing can describe it at all, one sees great bursts of flame & sparks … So we started at 6 & in 3 minutes were due in the front line – we only walked, this line was not expected to be held, in fact it hardly existed so knocked about was it by our guns, the wire too, which a few nights before had been 25 yds thick, had been hurled by our guns in all directions & caused us no trouble.[33]

The following wave closed behind Medlicott's first wave which was:

---

30   Liddle Archives: WW1/GS/1188: Papers of K.V. Norman, p.240.
31   Liddle Archives: WW1/GS/1188: Papers of K.V. Norman, pp.242–244.
32   Formerly PS/7092 Private Leonard Perkins, 21/R Fus, aged 28, and PS/6923 Private Reginald Conway Lloyd, aged 22. Both were gazetted into the Queen's RWS in August 1916 and are commemorated on the Thiepval Memorial. Gross is also commemorated on the Thiepval Memorial.
33   IWM: Documents 1708, 1985-11-05: Papers of W.B. Medlicott, book 2, pp.38–39.

... a much safer position for one thing, as you are close under your own barrage ... all you suffer from is machine gun & rifle fire ... The morning was still desperately thick & made clearing up very slow work. The Huns were coming out freely from their dug outs ... they put up no fight at all, poor chaps, the barrage had taken all the kick out of them...

... It must have been about 7 [a.m.] when we got rather a heavy firing from our left. I could see a group of Huns showing over the top of the parapet, they knocked out one or two round me & I got one in the right forearm myself, breaking the bone & causing a compound fracture, it didn't hurt much but I remember shaking it as if a wasp had stung me .... then I got another bullet thro' the right buttock which made me sit down rather quickly in a little shell hole ... while I was in this crater I got another bullet wound thro' the right shoulder.[34]

Medlicott was taken to the rear by his servant. Second Lieutenant William Parry (ex-21/R Fus) was mortally wounded, and Second Lieutenant Stanley Gregory (formerly 19/R Fus) was killed.[35] Six other ex-UPS men were killed with 24/R Fus. Elsewhere in 2nd Division, 22/R Fus and 23/R Fus were in action on 13–14 November and lost eleven more ex-UPS men killed. Another seven men from various UPS battalions were killed with 17/R Fus.

Herbert Vinden (2/Suffolks) bore a charmed life during the Somme fighting and went from the most junior subaltern to the most senior in just a few months. He missed one attack whilst he was replenishing provisions for the officers' mess at Doullens. Such was Vinden's pessimism at survival that he spent £60 over 48 hours during a short leave to Paris. He was appointed adjutant and served in battalion HQ for the remainder of the battle. According to Vinden on 13 November; 'The attacking troops [of 2/Suffolks] dragging their feet out of the mud moved at a snail's pace and were slaughtered.'[36] Second Lieutenant Keith Dingley had been separated from his brother William. Keith Dingley commanded an attacking platoon and recalled the conditions; 'The attack was made at dawn in a thick mist it being impossible to see more than 15–20 yards in any direction... The ground was very heavy and swampy; the mud reaching up to the men's knees in many places; this made progress slow...'[37] Dingley overcame a German strongpoint and continued to advance towards his objective until his platoon was reduced to three men. He fell into a shell hole on the lip of a German trench; 'Seeing Germans advancing down the trench I fired my revolver at them; before I was able to reload they rushed me...'[38]

Morgan Williams of 10/RWF (3rd Division) was also involved in the assault on Serre:

We went over the top at 5.45 a.m. on November 13th behind a most terrific bombardment ... we ran into uncut wire, and neither the Suffolks on our left nor the Royal Scots Fusiliers on our right succeeded in reaching their objective. Some of our men got as far as the fourth

---

34 IWM: Documents 1708, 1985-11-05: Papers of W.B. Medlicott, book 2, p.39–40.
35 Formerly PS/3128 Private William Henry Liddon Parry, aged 26, was gazetted into the RF in August 1916. He died on 29 November 1916 and is buried in le Treport Military Cemetery. Previously PS/416 Corporal Stanley Harris Gregory, aged 27, was from Hillingdon. Gregory enlisted in 19/R Fus in September 1914 and was gazetted into 15/R Fus in August 1916.
36 IWM: Documents 5565: Papers of F.H. Vinden, p.18.
37 TNA: WO 339/60080: K.M. Dingley Officer File.
38 Keith Dingley was captured at Serre on 13 November 1916 and was held in Osnabruck. He was repatriated in December 1918. TNA: WO 339/60080: K.M. Dingley Officer File.

enemy lines, where they were overwhelmed by [German] reinforcements … a couple of our wounded officers were taken prisoner … I was … cut off in a large shell-hole between the second and third German lines. For two hours ten men and I hung on to its lip warding off three determined counterattacks and by that time our number had dwindled to five, with a dozen disarmed prisoners sitting in the mud at our feet.[39]

10/RWF lost three ex-UPS officers; Arthur Edward Capell, Herbert Gordon Thomas and George Thomas; the latter two had served in 21/R Fus.[40] These ex-UP men were some of the last to be killed during the Battle of the Somme in 1916.

Some officers arrived in France too late. Second Lieutenant Donald Wright was awaiting his marching orders in Felixstowe. He wrote home; 'You will think I am a fraud as I am still in England – the ports are still closed as there are submarines about and mines are in the Channel. … We actually got on the boat this morning and were having lunch downstairs, when the order was given for us all to come on shore again!'[41] The men who missed the Somme would probably be better able to acclimatise as officers, train with their battalions, get to know their men and would hopefully survive longer in battle. They would still face challenges in spring 1917.

39 Williams, *From Khaki to Cloth*, pp.71–72.
40 Both Capell (ex-City of London School) and George Thomas are commemorated on the Thiepval Memorial; H.G. Thomas (ex-Llandovery School) is buried in Euston Road Cemetery, Colincamps.
41 IWM: Documents 17675: Papers of S.D. Wright, letter 9 November 1916.

23

20/R Fus and the Winter of 1916–1917

But take me away from the Somme!

Take me back to dear old Béthune
That's far away from here;
Take me back to dear old Beuvry,
Where I can get some beer (and your washing done)
Put me where there are no High Woods
And you get your tots of rum,
I'll stay in the Brickstacks as long as you like
But take me away from the Somme![1]
    33rd Division Song, Spring 1917

Conditions on the Somme in the Winter of 1916/17 were far from good. Both sides had more hardware to chuck around, the mud was appalling and… communication trenches were impossible … The whole battle area was a sea of mud and shell holes. Roads, farms and villages were all obliterated; not even a blade of grass remained and the only signs of former life were the scarred stumps of trees … The incidental wastage during those winter months was high, but perhaps the cold, wet and darkness was almost as bad as the man-made instruments of destruction and death.[2]
    Second Lieutenant Pease, ex-19/R Fus

The Battle of the Somme had first decimated 20/R Fus at High Wood before thinning its numbers further at Lesboeufs. Though the battle was over, 20/R Fus would still not free itself from the mud of the Somme for many months. It would be the winter that followed this battle that would be the making or breaking of the battalion. If it could absorb replacements, rebuild its cohesion, train and rest then 20/R Fus might rebuild its former morale and esprit de corps whilst otherwise improving its military effectiveness. A more cohesive battalion with longer traditions, and officers with wider influence, might reduce personnel departing for commissions and even encourage more UPS men back who had been temporarily incapacitated by wounds or illness. Failing in this task would encourage

---

1 IWM: E.F. Chapman Papers, 1799, p.55.
2 IWM: N.A. Pease Papers, 8230, p.36.

further wastage to commissions amongst the remaining UPS contingent. Having been at war for a year the old and new men of 20/R Fus knew that 1917 would bring further fighting and this would be the next major test their battalion would face. The surviving public school and university men in the battalion had experienced a winter in the trenches and knew that it would be arduous. Getting the battalion through this period of hardship, monotony and danger, whilst also rebuilding and improving it, would be difficult and would demand leadership from all officers and NCOs.

After Lesboeufs

On 9 November the battalion moved to Méaulte where billets were occupied for two nights. That night, the previous night and the next night, German aircraft conducted air raids and though the amount of damage was not recorded it had enough of an effect on Cecil Isom for him to record it in his diary. During the stay in Méaulte Captain Yorston was evacuated sick with trench fever and returned to the UK. Morale would have increased as the battalion entrained on the 11th bound for the rear areas. The men arrived at Airaines next day before marching to rest billets at Mérélessart. They battalion would reside here for almost three weeks. Here the surviving officers enjoyed a more luxurious time:

> The room I am sitting in now is beautifully furnished, with a lovely carpet, several easy chairs … after dinner every night we play at being back in the 18th Century; [Ashley] Gibson is the Marquise, the Padre [Mannering] is the Cardinal whilst I [Chapman] am the count. We drink red wine out of magnificent glasses … I am growing to love France; the country folk are delightful – gay and cheery.[3]

There had been considerable officer casualties at Lesboeufs in a battalion already undermanned in junior officers. These casualties had a detrimental effect on those remaining; Chapman and Gibson were the only two officers left in C Company though a reinforcement draft had replaced its losses in other ranks. Consequently, most of their evenings were taken up with censoring letters. Chapman's companion, Ashley Gibson, also remembered this time of officer shortages which meant that the length of periods on watch in the trenches increased and intervals between them reduced. In the line Gibson recalled never having more than two hours off duty.[4] Chapman was clearly relieved when he was given the prospect of several weeks away on a training course at the 33rd Division School at Hallencourt. Having been with the battalion whilst it suffered heavy casualties meant that his name rose to the top of the leave roster and he was home for eight days between 14 and 23 December. During December 1916 Gibson was also sent on a refresher course at Flixecourt; 'My CO and Adjutant had been very frank. "These courses are all poppycock," they said; "but we must send somebody, and you could do with a holiday, couldn't you?"'[5] However, some courses were of importance to better educate new officers to conditions in France; no sooner than he had arrived one fresh-faced subaltern was sent on a 33rd Division course from 9 December to New Year's Day.

3   IWM: E.F. Chapman Papers, 1799, p.37.
4   Gibson, *Postscript to Adventure*, p.156.
5   Gibson, *Postscript to Adventure*, p.157.

Though Somme veterans departed, there was also new blood arriving. During this break Second Lieutenants William Claughton and Henry Clark joined 20/R Fus. They were not archetypal officers as the UPS battalions had known previously. The former was a music teacher at Giggleswick School, aged 41, and had served in the Worcestershire Volunteer Artillery before being gazetted to a commission in the Royal Fusiliers. The latter worked for his father's tailoring firm but had served in the Gordon Highlanders Volunteer Battalion and London Scottish. He was trained at the Inns of Court before being gazetted to 6/R Fus in October 1916.

Original UPS men who had survived the Somme steadily drifted away from 20/R Fus. Hodson considered remaining with the battalion; 'My coming home for a commission … was the easier road, easier than staying on and being promoted sergeant. I had concluded that that winter, seen to its bitter end, would put out my light.'[6] Hodson, Skelton and others had seen the lights of many men extinguished during the recent fighting; men who might have lived longer if they had departed to commissions. Godfrey Skelton certainly believed he bore a charmed life during the Somme fighting. Once word of his commission came through, Skelton had a very nervous few days in the trenches awaiting his departure date. He finally left for the UK on 28 November 1916. James Hodson presumably knew of someone who had died so close to being sent home to temporary safety. His fictional protagonist, Hardcastle, 'died' in a bombardment the day before he was due to leave his battalion for a commission.[7] Whilst at Mérélessart he was likely appointed a job that improved his survival prospects:

> I've got a 'cushy' job: safe: in the orderly room. You remember how when I first joined the Lanchesters they kept shoving me into the orderly room to be a clerk and I kept fighting my way out. Times change! … Gus [Colonel Garnett] has signed my commission papers, and any day now I shall have to see the Brigadier to be O.K.'d.[8]

Hodson departed on Christmas Eve 1916.

December saw 20/R Fus still in rest billets. After the last tour of the trenches some recommendations for immediate awards had been submitted and PS/5361 Sergeant Arthur Murgatroyd was awarded the DCM and SPTS/4882 Private Thomas Prescott was awarded the MM. The former's citation read; 'For conspicuous gallantry in action. With a small party he established an advanced post under very heavy fire. He set a splendid example of coolness and courage throughout.'[9]

Before 20/R Fus departed Mérélessart, Colonel Bennett rejoined from the UK; with his return Colonel Garnett went back to England on leave. After a long break away from the trenches both packs and hearts were likely heavy for the men of the 20th Battalion as they disembarked their train at Edgehill Station at 2:00 p.m. on 8 December and set out to Morlancourt. They were going back to the Somme battlefields. After several nights in camp the battalion crept nearer to the front. On 14 December the battalion marched to Priez Farm (west of Rancourt) where 20/R Fus was to be in support. This was a new trench sector that the battalion had not served in previously. The front line was at Rancourt which lay to the southeast of Lesboeufs. The Rancourt and

---

6   Hodson, *Return to the Wood*, p.22.
7   Hodson, *Grey Dawn*, p.287.
8   Hodson, *Grey Dawn*, p.277.
9   *The London Gazette*, 10 January 1917.

Bouchavesnes sectors of trenches had been taken over by 19th Brigade around this time from 100th Brigade. The sector had been, until recently, occupied by the French and according to the 19th Brigade diary:

> This section of the line proved to be in a very bad condition. The front line consisted of a series of short strips of trench, with an average of over a foot of mud at the bottom, and very few dug-outs. There were no communication trenches, and no movement of any kind was possible by day within a thousand yards of the front line.[10]

On 18 December 20/R Fus moved into the right-hand subsector relieving 1/Cameronians. The battalion faced St Pierre Vaast Wood near Rancourt. 2/RWF were on the left with 9/HLI on the right. A Company, 20/R Fus, was on the right; C Company was on the left; B was in support and D Company was in reserve. On 18 December, during the relief, two officers, Captain Wylie and Second Lieutenant Morley, commanding C and A Companies respectively, were wounded.[11] They were hit by shrapnel before they even reached the reserve trenches; the former was wounded rather badly, the other was slightly hit. Wylie had taken over Chapman's company whilst the latter was on leave and had allegedly eaten a Christmas pudding sent to Chapman. Also on 18 December Lieutenant Thistle Robinson joined 20/R Fus; within a month he was appointed to command C Company, much to Chapman's chagrin, as it prevented him being formally appointed to fill this gap.[12]

20/R Fus would be involved in helping to improve the habitability of these trenches through laying trench boards. Those in support would also need to work on the communication trenches. Little rifle or MG fire was reported and 2/RWF certainly considered it a sector where the 'live and let live' ethos of trench warfare reigned. Between 18 and 22 December there was a hard frost followed by a thaw which made tracks impassable. To minimise the strain on the companies in the front line they were rotated every 24 hours. However, the slow movement meant that company reliefs took much of the night.

Meanwhile, on 18 December 33rd Division was instructed to form a Divisional Works Battalion which was to be commanded by Lieutenant Colonel Bennett. Each battalion in the division was to contribute a platoon of fifty men each and the division had been trawled for professional gangers from its battalions.[13] The formation of this additional battalion effectively side-lined Bennett and left Lieutenant Colonel Garnett to command 20/R Fus. As battalions

---

10   TNA: WO 95/2420: 19th Brigade War Diary.
11   Captain John Howie Wylie, aged 36, was from Oldham, Lancashire, and was a civil engineer. He originally enlisted in the Royal Marines in 1914 but this was never finalised. He was gazetted to a commission in the Royal Fusiliers in March 1915 and arrived in France in May 1916. He had formerly served as a platoon commander and Lewis Gun Officer with 20/R Fus. Second Lieutenant Charles Morley, aged 24, was from Highgate and had formerly been an estate agent's clerk. He had served in Malta and France with 2/Londons before being gazetted to a commission with 20/R Fus in July 1916. He suffered multiple wounds on 18 December 1916 and was later discharged unfit.
12   Former PS/2007 Private Thistle Robinson, aged 25, from Putney, had been educated at Corpus Christi College, Oxford, where he had joined the OTC. He enlisted in September 1914 and was gazetted as one of the first set of officers for 18/R Fus. He was promoted to lieutenant but had been posted to 28/R Fus. He later joined 17/R Fus and was wounded at Trones Wood. He recovered and served with 106/TRB and 5/R Fus before joining 20/R Fus.
13   TNA: WO 95/2408: 33rd Divisional Admin War Diary.

would only send their least effective personnel to the Works Battalion, its need to be formed from scratch, and its non-fighting status, the command appointment would not have appealed to Bennett despite its necessity. Colonel Bennett went sick the same day and was evacuated to the UK on 22 December.

On the night of 22–23 December 20/R Fus was relieved by 2/A&SH. According to the war diary December cost the battalion two officers wounded, one man killed and two missing, and thirteen men wounded and evacuated. The actual numbers of fatalities were nine.[14] On relief 20/R Fus moved to Maurepas Halte where the men were loaded in lorries and travelled to Camp 21 which lay to the north of Bray-sur-Somme. It was at Camp 21 that 20/R Fus enjoyed Christmas 1916 which was very different to the previous year. Gibson recalled; 'I was back [from his course] with the battalion just in time for its second Christmas in France. … We had our Noel celebrations and tried to make them gay, and on the morrow of Boxing Day were off again further south still…'[15] On Boxing Day the battalion marched nine miles to Camp 13 which was southeast of Morlancourt. On the 27th, 20/R Fus marched to Edgehill where they entrained for Pont Remy near Abbeville and on arrival marched nine miles to Yaucourt. Meanwhile the battalion transport left from Camp 13 on 26 December and after a stop in Argoeuves arrived at Yaucourt on 27 December. Though he failed to mention anything special happening at Christmas Cecil Isom recalled on 2 January; 'We drummers kept up New Year with a feed & sing song. Fine time.'[16] The New Year brought further morale for Edward Chapman who received a belated Christmas parcel. How he disposed of it highlights a lack of cohesion within the 20/R Fus officers' mess:

> We are having the pudding and mince pies warmed for dinner tonight. Our mess is crowded with some horrible new officers (ex-sergeants in this battalion), who are not gentlemen (or only half gentlemen, which is worse), and whose ignorance of their own defects simply staggers us all. Well, I swore I would keep your good things until they were well out of the way, and now, as luck would have it, they are all out to dinner. So I shall enjoy it with the Padre and two other real friends. I don't believe in throwing pearls before swine![17]

---

14 The following men were killed, died or were missing during December 1916; PS/9987 Corporal William Clifford Evans, aged 21, of Glanafon, near Newport, died of his wounds on 15 December and is buried in Grove Town Cemetery. GS/38033 Private Albert Rodwell was killed on 17 December and is named on the Thiepval Memorial. SPTS/4914 Private Peter Isherwood, aged 30, was killed on 18 December and is commemorated on the Thiepval Memorial. GS/63380 Private Charles Beckman was the assumed identity of Charles Watson, aged 35, from Shaftesbury Avenue, London; he was killed on 19 December and is buried in Grove Town Cemetery. PS/9352 Private Roger Gwynne Field, aged 20, from Oxford, was killed on 20 December; in 1920 his body was moved from a field burial into Assevillers New British Cemetery. GS/46406 Private Samuel Williams, aged 26, died on 6 December and is buried in Abbeville Communal Cemetery Extension. GS/63537 Private James Cranston, aged 20, died of wounds near Lincoln on 12 December. GS/47022 Private Reuben Hunt died on 31 December; he too is buried in Abbeville Communal Cemetery Extension. In addition, PS/8468 Private James Connolly, aged 19, was likely wounded at Rancourt and he went back to the UK on 2 January 1917 and died of his injuries in the UK in May 1917. He is buried in Coventry.
15 Gibson, *Postscript to Adventure*, p.158.
16 IWM: C.H. Isom Papers, 16336.
17 IWM: E.F. Chapman Papers, 1799, p.43.

Major General Reginald Pinney was educated at Winchester College and the Royal Military College Sandhurst. He joined the Royal Fusiliers in 1884 and after passing the Staff Course in 1890 fulfilled several staff appointments in India. He served in South Africa (1901–1902) and afterwards commanded 4/R Fus. In 1913 he took over a Territorial infantry brigade and in 1914 commanded 23rd Brigade in France. He was promoted to command 35th Division in January 1916 and swapped with General Landon.

The New Year Honours brought some more awards for 20/R Fus. The Adjutant, Captain Modera, was awarded the Military Cross though he was by this time employed as a major. Lieutenant Colonel Garnett, Captain Templar and Second Lieutenant Humphreys were MiD. Humphreys was long dead having been killed back in August. Quartermaster and Honorary Lieutenant Claude Carleton was awarded a higher rate of pay in the same list, presumably in recognition of his unsung work supporting the battalion. Captain Templar rejoined the battalion on 6 January after being sick with bronchitis since mid-December.

At midday on 13 January Medal ribbons were awarded to members of the battalion by Major General Pinney (who took over the division during the Autumn). These consisted of MC ribbons to Major Modera and CSM Armour, the DCM ribbon to Second Lieutenant Murgatroyd and the MM to Private Prescott.[18] According to Isom the 20/R Fus band, now a flute band, played a concert at No 2 South African Hospital near to Abbeville on 16 January. Afterwards the band had supper in the Sergeants' Mess.[19] On the 17th the rest of 19th Brigade moved off leaving 20/R Fus attached temporarily to 98th Brigade. The battalion followed them soon after with the Transport Section departing on 18 January and the remainder of the battalion departed at 4:00 a.m. on 19 January and entrained at Longpre. The railway destination was again Edgehill near Morlancourt for another period on the Somme. The battalion marched to Camp 12 near Bray and arrived at 2:00 p.m. and were joined later by the Transport.

---

18 It is presumed that WO Class 1 Raven had returned, and Armour had been relegated to CSM having only been RSM temporarily at High Wood.
19 IWM: C.H. Isom Papers, 16336.

Next day the battalion moved back to Camp 13. At this point Major R.H. Goldthorp of the 4/Reserve DWR was attached to 20/R Fus for instruction; presumably prior to his being allocated as a major to another battalion.[20] Several new subalterns joined 20/R Fus but they were not UPS officers having been posted from other Royal Fusiliers Reserve Battalions in the UK. Meanwhile the UPS Reserve Battalions had been subsumed by the Training Reserve and no longer directly supported 20/R Fus. Second Lieutenants Uriah Davis and Bertrand Whiter were commissioned into 10/Londons and Second Lieutenant Thomas Harrison had been gazetted into 25/Londons. Only Second Lieutenant Eric Powell had served with 21/R Fus before being gazetted into the Fusiliers. Despite the high numbers of ex-UPS officers serving in France by this time, few returned to 20/R Fus.

The next move to the trenches would see a further British 'side-step' to the south with 33rd Division taking over more trenches from the French, this time astride the River Somme. Within 19th Brigade the Cameronians and Scottish Rifles would go into the trenches first before being relieved later by 20/R Fus and the RWF. On this occasion the trenches were in good condition with good communication trenches and whilst the frozen conditions continued the trenches needed little work, nor could any work be done.

On 22 January 20/R Fus marched to Camp 18 which was just south of Vaux in the Somme Valley. Next day the battalion marched up to Howitzer Wood where it relieved 1/4 Suffolks in reserve to the right-hand brigade in the line. The Transport were stationed at Frise Bend (across the Somme from Curlu) with the QM's store being located to Vaux, presumably in Camp 18, whilst the surplus personnel, under the recently joined Second Lieutenant Harrison, were at Camp 19. Between 4:30 p.m. and 8:30 p.m. on 24 January, 20/R Fus relieved the Cameronians in the front right subsection of the right brigade with the RWF on the left and French troops on the right. Three companies were in the front line with a half-company on an island at Ommiecourt-les-Clery and a further half-company in reserve. On the evening of 25 January Second Lieutenant Harrison took the surplus personnel into the line. According to Cecil Isom this included all the new draft and all but six drummers.[21] They arrived after midnight and whilst the bandsmen went into dug outs the new draft went straight into the front line, presumably, to gain experience.

Lieutenant Colonel Crawshay injured

On 26 January another controversy occurred according to Robert Graves; 'Colonel Crawshay [commanding 2/RWF] was wounded while out on No Man's Land inspecting the [2/RWF] battalion wire: shot in the thigh by one of the 'rotten crowd' of his letter, who mistook him for a German and fired without challenging'.[22] The 'Rotten Crowd' is a reference to a previous letter to Graves by Crawshay regarding the retirement at High Wood.[23] Whilst challenging an enemy

---

20 Major Robert Goldthorp later commanded 2/10 Londons from September 1917 to February 1918 before being replaced.
21 This new draft consisted of a batch of men sent to join 1/R Fus but who were posted instead to 20/R Fus. The majority had arrived in France in early January. IWM: C.H. Isom Papers, 16336.
22 Graves, *Goodbye to All That*, p.196.
23 In this earlier context, this could refer to the Cameronians and Scottish Rifles, Quoted in Graves, *Goodbye to All That*, p.185.

was customary, another officer, experienced in patrolling stated; 'It is not exactly advisable to take people on trust in No-Man's-Land'; an error is equally possible whereby one inexperienced sentry did not inform his successor of the patrol.[24] Frank Richards mirrored and elaborated on Graves' accusation though, as a signaller, he was unlikely to have been an eyewitness. It could be argued that No man's land was no place for a battalion commander. Inexplicably, Crawshay took the battalion MO, Captain Dunn, with him according to Richards.[25] Dunn, possibly an eyewitness, and not one to pull punches regarding 20/R Fus, stated in his battalion history that Crawshay was shot by an enemy patrol and suggested that Crawshay was revealing himself to the enemy on a skyline. However, Dunn's brief account, which mirrored the war diary, would suggest he himself was either not present or did not want it known that he was. Neither the 19th Brigade, 20/R Fus nor 2/RWF diaries mention any details relating to this incident, suggesting the brigade did not want to draw attention to this event.[26]

The weather was dry, but cold, with crisp snow on the ground. The sedentary nature of trench warfare in this sector allowed hot tea to be brought up before dawn, hot food arrived after dusk with more tea arriving between 8:00 and 9:00 p.m. This all made the cold conditions slightly more bearable. The trenches were 80–130 yards apart which meant that the Germans were generally unwilling to shell the British trenches but employed rifle grenades and trench mortars (launching 'aerial torpedoes') which, according to Edward Chapman; '… were quite a new experience for me. They came very silently, with a just audible "swish". I prefer an honest shell, which makes a loud whizz and gives you a fair warning that it is coming! After a couple of days any sudden swish or rustle made me crouch down in a most undignified position!'[27] Edward Chapman had not experienced 'normal' trench warfare, unlike the surviving UPS veterans of the previous winter. Chapman chose to retaliate with some of his own rifle grenades but the Germans replied with aerial torpedoes and later British trench mortars were deployed. According to Chapman; 'We had a ripping dugout, the trenches were in good condition, and the men all had good dugouts too. We were all very jolly together until one night one of the other subalterns was killed, which put an end to all of our jollity. Poor chap – he was newly married and had only been in the battalion for a few days.'[28]

On 27 January was the Kaiser's birthday. Though the Germans were reportedly quiet there as still some mutual aggression. The officer Chapman referred to, Second Lieutenant Bertrand Whiter, was killed on 27 January 1917 along with two other men; a further man died of his wounds.[29] The battalion left the trenches on 28 January, having been relieved by the Cameronians, and went into dugouts in reserve but left one half-company, under Chapman,

---

24 IWM: E.E.H. Bate Papers, 747, p.51.
25 Frank Richards, *Old Soldiers Never Die* (Eastbourne: Anthony Rowe, 1933), pp.218–219.
26 TNA: WO95/2420: 19th Brigade War Diary; TNA: WO 95/2423: 2/RWF War Diary.
27 IWM: E.F. Chapman Papers, 1799, p.45.
28 IWM: E.F. Chapman Papers, 1799, p.45.
29 Whiter, aged 24, was born in Bethnal Green but became a farmer in Canada. He had served with the 104th Regiment of Militia and enlisted in the Canadian Expeditionary Force in September 1914 and served with 7/CEF in France. On returning to the UK for leave in December 1915 he applied for a commission in 10/Londons. He joined 20/R Fus on 21 January 1917. He is commemorated on the Thiepval Memorial. GS/63700 Lance Corporal Frederick Craig had joined as a replacement after High Wood. He is buried in Hem Farm Military Cemetery. GS/12488 Private Herbert Knapper was a former coal miner who had served at Gallipoli with 2/R Fus. He is buried in Peronne Communal Cemetery Extension. G/38146 Private Ernest Dawson, aged 34, is commemorated on the Thiepval Memorial.

in a post on an island in the river. During this stint in the line, in addition to the casualties above, 20/R Fus lost another man killed and ten wounded.[30] The weary men were greeted with braziers in their dugouts, hot food, blankets and a postal delivery. The rest of the month was spent attached to 98th Brigade as their reserve battalion.

On 4 February 20/R Fus was relieved by the Scottish Rifles and moved to Camp 19 where it returned to the command of 19th Brigade and conducted training whilst in divisional reserve. 8 February saw 20/R Fus back in the line after relieving 1/Queens and had 2/A&SH on the left and 2/RWF on the right. On this occasion Colonel Garnett remained with 'B' Echelon whilst Major Modera MC accompanied the battalion into the line. The Battalion was not entirely sedentary; on the night of 9–10 February Second Lieutenant Chapman and G/63352 Lance Corporal Charles Howes patrolled in no man's land and were complimented for their work by Brigadier General Mayne. Chapman's letters modestly fail to record this accolade.

Chapman was again commanding C Company after Lieutenant Thistle Robinson had been accidentally injured. Robinson had wished to use some rifle grenades to disperse an enemy wiring party at 1:45 a.m. on 11 February. He held the rifle at 45 degrees with a No 20 Grenade loaded and prepared to fire. Sergeant Russell, of C Company, pulled the trigger but the grenade detonated before the rifle grenade rod had left the barrel. Chapman, as bombing officer, compiled a report and believed Robinson had accidentally lowered the barrel of the rifle so the grenade had got stuck or grazed the parapet. The grenade discharged a foot from Robinson's head, and he was lucky to have only suffer minor injuries. The enquiry found that the rifle grenades stored for use had rusty rods, though the least rusty were cleaned and slid smoothly into the barrel. A small fragment of rust may have been in the barrel which likely caused the grenade to stick after firing and detonate unsafely. The grenade had allegedly been tested and the fuze could have been at fault in this instance. The accident was attributed to bad luck. However, this was another example of an ammunition or weapon accident that could, all together, bring into question the effectiveness and standard of weapons training in 20/R Fus. Though Sergeant Russell was the NCO on watch he had no idea about bombs and was likely being shown by Robinson when the accident occurred. However, it should be noted that this incident occurred *because* 20/R Fus was not indulging in a 'live-and-let-live' style of trench warfare. This was an attitude to the credit of the battalion.

Officer-man relations were good according to Chapman: 'The feeling of comradeship out here is a thing I have never experienced before. Social distinctions and so on do not count for very much when you are under fire.'[31] Likewise, as a paternalistic figurehead for his platoon, Chapman recognised; 'Many of my men are very poor and come from the poorer parts of London and I want to do all I can for them.'[32] Chapman asked his family to send him additional comfort parcels that he could share with his men.

30 L/15270 Private Frederick Gough was killed on 25 January 1917. He is buried in Peronne Communal Cemetery Extension. G/24721 Private George Coker, aged 28, left the unit on 25 January 1917, presumably having been wounded during this stint in the line. He died of his wounds in the UK on 12 July 1917; he is buried in Kensal Green. PS/5187 Private Leonard Lawrence, aged 23, also departed on this day; he was later discharged from the Army, sick, in June 1917. He had enlisted with 20/R Fus, in Manchester, on 8 September 1914. G/9132 Private James Oldbury also left on 25 January and was discharged in April 1918 due to wounds. GS/63551 Private Albert Jones, aged 33, was also sent home on 28 January 1917; he was discharged sick in August 1917.
31 IWM: E.F. Chapman Papers, 1799, p.46.
32 IWM: E.F. Chapman Papers, 1799, p.46.

On 12 February the battalion was relieved by the Cameronians and went into support; three companies were positioned in trenches in Road Wood and one company was in Man and Merlin Trenches. This situation remained until the 16th when 20/R Fus took over the right-hand sector again. Two companies were in the front line, one was in support and the fourth was attached to 5/Scot Rif who were the support battalion. The fourth Cameronian company was relieved by an RWF company. This elaborate organisation was intended to enable 20/R Fus and 2/RWF to attack the German trenches. This operation was postponed and was ultimately cancelled. Edward Chapman was far from being disposed to aggressive activity: 'Hurrah for peaceful trench warfare. I sadly lack the offensive spirit. If all were like me, we should never win this war. Incidentally, if all were like me there would never have been a war at all!'[33]

Up to this point there was a hard frost underfoot but on 17 February there was a thaw which made the ground soft and movement became difficult. On 19 February 1917 Lieutenant Colonel Garnett returned to 2/RWF to take over command; Major Modera MC took temporary command of 20/R Fus. However, a permanent solution was found, on 22 February, when Lieutenant Colonel Donald Whitley Figg DSO was appointed to command 20/R Fus.

An improvised tearoom at Clery during the winter of 1916–1917 seems to summarise the desolate landscape, destruction and poor weather. The artist, Captain Seton Hutchison (of 33rd Division) recorded that this scene was captured during 'dirty weather' on 20 February 1917.

33 IWM: E.F. Chapman Papers, 1799, p.51.

On paper Figg was an excellent candidate. Aged only 31, he was relatively young to command a battalion. He had arrived in France in March 1915 as a captain with 1/24 Londons and had been rapidly promoted. He was undoubtedly brave having been 'Mentioned' on 1 January 1916 and was awarded the DSO for his work at Givenchy on 25–26 May 1915 whereby, 'For seventeen hours his conduct was a brilliant example to the hard-pressed men around him…'[34] He was also a Chevalier of the *Légion D'Honneur*. As the battalion was becoming an increasingly 'London' unit (as the UPS contingent diminished), a London Regiment officer would be suitable to command it. As a Territorial Force officer he might be a more understanding commander when dealing with the remaining peculiarities associated with the Public School survivors of the battalion than a more rigid 'Regular' officer. Having two years of experience in France he was likely a good judge of men and their characters and would be qualified to further recommend UPS men for officer training. He certainly found favour with Chapman, 'The new C.O. is very young, but a fine soldier, with the D.S.O. and the Legion of Honour. I think we shall like him…'[35] Colonel Figg would likely make an excellent commander if he was able to effectively stamp his mark on 20/R Fus.

Early in the morning of 21 February the battalion relief with the Scottish Rifles was completed and 20/R Fus could retire. The state of the ground had delayed the handover to 2:05 a.m. 20/R Fus went into reserve at Howitzer Wood. The Battalion remained here until 23 February when 19th Brigade was relieved by 100th Brigade in the line. 20/R Fus moved back to a tented camp in Suzanne where it stayed until 3 March. This had not been an easy time in the trenches either in the front line, support or reserve. On 23 February 1917 Cecil Isom recorded it was the first wash and shave he had had in 14 days.[36] During February 20/R Fus suffered Lieutenant Robinson wounded, eighteen men wounded, two missing and three died of wounds.[37]

'An Unlucky Spell'

On 3 March 20/R Fus marched to Frise Bend and went into reserve for 100th Brigade. Next day, the battalion went into the right-hand sector opposite Clery with three companies in the line and the fourth in support. Captain Dunn of 2/RWF recalled, 'Enough of Clery is [still] standing to make it ghostly. A village razed is not so sad to see as roofless, windowless, sagging

---

34 *The London Gazette*, 1 June 1917.
35 IWM: E.F. Chapman Papers, 1799, p.51.
36 IWM: C.H. Isom Papers, 16336.
37 The following lost their lives in February 1917: GS/63444 Private Lewis Jeffery, aged 37, died of his wounds on 1 February 1917. G/50282 Private William Norton, aged 40, was killed four days after Jeffery. Both were buried in Bray Military Cemetery. G/50497 Private Reginald Blackman had attended Bithery Boys School, Goudhurst. He was killed on 16 February. His body lies in Bray Military Cemetery. G/46354 Private William Barrell, from Kingston, and L/12236 Lance Corporal William Chanin, from Hounslow, died on 18 February. The latter served with C Company and was a veteran of service in France in 1914; he died of his wounds and is buried in Bray Military Cemetery. Barrell was killed and interred in Peronne Communal Cemetery Extension. GS/6159 Private Albert Cross, aged 21, succumbed to his injuries on 21 February and is buried in Peronne Communal Cemetery Extension. He was a Gallipoli veteran who had joined 20/R Fus twelve days before his death. PS/6146 Private Gordon Selwyn Gilman had served with 18/R Fus. He was born in New South Wales, Australia, and lived in Reigate. He died of an illness in Dover on 6 February.

walls; they all give one creeps at night…'[38] This was expected to be another quiet period in the front line though one man was killed during the relief.[39] The diary recorded on 5 March; 'Lt Col D.W. Figg DSO, temporarily commanding the Battalion, was sniped at the point I7.c.5.4. with front line at 12:30 p.m. He died at 5:30 p.m. at Battn Aid Post.'[40] The site of his wounding was southeast of Clery-sur-Somme though where the Medical Officer, Captain McConnell, was located is unknown. Captain Hollingsworth, who had recovered from his wound at High Wood, took temporary command. Cecil Isom was a bearer at Figg's funeral the next day and noted that there was a firing party and that the Last Post was played. Isom also put a cross on the grave the day after and boarded the grave around; presumably to make it more durable.[41] Two quiet days followed but Edward Chapman wrote to his family that:

> We have had an unlucky spell. Our new Colonel fell by a sniper's bullet his very first day with us in the trenches. And our last day [8 March], at 4 a.m. the Boche raided one of our saps. My company front was not touched, but we came in for the barrage, which was intense. I was on watch at the time. We had no casualties from the barrage – a most miraculous thing. It was so sickening that the damned Boche got away safely. We didn't know of the raid until it was all over, as it happened some distance to our right. The amount of artillery preparation they had was enormous, considering what a little raid it was…[42]

Though the 20/R Fus War Diary omitted this raid the 19th Brigade War Diary stated, 'In the early morning, under cover of a very heavy local *minenwerfer* bombardment, a small party of the enemy succeeded in penetrating into one of our saps unnoticed, and secured two of our men, who are believed to have been wounded; a third man, who was also wounded, was captured, but dropped by the enemy in No Man's Land.'[43] During this raid two men of 20/R Fus were missing, believed captured according to Captain Dunn of 2/RWF: 'Indications of a raid woke me at 5 o'clock. It turned out to be on our neighbours, two [men] were lifted.'[44] At least one man was recorded as being killed.[45]

On 8 March 20/R Fus was relieved by the 17th Battalion Welsh Regiment (17/Welsh) and went into support for 119th Brigade (40th Division). The next night 20/R Fus was relieved by 19/RWF; 20/R Fus marched back to Suzanne where it stayed in a camp for the night and moved to Camp 12 on 10 March.

There followed three weeks out of the line in camp during which training in open warfare took place. Though relatively quiet, there were a few events of note. On 14 March Major Lionel Frederic Leader, King's Liverpool Regiment, was appointed as the commanding officer.[46]

38 Dunn, *War the Infantry Knew*, p.301.
39 GS/62484 Private Hamilton Street was killed on 4 March 1917. He had formerly served with 1/3 County of London Yeomanry at Gallipoli before being evacuated. He is commemorated on the Thiepval Memorial.
40 TNA: WO 95/2423: 20/R Fus War Diary.
41 IWM: C.H. Isom Papers, 16336.
42 IWM: E.F. Chapman Papers, 1799, p.53.
43 TNA: WO95/2421: 19th Brigade War Diary.
44 Dunn, *War the Infantry Knew*, p.302.
45 GS/63542 Private Harold Tomlin was an errand boy pre-war. He is commemorated on the Thiepval Memorial.
46 TNA: WO95/2421: 19th Brigade War Diary states 12 March 1917; TNA: WO95/2408: 33rd

Leader had embarked overseas with the Connaught Rangers to Mesopotamia in 1915. He was serving as 2IC of 4th Battalion The King's Regiment and had already stood in as CO for that battalion in November 1916 and in February 1917. The delay in taking over 20/R Fus was due to Leader awaiting the return from leave of the CO of 4/King's. At 47, Leader was older than his predecessor. However, what he lacked in youth and dynamic leadership would be mitigated by his considerable experience from a career in the Volunteers and Territorial Force spanning many years. He was clearly considered competent and capable enough and had proven his abilities sufficiently to be entrusted with a battalion. Unlike his predecessor, or successor, Dunn had little to write, positive or negative, about Leader.

'With great presence of mind …'

Another example of the dangers of training with live grenades was illustrated by the death of Second Lieutenant Ian Badenoch. The details were recorded in a citation for the Albert Medal written by Lieutenant Colonel Leader:

> At Camp 12, near Chipilly, on 19th March 1917 during bomb throwing practice 2nd Lieut. I.F.C. Badenoch, 20th Royal Fusiliers was in a bombing pit with No: [SPTS/]4905 Pte J. Ormerod of the same Bn. A live bomb thrown by Private Ormerod failed to clear the parapet and fell back into the bombing pit. With great presence of mind, and regardless of his own safety Lieut Badenoch at once rushed to pick up the bomb and throw it out of the pit. He collided with Pte Ormerod and was so prevented from throwing the bomb clear before it exploded. Owing to Lieut Badenoch's prompt action, Pte Ormerod escaped with slight injuries, while he himself was fatally wounded.[47]

The citation published in the London Gazette added that 'Lieutenant Badenoch's prompt and courageous action undoubtedly saved the man who threw the bomb from death or severe injury.'[48] Badenoch was copiously bleeding from several wounds in his left thigh and had slight wounds to his right leg and face. Ormerod had multiple wounds, most of which were superficial. According to Chapman, 'Officers in this battalion are very unlucky. Our bombing officer, who had only joined us recently, had an accident with a bomb and died yesterday of his injuries. He took over the job of bombing [officer] from me only the other day … It is rotten luck to be killed behind the lines like that…'[49] Not only had Badenoch only recently joined but he had been an officer less than a month. It suggested that there was a shortage of experienced junior officers in the battalion if a newly-appointed officer was chosen as bombing officer. This, like Thistle's and Coggin's earlier accidental wounding add to criticisms of training and weapon handling in 20/R Fus. March 1917 had seen Figg and Badenoch killed along with two other men. Eleven men

---

Division Admin War Diary records 11 March 1917.
47   TNA: HO 45/10890/355028: Albert Medal File for I.F.C. Badenoch.
48   Second Lieutenant Ian Forbes Clark Badenoch, aged 19, from Sentosa, Banff, was educated at Banff Academy. He enlisted in August 1915 with the A&SH and was gazetted to a commission in the Royal Fusiliers on 26 February 1917. He was awarded the Albert Medal, Second Class, in March 1918. He is buried in La Neuville Communal Cemetery Extension, Corbie.
49   IWM: E.F. Chapman Papers, 1799, p.54.

were wounded in action whilst two were accidentally wounded (one of whom was Ormerod), and two men were missing, believed to be POWs.[50]

The Spring brought a new feeling of positivity with the warmer weather. In late March Chapman enjoyed a show by the 33rd Division concert troupe, 'The Shrapnels' who provided 'Bursts of humour'. Chapman recounted; 'They sang all the good old BEF songs, the same that they sang last August. I have got to love these; they come out of the very soul of the 33rd Division. Everyone in the division appreciates them, and no one else possibly can …'[51] A long, hard, winter was over and Chapman noted his own increasing optimism in letters home:

> We marched through a small town [Doullens] on our way here, where we were billeted, also last Oct, in the earlier part of the month. There was my old hostess standing at the door. I think she recognised me. This is a *bon* life! We march through the most rich and lovely country, and some marches have been pure joy from start to finish. Our longest one was 11 miles. No one has fallen out from C Company … I wouldn't be a student again for worlds.[52]

If accounts like Edward Chapman's are to be believed some rejuvenation did occur after this tough winter. Chapman himself had steadily grown in confidence and developed good relationships with his men. He noted the quality of the battalion band, his love for his company, even the pride his company cooks took in their field kitchen. However, the extent of this revival of fortunes for 20/R Fus was soon to be put to the test. Though Spring was in the air the improving weather brought training and preparations for a new offensive. Just as 20/R Fus had started to overcome its crippling losses from the Somme fighting, now it faced a new 'Big Push' which might thin the ranks of both cockneys and UPS men irrespective of social background or education. It was during this winter that 20/R Fus further progressed from being a Royal Fusilier battalion comprised of Mancunian grammar school boys, Manchester University graduates and other ex-public school boys into a battalion with a new character. Though it still retained some of its previous members it became a more generic Royal Fusilier battalion; a melting pot of Londoners comprising an assortment of cockneys from the Territorials, New Army volunteers and conscripts.

The officer corps was also changing and the 'new' officers were trained in the OCB system and came with knowledge of the new weapon systems and tactics of trench warfare. The new techniques regarding platoon training, and the development of the platoon as a tactical group comprising bayonet men, Lewis gunners, rifle grenadiers and bombers, were also being pushed down from above. Officers returning from courses also brought back some of this knowledge. It could be argued that these new methods were introduced from the bottom-up, top-down and laterally. In addition, the lessons the battalion had learnt on the Somme might also contribute to future performance. Before 20/R Fus joined the Battle of Arras the experience of other UPS officers and men caught up in this fighting must be explored.

50   GS/63542 Privates Tomlin and GS/62484 Private Street were likely the two fatalities. GS/63435 David Duce died of his wounds on 12 March and was interred in Bray Military Cemetery. G/42519 Lance Corporal Sidney Chatfield was wounded and succumbed to his injuries on 25 March; he is buried in St Sever Cemetery Extension. G/46679 George Barnes died on 28 March 1917 and is buried in Bray Military Cemetery.
51   IWM: E.F. Chapman Papers, 1799, p.55.
52   IWM: E.F. Chapman Papers, 1799, p.58.

24

UPS Men at the Battle of Arras

I know a lone spot on the Arras road

I think it will be summer soon, and perhaps the war will end this year ...[1]

I know a lone spot on the Arras road
That murmurs with the moan of Memory's pain.
And I should grieve my heart with stifled sobs
If I could bear to walk that road again.
   Memory by Arthur Newberry Choyce[2]

Prelude

Spring 1917 might have been the end to a nightmare of appalling weather, and fostered the optimism of a fresh start, but it also saw more casualties. One such fatality was Second Lieutenant Leslie Ireland who was gazetted into the Manchester Regiment in September 1916. He joined 12th Battalion Manchester Regiment (12/Manch R) on 30 December. On 11 February he wrote home, 'Everything OK. Thaw just started after a month [of] continual freezing. Think we go back for a month rest in about a week…'.[3] Next day Leslie Ireland was killed by a piece of shell fragment whilst visiting the front line near Sailly-Saillisel. His Colonel wrote; 'He had not been many weeks with us. But even in that short time we had all got to like him so much, and we feel his loss, not only as an excellent officer, but as a friend. He was killed during a bad and very anxious time and bore himself all through as a courageous gentleman.'[4]

The British planned for a major ground offensive at Arras. However, the German withdrawal to the prepared defences of the *Siegfriedstellung* or Hindenburg Line seriously affected this scheme. There was, therefore, a need to rapidly pursue the Germans and beat them to

---

1 Robert Vernede, *Letters to his Wife* (London: Collins, 1917), p.219.
2 Arthur Newberry Choyce, *Memory Poems Of War And Love* (London: John Lane, 1918), p.13.
3 IWM: L.W.C. Ireland Papers, 14441.
4 IWM: L.W.C. Ireland Papers, 14441.

PS/6818 Private Leslie Woodhouse Cubitt Ireland, from Hampton Wick, was killed a month short of his twentieth birthday. He attended Kingston Grammar School and King's School, Rochester, where he was in the OTC for three years. He was a clerk for the Phoenix Assurance Company and was qualifying for actuary work. He enlisted with C Company, 19/R Fus and disembarked in France in January 1916. He returned to the UK for a commission in May 1916 which he received in October 1916. Ireland was awaiting a permanent commission in the Indian Army when he was killed. He was buried in Guards Cemetery, Combles. (L.W.C. Ireland papers)

their own defensive positions. A period of open warfare, a rarity since 1914, preceded the British Spring Offensive.

On 8 April 1917 Second Lieutenant Robert Vernede, now serving with 12/RB, 20th (Light) Division, wrote an up-beat letter to his wife describing his billets and the destruction left by the Germans during their recent retreat. He made no reference to his battalion going into action next day. He finished with the optimistic words, 'I think it will be summer soon, and perhaps the war will end this year and I shall see my Pretty One again.'[5] Vernede's wife would never hear from her husband again. At 11:00 p.m. on 8 April the battalion pushed its outpost line forward. A brother officer wrote:

> He ... went forward with a Yorkshire officer, who was in charge of the Coy. on his right, with his Sgt. and Cpl. and a couple of his men, and as far as I can gather, came right on top of an enemy machine gun and was very seriously wounded. His men got him back to the Aid Station, but he did not survive the journey there. ... his Corporal, who looked after him, said his last words to him were: "Send my love to my wife."[6]

Vernede was wounded in the stomach which proved fatal; he died on his way to the Field Ambulance.'[7] 7/KOYLI was on the right flank of 12/RB near the southwest corner of

---

5   Vernede, *Letters to his Wife*, p.219.
6   Vernede is buried in Lebucqueiere Communal Cemetery Extension. Vernede, *Letters to his Wife*, p.xiii.
7   TNA: WO 95/2121: 12/RB War Diary.

Havrincourt Wood and the battalion advanced 500 yards into the wood. During this operation Second Lieutenant Arthur Edward Foster (formerly 18/R Fus), aged 23, from Bedford Park, was also killed.[8]

9 April 1917

As with the First Day of the Battle of the Somme, the various British divisions making the attack each contained many ex-UPS officers and men. This next section will record some of their stories and examine some of their experiences as part of this milestone battle. From the Chemical Works at Roeux, to Bullecourt; from Monchy-le-Preux to Gavrelle, ex-UPS officers and men contributed to the British advance and many paid for this success with their lives.

One very young officer with 15th (Scottish) Division was Second Lieutenant Eric Edward Dale. Dale had served under-age with 19/R Fus before being gazetted into the Black Watch. He joined A Company, 9/BW, in France on 8 March 1917. Within a month 9/BW was fighting at Arras on 9 April 1917. At 5:30 a.m. the battalion was in the vanguard of the attack. Within fifteen minutes the battalion had reached the German trenches but lost five officers and about fifty men from the German barrage. The next phase, at 7:10 a.m., saw 44th Brigade held up near Hermes Trench by MG and sniper fire from the embankment of Railway Triangle. 9/BW eventually conquered Railway Triangle and reached the 'Blue Line' with the assistance of a re-scheduled bombardment, a tank named 'Lusitania', and their supporting battalion. The 9/BW attack was wholly successful having captured 200 prisoners but had a heavy price.[9] This was Dale's first offensive and it must have been a sobering experience for a newly-arrived subaltern. He survived two further offensive operations near Arras before he was sent home injured. He later underwent court martial for drunkenness and absence at which point it emerged that he was an Egyptian Prince, Said Fazil, who had enlisted under a false identity with the connivance of the War Office.

After 9th Division achieved some form of breakthrough, 4th Division was sent forward to exploit this success. Three ex-UPS men, all formerly of 19/R Fus, served with 1/SomLI, Stormont Hays Card, Eric Mawdsley Gardner and Alfred Cyril Parsons. Each had been trained at a different OCB but had nonetheless found their way to the same battalion. Parsons had arrived a month after being gazetted and returned to the UK wounded. Card was a graduate of Pembroke College, Oxford and was from St Albans; Gardner was aged 21 and from Ilfracombe, Devon. Card and Gardner were with the battalion in Hyderabad Redoubt, north of Fampoux, on 10 April 1917. That evening two platoon-sized patrols were sent out to reconnoitre but were hit by German MG fire and almost wiped out. Because both Card and Gardner were killed on 10 April, and their bodies were lost, it is probable that that they commanded these two platoons.[10] One lucky soldier on 9 April 1917 was Archie Brown of 4/R Fus; he had twisted his ankle during a fatigue party and missed the battle.

---

8   Foster served with 18/R Fus in France before undergoing officer training and being gazetted to a commission in the KOYLI in early August 1916. He is buried in Neuville-Bourjonval British Cemetery.
9   9/BW lost fifteen officer casualties; enlisted casualties were not recorded.
10 They are commemorated on the Arras Memorial.

Major Gerald Godfrey Knighton, aged 28, attended Southbourne-on-Sea Prep School, Aldenham School, Hertford, and Pembroke College, Cambridge where he studied science and joined the OTC. Employed as a science master at St Andrews College, Toronto, he returned to the UK in 1913 to read Holy Orders at Bishop's College, Cheshunt. Knighton enlisted in the UPS on 12 September 1914 and was gazetted to 9/OBLI in October. He is interred in Bucquoy Road Cemetery, Ficheux. (Author)

Three men of 5/OBLI (14th Division) were former soldiers of the UPS Brigade. James Stevens was gazetted to a commission in August 1916, having served with 18/R Fus. As a relatively new officer he may not have known that George Gray and Major Gerald Knighton, of the same battalion, had served for a short time with the UPS. Gray had already been wounded in the hand and thigh in November 1915 near Ypres and had returned to France on recovery. Knighton had trained troops in the UK until September 1916 when he went to France; by April 1917 he was commanding a company. The battalion attacked Telegraph Redoubt, south of Tilloy, on 9 April, and carried the German trenches. In the process Stevens was killed and Gray was mortally wounded; he died the next day.[11] On 28 April, 5/OBLI was in the line east of Wancourt; Major Knighton was wounded, along with another officer, and later died of his injuries.[12]

'Smashed by the appalling fire'

Over the winter of 1916–1917 Herbert Vinden had been appointed intelligence officer at 76th Brigade and even stood in as Brigade Major. Vinden was in action during the Arras fighting when he was sent forward to locate the positions of the forward troops:

11 Stevens is buried in Tilloy British Cemetery; Gray is buried in Warlincourt Halte British Cemetery.
12 He is buried in Bucquoy Road Cemetery.

I walked along the Arras-Cambrai Road and saw our cavalry coming into action. ... As I walked, a regiment of cavalry charged on the opposite of the road and were slaughtered. The ground was covered with dead men and dead horses. I was in this barrage and flopped down in a ditch as a crump [5.9-inch shell] ... fell just on the other side of the road. I was wounded in the hand, arm and leg, but fortunately the wounds were superficial. I continued and completed my task.[13]

76th Brigade were to attack Guémappe on 11 April. 8/KO and 2/Suff R were in the lead with 10/RWF in support. Morgan Williams (10/RWF, ex-21/R Fus) was involved in this 'dud' show where the preliminary German artillery fire was heavier than the British supporting bombardment. Williams wrote later:

We were now walking into a hell of machine gun fire, heavier even than at Serre and the air was alive with the "ping-ping" and the "zip-zip" of bullets. By half-past seven there were no troops in front of us, for the King's Own and Suffolks were smashed by the appalling fire. How they hung on as long as they did is a mystery to me. All that I could do was to reform the line and occupy shell-holes...[14]

One of his brother officers, Tom Rea (formerly of 21/R Fus), was wounded.

The UPS and 'Bloody April'

The Arras Offensive coincided with a resurgence of German air dominance. The ex-UPS men who had joined the RFC had a difficult time. Second Lieutenant Ronald Hume, an officer survivor of High Wood, was attached to the RFC as an Observer. After training in the UK he returned to France on 5 April 1917 and joined No 20 Squadron. He was shot down and killed the very next day near Polygon Wood whilst flying on a bombing mission in an FE2B. His death epitomises the severe casualties to members of the RFC and exemplifies the short life expectancy of aircrew.[15] Likewise, Leslie Graham Lovell (ex-21/R Fus) was gazetted to a commission in the RFC in August 1916. He joined No 48 Squadron on 7 April as an observer. His combat career was also tragically brief. On 11 April 1917 he went out on an offensive patrol

---

13 IWM: Documents 5565: Papers of H. Vinden, pp.19–20.
14 Williams, *From Khaki to Cloth*, p.93.
15 Ronald Hume was born in 1886 and was educated at St Ronans. He was formerly a lieutenant of the Royal Navy having joined in 1902. He resigned in 1910 to avoid a court martial after £150 disappeared from the wine fund he was managing. He enlisted at Leatherhead in February 1915, was a corporal by August and a sergeant by November. He was originally with the MG Section, D Company, 20/R Fus. He rejoined 20/R Fus in France as an officer in April 1916 and was slightly wounded in High Wood. He was made the intelligence officer in December 1916 and commanded A Company in January 1917. Hume transferred to the RFC as an observer and after training he joined No 20 Squadron RFC on 5 March 1917. He barely saw out the month being killed during 'Bloody April' on 6 April 1917. He is commemorated on the Arras Flying Services Memorial. He wrote in The Pow-Wow as 'Sinbad the Sailor' presumably because of his previous nautical career. He was probably the only member of the UPS Brigade to have served in the Royal Navy, Army, and RFC during his career.

and was shot down near Gavrelle and killed along with his pilot.[16] Cator Barclay Holland, aged 20, also of No 48 Squadron, was reported missing on 11 April but was captured. James Hesketh was formerly a chartered accountant who was gazetted to the RFC in September 1916 and went to France in February 1917 with 48 Squadron. On 22 April Hesketh and his pilot were shot down in flames whilst on photo-reconnaissance. Hesketh was killed but his pilot survived.[17] Hugh Cecil Patterson of 21/R Fus was considered a spoilt boy of a Kentish family according to Hugh Spurrell.[18] He was gazetted and attached to the same squadron on 18 April and flew Bristol Fighters. He was returning from an offensive patrol when he was killed in an air collision with a Sopwith Scout on 30 April.[19] Another officer, Walter Alfred Southey (ex-19/R Fus) arrived at 48 Squadron on 11 April before being wounded on 4 May.[20]

Despite the heavy aerial casualties, there were equally significant losses amongst the infantry.

Second Lieutenant Arthur Lawrence Piper, formerly of 19/R Fus, recalled the attack by 1/Essex at Monchy-le-Preux on 14 April:

> The Battalion attacked on the morning of the 14th of April 1917 from the village of Monchy, the objective being a line about 1,200 yards due east of Monchy. Having reached the objective, I dug in and passed word down to right and left for the next officer and found there were no more in touch. I had about 25 unwounded men with me… Very soon after, the enemy came down from … my left and occupied our jumping off trench. I thought there was a frontal counterattack developing when I was wounded by a machine gun bullet passing through my face. I then lost consciousness and remember nothing until I found myself lying outside a German dressing station.[21]

'I shall soon be going back to England …'

Elsewhere on the front, away from the Arras battlefields, Donald Wright was also involved in heavy fighting at this time. The 8/Bedfords war diary recorded on 19 April that; 'All ground previously gained was held despite many bombing attacks by the enemy and a very heavy hostile barrage. At night battalion was relieved … Relief was not completed until 6.30 a.m., 20 inst …'[22] Casualties at Maroc included Donald Wright who was badly wounded. He dictated an up-beat letter whilst in Hospital:

> My dearest Father, Thank you so much for the letters – I had an operation at Calais this morning & am much better. The opinion is that I shall soon be going back to England &

---

16   Lovell is commemorated on the Arras Flying Service Memorial.
17   Hesketh is commemorated on the Arras Flying Service Memorial.
18   IWM: Documents 2138, 1992-11: Papers of H. Spurrell.
19   He is buried in Warlincourt Halte British Cemetery.
20   Southey had worked in the timber trade before joining 19/R Fus (as PS/6741) in September 1914. He returned to France as a pilot and was awarded the DSO, DFC and Bar and was promoted to Captain. He left the RAF in March 1919 only to die in a motor accident in May 1920.
21   Piper suffered from a fractured mandible. After capture at Monchy he was interned in Holland in April 1918 and repatriated to the UK in July. TNA: WO 339/58910: A. Piper Officer File.
22   TNA: WO 95/1611: 8/Bedfords War Diary.

I think I shall. I hope you will not have had the wind up too much not having had a letter for so long. Much love to all, your devoted son, Donald.[23]

A note accompanying the letter from a nurse stated; 'Your son has dictated this letter to me and asked me to send it to you. I am grieved to tell you he is much worse than he thinks, he is very seriously ill indeed, he had a big operation this morning.'[24] Wright died during the night; 'He was restless and delirious most of the time, but several times spoke as though he was speaking to you, but most of the time he imagined he was in the trenches…'[25]

Second Battle of the Scarpe 23 April 1917

On St George's Day the offensive at Arras continued with another massive set-piece battle using artillery to batter the German defences. Again, this fighting would see ex-UPS officers and men engaged with most of the attacking divisions. In the very south the operation involving 33rd Division is of especial interest because several UPS men were officers in the division and 20/R Fus played a minor role. Their story provides a vignette for the wider British performance as battalions were sucked into the German 'elastic' defences, were fixed, were counter-attacked, and spat out back to their own lines having suffered grievous casualties, often for little gain.

'Shot Down Like Rabbits …'

On 23 April 33rd Division intended to capture sections of the Hindenburg Line, and the village of Fontaine-les-Croisilles, as part of a wider offensive. 19th Brigade was in reserve (the limited role played by 20/R Fus will be covered later). The attack was two-pronged with 98th Brigade attacking along the Hindenburg Line whilst 100th Brigade attacked the defences frontally. 100th Brigade employed 1/Queen's RWS, with two companies of 16/KRRC attached to provide carrying parties. Their advance was to be made using surprise and prior to 'Zero' the Queen's would sneak forward and form up astride the Croisilles-Fontaine Road. At zero, the waves of infantry were to closely follow the bombardment towards the German Trenches. Two tanks were to support 100th Brigade but they did not arrive. 1/Queen's comprised several ex-UPS officers; Captain Francis Ball, Second Lieutenants Robert Walker, Gordon Jacob, Harry Thompson, John Holliday, David Millard, and Gerald Burghope; they were all from different UPS battalions and had accumulated in 1/Queen's from consecutive drafts of replacements.

The heavy barrage, richly supplemented by massed indirect MG fire, was met with little return fire from the German front line. Few casualties were incurred by the infantry, but the strong Hindenburg Line wire entanglements caused delays. Three types of Germans were found in the trench; those that were dead, those that resisted (and were bayoneted) and those who surrendered. Owing to thick barbed wire between the front and support lines, the barrage was 'lost', and the advance was checked. Captain Ball established his company in the Hindenburg

---

23 IWM: Documents 17675: Papers of S. Wright.
24 IWM: Documents 17675: Papers of S. Wright.
25 IWM: Documents 17675: Papers of S. Wright.

Line with bomb blocks created on his left and right flanks and along the communication trenches. Between these companies a communication trench ran towards the Support Line; a short distance down it was a large dugout which was used as an HQ.

At 12:36 p.m. a message was sent stating there was no sign of 98th Brigade and that more artillery fire was needed. According to Captain Ball:

> At 2 p.m. our supply of hand grenades was nearly exhausted, and the position was becoming very difficult … At this point O.C. attack sent for company commanders in order to confer with them as to what steps might be taken to improve the situation. I went to the HQ dug-out with the other Coy commanders. At this moment the Germans broke through the bomb blocks and captured the whole of the front line. … We were entirely surrounded and without any hand grenades.[26]

The senior officer present ordered the surrender of the isolated party. The Germans had pushed back the blocks having bombarded the defenders with trench mortar bombs and rifle grenades; at this point the blocks also ran out of bombs. This forced a retirement along the trenches. The surviving men of the Queen's and KRRC withdrew at a run back across No man's land; these retreating soldiers were shot down 'like rabbits' according to one observer. Second Lieutenant Holliday, commanding B Company, recalled of this battle:

> At 1.30 p.m. the position was getting serious, casualties were serious and bombs and ammunition [were] nearly exhausted by continual counter attacks. Whilst a consultation was in progress at B.H.Q. the German attack broke C Coy on the right and occupied a position between B.H.Q. and the outpost line … I left B.H.Q. and rejoined my Coy where I was able to send some men back before the last gap in the wire was closed by the Germans. I was finally bombed out of a shell hole just inside the wire along with four privates of B Coy…[27]

1/Queen's blamed the failure on the tanks, the wire, the lack of bombs and the losses amongst officers. They lost three officers killed, including Gerald Burghope and David Millard, two wounded and eight missing including Ball, Holliday, Jacob, Thompson, and Walker.[28] Kenneth Norman commented; 'The 1st Queen's always finds itself in the thick of the fighting. All the officers and most of the men of my old company, 'A', were either killed, wounded, or taken prisoners… For the fifth time the 1st Queens were smashed.'[29] 16/KRRC was in support of the West Surreys; Lieutenant Guy Spreckley was killed on 23 April. He was already a Somme veteran who had been wounded near Delville Wood serving with 7/KRRC in September 1916. He had been educated at King Edward's School, Bromsgrove and was a farmer.

---

26 Captain Francis Shorland Ball was a 34-year-old solicitor from Sloan Square; he was educated at Repton School. He was commissioned in December 1914 having enlisted in B Company, 20/R Fus, in September 1914. TNA: WO 339/40500: F.S. Ball Officer File.
27 TNA: WO 339/60128: J. Holliday Officer File.
28 PS/144 Private Gerald Harry Vernon Burghope, aged 20, from Brighton, formerly of 19/R Fus, was gazetted to the Queen's RWS in early August 1916. PS/3665 Corporal Davis Edward Hall Millard was gazetted to 3/Queen's RWS in July 1916 having served with 19/R Fus in France. Both are commemorated on the Arras Memorial.
29 Liddle Archives: WW1/GS/1188: Papers of K. Norman, p.316.

Third Battle of the Scarpe 3 May 1917

The next phase of the fighting provides a further case study of ex-UPS officers and men through the focussed examination of 21st Division in action, again, in the Hindenburg Line. Field Marshal Haig wanted to maintain the Arras attacks to continue to aid the French offensive. There would be another general offensive on 3 May with the Third Army attacking along its front and the Fifth Army attacking Bullecourt. 21st Division would attack along the Hindenburg Line whilst 62nd (2nd West Riding) Division, on the right, attacked it frontally to the southeast. This 'hammer and anvil' approach would remove the Germans grip on this section of their defences. 110th Brigade contained several ex-UPS officers. Captain David Kelly, now with HQ 110th Brigade, on reconnoitring the brigade's new area, testified to 33rd Division's pyrrhic victory on 23 April near Croisilles; '… I turned up one of the communication trenches … and found it literally heaped with British dead. The troops we were relieving had been clearing the recently captured Tunnel Trench and had dumped all the corpses in this disused side trench.'[30] These bodies may have included some men of 20/R Fus.

The 62nd Division attack failed, and the fighting stalled, despite some skilful leadership by junior officers. Lieutenant Tremlett of 208th MG Company formed a group of infantrymen into a bombing party and through his leadership they routed enemy troops, cleared two MEBUs and captured 500 yards of trench.[31] Second Lieutenant Alfred Montford, also of 208th MG Company, was killed in this action. Their endeavours were unable to change the course of events; there would be no 'anvil' for 21st Division to strike.

21st Division attacked with two brigades. 64th Brigade would attack along the Hindenburg front and support trenches. 110th Brigade would start their attack at right angles to the Hindenburg Line and attack parallel to it across the open ground to the north. Zero hour was 3:45 a.m. and the attack proceeded in darkness. MG fire caused many casualties and the attacking troops quickly lost cohesion and were held by the German front line opposite Chérisy and Fontaine. Second Lieutenant Charles Boyd, attached 9/Leicesters, was wounded, and reported missing on 3 May.[32] Lieutenant Francis Pitts was wounded and captured; he succumbed to his wounds in German captivity.[33] The amateur war poet Second Lieutenant Arthur Newberry

---

30 Captain David Victor Kelly was born in Adelaide and educated at St Paul's School, London, and Magdalen College, Oxford. He enlisted in the UPS Brigade, aged 23, in September 1914. He was gazetted as a second lieutenant in the Leicestershire Regiment in December 1914 and went to France in July 1915. He was later ADC to GOC 110th Brigade. D.V. Kelly, *39 Months with the "Tigers"* (London: Ernest Benn, 1930), p.64.
31 Lieutenant Elias Tremlett, from Crediton, Devon, aged 26, was a law student at the University of London before enlisting in 19/R Fus in September 1914. He was gazetted to 9/Devons in December 1914 and went to France with the MGC in February 1917. He was told of being awarded the DSO for this bravery the morning before he was mortally wounded on 22 May 1917; he is buried in Mory Abbey Cemetery.
32 PS/6049 Private Charles Gordon Boyd, aged 24, from St Peters in Thanet, was a civil servant. He enlisted in January 1915 in 18/R Fus. After enlisted service in France he was gazetted in August 1916 in 7/Notts&Derby. He is buried in Héninel-Croisilles Road Cemetery.
33 PS/3098 Private Francis Burton Pitts, aged 27, from Loughborough, was educated at Loughborough Grammar School and Magdalen College Oxford; he described his profession as 'gentleman'. He enlisted in 21/R Fus in September 1914 and joined No 2 Company. After service in France with 21/R Fus he was gazetted into 3/Leicesters in August 1916. He is buried in Cabaret-Rouge British Cemetery.

Choyce (formerly 28/R Fus) was shot in the arm whilst advancing. He continued forwards but was wounded for a second time; he lay in No man's land for twenty hours before he painfully crawled back 600 yards towards British lines. Reginald Minnis, 9/Leicesters, was wounded in the head but survived.[34] Previously 6/Leicesters had attracted ex-UPS men as officers. Amongst these were David Kelly, George Gillet, James Burdett, and Frank Curtis; all four were gazetted in late 1914 or early 1915. Several other ex-UPS men were with 110th Brigade; Second Lieutenant James Harratt, 6/Leicesters, luckily survived this attack.[35] Henry Duvall, of 8/Leicesters, and Alfred Pickard, with 6/Leicesters, were both formerly of 18/R Fus; Pickard was later wounded on the Aisne in 1918. On 4 May Captain James Burdett rejoined 6/Leicesters. As a result of casualties, he was rapidly promoted and in August 1917 he became battalion 2IC. In August 1918 he became the commanding officer of 6/Leics R and held the post until the end of the war.[36]

George Eyston was an artillery officer with 21st Division; he recalled conducting a reconnaissance and laying a telephone wire during the attack. He was shot in the leg and after completing his mission he crawled beneath a tank for shelter. This was a gamble; though the engine was running he believed that the transmission was damaged. The protection of this vehicle prevented him being killed or captured.[37]

This may have been tank 784, commanded by Second Lieutenant Harold Puttock (ex-18/R Fus), which was supporting 21st Division. It crossed the British trenches at Zero but the infantry did not follow.[38] The tank turned back to find them but when they did advance the tank soon became immobilised. Sadly, Puttock had contributed little to assisting 110th Brigade. Tank No 554 was commanded by Second Lieutenant Horace Richards (ex-21/R Fus).[39] It was engaged by enemy grenades and MG fire with armour-piercing bullets ripped through it wounding several crewmen and perforating the radiator. The tank limped back but had to be abandoned.[40]

Lieutenant Nicholas Pease recalled 8/E Surrey (18th Division) attacking further to the north:

> Just before dawn the guns opened up with a roar and over the top went the front line infantry, followed by their supports. I think I was due to go into action at zero plus 15

34 PS/1158 Private Reginald Ernest Minnis, 19/R Fus, was aged 19 on enlisting on 12 November 1914. He was from Streatham, educated at Merchant Taylor's School and was employed as a clerk in a jewellery manufacturing firm. After enlisted service in France he was gazetted in August 1916 and later joined 9/Leicesters.
35 Formerly PS/2809 Private James Horace Harratt, of 21/R Fus, a schoolmaster from Ibstock. He was awarded the MC in 1918.
36 Burdett had enlisted in 21/R Fus in August 1914 having been a hosiery manufacturer. He played rugby for Leicester and the UPS and was promoted corporal in December 1914; in January 1915 he was gazetted to 6/Leicesters to join his brother.
37 Eyston, *Safety Last*, pp.36–37.
38 PS/6034 Corporal Harold Harvey Puttock, aged 20, from Guildford, was educated at Archbishop Abbot's School, Guildford, and was employed as a chemist. He and his brother, Arthur, enlisted in November 1914 and served in the battalion until he was gazetted in August 1916. Puttock was transferred to the tanks in December 1916; he saw action at Bullecourt on 11 April 1917, Fontaine on 3 May and St Julien on 27 August; he was gassed on 10 October near St Julien.
39 Formerly PS/8216 Private Horace Leslie Richards, of 21/R Fus, was from Bournemouth. He was gazetted to a commission in the MMGC in April 1916. He initially served in D Battalion and later 19th Tank Battalion.
40 *D Battalion Tank Corps War Diary* <https://sites.google.com/site/landships/home/narratives/1917/battleofmessinesridge/arrasnarratives/10company3may1917> (accessed 8 May 2018).

minutes, so off we went ... The attack had evidently gone well – the enemy had fled without much resistance leaving plenty of pickings in the form of food, cigarettes, cigars and a few highly prized automatic pistols. There were, of course, a number of dead and wounded...It was a boiling hot day and in spite of a good deal of enemy shell-fire the situation appeared satisfactory ... By mid-day the enemy artillery fire increased somewhat and was now being punctuated with machine gun fire, and on closer observation I distinguished a line of infantry in extended order advancing...[41]

This was the German counterattack which forced the British infantry back to their original front line.

Further north, PS/10149 Private Albert Clayton, aged 22, formerly 29/R Fus, was with A Company, 8/R Fus (12th Division). He recalled his part in the attack:

The night was far too dark, however, for any kind of formation to be maintained for long, and after the scramble over our own front line all semblance of order was lost ... the affair took on more the likeness of a migrating swarm of locusts. ... The crowds of men seemed to float onwards in groups, like wisps of fog blown along by the wind.[42]

Clayton and a few comrades stopped 30 yards short of their objective which was strongly defended. By daylight it was clear that the Germans had re-occupied their front line, trapping this depleted spearhead of British infantrymen who sheltered in shell holes and exchanged rifle fire with the enemy once it was daylight. Eventually, about thirty men decided to retire across the open ground:

I was progressing well, I thought, putting a safe distance of undulant ground between me and the rattling [machine] gun, when suddenly ... something seemed to hit my left foot with the force of a flying brick ... and I rolled head over heels into a shell hole ... large numbers of Germans were now standing out in the open firing their rifles from the shoulder. I examined my right foot but, to my great surprise, found the blood now staining my puttees had its source at the knee...[43]

Most of Clayton's party were shot down and killed. Clayton passed out in a shell hole but was eventually forced to surrender. The Arras offensive petered-out though there were small-scale attacks in late May and June. The specific role of 20/R Fus throughout this offensive must now be examined.

---

41   IWM: Documents.8230: Papers of N. Pease.
42   Clayton, *Long Before Daybreak*, p.185.
43   Clayton, *Long Before Daybreak*, p.188.

25

20/R Fus and the Arras

The Chocolate Soldiers went well and suffered for it

> *We had become so callous in regard to death. I seemed to shrug it off somehow …*[1]

Having endured the harsh winter of 1916–1917 the improving weather and time away from the trenches was welcomed. Spring was in the air but with it came the prospect of another offensive. 20/R Fus spent three weeks training at Camp 12, north of Chipilly; optimistically, that training was for 'open warfare'. The endless field days and manoeuvres at Epsom, Clipstone and Tidworth were likely distant memories for the few surviving UPS officers and men. It is unlikely that the battalion had engaged in similar training for some time and it presumably needed an opportunity to train collectively as a battalion. Lieutenant Edward Chapman, of 20/R Fus, wrote on 31 March that:

> It may be some time before we go into action again. This will depend on the success of other troops. I hope it is not the case of "out of the frying pan into the fire"! The 33rd Division is never put in a really soft sector, for we are one of the crack divisions, and the Divisional General [Pinney] is full of the "offensive spirit".[2]

This training continued until 2 April when the battalion moved to Corbie where it was billeted before moving to Bertangles via Querrieu, Allonville and Coisy. Second Lieutenant Roy Greenwood joined the battalion on the 2nd.[3] From Bertangles, the battalion marched via Villers-Bocage to Beauval. During this latter journey 20/R Fus likely marched past their corps commander, Lieutenant General Sir Ivor Maxse (commanding XVIII Corps) and their divisional commander, Major General Reginald Pinney. Siegfried Sassoon, a subaltern in 2/RWF recorded that as his battalion approached a group of mounted officers the word spread that this was their corps commander. Lieutenant Colonel Garnett, now commanding 2/RWF, rode

1 LA: WW1/GS/1299: D.J. Price Papers, pp.20–21.
2 By April 1917 Chapman was an acting captain commanding a company. See IWM: E.F. Chapman Papers, 1799, p.56.
3 Second Lieutenant Roy Osborne Greenwood, aged 22, had been gazetted into 20/R Fus in April 1915 but had previously been a subaltern in the 5th Dragoon Guards before joining 20/R Fus. He went to France in November 1916.

forward and saluted Maxse only to undergo a tirade for not dismounting and regarding the state of his battalion. It is not recorded what Maxse thought of 20/R Fus, if they passed him, but he was known to have had very high standards. Passing such an important figure did not warrant recording or being remembered by Cecil Isom, Donald Price, or Edward Chapman.

A large reinforcement draft comprising ex-ASC men arrived on 4 April 1917; they had been intended for 32/R Fus. Receiving these replacements so soon before a battle (and after a training period) was aggravating as they were an unknown quantity, were likely under-trained and would not have known their officers and SNCOs. Though the quality of this draft was not recorded these men needed to be integrated into the battalion. Their arrival may have necessitated a day of rest on 6 April whilst they were incorporated. From Beauval the battalion marched to Lucheux where it rested for a day before moving onwards to Humbercamps on 7 April, and to billets split between Bailleulval and Bailleulmont next day. Here the battalion engaged in another few days of training in open warfare. On 11 April 20/R Fus was on 10-minutes notice to move and at 4:00 p.m. marched off to billets in Boisleux-au-Mont.[4]

Before describing the role of 20/R Fus in the successive phases of the Battle of Arras, the circumstances under which 19th Brigade was to be engaged must be examined. These will highlight the difficulties that 20/R Fus was to face.

Storming the Hindenburg Line

The Battle of Arras opened on 9 April 1917. In many areas, such as on Vimy Ridge, success was achieved. There were considerable delays on the southern end of the battlefront. Generally, the further south from Neuville-Vitasse, the greater the failure for the attacking troops against the defences of the Hindenburg Line or *Siegfriedstellung*. 30th Division failed with heavy losses and 21st Division was unsuccessful except for a tenuous lodgement in the German trenches at the top of Henin Hill. These troops were evicted the next day. Bombing attacks along the Hindenburg Line slowly 'unlocked' the German grip on the trenches opposite the 30th Division but not before another frontal attack on the Hindenburg Line by both divisions failed on 11 April. Consequently, 18/Manch R (30th Division) was ordered to enter the Hindenburg Line and to continue the advance from the Cojeul River to capture Henin Hill in the 21st Division sector. Against the odds 18/Manchesters stormed German trench barricades and using good low-level tactics they captured 1,700 yards of trench, and Henin Hill, despite strong German resistance. One of the company commanders involved in this action was Second Lieutenant Benjamin Augustus Westphal (formerly 21/R Fus).[5] Patrols from 21st Division then re-entered the Hindenburg Line and at 9:55 a.m. on Friday 13 April two battalions of 21st Division launched further attacks along the Hindenburg Line. This operation culminated with no ground gained and these positions were consolidated.

Meanwhile, 20/R Fus was moving closer to the battle. On 11 April 33rd Division joined VII Corps under Lieutenant General Sir Thomas D'Oyly Snow. At 6:00 p.m. on 12 April 20/R

---

4   Isom stayed the night at Mercatel.
5   Second Lieutenant Benjamin Augustus Westphal, aged 20, was born in Salem, Jamaica. He had attended Fulneck School, Leeds, and Manchester University and enlisted in the 21/R Fus in November 1914. He was gazetted into the Manchester Regiment in July 1916.

Fus followed the route cleared by 18/Manchesters along the Hindenburg Line and relieved troops from 30th Division. 19th Brigade was placed under the orders of 21st Division on 13 April. Once the 21st Division attack commenced the Cameronians and 20/R Fus took over the 'jumping-off' trenches and an outpost line running southwards from Henin Hill almost as far as Croisilles. Though this line was held for only twelve hours the battalion lost five men killed and twenty-one wounded, presumably from shellfire and sniping.[6] Second Lieutenant Arthur Murgatroyd was wounded in the right arm by shell fragments, buried and deafened by the same shell on 13 April.[7] At 7:00 p.m. 20/R Fus withdrew to the Hindenburg Support Line which the battalion held from the Henin–Héninel Road to near the top of Henin Hill.

19th Brigade was to resume the attack the next morning, the 14th. At 6:00 p.m. on 13 April Field Marshall Haig visited HQ 33rd Division which showed the importance of this operation. The attack was to be made by 1/Cameronians and 20/R Fus with 5/Scottish Rif in support. At 10:30 p.m. on 13 April the written operations orders arrived at HQ 5/Scottish Rif which stated they would attack through 20/R Fus instead. The 1/Cameronians would attack along the Hindenburg Line trenches whilst 5/Scottish Rif would attack over the open ground to their left.

The German blocking positions in the Hindenburg Line were a difficult prospect to attack. The Germans had strengthened their communication trenches to create new lines of defence facing westwards. Two battalions, attacking side by side, would face two very different tactical problems. Troops attacking in the open ground north of the Hindenburg Line would attack a German trench over open ground requiring speed and surprise. The other force would have to conduct slow methodical bombing attacks to storm the length of a complex German trench system with extensive dugouts. Both attacks would be difficult to coordinate and might end up being executed in isolation despite their proximity.

5/Scot Rif deployed onto the western slope of Henin Hill with two companies positioned to make the advance. One company of 1/Cameronians was to attack along the Hindenburg Line with two platoons attacking up each trench and another company was supposed to attack across the open ground to the north of the Hindenburg Line but failed to get into position in time; another company took over at short notice. The objective was the sunken road which crossed the Hindenburg Line south of Fontaine. Early on 14 April, the outpost line was re-occupied by 20/R Fus with the battalion being notionally in brigade support. Cecil Isom merely noted; 'In dugout 6 kilos[kilometres] long. Fritz at each end.'[8]

The attack commenced at 5:30 a.m. on 14 April with an artillery barrage that sounded to critical observers like it came from little more than an artillery brigade.[9] The Cameronian advance in the Hindenburg Line was stopped by enemy blocks having advanced 100 yards down the support line and 150 yards along the front line.[10] The company that attacked across the open

---

6 The following were killed on 13 April 1917: PS/9347 Private Harry Spencer Parrott, aged 20; GS/46179 Private William Cook; GS/63466 Private George Ansell, aged 26; and GS/63586 Private Alfred Young. All four were buried in Cojeul British Cemetery. G/32830 Private Richard Arnell, aged 30, is commemorated on the Arras Memorial.
7 Murgatroyd returned to duty two weeks later but suffered hearing and nasal problems, likely due to the blast.
8 IWM: C.H. Isom Papers, 16336.
9 Martin, *Fifth Battalion, Cameronians*, p.108.
10 Navigation and map-reading were difficult in trench systems at night and several sources give erroneous or conflicting trench map references.

was stopped by MG fire but dug in having advanced the line slightly. The positions gained were consolidated. The companies of 5/Scot Rif formed up and advanced at zero hour. As they topped the crest of the spur pointing towards Henin they were hit by German artillery fire and MG fire. The advancing waves pushed forward over the trench occupied by 20/R Fus and advanced about 1,000 yards. Enemy outposts were seen withdrawing towards Fontaine but nothing was seen of the Cameronians on the right. The 5/Scot Rif companies suffered heavy losses especially from enfilading fire from the left. According to HQ 33rd Division; 'The [19th Brigade] attack was unsuccessful owing to Division on the left [56th Division] not advancing…'[11] The barrage was so weak that the rapidly advancing Scottish Rifles passed through it without realising. The advance petered out and the Scottish Rifles dug in on a reverse slope. 20/R Fus moved to relieve them during the night 14–15 April. The Fusiliers would be next to attack. During the night 21st Division was relieved by 33rd Division; the latter would organise the subsequent attacks. 14 April cost 20/R Fus a further seven fatalities.[12]

33rd Division takes over

The war poet, Siegfried Sassoon, of 2/RWF, occupied trenches in the Hindenburg Line along with his company. Once his men were positioned, he investigated the dispositions of the company on his left:

> Poor devils – they belonged to an amateur battalion who suffered badly in the dud attack 36 hours later. I came around a corner and found a sort of panic–party going on at a point where the trench was like a wide nullah. A platoon (who must have been more than half as numerous as my whole company) had taken alarm. NCOs and men were jostling one another in their task to vanish through a narrow doorway which led to the bowels of the earth … one of them, panting excitedly, told me that "the Germans were coming over."[13]

Two officers were involved in this panicked display but there was no sign of an enemy attack. 20/R Fus was in brigade support in the Hindenburg Line and it is presumed that Sassoon was referring to 20/R Fus though he may have added some exaggeration concerning the extent of this display. He did not mention it in his contemporary diary.[14] The conditions in the trenches

---

11 TNA: WO95/2406: 33rd Division War Diary.
12 PS/5310 Private George Wilfred Meldrum, aged 25; GS/63480 Private Frederick Clark and GS/62841 Private Edward Laver, are commemorated on the Arras Memorial. PS/5718 Sergeant Frederick William Sutcliffe and GS/62828 Private Sidney Golding were buried in buried in Cojeul British Cemetery. GS/53025 Private Philip John Ingerson is buried in Héninel-Croisilles Road Cemetery. GS/63487 Private Frederick Chipperfield, aged 24, lies in Wancourt British Cemetery. On 15 April 1917 PS/8718 Private Alfred Joseph Brown, aged 21, from Addiscombe, was killed. He was educated at St Dunstan's College, Catford, and was employed as a bank clerk. He went to France with 20/R Fus in November 1915 and returned to the UK in late July 1916 having presumably been wounded at High Wood. Brown rejoined 20/R Fus (GS/47623) on 9 February 1917. He is commemorated on the Arras Memorial.
13 Dunn, *War the Infantry Knew*, pp.316–317; Siegfried Sassoon, *Memoirs of an Infantry Officer* (London: Faber, 1993), p.157.
14 Rupert Hart-Davis (ed.), *Siegfried Sassoon Diaries* (London: Faber, 1983), p.154.

around this time were unpleasant and morale-sapping as Donald Price remembered; '... We had become so callous in regard to death. I seemed to shrug it off somehow ... One lived with death and one became used to it.'[15] That apathy went hand in hand with periods of panic and fear; 'When I was in the agony of "Wind Up" – I don't think I could think of anything other than the terrible panic and fear of cracking up.'[16] Sassoon's own morale was also suffering from his proximity to dismembered British dead.

The 16 April assault – 'Another dud show'

After some delays the French Spring offensive was to commence on 16 April; a week after the first British attack. The British Army needed to exert further pressure to assist the French. On the evening of 14 April Allenby issued orders for the continuation of the offensive before being informed by Haig that there would be a week-long pause; theoretically this ended the First Battle of the Scarpe.[17] Nonetheless, on the morning of 16 April 19th Brigade was to launch an attack down the Hindenburg Line with Fontaine-les-Croisilles as the objective. Meanwhile, 100th Brigade was to advance to face the Hindenburg Line to the northeast of Croisilles and to consolidate their position. 1/Cameronians were to advance with one company down each trench in the Hindenburg Line (front and support) with the attack starting at 4:15 a.m. On the left 20/R Fus was to attack across the open ground in the same manner that the Scottish Rifles had done. The Cameronians were to lead the assault but were to be supported by a bombing party from 2/RWF under Siegfried Sassoon. It had been intended for this party to conduct a bombing attack along the tunnel concurrently with the attack along the trenches but this was cancelled. Sassoon recalled that the Cameronians were confident and already knew the ground having attacked on 14 April. However, the Cameronian plan may have been inadequately conceived and articulated because Sassoon returned from their briefing without much idea of his role.[18] The Cameronian attack the day before had also been imperfectly coordinated and executed which was not auspicious for the 16 April operation. On the 15th it rained and the trenches were filled with glutinous mud. Parties from 1/Cameronians carried forward and stockpiled bombs, along with other stores, ready for furthering the offensive the next day.

20/R Fus Attack Across the Open

On the left 20/R Fus was ready to attack by 2:45 a.m. on 16 April having taken over the positions reached by 5/Scot Rif on the reverse slope of the spur during their attack two days before. This was a trench roughly running to the northeast from the area of the forward trench barricade in the Hindenburg Line around Friedrich Trench. The Scottish Rifles, in gaining this line the day before, had already picked the low hanging fruit. The difficult section of the advance would take place next

---

15   LA: WW1/GS/1299: D.J. Price Papers, pp.20–21.
16   LA: WW1/GS/1299: D.J. Price Papers, p.21.
17   This may have been brought about by three of Allenby's divisional commanders petitioning Haig to halt piecemeal attacks. Cyril Falls, *The History of the Great War, France and Belgium 1917* (London: Macmillan, 1939), p.378.
18   Dunn, *War the Infantry Knew*, p.318.

20/R Fus and the Arras 379

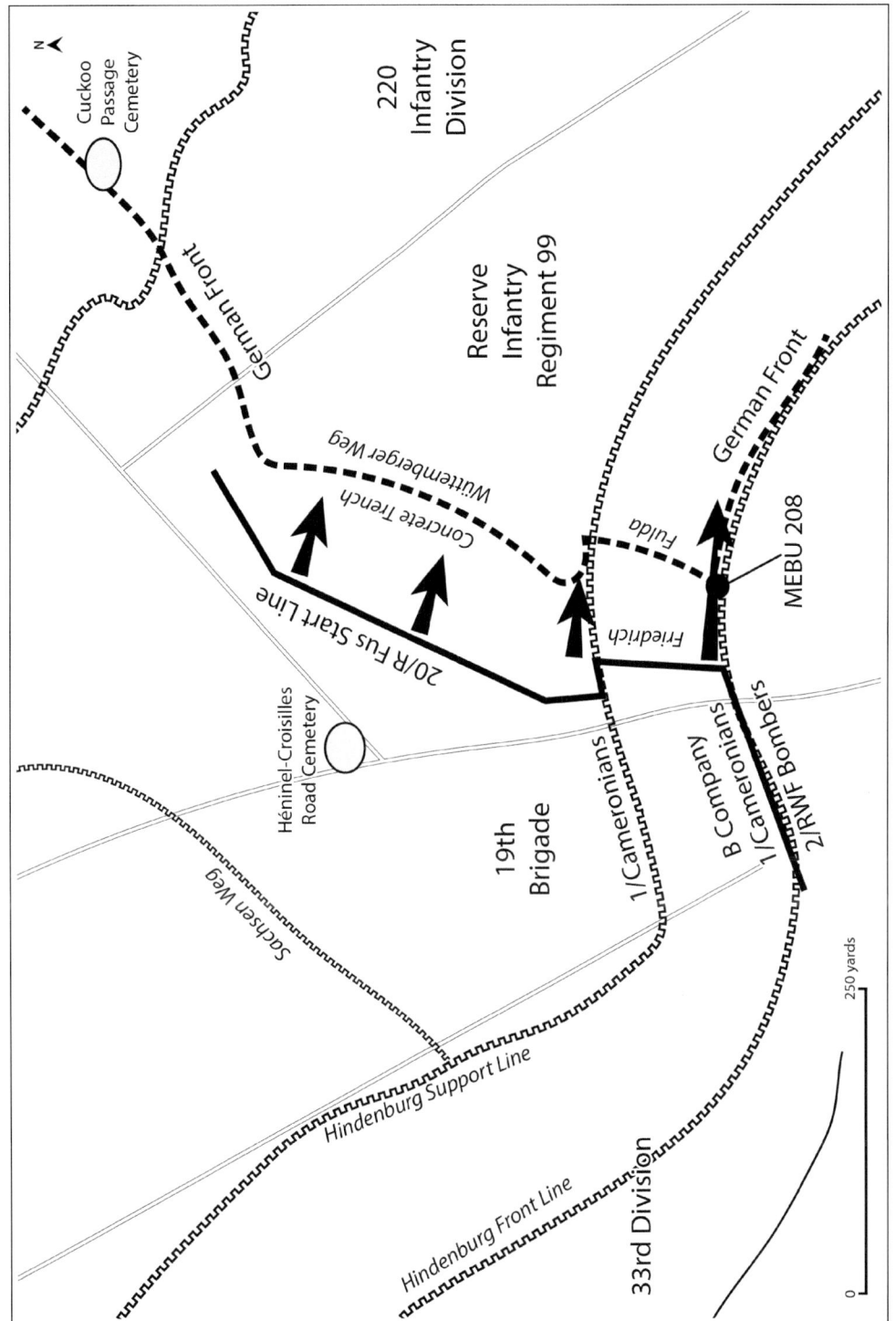

Map 9 Concrete Trench, 16 April 1917.

as 20/R Fus had to advance over the crest of the ridge against German positions dug in on the reverse slope. These German trenches could not be seen from the British line and the Fusiliers would be exposed to their fire at relatively sort range as the crested the rise. In addition, if German positions on either flank were not eliminated by the British barrage or kept busy by the Cameronians on the right or 56th Division on the left, German MGs would also rake 20/R Fus from the flanks. This was an attack where there was limited scope for tactical flair, as it was part of a set-piece operation, and much would depend on the effectiveness of the barrage and the success of flanking units.

At 4:15 a.m. the Cameronians attacked down the Hindenburg Line from Friedrich Trench. The commander of the Cameronian company in the support line failed to start at the correct time and did not make any progress. The Cameronian company in the front line advanced 500 yards but were held by German reinforcements. This watershed coincided with a shortage of bombs and this company was forced back to their start line. The Cameronian regimental history stating that; '…this lack of synchronization contributed to the success of the German counter-attack.'[19] However, 20/R Fus was not to know that the Cameronian company on their right had not succeeded. They would soon experience the consequences.

A and D Companies, 20/R Fus, were in their front line trench with B in the second line and C Company in the trench behind them. At Zero Hour, A, D and B Companies were to get into formation and advance. This was accomplished, but after covering only 100 yards they came under heavy MG fire from their front and flanks. This presumably occurred as the leading waves crested the ridge of the low ridge. From this point onwards progress was small and the advance inclined to the right. The attack was ceased having failed due to the severity of the enemy MG fire. In the words of Captain Dunn of 2/RWF; 'This was another dud show. The Chocolate Soldiers [20/R Fus] went well and suffered for it …'[20] This is interpreted to mean that 20/R Fus advanced with spirit but consequently lost heavily in casualties in this futile attack.

In light of the Cameronian failure, Sassoon was therefore ordered up to restore the situation though there is some confusion as to which trench he advanced up. Sassoon, and a Cameronian NCO advanced to their objective, where Sassoon was wounded.[21] Apparently the position was later handed back to the Cameronians and the trenches were lost again. According to the Cameronian regimental history 2/RWF sent fifty bombers to reinforce the front line but they arrived too late.[22] According to Private 'Dick' Richards the attack was successful but the gains were lost to a German counterattack. According to the Germans, *Reserve Infanterie Regiment 99.* bore the brunt of the British attack. They reported that a British battalion attacked down the *Siegfriedstellung* and a fierce bombing battle raged around one of the big shelters, presumably around MEBU 208, this ended in the favour of the British but the shelter was snatched back from them later.[23] Sassoon's vividly-described and self-deprecating attack may attract the attention of historians. However, no sooner had the first Cameronian attack been blunted than the 20/R Fus attack was doomed to failure. Despite the success of Sassoon's party the damage was already done for 20/R Fus. The battalion was left depleted in numbers, lying in the open on the hillside under enemy observation and unable to disengage, reorganise and press home

---

19 Story, *History of the Cameronians*, p.163.
20 Dunn, *War the Infantry Knew*, p.329.
21 Hart-Davis, *Sassoon Diaries*, p.155.
22 Story, *History of the Cameronians*, p.163.
23 *Die Osterschlacht bei Arras 1917. II Teil.* p.117.

their attack. The surviving men in the leading waves could only stay in cover and await darkness before returning to their own lines. It was not just the Cameronians' fault. On the left flank the troops of 56th Division had to advance rapidly across long swathes of open ground against heavy MG fire; they fared no better than 20/R Fus. 56th Division had suffered significant casualties since attacking at Neuville-Vitasse a week before. It was relieved by 30th Division.

Cecil Isom recorded his brief memories of 16 April 1917:

> Batt[alion] in attack. Unsuccessful. Out all night fetching in wounded from between the lines. Mud knee deep.
> April 17th Move out from trenches. In dugouts near Henin. We four H.Q. S.B.s went up line again and fetched more wounded in from front line. Third night without sleep.
> April 18th Returned to dugouts nr. Henin.[24]

Second Lieutenant Harry Clark, commanding C Company in Chapman's absence, showed his mettle during the attack on 16 April and was later awarded the Military Cross with the following citation: 'For conspicuous gallantry and devotion to duty. The attack being held up he got his men dug in and stuck to his position until relieved twenty-four hours later, when he withdrew the companies in good order. He set a splendid example to his men.'[25] Clark's MC was awarded for making a judgement that a further advance was likely futile and for holding the ground gained with the survivors. Luckily Chapman had been left out of the battle; he wrote afterwards:

> I had no say in the matter and was not particularly glad or sorry. So I have been out of it all. They have had a very bad time and I have lost a lot of my best friends, killed and wounded. How many I don't yet know. So I am not feeling very jolly. My best subaltern [Soro] is killed. … God knows how many of my NCOs & men are dead.[26]

The 20/R Fus suffered severe losses in this action especially amongst junior officers. Second Lieutenants Davis, Soro, Scott and Powell were killed; the Adjutant, Lieutenant Thomas Millard, was 'dangerously hit' and Lieutenant Colonel Leader and Second Lieutenant Archibald Graham were lightly wounded. Total fatalities for April were seventy-one men killed and

---

24 IWM: C.H. Isom Papers, 16336.
25 MC Citation, The London Gazette 30 July 1917.
26 In C Company Chapman lost Soro and three new officers (Davis, Scott and Powell) killed, as well as PS/9727 Sergeant Cross, a Corporal and about ten men. Lieutenant Uriah Philip Davis, aged 32, from Wandsworth, was gazetted into 3/10 Londons and joined 20/R Fus in January 1917. Second Lieutenant Eric Layton Powell was from Brabourne, Ashford, Kent, and had served in France with 21/R Fus from November 1915 before he was sent to an OCB for officer training. He was gazetted in August 1916 and joined 20/R Fus on 20 January 1917. Davis and Powell are named on the Arras Memorial. Lieutenant Ralph Quintus Scott, aged 32, arrived in France in December 1916. He had formerly been a captain in 3/E Surrey but resigned in 1909. He was in Australia early in the war. He joined 20/R Fus on 18 December 1916. He attended a Lewis gun course at le Touquet from 27 December to 14 January and was attached to the 33rd Division Works Battalion from 3 February to 10 March 1917. Second Lieutenant William Soro was aged 27 and from Reddish, Stockport. He was educated at Manchester Secondary School and Metropolitan College, St Albans, and was employed as a clerk. Soro had enlisted in September 1914 and served with B Company initially, and later as the MG Sergeant. He was gazetted in 20/R Fus on 18 December 1916 having been recommended by the CO. He attended a Lewis gun course on 31 December 1916 before going on leave 18–28 January 1917. Scott and Soro are buried in Héninel-Croisilles Road Cemetery. IWM: E.F. Chapman Papers, 1799.

Sketch by Cecil Isom depicting a Tommy and a sullen German prisoner. Though rendered months after the event, it was likely intended to trigger Arras memories. Unfortunately, 20/R Fus did not obtain many prisoners for the high casualties they suffered. (Reproduced with the permission of Pat Isom)

eighteen were missing.[27] Another 111 were wounded, which including Lance Corporal Harold Tyson who was sent to hospital with concussion.[28]

27 The following UPS men were killed on 16 April: PS/1308 Private Bernard Hepworth Brown, aged 31, was educated at Haileybury School. PS/8176 Lance Corporal Wilfred Lionel Coulshaw, from Muswell Hill, was an apprentice optician; PS/10134 Private William Darbyshire had attended a Lewis gun course. They were all buried in Héninel-Croisilles Road Cemetery. PS/4829 Lance Corporal Harry Fletcher, aged 21; PS/9727 Sergeant Francis Cross, aged 36; PS/10164 Private Frank Bartram, aged 20, of A Company; and PS/10059 Lance Corporal Wenden Ray Wilde, aged 24, had attended Manchester University. All four were commemorated on the Arras memorial. PS/5686 Sergeant Thomas Richard Sturley, aged 34, was an assistant teacher. PS/8274 Lance Corporal Geoffrey Thurnam Fairbank, aged 19, was educated at Haileybury. Sturley and Fairbank were buried in Wancourt British Cemetery. Two old boys from St Olaves School were killed and were buried in Héninel-Croisilles Road Cemetery; GS/53109 Private Wallace Gray, aged 26, was an apprentice clerk for a confectioner and PS/10635 Private Harold William Baker, played cricket and rugby at St Olaves and was a Cadet Corps NCO. Several veterans were killed: L/15052 Sergeant Fred Palmer had served with 2/R Fus at Helles, Gallipoli, and was awarded the DCM on 25 April 1915. L/14844 Corporal William Jolly was belatedly awarded the MM on 17 June 1917. G/13943 Private Albert Bilton had served with 2/R Fus at Gallipoli from September 1915. GS/15421 Private Thomas Metcalf had formerly served in the Balkans. All four were buried in Héninel -Croisilles Road Cemetery. L/15509 Sergeant Henry Sear, aged 24, had arrived in France in September 1914 with 1/R Fus; he is named on the Arras Memorial. Several 20/R Fus soldiers had previously served with other R Fus service battalions: G/50286 Private Herbert Penton, aged 31; G/47771 Private Frederick Morrison, aged 23, and GS/63058 Private Charles Barrett; were buried in Héninel-Croisilles Road Cemetery as was SPTS/4703 Private Frank Hollingworth, aged 24. SPTS/4882 Corporal Thomas Prescott, aged 21, of B Company, was awarded the MM for good work 3–8 November 1916; he is buried in Wancourt British Cemetery. The following served in the ASC before they joined 20/R Fus on 4 April 1917: GS/53264 Private Bernard Summerfield; GS/53254 Private William Payne, aged 24, of D Company; GS/53242 Private Alfred King, aged 24; GS/53234 Private Stanley Haill, aged 23; GS/53233 Private William Gardiner, aged 20; GS/53202 Private Bertie Huggins; GS/53200 Private James Hastie, aged 32; GS/53120 Private Thomas Wade, aged 32, of A Company; and G/47244 Corporal Robert Proctor, aged 32. All nine were buried in Héninel-Croisilles Road Cemetery. GS/53257 Private James Rushton's name is inscribed on the Arras Memorial.

The following formerly served with the Middlesex Regiment before they joined 20/R Fus in March 1917: GS/62811 Private Frederick Bishop, of B Company; GS/62830 Private Albert Havis; GS/62856 Private William Springall, aged 35; GS/62863 Private George Weeks, aged 31, reside in Héninel-Croisilles Road Cemetery. GS/62789 Private Walter Samuel Malpas; GS/62832 Private John Hale; GS/62849 Private James Radford; GS/62854 Private George Sellen, of B Company, are commemorated on the Arras Memorial. Of the remainder the following are named on the Arras Memorial: GS/53207 Private William Hastings, of A Company; GS/60926 Private Charles Godbold; and GS/63332 Private Solomon Cornblatt; GS/9436 Private Abraham Gregory, an errand boy for a draper; L/14675 Lance Corporal Edward Sacre, aged 21, was educated at the Central London District School; L/16220 Lance Corporal Thomas Lippy, aged 19, of B Company; G/24244 Private Benjamin Thomas, aged 22; G/38675 Private Frederick Tyler; GS/52825 Private Israel Goldman, of B Company; GS/53038 Private John Hewitt, aged 39. The following were buried in Héninel-Croisilles Road Cemetery: GS/7613 Private Henry Kirby; G/51259 Lance Corporal Owen Miles, of A Company; G/16645 Private George Irwin, aged 28; GS/53051 Private Frederick Bates, aged 26; GS/63518 Private Albert Radwell; G/743 Private Robert Edwards, aged 21; G/1317 Private Peter Davies; G/3285 Private Leonard Steel, aged 33, B Company, MG Section; G/3977 Acting Sergeant Alexander Brown; G/6056 Private William Rawlinson, aged 21; GS/9518 Private Richard Edmonds; G/26312 Private Frank Rose, aged 23; G/35375 Private Edward Sykes; G/36847 Private Henry Matthews; G/41787 Private Ernest Bird, aged 23; GS/43520 Private Herbert Smith; G/50184 Private Robert Dumbrill, who was employed at Christ's Hospital School, near Horsham; GS/53117 Private Thomas Roe; GS/53194 Private Percy Clarke; GS/62851 Private Wilfrid Rolfe, aged 20. GS/4960 Lance Corporal William Stanbrook, aged 21, is buried in Wancourt British Cemetery.

28 The following men died next day: PS/8055 Corporal John Edward Barlow, aged 33, of B Company; along with GS/53191 Private Samuel Carnell, were buried in Warlincourt Halte, British Cemetery. GS/62791 Private Thomas Mitchell, aged 38, is buried in Gouy-en-Artois Communal Cemetery Extension. GS/42529 Private James Walpole, aged 27, a bookstall manager from East Dulwich, and GS/53212 Private Stanley Kennaugh, aged 27, died on 19 April and were buried at Warlincourt Halte.

20/R Fus had indeed 'gone well' into this action and the failure of this attack could easily be blamed on the Cameronians. On 23 April 1/Mx and 2/A&SH attacked over the same terrain as did 2/RWF later in the day. They were unsuccessful; on the brow of the hill where the German trenches came into view there was; '… nearly a solid hurdle about thirty yards log, a swathe of Royal Fusilier and some Middlesex dead, lying where the machine-guns swept the brow of the rise – all shot in the head or chest …'[29] This 'hurdle' ultimately consisted of 20/R Fus, 1/Mx and 2/RWF dead.

Dunn and Sassoon suggested that 20/R Fus was not an effective unit. Chapman, a company commander, had a sanguine view of the abilities of his battalion; 'The battalion is going strong. It is a better battalion now than it was 6 or 7 months ago.'[30] 19th Brigade had not significantly advanced the line and according to rumours recounted by Captain Dunn the GOC was unhappy with the brigade's performance.[31] In the belief of the higher echelons of the British Army these attacks were chasing a defeated enemy but this was clearly incorrect. The brigade was committed piecemeal into a battle over which it initially had little control. It largely repeated the previous attacks made by 21st Division though the situation was not one where there was much scope for manoeuvre. The brigade, and 20/R Fus, had fought against a determined enemy in the confusing and difficult terrain of the Hindenburg Line. This sector was the linchpin of the German defences in the area and where the enemy was at their strongest. 5/Scot Rif, and later 20/R Fus, advanced across the open and suffered high casualties from enfilading MG fire. It was true to suggest that 19th Brigade had not been fully prepared for bombing operations and some units had mastered the techniques for fighting in the constricting trenches of the Hindenburg Line better than others. Sassoon was critical of the staff-work at brigade HQ which had exacerbated problems and which inhibited the movement and operations of the battalions. However, the planning and execution at battalion level had also been imperfect in some cases.

19th Brigade was relieved by 98th Brigade, which had been in support and were now to hold the line. That night the Cameronians and RWF were relieved though nothing went smoothly. 20/R Fus was supposed to be relieved by 1/Mx but the guides became lost and 1/Mx wandered all night before daylight enabled them to right themselves. The two companies of 20/R Fus in isolated positions had to await nightfall on 17 April before being relieved.

The French offensive had started on 16 April. Sir Douglas Haig, after initially receiving positive reports at GHQ, recorded in his diary that though the French claimed to have secured 10,000 prisoners their much-claimed victory had not been realised. He was unable to get firm details on the progress of the French attacks and remarked; 'This is always a bad sign, and I fear that things are going badly with their offensive.'[32]

The rest for 20/R Fus was brief. The battalion relieved 1/4 Suffolks (who had taken over from the Cameronians) on 20 April to hold the line whilst 98th Brigade prepared for the next phase of the Arras Offensive. C Company held the Hindenburg front line including the trench block; D Company were in the support line also holding a block; B Company were holding a short trench north of the Hindenburg Line whilst A Company were in the Hindenburg Support

---

29 Dunn, *War the Infantry Knew*, p.337.
30 IWM: E.F. Chapman Papers, 1799, p.56.
31 Dunn, *War the Infantry Knew*.
32 Robert Blake (ed.), *The Private Papers of Douglas Haig 1914-1919* (London: Eyre & Spottiswoode, 1952), pp.217–218.

An idealised view of a bomb fight published in *The War Illustrated*. Though depicting an engagement near Oppy Wood, it amply demonstrates a bombing party moving along a trench with both bombers and bayonet men advancing in unison. They would be supported by Lewis gunners and rifle grenadiers who would suppress the activities of enemy bombers. Not depicted are and the numerous bomb carriers required to supply the operation.

Line. At 6:00 p.m. on 21 April 1/4 Suffolks took over again and 20/R Fus withdrew and moved to billets at Boyelles. During this move Second Lieutenant Charles Aberdein was wounded in his left leg and right hand, presumably by shellfire.[33] Another man was killed on 21 April.[34]

St George's Day – Second Battle of the Scarpe

The 33rd Division plan for the next attack on the Hindenburg Line and Fontaine-les-Croisilles called for a two-brigade assault. 100th Brigade were to attack on the right facing the Hindenburg Line, 98th Brigade advanced on the left along the trenches. 19th Brigade held the line prior to the attack and were in reserve. On the left 30th Division was attacking. Fifth Army, on the right, would not advance. The artillery of three divisions supported the operation with a 'rolling' or 'creeping' barrage to help the infantry forward and smoke was to be used to screen the very right flank of 33rd Division from observation. Between 13 and 20 April the artillery of VII Corps fired a total of 148,736 shells and between 20 and 27 April 392,016 shells. Edward Chapman observed the artillery barrage; 'I shall never forget the preliminary bombardment.

---

33 Second Lieutenant Charles Aberdein, from Sefton Park, Liverpool, was gazetted to a commission in the RF on 2 June 1916. After being wounded on 21 April 1917 he was discharged on 1 August 1918 due to wounds.
34 G/53121 Private James White, from Limekilns, Dunfermline, is buried in Cojeul British Cemetery.

We stood and watched it at 5 o'clock in the morning, just as it grew light. It was more intense than anything I saw on the Somme'[35]

Despite the plan, and strong artillery support, the attacks went badly at 4:45 a.m.; after initial success the spearhead of 98th Brigade in the Hindenburg Line, 1/4 Suffolks supported by 2/RWF and two tanks, were forced back to their start line. Two battalions which attacked across the open ground north of the Hindenburg Line failed, like 20/R Fus had a week before, due to frontal and enfilading MG fire. Though elements of 1/Mx and 2/A&SH penetrated the German front line they were cut off behind it. The 100th Brigade attack by 1/Queen's RWS also failed after some lodgements were made. Taking advantage of British casualties and confusion German counterattacks on the British Hindenburg Line barricades almost led to a British collapse. In the confused fighting much of 19th and 98th Brigades were committed to this defence of the Hindenburg Line and a further attempt to advance into German territory in the evening, which also failed. Lieutenant General Snow wrote; 'We are having a deuce of a fight today everywhere. – Success somewhat variable … we have not made the headway we hoped. However, perhaps we shall by evening.'[36] Though the Germans had seemingly blunted all British attacks the existence of the 1/Mx and 2/A&SH companies trapped in their rear dislocated their defence and the Germans withdrew their forward troops during the night.

The 20/R Fus remained in bivouacs in Boyelles but was held at ten-minutes notice to move all day on the 23rd and were about the only battalion in 33rd Division that was not engaged. Whether this was to preserve it as a reserve by the divisional commander or due to concerns concerning its ability are unknown. At 10:00 p.m. it moved forward to shelters in the sunken road southeast of Henin.

Throughout the night flares went up from the German lines and No man's land was swept by traversing machine guns. Just before dawn a wounded highlander limped through the German wire to the British Trenches held by 2/RWF. He reported that the enemy had vacated their positions. Orders were received at 6:30 a.m. on 24 April that 98th Brigade were to be relieved by 19th Brigade. At 7:00 a.m. 1/Cameronians relieved 5/Scot Rif and 1/4 Suffolks in the forward positions. At 9:30 a.m. on the 24th 20/R Fus moved forward from its overnight position and relieved the remains of 2/RWF by 1:00 p.m. When it was known the Germans had withdrawn the Cameronians and 20/R Fus sent out patrols and companies advanced behind them to determine the extent of this withdrawal. Patrols advanced as far as where the Hindenburg Line crossed the St-Martin-sur-Cojeul road where they came under enemy fire. The leading troops erected barricades at the extent of this advance and a line of posts was established. The ground disputed the previous day was taken without a fight. The 20/R Fus war diary states that they pushed out patrols as far as the Sensée.

Ultimately, 20/R Fus took a position north of the Hindenburg Line to face the new German positions that they had withdrawn to. This line consisted of ten strongpoints running north eastwards from the Hindenburg support line. A Company was on the right; D Company was on the left. B and C Companies 20/R Fus were in support to these new positions. Two companies of pioneers dug a new trench to the north of the Hindenburg support line which was occupied by 1/Cameronians – this line was continued to the left by 20/R Fus. Edward Chapman arrived on the battlefield after the attack:

---

35   IWM: E.F. Chapman Papers, 1799, p.62.
36   IWM: Documents 7274, 1976-03: Papers of Sir Thomas Snow.

... before long I found myself in an old Boche Trench, and he had evidently left rather hurriedly. We came back home laden with souvenirs! ... The ground there was simply covered with dead – our dead. We buried a lot of our own, who had been killed the week before. The thought of all those fellows knocked out makes you fed up. Of course there are a good many dead Boche in their old trench.[37]

All Cecil Isom recorded on 23 April was; 'Big attack on. Battle raging <u>all day</u>. Moved forward to Sunken Rd, Near Henin.' Next day he moved into the support trenches in the old front line where his company relieved the RWF.[38] Lieutenant General Snow wrote to 33rd Division; '... the Corps Commander wishes to record his appreciation of the high fighting qualities of your Division. The fighting of the last few days has been severe, and in open fighting your Division, in a difficult situation, outfought, and inflicted severe losses on the enemy.'[39] Snow wrote in his own diary; '[The Germans have] ... now had enough of it in front + has retired and we are pressing on.'[40]

This advance and patrolling would have taken place in view of German MGs and artillery spotters. Three men were recorded as killed on 24 April. These included PS/5404 Private Alfred Ormerod who was struck in the stomach by a shell fragment and died almost immediately. Ormerod was running to bring a stretcher to give aid to his platoon sergeant who had been seriously wounded. His sergeant was likely GS/12220 Lance Sergeant William Harwood who was also killed.[41] A further five men were killed the next day.[42]

At 11:00 p.m. on 25 April, 20/R Fus was relieved by 10/KOYLI and withdrew to billets at Boiry-Becquerelle. At 5:00 p.m. next day the battalion moved to Boyelles and on the 27th marched back to Bailleulval to rest and train. The remainder of the month was spent in training for open warfare. Special classes were run for scouts, signallers and map reading for NCOs. One non-battle casualty on 26 April 1917 was Cecil Isom who was taken ill with Pleurisy and was sent 'down the line' to hospital with a temperature of 102 degrees. He was despatched to the UK on 13 May 1917 and would not return to the 20/R Fus.

Between 22 and 26 April an estimated 212 officers and 5,739 men of 33rd Division became casualties. Edward Chapman summed up the situation succinctly; '... we are winning – at least the newspapers tell us we are'.[43] On 2 May the battalion marched to bivouacs at Adinfer where open warfare and specialist training continued. On 3 May there was another major 'push'

---

37 IWM: E.F. Chapman Papers, 1799, p.62.
38 IWM: C.H. Isom Papers, 16336.
39 TNA: WO95/2426: 2/A&SH War Diary.
40 IWM: Snow Papers, 7274.
41 PS/5404 Private Alfred Ormerod, aged 34, was educated at Manchester Grammar School and the University of Durham. He was about to be ordained by the Bishop of Southwell when war broke out. He is buried at Wancourt British Cemetery. GS/12220 Lance Sergeant William John Harwood had formerly served with 11/R Fus and GS/62816 Private Joseph Banks, aged 35, had previously been with the Middlesex Regiment. Banks and Harwood are commemorated on the Arras Memorial.
42 The following men were killed on 25 April 1917 and are commemorated on the Arras Memorial: GS/63376 Private Frank Johnston, aged 22, of D Company; G/36349 Private Jesse Foster, a baker and confectioner; G/42226 Private Oliver Ablitt, aged 20; GS/62843 Private Thomas Manchip, aged 35; GS/62847 Private Edward Pye, aged 26, of B Company. GS/63562 Private Frederick Goodling died of his wounds on 29 April and lies in Warlincourt Halte British Cemetery.
43 IWM: E.F. Chapman Papers, 1799, p.63.

planned for the whole Arras front but this would not involve 33rd Division. On returning to the line 20/R Fus would find little change to the forward positions beyond the increase in bodies.

On 4 and 5 May the training programme changed to trench-to-trench attacks and troops were exercised in the drills for bombing along communication trenches. In addition, five officers and five NCOs per company were practiced in night marching using compass bearings which suggested that lessons from the 3 May attack had been learnt and promulgated. Training on 6 and 7 May took the form of moving as a battalion in artillery formation and practicing deployment from that formation. After this period of training there was a quieter day on the 8th when there was a divisional sports day held 2km northeast of Douchy-les-Ayette. Training continued on the 9th, 10th and 11th with further NCO map reading classes and regimental sports held on the 10th. At 6:00 a.m. on 12 May the battalion moved to bivouac in Moyenville.

On 15 May this rest period came to an end; 20/R Fus relieved 1/9th HLI in six outposts which covered the British front from just east of Croisilles to just south of the forward British positions in the Hindenburg Line. D Company held the right-hand three posts whilst C Company held the three on the left. B Company was in support and A Company was in reserve; Battalion HQ was in a sunken road just northwest of Croisilles.

Around 15 May there was a suspicion of a further German retreat and patrolling was increased in the following days to ascertain if this was true. On 16 May 20/R Fus described patrolling as 'intense' around this time and it is presumed that Second Lieutenant Hill and the Observer Section and Snipers were active. To conserve the morale and comfort of the battalion for future operations there was a company relief on the night of 17–18 May whereby A Company took over Posts 1–3 and B Company occupied Posts 4–6. C Company went into support and D Company into reserve.

Attacks on Tunnel Trench – 20 and 27 May 1917

Though the 3 May 'push' had cost the British Army considerable casualties, the offensive was not discontinued. A strategic problem was brewing; in the matter of a few weeks the British Army had gone from the supporting act to now being a one-man show after the failure of the French offensive. Field Marshal Haig was now planning a British 'knock-out blow' in Flanders that would become The Third Battle of Ypres. This required the centre of gravity for the BEF to be shifted northwards; thereby denuding the Arras offensive. Attacks at Arras would maintain pressure on the Germans. Haig hoped that, by attacking limited objectives, this pressure could be economically continued without the enemy realising that the British plan had changed. Additionally, there were several kinks in the front line still needed to be straightened-out. In VII Corps, only 33rd Division was destined for offensive action in late May.

These operations were planned by Major General Pinney and his headquarters. A warning order was issued to 100th Brigade on 15 May for an attack on 17 May which was postponed to 20 May. Even more troops were to attack the Hindenburg Line this time. 100th Brigade would assault the trenches south of Fontaine between the Croisilles-Hendecourt and Croisilles-Fontaine roads with three battalions; in addition, 5/Scot Rif would attack a feature called 'The Hump' assisted by a carrying party from 20/R Fus under Second Lieutenant Morison. Concurrently, 98th Brigade would attack down the Hindenburg Line and link up the two lodgements in the German Trenches. The Cameronians and 20/R Fus were in reserve to potentially

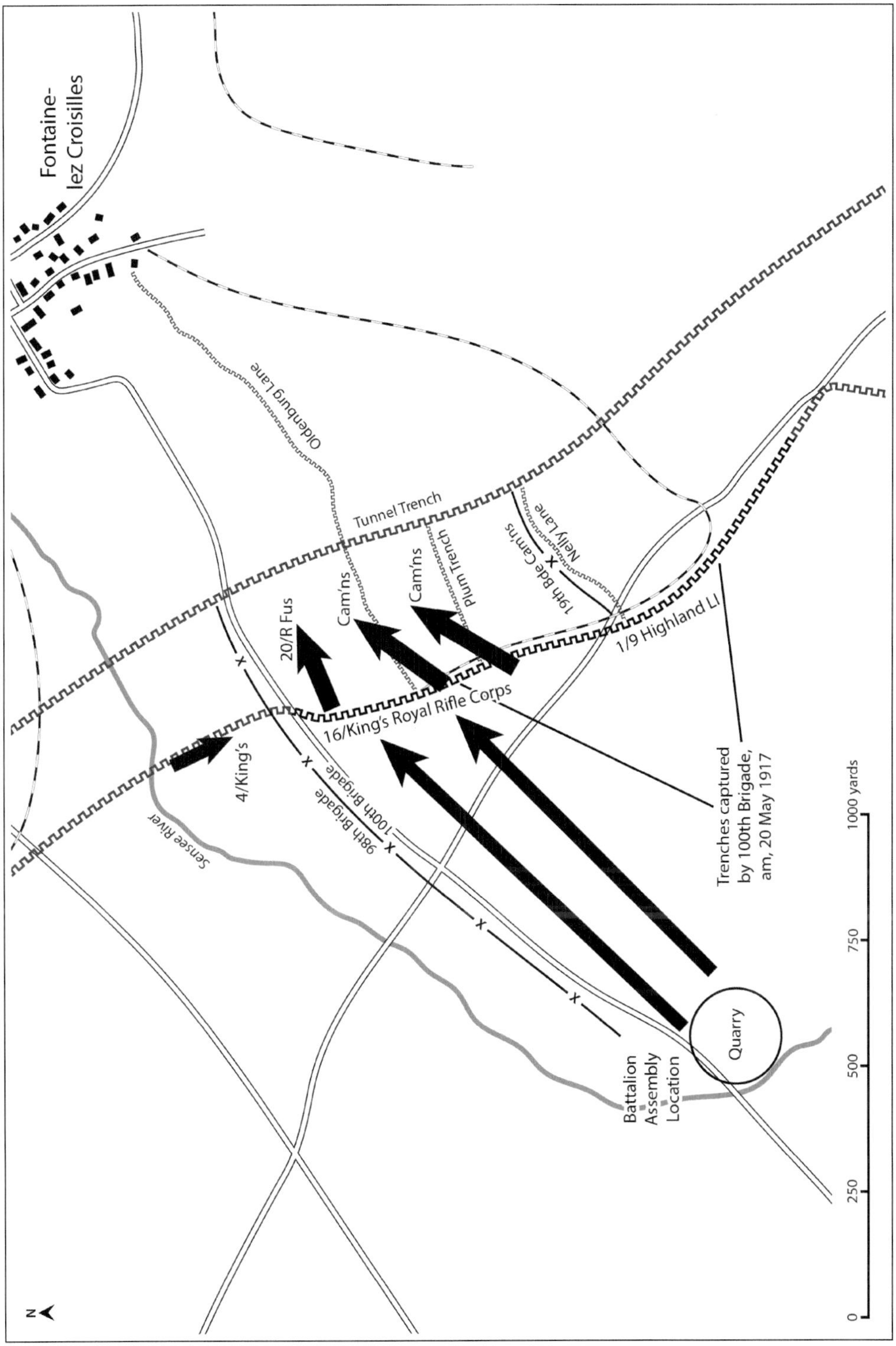

Map 10 Tunnel Trench, 20 May 1917.

attack through 100th Brigade and capture Tunnel Trench. Ultimately, the attack was similar to 23 April but employed a greater mass of infantry.

100th Brigade were to attack with 2/Worcs R on the right; 1/9 HLI in the centre and 16/KRRC on the left. 100th MG Company positioned a 12-gun battery in a camouflaged trench 80 yards long with two further four-gun batteries elsewhere and dummy gun positions to draw retaliatory German fire. On 20 May Allenby and his entourage watched the attack from some high ground near St Leger and Sir Arthur Conan Doyle recorded that 100th Brigade were; '… advancing as upon an Aldershot field day.'[44]

The 100th Brigade attack resulted in mixed success. The brigade captured the first objective without much resistance on account of the mist. The advance against the support line failed and those survivors dug themselves in forward of the Hindenburg front line. There were several bombing fights along the trenches captured but the Hindenburg front line was relatively secure. 5/Scot Rif, though it suffered heavy casualties did not secure 'The Hump'. The 98th Brigade attack along the Hindenburg Line on 20 May achieved partial success. Between 7:00 a.m. and 8:00 a.m. reports were received by 33rd Division that the party of 4/King's in the Hindenburg front line had taken their objectives and linked up with 100th Brigade. They had advanced to the River Sensée but had not reached 100th Brigade.

Tunnel Trench needed to be captured if the 100th Brigade lodgement in the Hindenburg Line was to be secured. The two battalions of 19th Brigade would need to capture Tunnel Trench to resolve the situation.

'The End of a Perfect Day' – The Evening Attack

At 5:20 p.m. 1/Cameronians received orders to attack the Hindenburg support line in conjunction with 20/R Fus. Time was short to make the necessary preparations for an attack through the battalions of 100th Brigade. Though 20/R Fus was in reserve the battalion needed to extricate itself from the posts it was holding. The four companies were to abandon their outposts, which were now redundant, and rendezvous in the Quarry. The Battalion would then take up positions for the attack. 20/R Fus was to attack on the left of the Cameronians in four waves with A Company on the left and B Company on the right and with C and D Companies in support on the left and right respectively. Their objective was Tunnel Trench from the Croisilles-Fontaine Road to the end of Oldenburg Lane. The Cameronians advanced with two companies in front and two behind; their objective was Tunnel Trench between Nelly Lane and Oldenburg Lane. Zero hour for the attack was at 7:30 p.m. and guides were supplied by 1/9 HLI.

At 7:15 p.m. the two battalions advanced covered by the Lewis Guns of 16/KRRC which had been pushed forwards and to the left flank. The British artillery barrage commenced at 7:30 p.m. and German counter-fire landed in the Quarry and the valley to the rear. Lieutenant Edward Chapman, leading C Company was severely wounded in the head during the 20/R Fus advance from the Quarry to the Hindenburg Line.

When the Cameronians reached the Hindenburg front line the British barrage remained in place 60 yards in front of that position. When this barrage lifted the Cameronians advanced

---

44 Sir Arthur Conan Doyle, *The British Campaign in France and Flanders 1917* (London: Hodder and Stoughton, 1919), p.87.

and were met by 40–50 Germans who surrendered at the next trench which the Cameronians mistakenly believed was Tunnel Trench. By this stage most trenches have been somewhat obliterated. Because the position was so hard to locate on the map no rations, water or ammunition arrived to aid in the consolidation of the position. The next day the Cameronians found that they were in a different trench fifty yards short of Tunnel Trench between the river and the road.

Donald Price was a battalion scout with 20/R Fus and accompanied the Scout Officer, Second Lieutenant 'Bill' Hill:

> I was always in attendance near my officer, and as we were in single file I was probably 2 yards from him as we went forward. We were being shelled with whizz-bangs and shrapnel, when a dud whizz-bang fell between Mr Hill (officer) and myself. I remember he went deathly white. I naturally became very "windy", but did not panic, but kept my head down and kept on crouching.[45]

The 20/R Fus attack did not go well; the Cameronians strayed to the left and crossing into the advance of the Fusiliers. 20/R Fus was halted 100 yards short of Tunnel Trench by MG fire and could not advance. The battalion ended up holding a 180-yard section of trench and a line of shell holes to the right of it. These positions existed between the two main trench lines. Private Donald Price, with Second Lieutenant Hill, recalled this attack:

> We attacked that night, but I was with Bill Hill, I never went over in the attack. ... the attack went off and there were flares and bombs flying around and bullets and Christ knows what. Eventually he said to me 'look there's a message, I want you to go out there and see what's happening', and he gave me a message, and I'd got to go and crawl over and find them about 100 yards in this German trench which they were trying to capture. And I went over there crawling ... When I eventually got there – the lights and everything else showed me where the attack was. I handed this piece of paper to a soldier and said to give that to an officer ... So I started on the way back [which] wasn't more than 100 yards. And I looked back and thought where I'd come from and I'd missed my way. And the next thing I found myself about 100 yards into another battalion. They thought I was a spy you see. ... I think they were KRRs or something. ... They took me down to the[ir] officer ...[46]

The 16/KRRC officer sent Price back to his battalion. On his return journey Price fell into a river, presumably the Sensée, and finally found Lieutenant Hill in a dugout where Price reported, soaked to the skin.[47] A 16/KRRC rifleman recalled; 'After dark a Battalion of the Cameronians ... advanced through us to attack the German second and third lines, but this only met with partial success. I remember a wounded Officer of that regiment being carried through our line and he was singing that "This is the end of a perfect day."'[48]

45 IWM: Donald Price Sound Recording 10168.
46 IWM: Donald Price Sound Recording 10168.
47 Second Lieutenant William Ernest Hill was OC D Company, 20/R Fus. He was from Tyldesley, Lancashire, enlisted in 20/R Fus. He went to France with the battalion in November 1915 and served with them until commissioned in December 1916.
48 IWM: Documents 01/36/1: Papers of H. Gore, p.58

Though drawn in October 1917, this Bairnsfather style sketch by Cecil Isom sums up the experience of 20/R Fus Tommies at Arras. The title also chimes with the words of a wounded Cameronians officer following the failed assault in May 1917. (Reproduced with the permission of Pat Isom)

20/R Fus lost seven officers wounded: Captain Hodgson-Jones, Captain Chapman, Second Lieutenants Graham, Bryant, Davis, Garrity and Morison. Chapman lost an eye and would never soldier again on active service; he spent the remainder of the war training recruits.[49] Hodgson-Jones suffered a fractured right arm above his elbow.[50] Graham received a bullet wound to his left calf alongside a bout of neurasthenia and did not return to France until November 1918. Little is known of the actions of Second Lieutenant Morison's carrying party except that he and twenty men went forward with bombs in support of the Scottish Rifles at about 7:00 a.m. Morison was wounded whilst leading these men. Detailed other rank losses were not recorded but the 20/R Fus war diary stated that during May 1917 the battalion lost sixteen men killed, twenty-nine missing and seventy-three men were wounded.[51] The Cameronians had lost three officers and fourteen

---

49 When Edward Chapman recovered, he was attached to 53/R Sussex (Young Soldiers) for training recruits.
50 Hodgson-Jones was MiD on 20 May 1917. On recovery, and after a period with 5/R Fus, he transferred to the Railway Operating Division of the RE. He died in hospital on 13 February 1919 whilst based at the Railway Training Centre at Longmoor.
51 The following men were killed on 20 May and are commemorated on the Arras Memorial unless otherwise stated: PS/4662 Sergeant Archibald Stanley Pearse Cordingley; L/13532 Sergeant Thomas Harris, aged 33, from Limehouse, had landed at Helles, with 2/R Fus on 25 April 1915; GS/63326 Sergeant Archer Smith, aged 20, had served with 1/Londons at Gallipoli; GS/63504 Private Alfred Field, of B Company, deserted and was recaptured and convicted in November 1916; his death sentence was commuted to penal servitude as a suspended sentence; STK/651 Lance Corporal Thomas Cunliffe, of D Company; G/2660 Sergeant Arthur Slatter, of C Company; GS/11201 Private Amos Bell; GS/14234 Lance Corporal Ernest Fripp, of D Company; G/24741 Private James Hardy, of C Company; G/25049 Private Harold Burrows, aged 23, of C Company; G/29857 Private Ellis Stevens, of B Company; G/41986 Private Edwin Thrower, of C Company; G/42243 Private Harry Gosling, aged 35, of B Company; GS/60924 Private Thomas Davis, aged

Eric Alexander Ogilvie Durlacher (2/Worcs R) was a solicitor's articled clerk who attended St Benedict's School Ealing, St Augustinius College Ramsgate and St Pauls School where he was a member of the OTC. He was killed whilst attacking the German trenches. Durlacher, Forrest and Corbridge had previously served with 18/R Fus. (Reproduced with permission of the Sutton Archives)

men killed, 100 wounded and sixteen missing. They had captured two officers and fifty-two PoWs. The 33rd Division attack on 20 May fell on *Reserve Infanterie Regiment 225*. (*49. Reserve Infanterie Division*); their defence crumbled due to the nature of the attack on two sides. The regiment was relieved on 23 May having suffered 113 men killed, 482 wounded and twenty-three missing whilst holding the line during the Arras fighting. This was a local, incomplete and largely pyrrhic victory for 33rd Division. The CO of 20/R Fus, Lieutenant Colonel Leader, summed up; 'That awful attack on 20th May was a bad job. The attack was as successful as possible & nobody could have done more.'[52] A further attack was required on Tunnel Trench and to resolve this 'untidy' situation.

36; GS/60941 Private Alfred Thompson, of C Company; GS/62858 Private William Tassell, of D Company; GS/63579 Sergeant Arthur Briers, aged 34, a fishmonger's assistant; GS/63626 Private Charles Ives, aged 23, joined C Company; 228057 Private George Burley, aged 21. The following were killed on 21 May: PS/4460 Lance Corporal Marmaduke Wellesley Bannister; GS/12672 Lance Sergeant George Tuersley; GS/63315 Sergeant George James, aged 20; GS/9702 James Leat, aged 24, is buried in Bucquoy Road Cemetery. SPTS/1726 Lance Corporal Julien Walton; SPTS/4614 Private John Griffin, was killed the next day. GS/53206 Private John Hale; GS/53245 Private Arthur Mullens, a baker's van boy, and GS/53246 Private Sidney Moore, aged 31, had all served with the ASC before joining 20/R Fus. All three were killed on 21 May. The following died of their wounds later: GS/63357 Private Percy Smith, aged 19, died of his wounds on 24 May and is buried in Sunken Road Cemetery, Boisleux-St-Marc. GS/53261 Private Percy Richardson, aged 25, a pastry cook, died of his wounds on 25 May, he is buried in Sunken Road Cemetery. GS/62470 Private Edward Burrington, aged 26, died on 27 May and is named on the Arras Memorial. Stk/2020 Private John Jones, aged 23, died of his wounds on 29 May; he is buried in Sunken Road Cemetery. A handful of men died before these attacks: SPTS/4902 Private Sydney Stringer, aged 22, died of his wounds on 10 May and is buried in Mont Huon Military Cemetery, Le Treport. G/46691 Private John Tuffs died of wounds on 17 May; he is buried in Bucquoy Road Cemetery. SPTS/5296 Private George Funge died on 17 May and resides in St Leger British Cemetery. TNA: WO95/2423: 20/R Fus War Diary.

52 IWM: E.F. Chapman Papers, 1799, p.68.

20/R Fus held their new trench throughout the next day and were relieved at midnight 21/22 May by 16/KRRC and moved to bivouacs in a sunken road west of Croisilles. They arrived at 3:00 a.m. and spent three days resting and cleaning uniforms and equipment. On 21 May news of further decorations was received; Second Lieutenant Clark was awarded the MC for the 16 April attack and Major Modera, Captain Hollingsworth, Captain Hodgson-Jones and the Regimental Sergeant Major, PS/8860 Warrant Officer Class 2 Leonard Raven, were all MiD. Later in the month Hollingsworth was also awarded the *Croix de Chevalier* of the *Legion D'Honneur*.

There were other ex-UPS officers in action near Tunnel Trench on 20 May. Lieutenant Laurence Bernard Forrest (16/KRRC) was killed during this operation; his body was never identified, and he is commemorated on the Arras Memorial. He was a bank clerk in Paris and had been educated at Denstone College. Arthur Corbridge, a bank clerk from Blackburn, was killed whilst 4/King's attacked along the Hindenburg Line.

Whitsunday Attack on Tunnel Trench on 27 May

Though 20/R Fus had completed its last attack in the Croisilles Sector there was still a last job for 33rd Division to complete. One company of 20/R Fus, by then reduced to three officers and 85 men, was attached to 1/9 HLI to hold the line from 9:00 p.m. 26 May onwards. Members of 20/R Fus would be spectators to a further disaster to befall 19th Brigade. In order to maintain pressure on the Germans a further attack was planned for 27 May to capture 800 yards of Tunnel Trench from Plum Lane to the Fontaine-Croisilles Road. The plan was employ the element of surprise; the operation would commence at 1:55 p.m. because the Germans were less likely to be alert at this time. 2/RWF and 1/Cameronians would make the attack and would experience similar difficulties to 20/R Fus. The Hindenburg front line further to the right was held by 20/R Fus and they would have a front row seat for this operation.

Tunnel Trench was by now almost obliterated and opposite the RWF it was invisible on the reverse slope of a small ridge. Between the crest and Tunnel Trench the Germans had linked several shell-holes together as a screen about 120 yards in front of Tunnel Trench. The attack started well with few casualties until the attacking companies advanced over the ridge. The intermediate shell hole line, weakly manned by the Germans, absorbed the British impetus and once the British barrage lifted from Tunnel Trench the Germans manned the parapet in force and were able to stop any advances beyond the intermediate position. As a result of an unknown person shouting 'retire!' on the left flank both attacking battalions withdrew with heavy casualties. One of the RWF companies was to concurrently bomb their way along Plum Trench. They met with some success until stopped by a barricade eighty yards from Tunnel Trench. They were forced to retire when the general withdrawal coincided with a German attack from the open ground on the right which entered Plum Lane. This German move cut off some members of RWF. The rest of this company withdrew to their original block and engaged in a stiff bombing fight. During this action a 20/R Fus Lewis gun team assisted the RWF by shooting at German troops crossing the open ground.

At 1:00 a.m. on 29 May the 20/R Fus company was relieved and returned to 20/R Fus who were still in the sunken road. Next day 20/R Fus was relieved by 13/NF (21st Division), and moved to Moyenville. On 31 May 20/R Fus marched to billets in Bellacourt. The first three days of June were spent in reorganising companies and conducting inspections. In late May Second Lieutenant Joseph Finch joined the battalion, presumably as a replacement for the casualties earlier in the month. He was not an archetypal infantry officer being aged 29 and having been a furniture clerk from Cheltenham who had been a sergeant with the Army Ordnance Corps since late 1915. He recalled his first few months as a commissioned officer:

> Since I left Oxford things have changed and varied somewhat. I was home for nearly 5 weeks before being posted to the 20th Royal Fusiliers and then was at Dover for a fortnight and eventually crossed to France on May 10th last … From landing in France I was about a fortnight at the base and eventually reached the line the last week in May…[53]

Finch would have a steep learning curve. Training for the battalion began on 4 June from 6:15 a.m. to 4:00 p.m. each day. On 5 June the battalion was inspected by Lieutenant General Snow (VII Corps). 6–10 June were spent with further training including range practices. The next week involved 'advanced training' whereby there were inter-company and battalion tactical exercises. A brigade church parade took place at Basseux on 17 June and General Snow returned to present medal ribbons. Captain Hollingsworth was presented with his *Legion d'Honneur*; L/17164 Company Sergeant Major Burns received the DCM; Second Lieutenant Clark his MC and PS/5428 Sergeant Page the MM.

On 18 June the battalion marched to a camp in Moyenville. Meanwhile the commanding officer and company commanders were collected by bus and were taken forward to reconnoitre the sector of front line they were to next occupy. At 7:00 p.m. on 19 June 20/R Fus left camp and marched to the trenches to relieve 8/Leics R. They were to occupy the sector south of the Croisilles–Fontaine Road; their right was on the remains of Marmalade Lane and their left was at King's Point. The latter post was a dangerous position consisting of a trench block in the Hindenburg Support Line beyond which the Germans were holding Tunnel Trench. This was the linchpin of the British defences; the problem lay in the proximity of the German trench block and the vulnerability to enemy attacks. The British field and heavy artillery routinely shelled Tunnel Trench twice a day at 11:00 a.m. to noon and 7:00–8:00 p.m. in bouts of daily 'hate'. During these periods the 20/R Fus withdrew non-essential personnel and held Humber Trench and Lump Lane with a skeleton force due to the proximity of Tunnel Trench; Burg Trench remained strongly held. It is presumed that the German garrison in Tunnel Trench withdrew their men into the Hindenburg Line Tunnel.

The timing of this stint in the line was unfortunate as the rest of the 33rd Division enjoyed a morale boost on 20 June when a race meeting was organised by the divisional staff. This was to provide a distraction from the recent offensive and to break the monotony of the period holding the line. A lot of effort and improvisation was put into this event and it was a shame that the majority of the men of 20/R Fus could not attend. This also provides an apt moment to wonder how many of the original members of the battalion who had attended the race meeting at Epsom were still on the strength.

On 24 June the Cameronians took over from C Company, 20/R Fus, in Lump Lane, to conduct another attack on Tunnel Trench. The Cameronians advanced from King's Point along the Hindenburg Line to capture the trench block and Tunnel Trench. According to Captain

---

53 Letter from J.H. Finch, 7 August 1917. (Author).

Dunn: 'They didn't. At midnight they bumped on to manned shell-holes in front of uncut wire. Fritz has become an artist in shell-hole defence…'[54] The attack having failed, the Cameronians were relieved again by C Company, 20/R Fus, next morning. 20/R Fus was relieved by 1/Queen's RWS and 2/Worcs R with the relief complete by 1:30 a.m. on 25 June.

During this tour there were two fatalities and eighteen wounded (of whom three remained at duty). One of those killed was PS/5428 Sergeant Victor Page was a company quartermaster sergeant and an original member of the battalion who had been presented with the MM days before his death.[55] In addition, Second Lieutenant Mark Gilchrist Whyte was slightly wounded on 20 June 1917 but only had to return to the transport lines to recover. Next day, Second Lieutenant Jones was wounded and two days later Second Lieutenant Geoffrey Colbourne was wounded in the right hand by a grenade.[56]

Though 20 May was the last attack by 20/R Fus this last period in the line marked the culmination of the part 20/R Fus played, as a battalion, in the 1917 campaign at Arras. It had been a costly affair. The battalion war diary recorded that casualties sustained from April to June had been four officers killed and thirteen wounded and 90 men killed, 198 wounded and 47 men missing. The death toll later rose to 140 men killed. Whilst the battalion had performed well, these casualties had further eroded the effectiveness that it had built up since the disaster at High Wood. Further men from the hard-core of the battalion, the UPS men who had originally comprised it, had been either killed, wounded or had sought an escape through commissions. The members of the battalion hoped for another stretch in some comfortable trenches rather than another bout in the line like they had experienced near Arras.

Whilst in reserve at 'C' Camp at Moyenville, 20/R Fus continued training but with less intensity than before. On 29 June the battalion, and 19th Brigade, moved to Monchy-au-Bois to conduct further training. From here between 1 and 5 July 20/R Fus marched from Monchy to Airaines, via nights in Arqueves, Naours, Yzeux.[57] Once at Airaines, training was conducted for the remainder of the month.

It could be argued that during these operations there were complaints of some poor understanding of conditions, planning and staff-work by staff officers at brigade and divisional HQs. However, there were some capable staff officers conducting this planning; the experienced Major Twiss was still the brigade major at 19th Brigade. Meanwhile, the 33rd Division GSO2, was Major Bernard Montgomery who would later become a field marshal.[58] Many staff officers were experienced regimental officers.

---

54 Dunn, *War the Infantry Knew*, p.359.
55 The 20/R Fus war diary states three killed and fourteen wounded; the 19th Brigade diary stated two, and only two registered war dead could be found for 20/R Fus. L/16047 Lance Corporal Edward Goff, aged 21, is commemorated on the Arras Memorial. PS/5428 Sergeant Victor Page MM, aged 28, had been employed as a traveller for a firm of shop fitters; he is buried in Croisilles British Cemetery. PS/4648 Private Henry Rayner Claye, aged 46, of A Company, died of wounds on 27 June 1917 and is buried in Dunkirk Town Cemetery. GS/53293 Private Ernest Whitbread died of his wounds and is buried in St Sever Cemetery.
56 Colbourne had only crossed to France on 11 May 1917 and Whyte was gazetted to a commission on 29 March 1917.
57 The 20/R Fus war diary for 1–15 July is missing.
58 In early July Major Montgomery departed to a new post at a corps HQ.

26

Trenches at Nieuport

The veriest contrast of pleasantness and unpleasantness

The veriest contrast of pleasantness and unpleasantness[1]

We were sent up to the coast to a place called Nieuport. We all called it a rest but actually it wasn't a rest ...[2]

Most of July 1917 saw 20/R Fus at Airaines conducting training; some of this involved trench digging practice. However, up to 15 July, the battalion war diary is missing.[3] More widely during this period training took place, battalion sports were held and further drafts joined the battalion; amongst those rejoining was Captain George Francis Jones-Williams. On 7 July a reinforcement draft joined 20/R Fus which consisted of personnel who had returned to France having served with other Royal Fusilier battalions. Another draft arrived shortly after consisting of men bound for the 8th Battalion of the Border Regiment (8/Border) and another draft arrived on 21 July comprised of men formerly of the Northamptonshire Regiment. Integrating these men would further dilute the identity of the battalion and might reduce the effectiveness of training for the battalion.

Captain George Ziegler rejoined on 12 July and his diary fills a couple of days of this gap. After arriving he had tea at battalion HQ and took one of the battalion horses over some jumps. Having been placed in charge of C Company, replacing Chapman, he became a mounted officer who needed to be able to control a horse on parade and on the march. He also ordered some riding breeches to match his new status. Next day a 5:00 a.m. rise was required for a route march and that afternoon a bathing parade took place in a muddy, shallow stream nearby. On 14 July training recommenced but heavy rain prevented this until the afternoon. That evening the officers practiced for a rifle competition. Next day was a church parade at the camp cinema with censoring of letters during the afternoon.

---

1 O'Neill, *The Royal Fusiliers*, p.394.
2 IWM: Donald Price Sound Recording 10168, Reel 11.
3 TNA: WO 95/2423: 20/R Fus War Diary.

A group of new officers joined on 29 July; these were Second Lieutenants Harry Chapman, Cecil Butler, George Cook, Harold Kirk DCM, William Lewis MM and Ronald Barker.[4] Though only one was an ex-UPS man the rest had considerable front-line experience. This period was a welcome break after the battalion having spent a lengthy period in the Arras area since mid-April. Captain Dunn, of 2/RWF, billeted nearby, considered that the weeks at Airaines passed very quickly. The village boasted a cinema and there were a few visits by the divisional band which played Gilbert and Sullivan.[5]

On 17–18 July there was another unusual divisional event; 33rd Division held a horse show at Cavillon which was carried out in a similar scale to the summer race meeting. According to Captain Dunn (2/RWF) this 'stunt' was instigated by Major General Pinney and it took; 'forethought and two weeks of painstaking preparation.' The event was well-organised and enjoyed by many. However, Dunn's account implied a level of incongruity between such a spectacle and some of the poor preparation and divisional staff-work that 2/RWF, and 20/R Fus, had endured at Arras.[6] 20/R Fus presumably had a reasonable set of animals under the Transport Officer (Captain Jones-Williams).[7] Two light draught mules of the battalion, though not winners, were commended by the judges. The battalion did win the competition for the best turned-out cooker. This was not a prominent prize but one that implied a high level of equipment husbandry.

An un-named officer of 20/R Fus summed up his thoughts on what would happen next:

> For days past, nay for weeks past, the rumble of the guns in the north had foreshadowed that there was to be trouble for the Hun before long; the official communiques in the daily papers spoke continuously of heavy bombardment in that neighbourhood, so that when on July 29th we heard that we were to leave Airaines, where we had spent a very pleasant few weeks in rest billets, conjecture ran wild as to what was to be our destination. We were not

4   Second Lieutenant Harold Kirk, aged 37, was born in Hull and was an actor at Daly's Theatre, London. He enlisted on 18 June 1915 with 24/R Fus in London and was SPTS/3592 Private H. Kirk. He later served with 30/R Fus and joined 23/R Fus before going to France with them on 16 November 1915. Kirk was awarded the DCM for bravery on 26 January 1916 (gazetted 15 March 1916). The citation read, 'For conspicuous gallantry in crawling across the open within 100 yards of the enemy to fetch water for a wounded comrade. Later he went out again across the open to fetch a supply of bombs and succeeded in bringing them back although fired at on both journeys.' He suffered a wound to the ear as a lance corporal on 27 May 1916 but returned to duty before being wounded more severely in the head on 27 July which necessitated his convalescence. He joined No 6 OCB on 3 January 1917. He was gazetted on 30 May 1917. Second Lieutenant William Lewis was formerly a sergeant with 17/R Fus (E/403). He had been awarded the MM and had embarked for France back in November 1915. He was gazetted on 30 May 1917. Second Lieutenant Harry Denington Chapman was from Horsham, Sussex and was educated at Cranleigh School. He enlisted in 18/R Fus and went to France in November 1915. He returned to the UK, likely with the disbandment of his battalion. He was gazetted in May 1917. Second Lieutenant Cecil Talbot Butler, aged 19, was a clerk for Lloyds Bank. He was formerly a sergeant in 10/R Fus (STK/2114) who went to France on 16 August 1916 until 28 December 1916. He was gazetted on 30 May 1917. Second Lieutenant George Cook was formerly a corporal with 17/Londons who went to France on 9 March 1915. He was gazetted on 29 May 1917. Second Lieutenant Ronald Alfred Barker, from Chelmsford, was gazetted on 30 May 1917.
5   Dunn, *War the Infantry Knew*, p.364.
6   Dunn, *War the Infantry Knew*, p.367.
7   Captain George Francis Jones-Williams. He was gazetted to 20/R Fus in July 1915; he was MiD on 11 February 1917.

long left in doubt and learnt that we were off to Dunkerque. Dunkerque – why? Perhaps we were off to England, not likely. What could be afoot?[8]

On 30 July orders were received to entrain for Dunkirk. All the practice trenches that had been dug were filled in which must have seemed galling work. Kit was packed. A billeting party of four men under Second Lieutenant Arthur Murgatroyd was sent on ahead to Bray-Dunes. Next day D Company marched out at 4:00 a.m. and entrained at Pont-Remy at 8:35 a.m. At Dunkirk, D Company travelled up the canal in barges to Bray Dunes. The remainder of the battalion departed at 7:00 a.m. and entrained at 11:00 a.m.:

> No sooner had we entrained than down came the rain, and it rained persistently for the rest of the day, and for several days following – a striking contrast to the weather of the past weeks, which had been magnificent. It was, indeed, unfortunate, as we afterwards learnt, that the opening of the great fight for the Paschendaele [sic] Ridge [Third Battle of Ypres] should have been so visited with such an upheaval of weather…[9]

This battle started on 31 July and though it was far away it might yet involve 33rd Division and 20/R Fus. However, their current journey took them by rail. As the train passed a nondescript Belgian village; '…two tremendous explosions occurred, and the carriage windows rattled. We jumped up and seized rifles and revolvers, thinking that the train was being bombed….'[10] This was in fact Belgian engineers blowing stone for quarrying though it was close enough that many stone chips hit the wagons. The train arrived at Dunkirk at 9:00 p.m.; the transport was unloaded and despatched onwards by road but did not arrive at its destination until 4:30 a.m. The journey from Dunkirk for the main body was via barge and with the barges to depart next morning the men slept relatively comfortably on them overnight. By midday the three companies had all arrived at Bray Dunes.

Here in Bray there 'were comfortable billets for officers and men, and we hoped to be allowed a day or two to enjoy them. At this time there was a considerable amount of hush-hush, and we did not know how or when we were to be employed.'[11] Bray Dunes resembled Deal or Sandwich in the UK. However:

> … the hotels and houses on the seashore had not their pre-war usage and were occupied mostly by military staffs. The sand dunes had been placed in a state of defence, with trenches dug and much barbed wire erected. Batteries were here and there, and there was an elaborate defence scheme, which we all had to study and know, so that in case of emergency each man knew his appointed task.[12]

Battalion HQ was opposite the church and was billeted nearby along with B Company; the remainder of the battalion was under canvas. Whilst here the battalion had to be prepared to

---

8 O'Neill, *The Royal Fusiliers*, p.393.
9 O'Neill, *The Royal Fusiliers*, p.393.
10 O'Neill, *The Royal Fusiliers*, p.394.
11 O'Neill, *The Royal Fusiliers*, p.394.
12 O'Neill, *The Royal Fusiliers*, p.394.

man seaward facing defences in case of an enemy amphibious landing; these defences consisted of a series of strongpoints.

During July 1917 the battalion had lost four men wounded to accidental injuries. However, training continued in this new location until the middle of the month using the beach and dunes 'the men used to parade there each morning, carry out manoeuvres, drill, rifle and bombing practice, and then take a bath before marching back to dinners. The afternoons were spent in recreation…'[13] The weather was good and training areas were satisfactory, if limited. The Brigade benefited from some good training. The officers rode a good deal and arranged horse races across the sand for their chargers; Captain Jones-Williams and his grey mare won on several occasions. The junior officers had to endure or enjoy equitation lessons though, to save the energy of the horses, these took place on mules and resulted in several of them being involuntarily unhorsed. There was also time to visit a neighbouring aircraft squadron and invite its members to dinner.

Second Lieutenant Joseph Finch took the opportunity to write home to record his activities over the past few months and some close shaves since late May:

> [I] … have been amongst one or two "do's" and have managed to get through so far with a small wound from shrapnel in my right hand, nothing, but just enough to draw blood and have had a lucky escape from snipers, having had my tin tad badly dented on one occasion. Out of the line [I] have been able to have rather a decent time and on one or two occasions have visited some good French towns and altogether things are great.[14]

Finch's luck would not last. Donald Price, with hindsight, recalled; 'We rested at Dunkirk [sic] before we collected ourselves really, we got all our equipment right and we were all refitted out … to sort of start again really. I think they [The Staff] had got another battle in mind so they wanted to spruce us up a bit.'[15] He realised that this break, and being sent to a 'quiet sector', was only going to be temporary and that they were being readied for the Third Battle of Ypres. The account of their time here demonstrated the contrast between the pleasant nature of this seaside sojourn and what was to come was summed up thus, 'the fact being that we remained in Bray Dunes until August 15th and then spent twelve days in the Nieuport sector of the line, the two periods, namely, August 1st–15th and 16th–27th, firmly impressing themselves on the minds of all those who live to remember them as presenting the veriest contrast of pleasantness and unpleasantness.'[16]

On 15 August 33rd Division commenced relief of 32nd Division and as a precursor 19th Brigade relieved 147th Brigade (49th (West Riding) Division) which was in reserve. At 4:10 a.m. on 15 August 20/R Fus marched via La Panne and Coxyde to huts at Kuhn or Queensland Camp 500 yards southwest of Oost Dunkirk. This camp was well appointed; '…the last thing in comfort in the way of camps, every hut being lavishly fitted with electric lights, and wire beds and bunks for almost every man; tables and comfortable armchairs adorned the officers'

---

13   O'Neill, *The Royal Fusiliers*, p.395.
14   Letter from Second Lieutenant J.H. Finch, 7 August 1917 (author's collection).
15   IWM: Donald Price Sound Recording 10168, Reel 11.
16   The older spelling of 'Nieuport' (as seen on trench maps) is used instead of the more-modern 'Nieuwpoort'. O'Neill, *The Royal Fusiliers*, p.394.

quarters.'[17] The transport stayed at Jenniot or 'Canada' Camp just west of Coxyde.[18] At 9:00 a.m. the minimum reserve, including the commanding officer, was separated and marched to the Divisional Depot Battalion at Ghyvelde. This consisted of twenty-five men per company and five officers. Donald Price summed up his memories of this period:

> We were sent up to the coast to a place called Nieuport ... we all called it a rest but actually it wasn't a rest. It was a rest from really hard battles ... we were in trenches and had to do our stint in the trench of course, but there weren't any attacks going on or anything like that. ... It was quite a rest really but we were still getting shelled and we were still getting casualties all the time.[19]

This period was more restful for Price; he was granted leave to the UK from 7–17 August. On the afternoon of the 15th the officers conducted a reconnaissance of the trenches they were to take over near Nieuport:

> Away, then, we went to reconnoitre, and to see our opposite numbers, so to speak, from whom we should take over. We rode to the outskirts of Nieuport, and then proceeded on foot. Nieuport is a largish town, which had recently suffered considerably from bombardment. Though the streets were quite intact, they contained a good deal of debris, and covered in "ways" had been made along the sides of the streets; in almost every case these "ways" had been dug for a few feet, so that one was, so to speak, half above and half below ground level. Everywhere there were gas gongs and rattles, for only recently had the town received a goodly libation of the Huns' new mustard gas, and in various parts of the town the smell of this gas was still fairly potent. The brigade headquarters was situated in a cellar in a street in the town, and having reported there, we proceeded to the headquarters of the battalion which we were to relieve.[20]

They were to take over as the right-hand front-line battalion in the Lombartzyde sector. The area had increased in prominence and value since the German attack on 11 July where they had captured the whole of the Dunes Sector. All that remained of the British Bridgehead over the Yser Canal at Nieuport was the 1,500-yard wide and 1,000-yard deep Lombartzyde Bridgehead. On the right it was abutted by the Bamburgh Polde and on the left the flooded Geleide Stream secured the left flank. Trenches mostly consisted of 'breastworks', due to the poor drainage, and many were in a poor state of repair after recent heavy shelling. There were only two communication trenches in the sector, one called Nasal Alley and the other was *Petit Boyau*.[21] There were only three wooden bridges across the canal to the bridgehead; German shells often destroyed them during the day and the RE rebuilt them at night. The liaison with the unit to be relieved took place and the recce party returned safely. The next night at 7:30 p.m. the remainder of 20/R Fus marched, via Wulpen and Nieuport, and took over the Lombartzyde

17 O'Neill, *The Royal Fusiliers*, p.395.
18 There is still a military camp on this site.
19 IWM: Donald Price Sound Recording 10168, Reel 11.
20 O'Neill, *The Royal Fusiliers*, p.396.
21 The latter trench was unofficially known as *Toute Suite* Alley due to the speed men moved along it. P.G. Bales, *The History of the 1/4th Battalion Duke of Wellington's Regiment 1914-1919* (London: Edward Mortimer, 1920), p.148.

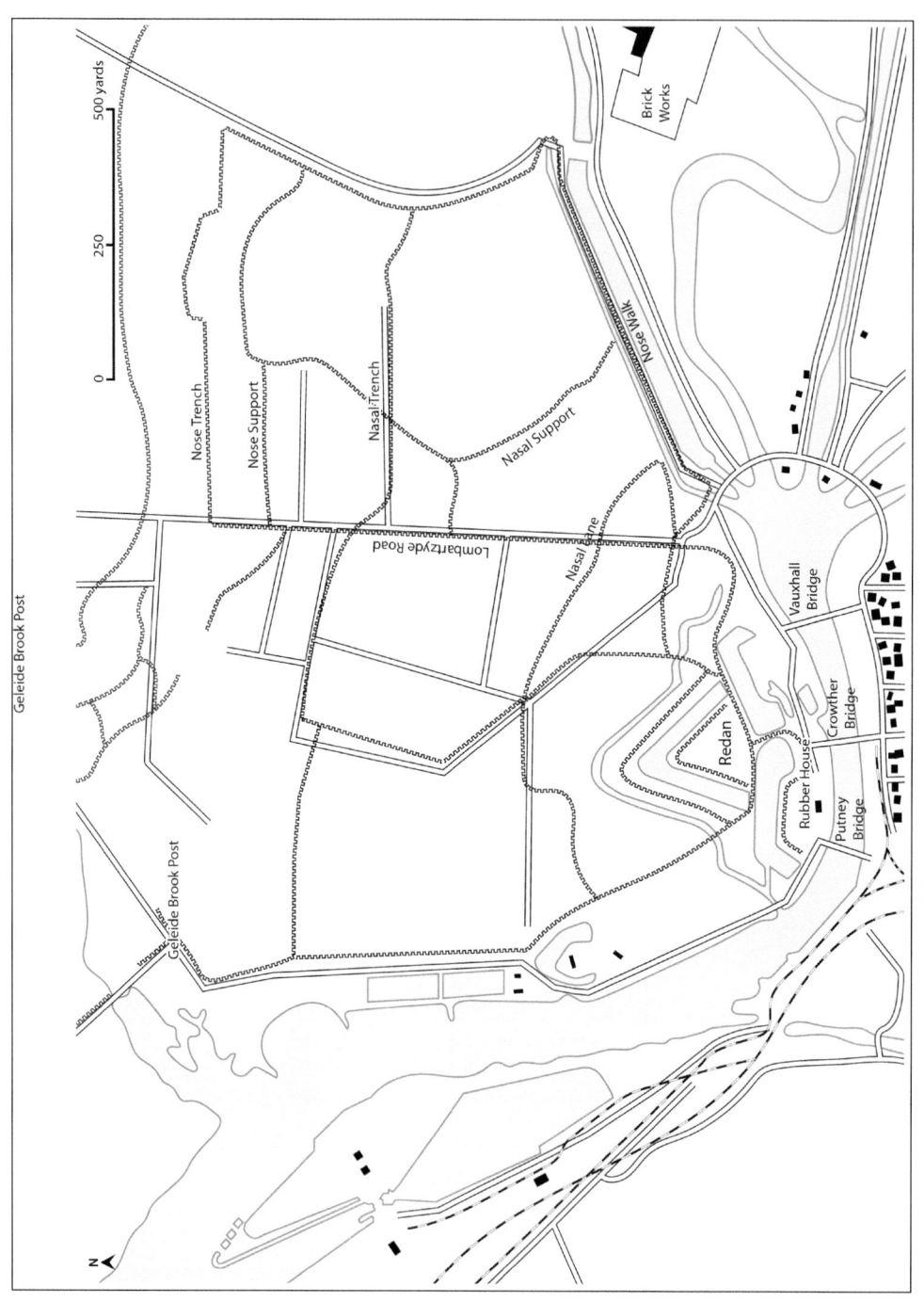

Map 11 Lombartzyde Sector, Nieuport, summer 1917.

sector from 1/4th Battalion West Riding Regiment (1/4 DWR). Captain Ziegler recorded the journey to the trenches:

> I set off to go up the line to take over. We had to go some way out of our way [and] finally through Nieuport and up over the bridges to the line. It was worse than Festubert as we had to go up over the open through a hell of a barrage which they kept up the whole night blowing all the lines, such as they were, to bits. The Reserve line had several casualties amongst NCOs as they were all in one concrete dug out which was blown in.[22]

20/R Fus apparently only lost one man during the relief but the Germans bombarded the British trenches for half an hour early in the morning on 17 August which killed six men when a shelter received a direct hit.[23] The 'Redan' was an old bastion of the historic Vauban defences and at its centre was the 'India Rubber House' – so called because of its survivability to German shells having a concrete roof 2ft thick. This was the headquarters for two battalions and a shelter for many troops. The large building had cubicles and electric lighting.

View from the site of the forward trenches at Lombartzyde back towards the Nieuport Memorial which is inscribed with the names of the missing from this sector. This image depicts the relatively small bridgehead here and the very flat nature of the ground. (Author)

22   FML: RFM.ARC.3015.2: Diary of Captain Ziegler, p.16.
23   PS/3898 Corporal George Victor Rushton, aged about 20, from Snodsbury, Manchester; had enlisted with 18/R Fus in 1914; GS/63322 Lance Corporal Charles Turner, aged 23; GS/63712 Lance Corporal Charles Woollaston, aged 28 and GS/66431 Private Joseph Lycett, were commemorated on the Nieuport Memorial. GS/46404 Private Sidney Anthony, aged 26, was buried in Adinkerke Military Cemetery. The sixth man is unknown. The following men, likely of C Company under Captain Ziegler, were wounded: GS/63301 Private James Allen; GS/63644 Sergeant Henry Woollven; GS/66445 Private Cledwyn Price; GS/29993 Private Stanley Goodson and GS/38672 Private Bertie Williams. See O'Neill, *The Royal Fusiliers*, p.398.

Once they had taken over 20/R Fus was holding the right front, 2/RWF held the left front. 1/Cameronians were in support and 5/6 Scot Rif were in reserve. One platoon of C Company, 20/R Fus, held the Front Line in Nose Trench, a further platoon held Nose Support with two platoons and Company HQ in Nasal Trench. A Company was in support with two platoons and Company HQ in Nasal Support and two platoons in Nasal Walk. D Company was in Reserve with two platoons and Company HQ in Nasal Lane. The remaining two platoons of D Company were attached to 57th Field Company and moved to dugouts in Nieuport. B Company were quartered in the Tunnel at the Redan; they provided carrying parties to the front line company. This was because the Transport could only get to 'Arch Bridge'. The Transport and QM's Stores had moved to Coxyde Bains. During this period Major F.S. Modera MC commanded the Battalion in the trenches whilst Lieutenant Colonel Leader was with the Transport.

20/R Fus names inscribed on the Nieuport Memorial to the Missing. (Author)

Donald Price was now working as a sniper with D Company. His sniper's post was built into one of the breastworks and must have been structurally well built. The post was still there when he re-visited the battlefields in 1923. His memories of this sector were that:

> It was quite peaceful except for when you were bringing up rations at night as soon as "Stand To!" was on he [the Germans] started shelling the [canal] crossing and you'd got to watch the shells and dodge between the salvos to get across for rations or anything else … When we were relieved you had to be damned careful to get out… But when you got in [the trenches] there didn't seem to be much shelling about. … Whether it was a sort of reciprocation; we didn't shell them and they didn't shell us, I think that might have been something because it was very quiet really there and that's why we called it a rest; but you were still getting shelled. It was still dodgy to start wandering around over this [canal] crossing.[24]

20/R Fus engaged in patrolling, but this was curtailed by the amount of water between the trenches. All was quiet though a German patrol was spotted and was dispersed with fire on one night. On 19 August A Company and C Companies relieved one another in the trenches.[25] On the night of 21/22 August 20/R Fus was relieved by 5/6 Scot Rif and moved predominantly into the Tunnel in the Redan. Battalion HQ moved to the India-rubber House near Putney Bridge and two platoons of D Company were still in dugouts in the town. 20/R Fus provided working parties for the left and right subsectors as well as defending the Redan.

During the night of 22–23 August 1/Cameronians relieved 2/RWF in the left sector. No sooner was the relief complete than the enemy successfully raided the Geleide Brook Post held by 1/Cameronians who had just taken over from 2/RWF. Hand-to-hand fighting took place during the struggle for this post though Dunn recorded two Cameronians were lost for four enemy killed.[26] The Cameronians were forced to withdraw from the post, due to the SOS bombardment called down by 2/RWF who were notionally still in charge of the sector. In addition, during a heavy German bombardment of Nieuport a shell hit the building housing brigade HQ.

Plans were developed for the Cameronians to launch a raid on 24 August 1917 and 20/R Fus would launch an enterprise on 25 August. The Cameronian raid was successful, and the 20/R Fus raid plan was finalised. However, it did not come to fruition; on the evening of 25 August the Cameronian gains were lost to a German counterattack which recaptured the post in question after heavy fighting. B Company 20/R Fus was rushed forward to Nasal Trench which they garrisoned and held two posts on the Geileide Stream. B Company was the one which had prepared the raid. There were minor attacks and operations in this area involving 1/Cameronians over the next couple of days and B Company, 20/R Fus remained in support. They were relieved and withdrawn on the night 26/27 August and returned to the Redan tunnel. 20/R Fus lost

---

24 IWM: Donald Price Sound Recording 10168, Reel 11.
25 GS/48225 Private Frederick Bunyon, aged 19, was killed in action on 19 August and is commemorated on the Nieuport Memorial.
26 Dunn, *War the Infantry Knew*, p.378.

three men killed 20–24 August.[27] On 25 August 20/R Fus lost four men – presumably during the intensified shelling.[28]

On the night 27/28 August 20/R Fus was relieved by 1/Dorsets; the relief was complete by 3:15 a.m. on 28 August 1917. So ended a hard stint in the trenches:

> We had been in a good many parts of the line, but it was generally agreed that this particular tour was to be remembered as one of the most unpleasant; we had sustained a good number of casualties, and the Hun artillery had been very active with shells of every calibre and with plenty of gas shells. Our consolation, however, was that our guns were very active, too, and the best feature was the weather, which was on the whole good.[29]

The battalion marched to Kuhn Camp and once the battalion was present it moved off at 9:30 a.m. to billets at La Panne; here the Transport and QM's Stores rejoined. Next day the battalion departed to Petit Synthe near Dunkirk in buses; the transport travelled by road. The Minimum Reserve rejoined on 30 August. The next day the battalion departed via buses to billets in Houlle.

During August 20/R Fus lost twelve men killed, 49 wounded and two missing. Major Hickley was attached to Fourth Army HQ and departed. Second Lieutenants William Gould, George Cook and Geoffrey Colbourne were admitted to hospital during the month. Second Lieutenant Alfred Hewlett Smith was stuck off strength as being permanently assigned to a base unit. Further officers joined during the month; Second Lieutenants Leonard Brooke, William Miles, John Smith and William Stokes.[30] The next task would be to march southwards and take part in the Third Battle of Ypres. This would be a stern test for the battalion.

---

27   G/63503 Private Matthew Fitzpatrick had previously served at Gallipoli with 1/10 Londons. He was buried in Ramscappelle Road Military Cemetery. GS/21435 Private Philip Fernihough, aged 22, and GS/5568 Private Charles Storey are commemorated on the Nieuport Memorial.
28   G/5115 Private Harold Cockfield, aged 23; GS/63670 Private Edward Chaplin, aged 23; GS/63689 Private George Kelton, approximate age 22; and GS/66415 Private Thomas Gelson, aged 21, were all commemorated on the Nieuport Memorial.
29   O'Neill, *The Royal Fusiliers*, p.400.
30   The following officers joined 20/R Fus 10–20 August 1917. Second Lieutenant Leonard Brooke was formerly a private with 17/R Fus and went to France in November 1915. He was gazetted to a commission with 6/R Fus on 25 April 1917. Second Lieutenant William Ewart Stokes was from Derby. He had enlisted service with the RAMC and went to France in January 1915. He was gazetted to a commission on 10 July 1917. Second Lieutenant John Thomas Smith was formerly a Grenadier Guardsman and went to France in February 1915; he later joined the Guards MG Regiment. Second Lieutenant William Stephen Miles was from Kew. He enlisted with 18/Londons and went to France in March 1915. Both were gazetted on 26 June 1917.

27

Ex-UPS Men at Ypres, June to November 1917

A ghastly, decaying, body-strewn, tree-less corridor of death

---

We have gone through ten of the most awful days I have ever spent & even now I do not seem to be able to get my mind to bear on anything'[1]

A ghastly, decaying, body-strewn, tree-less corridor of death that was No man's land.[2]

Messines – 'Another big fight'

The Battle of Arras had been a coalition battle in conjunction with the French. An offensive from the Ypres Salient was what Field Marshal Haig had long been pushing for. The 'curtain-raiser' to the main battle was a British offensive against the Messines Ridge. This would see the execution of a long-running engineering plan to mine beneath the German defences at Messines. This would be combined with the methodical operation by General Plumer (Second Army) for a well-coordinated set-piece infantry attack. This would commence at 3:00 a.m. on 7 June 1917 when the mines would detonate under the German strongpoints and a hurricane bombardment would begin the attack by nine divisions. Though 20/R Fus was still in the vicinity of Croisilles there were many ex-UPS men within Plumer's forces though this section will only cover a few.

In the north of the offensive was 24th Division. Andrew Buxton (3/RB) wrote home on 6 June; 'Just a line written from a tent. Life is very big and interesting. … out here the Bosch must be having a real bad time.'[3] Alongside Buxton were three other ex-UPS men; Edward Patey (who had served with 18/R Fus) and Peter Adam and Charles Goody (formerly of 19/R Fus). Patey was a schoolmaster who had been educated at Norwich School and Hertford College, Oxford. Adam was educated at Leys School (where he joined the OTC) and Cambridge University; he was an engineering student. Goody had been taught at Merchant Taylors' School and Keble College Oxford and was a Gold Coast civil servant.

---

1 FML: RFM.ARC.2482.47: Documents of H.D. Etheridge, letter 17 June 1917.
2 Hiscock, *The Bells of Hell*, p.27.
3 Woods, *Andrew, Andrew Buxton*, p.284.

The battalion attacked next day at Messines and reached its objectives. Buxton took forward a carrying party with supplies and when these were delivered he went to visit his company but was killed by MG fire on the way. Lieutenant Colonel Pigot, commanding his Battalion, wrote:

> He was just coming back from the front line after an attack yesterday when he was hit by a bullet and died almost at once. I can't tell you how much I deplore his loss. He had been with us a long time and on ever so many occasions had shown himself a very brave man. Everyone loved him, and all the men of his Company will, I know, regret his loss. He was always doing his best to make his men comfortable …[4]

Buxton was buried in Oostaverne Wood Cemetery. Charles Goody had been wounded in the leg in January 1916 and was hit again on 8 June 1917. A bullet or shrapnel ball hit him in the face whilst his mouth was open removing part of his tongue, shattering ten teeth and damaging his jaw. Despite this he could later count himself lucky having received this wound. He was the second of three brothers to be injured or killed. Time would tell as to whether Patey and Adam would also survive the campaign.

Carrying parties of 12/R Fus followed behind 3/RB. The former suffered a steady loss of men and later took to the offensive to capture the dugouts in Battle Wood on 14 June. PS/1462 Private Hugh Dimsdale Etheridge, formerly of 18/R Fus, was a runner with 12/R Fus. He later wrote home to his family:

> … we have been in the thick of all this fighting & only came out yesterday. … We have gone through ten of the most awful days I have ever spent & even now I do not seem to be able to get my mind to bear on anything, but I am all right not [hurt] a scratch. Most of the prisoners we took had only been on the western front for a few days, one lot from Russia & another from Serbia … The first four days we took all objectives & then we came back for 24 hours rest & then we went up & took a place that another division did not quite take, we were there for 4 days & have just come out again. … You will be glad to hear that I have been recommend[ed] for the same honour probably the Military X, my name was sent in yesterday.[5]

He was awarded the MM, 'this man displayed great bravery and coolness under heavy machine gun fire, in bringing back messages from the Outpost Line to Headquarters. It was greatly owing to his determination that information was kept up between forward post[s] and [the] Company Commander during the operations.'[6] The battalion considered; 'The work of the Runners during the operation was exceptionally good, they had to take risks which they did, in performance of their duties, and all messages brought back from the line of Outposts had to be carried in full view of the enemy … the runners did splendid work.'[7] Etheridge wrote a few days later; 'You will be glad to hear that it came through in Battalion orders last night that the

---

4   Woods (ed), *Andrew*, p.289.
5   FML: RFM.ARC.2482.47: Documents of H.D. Etheridge, letter 17 June 1917.
6   FML: RFM.ARC.2482.47: Documents of H.D. Etheridge, MM citation.
7   TNA: WO 95/2208: 12/R Fus War Diary.

PS/1657 Private Edmund Thornber Hussey, aged 21, was from Vanbrugh Hill, Blackheath. He enlisted with 18/R Fus and was gazetted to 4/RMF; he was attached to 1/RMF in France in 1916. Hussey was gassed in September 1916 but recovered and returned to the front. He was buried in Irish House Cemetery. (Reproduced with permission of Sutton Archives)

general had given me the Military Meddle [sic] [,] only wish they would give me a few day's leave…'[8] He was starting to feel the pressure from this fighting; 'You will be glad to here [sic] that I have come out safe again out of the hottest fighting I have yet been in … after the incessant shelling we have been through I do not think my nerves would of stood it much longer. We have been hard at it since the 7th…'[9]

On 7 June, the next division in line to the south was 23rd Division which would attack at Hill 60 and Zwarteleen. The night before the attack Second Lieutenant Vincent Edwards, of 10/DWR (formerly 19/R Fus), conducted a reconnaissance of the German wire in No man's land. He had to crawl, along with one of his men, through the darkness to about 30 yards from the enemy trenches to check the state of the enemy barbed wire. Luckily for Edwards and his battalion the artillery had done a good job of cutting the wire and they were able to return safely to their own lines without being seen or shot at.[10]

Next morning, at about 4:00 a.m., the mines under Hill 60 were detonated and Edwards' battalion advanced. He was in the first wave and they followed closely behind the British artillery barrage. When they entered the German trenches; the enemy were either dead or had withdrawn. There was no major counterattack to try and recapture these gains. Also with 10/DWR was Arthur Calvert Tetley, from Rhyl. Though Edwards survived this action Tetley was killed during the day.

Further south still, 16th (Irish) Division attacked facing Wytschaete. Edmund Thornber Hussey commanded Y Company, 1/RMF, when it attacked on 7 June. As the battalion reached the Oosttaverne–Wytschaete road several prisoners were captured in dug outs. One flank of Hussey's company pushed forward too quickly and found themselves under fire from their own creeping barrage. Hussey dashed forward to move them back but was killed by British shellfire.

8   FML: RFM.ARC.2482.47: Documents of H.D. Etheridge, letter 25 June 1917.
9   FML: RFM.ARC.2482.47: Documents of H.D. Etheridge, letter 28 June 1917.
10 Liddle Archives: WW1/GS/0505: Papers of V. Edwards, p.4.

Though only 21, Hussey was a Somme veteran who had been gassed in September 1916; his loss to the battalion was as great as that to the Church of England, in which he had chosen to serve but for the war.

The Third Battle of Ypres was to start on 31 July 1917. Though there was a gap between the two offensives there were still numerous raids, bombardments, and patrols. One raid was conducted on 28 June 1917 near Hill 70, Loos, by 11/Essex. Frank Bernard Wearne had served with 18/R Fus and had been wounded in four places on 3 July 1916 whilst with 10/Essex. On 28 June he commanded a blocking party and the circumstances of his death were written into a citation for the Victoria Cross:

> For most conspicuous bravery when in command of a small party on the left of a raid on the enemy's trenches. He gained his objective in the face of much opposition and by his magnificent example and daring was able to maintain this position for a considerable time, according to instructions. During this period 2nd Lt. Wearne and his small party were repeatedly counter attacked. Grasping the fact that if the left flank was lost his men would have to give way, 2nd Lt. Wearne, at a moment when the enemy's attack was being heavily pressed and when matters were most critical, leapt on the parapet and followed by his left section, ran along the top of the trench, firing and throwing bombs. This unexpected and daring manoeuvre through the enemy off his guard and back in disorder. Whilst on the top of the trench 2nd Lt. Wearne was severely wounded but refused to leave his men. Afterwards he remained in the trench directing operations, consolidating his position, and encouraging all ranks. Just before the order to withdraw was given, this gallant officer was again severely hit for the second time, and while being carried away was mortally wounded...[11]

On the evening of 17 July Lieutenant William Bentley (formerly 19/R Fus), his company commander and a guide (all from 9th Battalion North Staffordshire Regiment) went into No man's land southeast of Wytschaete to mark positions for a works party. Enemy shelling killed or mortally wounded the other two and wounded an officer leading the works party. Having tried to help his comrades Bentley took over the works party and arranged the completion of Dorset Trench before withdrawing. For this feat of endurance Lieutenant Bentley was awarded the MC.

Flanders Offensive

Many battalions were heavily shelled whilst awaiting the offensive on 31 July. An officer of 2/Northants arrived at battalion HQ near Zillebeke Lake on 27 July, 'I found utter confusion. The surface shelter which housed the HQ had sustained a direct hit and the adjutant, the intelligence officer, the signals officer and the chief clerk had all been blown up and killed or evacuated. Even the leather-faced old RSM looked off colour.'[12] Second Lieutenant Frederick Mervyn Hills, the intelligence officer for 2/Northants (formerly of 18/R Fus) was killed when a German

---

11   Wearne is commemorated on the Loos Memorial. *The London Gazette*, 31 July 1917.
12   Hubert Essame, *Passchendaele 1917*, War Monthly, p.37.

8-inch shell hit the mess shelter he was in.[13] He had been educated at Tonbridge School and had played rugby for his UPS company. Others were victims of this artillery fire. Thomas Johnson, formerly of 19/R Fus, was serving with 217th MG Company, 20th Division. On 27 July he was badly wounded by a shell whilst returning from Yorkshire Trench south of Boesinghe. He died of his wounds on 5 August.

Third Ypres

The main part of the British offensive at Ypres opened on 31 July 1917. From the north to the south of the battlefront ex-UPS officers and enlisted men, saw action and contributed to this operation. Many were wounded or lost their lives.

Over this period 3/RB (24th Division) operating near Shrewsbury Forest, suffered two more ex-UPS officer casualties. On 31 July, Peter Adam MC and Captain Edward Patey were killed, the latter by MG fire, early in the morning of 2 August, along with two men. Lieutenant William Folds Cooper, of 12/R Fus, was also killed during the attack on 31 July 1917. His body was found in Shrewsbury Forest near Jehu Trench; he was reburied in Hooge Crater Cemetery. On 3 August PS/6626 Private John Clegg, a butcher from Rochdale (formerly of 21/R Fus) and PS/7367 Lance Corporal Leslie Foot (formerly 19/R Fus) died whilst serving with 12/R Fus as was GS/47407 Private Harry Oldfield who had survived High Wood with 20/R Fus. Private Etheridge was again in harm's way as a runner and carried messages under intense MG and artillery fire back; he was awarded a bar to his Military Medal.[14] These casualties, and those amongst officers, in 12/R Fus meant that Major John Hartley was posted in as 2IC having been attached to a training battalion in the UK. He was soon re-assigned to command 8/E Kent. Hugh Etheridge realised his survival chances as an infantryman were slim and renewed his interest in a commission.[15] He did not get his papers signed by his brigade commander until November.

26/R Fus (41st Division) was in action during The Battle of the Menin Road and had several ex-UPS officers and men in its ranks. On 20 September 1917 26/R Fus suffered heavy casualties when assaulting Tower Hamlets east of Ypres. Within ten minutes of advancing only one officer remained unwounded in the rifle companies. Casualties included the commanding officer. Major Allen Maxwell (formerly 20/R Fus) was serving with 26/R Fus; he took command and was awarded the DSO for his leadership.[16] Second Lieutenants Eric Woodville-Morgan, John Jackson and Sydney Smith were three of the junior officers killed; all were former UPS men.[17]

---

13   Hills is buried in Perth Cemetery (China Wall).
14   FML: RFM.ARC.2482.47: Documents of H.D. Etheridge, MM bar citation.
15   FML: RFM.ARC.2482.47: Documents of H.D. Etheridge, letter 30 September 1917.
16   Maxwell's citation stated; 'He took command of his battalion during an attack when his commanding officer became a casualty, and with great energy and determination led his men forward under heavy and continuous fire. Throughout the attack and during the enemy counter-attacks which followed the capture of the position he set a magnificent example.' *The London Gazette*, 22 March 1918.
17   Formerly PS/1088 Private Eric Theodore Woodville-Morgan, aged 23, enlisted in 19/R Fus. He was from Teignmouth and was educated in Lausanne and at Clifton College, Bristol. He was gazetted to a commission on 24 January 1917. PS/5131 Private John Jackson, formerly of 20/R Fus was from Blackpool. He served with the battalion from November 1915 before returning to the UK on 14 January 1917. He had only been gazetted to a commission on 30 May 1917. PS/7811 Private Sydney

Three ex-UPS enlisted men, were also killed and their bodies lost; PS/10765 Private Gwilym Evans, PS/8347 Private Charles Turner and PS/2470 Private Claude Briggs.[18] Claude Briggs was an original UPS man who had served with 21/R Fus since enlisting in September 1914. He was not considered fit enough for training at an OCB but instead went to 26/R Fus. He was given clerical work until November 1916 but after this he was sent to his battalion. He broke down with rheumatism and worked in the battalion Orderly Room until again being employed in a rifle company in March 1917. Briggs served in the trenches from then on despite having a slight limp, until he was reported missing on 20 September 1917.[19] Also serving with 26/R Fus was Lieutenant Francis Steinthal who had formerly served with 29/R Fus.

On 23 September Captain Robert Hammond, from 32/R Fus (formerly a company commander with 19/R Fus), was appointed as CO.[20] Another replacement was Second Lieutenant Harry Vander-Weyden who returned having recovered from a wound in January 1917.[21] On 28 September a German aircraft dropped a 'lucky' bomb outside the HQ of 26/R Fus. The detonation mortally wounded Lieutenant Colonel Hammond and his adjutant and wounded four other officers. Next day Major Henry Tuite arrived to take command; Tuite had also commanded a company with 19/R Fus and had instructed at an OCB since May 1916. On 30 September Hammond died of his wounds and was buried in Zuydcoote Military Cemetery; he was awarded the Military Cross after his death.[22] Another replacement who arrived after these events to increase the manning of 26/R Fus was Private Cecil Isom who had recovered after being hospitalised whilst with 20/R Fus back in April. In his diary he recorded air raids throughout this period. Within a few short days 26 R/Fus had suffered severe officer casualties. Experienced former-UPS officers had stepped-in to provide leadership and had helped sustain a sister battalion through a difficult period.

Second Lieutenant Arthur Drumgold, 11/Queen's RWK (41st Division), was missing after an attack on the notorious Tower Hamlets spur. He and a handful of men were found to have held a position in advance of the most forward troops. He was subsequently awarded the DSO for the period 20-23 September 1917, 'He led his company forward and gained his objectives in the face of heavy machine-gun and rifle fire from enemy strong points. With eight men he remained in an isolated position without food or water for thirty-six hours under very heavy artillery fire,

    John Smith, of 18/R Fus, was aged 36, from Lichfield; he had served with 1/R Fus before being gazetted to a commission on 26 July 1917. Woodville-Morgan is buried in Oxford Road Cemetery. Jackson and Woodville-Morgan are commemorated at Tyne Cot.
18 PS/8347 Private Charles Turner, aged 21, from Ashton-under-Lyne and PS/10765 Private Gwilym Evans, aged 20, from Clynderwen, Carmarthenshire, are both commemorated on the Tyne Cot Memorial.
19 Brigg's body was later found and buried in Bedford House Cemetery. See IWM: Documents 21795: Papers of L.K. Briggs.
20 Formerly PS/432; Robert Whitehead Hammond, aged 41, was a consulting engineer. He had enlisted in London in September 1914 before being one of the original officers gazetted into 19/R Fus in October. He later commanded C Company. He returned to France and joined 32/R Fus on 23 July 1917.
21 Formerly PS/3960 Private Harry Vander-Weyden, of 18/R Fus, was aged 19, and formerly a student educated at West Heath School, Hampstead.
22 His MC citation stated, 'He rallied men of another unit and consolidated the captured position. Later, he went forward under heavy fire and warned detached posts of an impending counterattack, and organised various parties for defence. He set a splendid example of courage and determination.'

PS/279 Private Arthur Drumgold, formerly of 19/R Fus, was from Shanklin, Isle of Wight, and attended the City of London School. He trained at an OCB from May 1916 and was gazetted into the Queen's RWK in October. (Reproduced with permission of Sutton Archives)

holding his position until relieved. He set a splendid example throughout the engagement.'[23]

The next stage of the campaign was the Battle of Polygon Wood from 26 September to 3 October. Archie Brown, of 4/R Fus, attacked in daylight and the men were engaged by artillery and MG fire, which caused many casualties. He remembered it as; 'a frightening ordeal'. Despite these setbacks, the battalion reached its objective. Brown leapt into a gun pit near the objective and took cover until the enemy artillery-fire ceased. They later took up a new position and consolidated some trench lines from a series of shell holes. Accordingly, 'the battalion lost 205 dead and wounded and among these were some who had served with me in the old 28th Public School Battalion …'[24]

Gerald Drew, of 8th Battalion Devonshire Regiment (8/Devons), formerly 19/R Fus, went into the attack on 25 October:

> I, with two platoons on the right of my battalion, was detailed to mop up behind the first wave. Having reached the first objective, I gave the order to the remnant of my men to re-organise and prepare to carry on to the next objective. Before the time arrived, something happened on my left, & at the same time many enemy appeared on the right, where there was a large gap… I found it impossible to move from my position owing to intense machine gun fire from all sides. At night, I tried to get back & while helping an officer of the 9th Devons, who was shot through the knee, to get to our own lines, we were surrounded…[25]

23　*The London Gazette*, 18 March 1918, p.3416.
24　Brown, *Destiny*, p.21.
25　TNA: WO 339/60084: Drew Officer File.

Drew was captured and spent the rest of the war as a POW. On 26 October Herbert Edward Fry, of 1/Queen's RWK was in action near Gheluvelt:

> We moved forward to the attack on a position to the north of Gheluvelt off the Menin Road at 4-40 on the 26th Oct 1917. D + B Companys [sic] moved forward in 4 waves D Company leading. Within a few minutes of zero hour the enemy began a very heavy barrage, including heavy machine gun fire, with the result that both companys suffered heavy casualties. The officer commanding the leading company was killed + all his subalterns wounded + my own subalterns were hit directly after leaving the jumping off trench. My C.H.Q. being wiped out with the exception of one runner whom I sent back to B.H.Q. I then went forward with the remnants of the 2 companys to my first objective + was wounded in my right wrist, the bone being broken. On obtaining my objective the enemy immediately counter-attacked very heavily + the 2 companys at this time numbered about 40. The position was held until nearly all had become casualties. My right thigh was badly shattered, the flanks retired, and I was taken prisoner 2 days later.[26]

Fry had his leg amputated due to frostbite whilst a POW and he was repatriated in May 1918.

The Third Battle of Ypres ended on 10 November in appalling conditions near Passchendaele. Though individual phases of the offensive had been successful it had been very costly for the whole British Army and equally so for former officers and men of the UPS Brigade. Whilst numbers of UPS-men had played parts in the battle the true extent of their service cannot be determined from this short chapter. Nicholas Pease, and 8/E Surrey, remained in the Ypres Salient after the battle was over; 'From November until February 1918 we remained in the Passchendaele Houthulst Forest region holding the line. The winter was hard; the frost mitigated the mud problem from time to time but produced many casualties from frostbite and some men even died of exposure.'[27]

Another ex-UPS man with 26/R Fus, Eric Hiscock, remembered his induction into the front line immediately after this great battle, we filed obediently into a deeply-cut trench in the Ypres Salient ironically signposted 'BELLEVUE' a sniper's bullet fired from the other side of a ghastly, decaying, body-strewn, tree-less corridor of death that was No Man's Land, penetrated the skull of the sergeant-in-charge a few feet from me ... his lifeless body fell to the sodden duckboards ... above the rat-infested slime.'[28]

The 20/R Fus experienced a similar depressing wasteland and would also have a hard winter at the eastern extremity of the salient. Before examining this, battalion performance aduring the autumn must be explored.

---

26 TNA: WO 339/60774: H.E. Fry Officer File.
27 IWM: Documents.8230: Papers of N. Pease, p.44.
28 Hiscock, *The Bells of Hell*, p.27.

28

20/R Fus and the Third Battle of Ypres

It was hell's delight

> *It was Hell's Delight; there was not a soul about, not a tree, not a mark, nothing. … the whole place was dreadful, evil the whole place … pieces of equipment, pieces of mules, pieces of men's arms, you have no idea …*[1]

There was a relative calm before the storm of the Third Battle of Ypres. For the first half of September 1917 20/R Fus was at Houlle (near St Omer) where they were able to conduct training in semi-open warfare.[2] On 2 September Major General Pinney was sent to hospital in Boulogne to get a leg injury treated. He was temporarily replaced by Brigadier General P.R. Wood CB CMG who had formerly commanded 43rd Brigade, 14th Division. Within three days Wood was visiting the forward areas at Ypres; presumably in readiness for the next offensive by 33rd Division. In the process, however, the GSO1, Lieutenant Colonel Forster DSO, was wounded. The GSO1 was a vital staff role in a divisional HQ and losing Forster would significantly impact on the efficiency of HQ 33rd Division. Dunn clearly respected Forster; 'He was a sad loss. Not helped by some puritan stodginess, he was an earnest, capable worker in a Division whose General Officers, many of us thought, were not notable for earnestness or capability.'[3]

On 13 September 1917 a further thirty-one soldiers joined 20/R Fus from 39 IBD having previously served with various other Royal Fusilier battalions; a couple had transferred from other regiments. None were original UPS soldiers. The majority were likely to be conscripts and for many it was their first time in France. They needed to be assimilated quickly as the battalion was likely to go into battle again relatively soon. Next day the transport was inspected by Lieutenant Colonel James Chaplin DSO, Commanding Officer of 1/Cameronians, who was acting as the Brigade Commander.

At 10:00 a.m. on 15 September 20/R Fus moved off with the rest of the Brigade to billets at Lederzeele. The next day the Battalion marched to Steenvoorde and the next to Thieushouk;

---

1 IWM: Donald Price Sound Recording 10168, Reel 11.
2 During this training GS/66401 Private Wilfred Coltman died of wounds and is buried at Longueness (St Omer) Souvenir Cemetery. He was formerly a coal miner and a soldier of the ASC. The 19th Brigade reported one casualty the day before, presumably due to a training accident.
3 Dunn, *War the Infantry Knew*, p.382.

the battalion getting ever nearer to Ypres. 20/R Fus remained in billets and tents on 18 and 19 September engaging in musketry training and close order drill. On 20 September the battalion moved to tents northwest of Westoutre where 33rd Division was concentrating. At this time the division was in X Corps reserve whilst there was a major attack astride the Menin Road on 20 September; this included 23rd Division (which 33rd Division would relieve). Despite training in readiness for going into action the battalion was still absorbing replacements; another 100 men arrived on 21 September. However, this draft, presumably not suitably trained, were sent with the Minimum Reserve to Mont-des-Cats where the Divisional Depot Battalion resided. The Minimum Reserve consisted of 25 men per company and five officers and would have included sufficient specialists to reconstitute the battalion in the event of heavy casualties.

With offensive action potentially very close there must have been some consternation when Captain Arthur Murgatroyd was accidentally burned on the evening of 21 September. He was bending over to pick up his stick at 7:00 p.m. when two boxes of matches in his trouser pocket caught fire (presumably a match sticking out of one packet struck the other box). Luckily Captain Henry Clark and Captain Ziegler were on hand to smother the flame and throw water on the affected areas. Ziegler was to be assigned to the Minimum Reserve and recorded:

> I was handing over [the company] to Murgatroyd when he found himself suddenly on fire with a box of matches, he was badly burnt as it burnt through his pocket and caught at his knee and he went to hospital and poor Greenwood was selected to take the company up the line. I felt very sorry for him as he is rather worried over his wife and also he had been promised a job on [the] Brigade [staff] in a few weeks.[4]

Though the burns to his abdomen and right thigh were relatively superficial they were severe enough that he was still sent to hospital by Captain McConnell, the MO.[5] No blame was attached to Murgatroyd for this accident. Murgatroyd was evacuated to the UK three days later; though his burns would quickly heal the damage to his ears and nose, from Arras in April 1917, meant that he would not return to France.

The onerous task of working parties also commenced, and two officers and 160 men were sent to Fitzclarence Farm to bury cables going forward to the front line at 4:30 p.m. on 22 September; they returned, presumably exhausted, at 7:00 a.m. next day. The new draft was presumably put through their paces and sorted out on 22 September and forty were considered well trained enough, or with sufficient previous training, to be despatched to the battalion at midday next day. A further works party was required on 23 September – four officers and 300 men departed at 2:00 p.m. for the same task but did not return until 9:00 a.m. next day. Both working parties were close enough to be exposed to poison gas, presumably Mustard Gas, and 66 men were affected by gas poisoning.[6] Meanwhile on 24 September 1917 33rd Division commenced relieving units of 23rd Division in the line north of the Menin Road. 19th Brigade

---

4   FML: RFM.ARC.3015.2: Diary of Captain G.G. Ziegler.
5   TNA: WO 339/83719: Officer File A. Murgatroyd.
6   G/53203 Private Harold Holmes, aged 25, from Manthorpe, Lincolnshire, had formerly been an apprentice baker. He was killed on 24 September 1917, presumably by shellfire whilst on one of these working parties. He formerly served with the ASC before joining 20/R Fus on 4 April 1917. He is commemorated on the Tyne Cot Memorial.

was to be in divisional reserve with Brigade HQ, the Cameronians and 5/6 Scot Rif at Bedford House. The remainder of the Brigade remained at Westoutre.

There was little time for these working parties to rest; at 5:45 a.m. on the 25th 20/R Fus, minus the Transport and B Echelon, marched to Railway Dugouts along the railway embankment to the southwest of Zillebeke Lake and Transport Farm. 19th Brigade was concentrating near Bedford House. The battalion arrived at around midday. As soon as they did, officers and a pair of runners from each company departed to reconnoitre the tracks leading towards future positions near Stirling Castle, Clapham Junction and Sanctuary Wood. The Transport moved to just east of Dickebusch Lake and the Minimum Reserve joined them nearby to the southwest of the lake later that day. At 11:30 p.m. the battalion commenced moving to reserve position around Jam Row southwest of Stirling Castle. During the day Second Lieutenant Joseph Finch was severely wounded near Zillebeke Lake. He was hit in the face by a shell fragment which resulted in him losing his left eye and having his forehead smashed.

For the coming offensives 19th Brigade would be divided up across 33rd Division. 100th Brigade was attacking on the right and 1/Cameronians was attached to them as a reserve. 98th Brigade were attacking on the left and 5/6 Scot Rif would bolster them. 2/RWF and 20/R Fus were 'queued-up' in divisional reserve and would be employed wherever the divisional commander decided. Though 20/R Fus was in reserve they were held in readiness to support or reinforce the attacking battalions as required. Zero hour for 98th and 100th Brigades was 5:50 a.m. on 26 September.

At 8:00 a.m. 2/RWF were placed at the disposal of 98th Brigade. 20/R Fus would therefore be next in the queue to be assigned to support one of the attacking brigades; 20/R Fus moved forward to the positions vacated by 2/RWF. However, this was not to be. PS/5473 Private Donald Price, now a runner, recalled returning to the forward trenches, near Stirling Castle:

> … we got to a place called Sanctuary Wood and were making our way up to the front line and it was Hell's Delight; there was not a soul about, not a tree, not a mark, nothing. … the whole place was dreadful, evil the whole place … pieces of equipment, pieces of mules, pieces of men's arms, you have no idea…. The sergeant major said to me, take this [message] back to the officers' mess; we were getting plastered [it was] bloody terrible. Anyhow, Jock Muir was with me and we had to run along one of these duck boards … there was no path, no nothing, just these duck bards we'd put down… We were running down this [duck board track] in the dark … poor old Jock, a shell came and knocked him out, killed old Jock… So I dashed on, with this message, to the dug-out … I slid down this dugout and I shall never forget this bloody officer, I was absolutely knackered, … I gave him this message… This officer said; "Right, get back as quick as you can". I could have shot him …[7]

After the arduous journey Price got back to his trench and sat down, fatigued; 'All of a sudden, "poof", he [the German gunners] started again and I got a real beauty … a mustard gas shell burst on the back of the trench and plastered my back with mustard gas. Anyhow he [the sergeant major] says; "off you go"…'[8] Price was also struck on the left knee by a shell fragment. He next remembered being at Boulogne with painful blisters on his back and a cut to his knee.[9]

---

7   IWM: Donald Price Sound Recording 10168, Reel 11.
8   IWM: Donald Price Sound Recording 10168, Reel 11.
9   Price was back with his battalion two weeks later.

At 9:30 a.m. on 26 September a gas shell hit battalion headquarters, likely from the same barrage of gas shells that affected Price. This incident rendered most of the senior personalities of the battalion *hors de combat*. The commanding officer, Lieutenant Colonel Leader; the adjutant, Captain Clark; the MO, Captain McConnell, and Second Lieutenants Hill, Philip and Lewis were all affected by gas.[10] Lieutenant Roy Greenwood was also sent back as a casualty.[11] Clark's symptoms were, 'went to sleep for 8 hours, then loss of vision, drowsiness, & severe inflammation of eyes, with cough & expectoration.'[12] These appear consistent with Mustard Gas. If 2/RWF had not been sent forwards, that battalion might have suffered similarly. The RWF did not benefit from this change of location; their commanding officer and adjutant were both killed by shellfire after moving.

In the immediate aftermath of the gas attack Second Lieutenant Harry Chapman was sent to 98th Brigade as a liaison officer at 2:00 p.m. By 3:00 p.m. Captain John Templar had taken command of the battalion with Second Lieutenant Percival James Tomkinson as adjutant. Tomkinson had formerly been a sergeant with the Battalion until gazetted in December 1916. During the latter half of the day Second Lieutenant Chapman also became a gas casualty. In the absence of an MO the padre, Reverend Ernest Mannering, volunteered to take command of the battalion stretcher bearers until a new doctor arrived.[13]

Orders were issued by Major General Wood at 4:10 p.m. that the battalion was to be attached to 98th Brigade and sent further forward to trenches near Stirling Castle where it would be in reserve. B Company was to support 5/6 Scot Rif and moved to the area of the road junction on the south side of Glencorse Wood. Further orders from General Wood to 98th Brigade placed 20/R Fus at the latter's disposal 'for the purposes of supporting your own front line and enabling you to maintain it.'[14] At 8:00 p.m. that evening 19th Brigade received orders that the division was to be relieved by 23rd Division over the next two nights.

After a few relatively quiet days with the 20/R Fus 'minimum reserve' George Ziegler wrote in his diary; 'In the evening Hollingsworth had orders to go up to the Battalion. We heard that the Colonel had been gassed, Finch wounded (lost an eye), Hill badly gassed, Lewis and Philip gassed. Later, about 11 p.m., [Second Lieutenant] Cook was sent for. We then retired to bed to be woken up by bombs being dropped all round the camp.'[15] Just before midnight Captain Hollingsworth arrived with the battalion from the 'minimum reserve' to take over as commanding officer 20/R Fus. Ziegler avoided going up as he was to attend a course on 28 September.[16]

---

10   Lieutenant Colonel Leader recovered from the effects of this gas. He later commanded 2/King's Regiment in France. Acting Captain Henry Clark joined 20/R Fus on 25 November 1916 and was awarded the MC in May 1917. Second Lieutenant Douglas Campbell Philip had enlisted in the Canadian Army and arrived in the UK as 463538 Private Philip of 62nd Battalion CEF. He was gazetted into the Royal Fusiliers in April 1917 and was a barrister by trade. Second Lieutenant William Lewis had formerly served as Sergeant in 17/R Fus. He had been in France since November 1915 and had been awarded the Military Medal. He was gazetted to a commission in May 1917. After recovering from the effects of gas he served with 3/R Fus. Second Lieutenant William Ernest 'Bill' Hill from Tyldesley, Lancashire, aged 24, had previously served as PS/5003 in 20/R Fus.
11   Lieutenant Roy Greenwood was recorded as Not Yet Diagnosed Nervous (NYDN); i.e. shell shock.
12   TNA: WO339/61692: H. Clark Officer File.
13   IWM: Reverend E. Mannering Papers, 6756.
14   TNA: WO95/2406/3: 33rd Division War Diary.
15   FML: RFM.ARC.3015.2: Diary of Captain G.G. Ziegler.
16   Ziegler was lucky. He returned from course in early November just in time to go on leave to

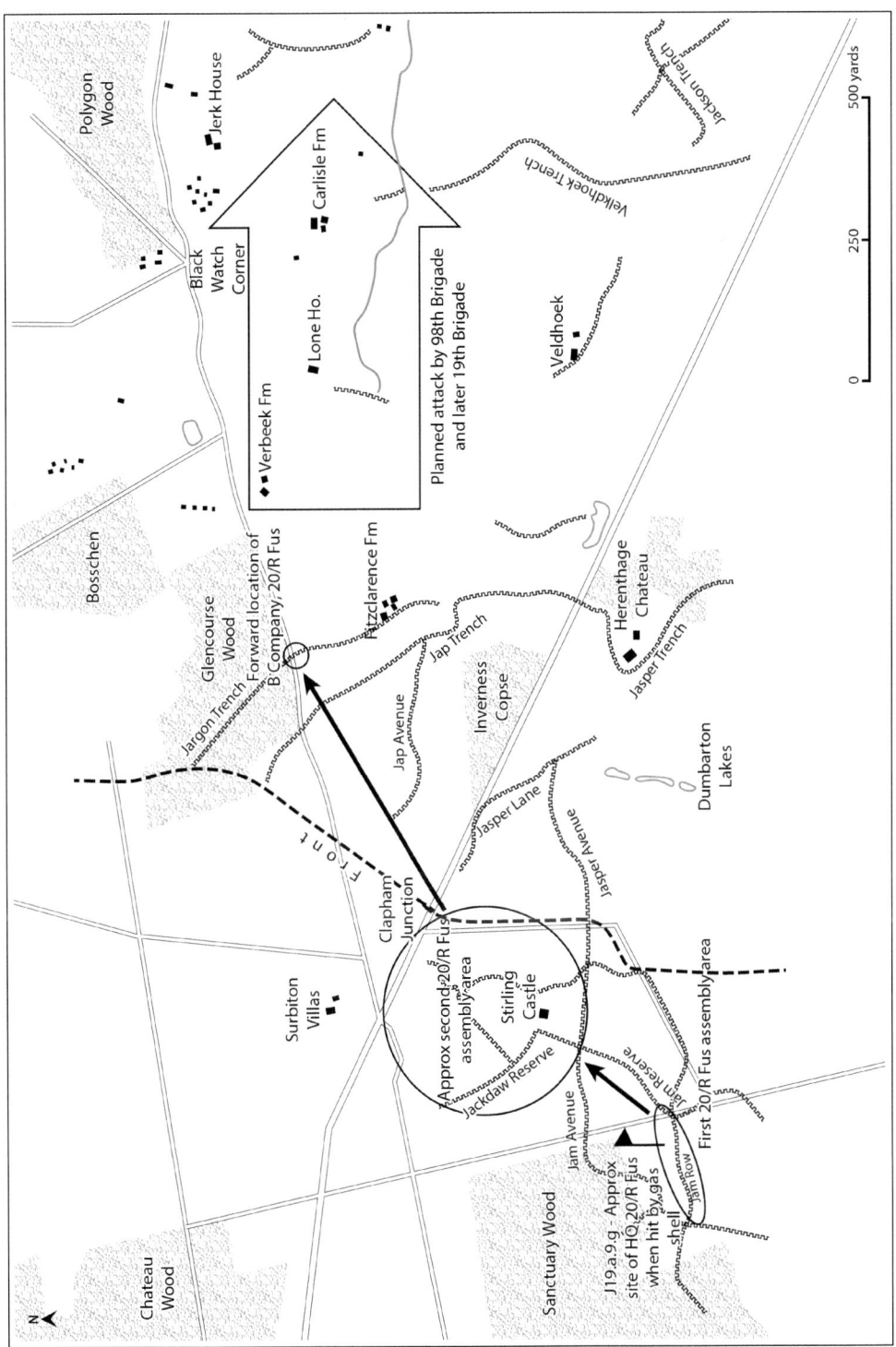

Map 12 20/R Fus operations, Third Battle of Ypres, September 1917.

The battalion remained in reserve, largely unused. This may have been recognition that the unit would not function effectively due to its officer casualties. On 27 September 20/R Fus was attached to 69th Brigade (23rd Division) at 4:00 p.m. That evening (10:00 p.m.) B Company was returned to 20/R Fus from 5/6 Scot Rif. 69th Brigade made no mention of 20/R Fus; presumably they released them quickly back to 19th Brigade. The battalion went back to camp in No 1 Area west of Dickebusch Road at 9:00 a.m. on 28 September.

During the day Major Modera returned from leave and took over the battalion with Captain Hollingsworth then becoming adjutant. Since moving into the line 20/R Fus had been commanded by four different officers and had four different adjutants. The role of MO was taken over by Lieutenant J.H. Boag on 1 October.[17]

The same day the transport and minimum reserve moved to Le Croquet and both were there by 6:00 p.m. To add insult to injury the camp near Dickebusch was bombed by enemy aircraft; one man was killed, and one was wounded.[18] The next afternoon the battalion left for Ouderdom; it then entrained and arrived at Ebblinghem at 8:30 p.m. After a short march it finally reached billets at Le Croquet at 10:30 p.m. The next few days were spent in cleaning up but on 2 October rehearsal parades were conducted ready for the inspection of the battalion by Field Marshal Haig. The whole of 19th Brigade was inspected; what he thought of the Brigade, and 20/R Fus in particular, were not recorded. Captain Dunn of 2/RWF recorded that 19th Brigade paraded at 7:30 a.m. and was inspected by battalion and brigade officers before; 'The C-in-C rode on to the ground at 12.30, twenty minutes late. After pinning ribbons on a few he remounted and passed along the lines of Infantry. Then we marched past, uninspired, on our way back to billets.'[19]

Casualties for September for 20/R Fus were nine officers wounded, two men killed, thirty-four wounded, seventy gassed and one man missing.[20] These were significant losses for a battalion that had only had one company engaged. By comparison 2/RWF had lost ten officers and 229 men, 1/Cameronians six officers and 104 men and 5/6 Scot Rif eight officers and 201 men.

the UK 6–20 November. FML: RFM.ARC.3015.2: Diary of Captain G.G. Ziegler and TNA: WO339/67877: G.G. Ziegler Officer File.

17 Lieutenant John Hamilton Boag was an experienced peacetime doctor and MO. He was born in 1893 and had lived in Edinburgh. Whilst at the University of Edinburgh he was a member of the OTC Artillery (1908-1912) and was an RAMC Cadet. He arrived in France in May 1915 and served with 29th Field Ambulance. In October 1915 he became MO for 12th Battalion Royal Scots and later served with 20th Sanitary Section RAMC. He was MiD in January 1916. After he joined 20/R Fus he was promoted to captain. He was again 'Mentioned' in December 1917 and was awarded the MC in September 1918 for carrying casualties whilst under machine-gun fire.

18 G/66408 Private John Fraser was reportedly killed in a hostile air attack on 30 September 1917. He had served with the ASC in France since February 1915. He is buried at Longueness (St Omer) Souvenir Cemetery.

19 Dunn, *War the Infantry Knew*, p.406.

20 Holmes was presumably the soldier recorded as missing. The soldier killed was likely GS/20481 Private George Whittle, aged 26; he only served with 20/R Fus for sixteen days. He died of his wounds on 28 September 1917 and is buried at The Huts Cemetery. In addition, S/8145 Acting Corporal Thomas Hughes died of his wounds on 29 September 1917 and is buried at Lijssenthoek.

29

20/R Fus October–December 1917

A plunge into a slough of filthy shell holes

The shell had blown him, and everything else, to pieces …[1]

Entry into the Salient was nothing more or less than a plunge into a slough of filthy shell holes, amongst which hundreds of unburied dead grinned and gesticulated from amongst miles of tangled wire and stunted trees.[2]

G.S. Hutchison

Autumn in the Ypres Salient was a series of bloody battles to wrest difficult terrain off the German Army. Despite this battle of wills, this battle of men against materiel, the involvement of 20/R Fus was peripheral. Though not central to military operations they endured the hard conditions, and a tough workload, to indirectly assist the campaign. During this period, more ex-UPS men died with other battalions than with 20/R Fus. October was to be a difficult month for 20/R Fus. A sign of the work ahead was evident when a 19th Brigade Pioneer Company was formed and Second Lieutenant Harold Kirk and twenty-two men helped comprise it.

On 4 October 20/R Fus departed Le Croquet at 11:30 a.m. and marched to Ebblinghem Station and caught a train to Vlamertinghe arriving at 6:00 p.m. The Transport travelled direct and remained there. The rest of the Battalion marched to a field near Goldfish Dugouts arriving at 7:00 p.m. and set up camp for the night. Tents did not arrive until the next day and the battalion presumably had to rough-it without proper shelter. Here 20/R Fus came under 3rd Australian Division for the purposes of administration and provided labour for 2nd Canadian Railway Troops (2/CRT) who were constructing light railways in the Ypres Area. 20/R Fus reported along with three other battalions (8/KOYLI, 2/A&SH and 10/DWR). All four were recorded as supposedly being out of the line 'on rest'. However, the requirements of battlefield logistics, combined with the hampering effects of the wet weather, meant that this work was vital for maintaining the fighting troops in the front line.

---

1 IWM: Donald Price Sound Recording 10168, Reel 12.
2 Hutchison, *Thirty-Third Division*, p.76.

The infantry would provide the labour to accomplish the extension of light railways eastwards under the expertise of 2/CRT. At 11:30 a.m. on 5 October, 237 men and five officers paraded and started work with 2/CRT at midday. They were involved in ballasting, grading and laying rails on light railways northeast of Wieltje. The next day work was arranged on a two-shift basis with the first shift, consisting of a half-battalion of 20/R Fus starting at 6:15 a.m. and finishing at midday. A second shift of similar numbers began at midday and worked until 6:30 p.m. For example on 6 October five officers and 234 men worked in the morning and five officers and 215 men worked in the afternoon. The next day the other half of 20/R Fus presumably worked the morning shift or afternoon shift. From 7 October C and D Companies provided the morning manpower each day and A and B Companies the afternoon parties.

The shell-pocked terrain made the line-laying a slow process. As the railway approached the rear of the battle zone progress became more dangerous and more horrible. Near Pommern Castle; 'The ground is covered with dead bodies, making the work of grading very unpleasant. The block houses are also filled with dead Germans…'[3] On 8 October 20/R Fus parties assisted in unloading ammunition at Oxford Road, the Becklington Line and Bavaria House. These labour tasks continued until 11 October, when the companies swapped over. There was a rest day on 10 October.

On 11 October Major F.S. Modera MC took over command of 20/R Fus and became an acting lieutenant colonel. In addition, Second Lieutenant Richard Owen Darker joined 20/R Fus on 8 October having been 'gazetted' two months before.[4] 12 October saw a smaller working party being provided by the battalion; this party moved eastwards to a new tented camp just northeast of Ypres. This change would presumably eliminate time slogging back and forward through the mud for tired working parties. From here, for the next week the battalion provided six officers and 450 men each day from 6:45 a.m. to 11:00 a.m. 20/R Fus was relieved here by 1/Cameronians on 17 October having provided a working party that morning. A Cameronian officer described his experiences doing the same job; 'We came back from our labour job [repairing roads and railways] … and I wasn't sorry, as with the weather we had, living in tents wasn't an ideal existence, to say nothing of the daily and nightly [aerial] bombing…'[5]

The men of 20/R Fus were taken in buses to Shankill Camp just east of Neuve Eglise and there they rejoined 19th Brigade. The interlude of labouring that 20/R Fus underwent was not necessarily because they were poorly thought of in the Brigade, they just did it first. All four battalions were sent over the coming weeks. Whilst 20/R Fus had been away, 19th Brigade had been holding the line east of Messines. This was the front line secured after the capture of the Messines Ridge. The next day was devoted to cleaning up in camp and getting the soldiers bathed. Further cleaning and kit inspections followed on 19 October. Small parties were found for working in nearby camps but training kicked-off on 20 October and continued afterwards, albeit with a break on Sunday 21 October, until almost the end of the month. Subjects covered included training of specialists, digging and revetting trenches and even some ceremonial drill. However, there were soon plans afoot to return to the front line and the commanding officer and five other officers went forward to reconnoitre ground.

---

3   TNA: WO 95/4065: 2/CRT War Diary.
4   Second Lieutenant Richard Darker had formerly served with 16/Mx (PS/1192) and went to France in November 1915. He was gazetted to the Royal Fusiliers in August 1917.
5   Hugh Maclean and John Baynes, *A Tale of Two Captains* (Edinburgh: Pentland Press, 1990), p.121.

Meanwhile, several working parties had to be found which included one officer and eighty men to a nearby camp; ten men to assist 290th Army Troops Company RE; 22 men to the main divisional RE dump and an officer and eighteen men reported to the PoW Cage at Bailleul. These parties were all back by 28 October.

On 29 October the Drums and Fifes left camp for Penzance Lines and Second Lieutenant Richard Worley proceeded to take over the bands, drums and pipes of 19th Brigade.[6] Casualties were relatively light during October with only one man killed and ten wounded; of these, three remained with the battalion.[7]

On the evening of 30 October 20/R Fus moved out again, via Wulverghem, to 'Bristol Castle' where they took over reserve positions from 1/Mx. A, B and D Companies were dispersed in trenches to the southwest of this location. C Company were under the orders of 1/Cameronians who were holding the front line. Next day, working parties began with Major D.W. Hollingsworth, six officers and 200 men digging a support trench. The situation and working parties remained the same over the next two days.

On 2 November 1917 the company commanders reconnoitred the front line; during this activity Captain Cecil Butler was wounded.[8] This activity was in preparation for the relief of 1/Cameronians in the right-hand sector of the front line on the night of 3–4 November 1917. By 12:30 a.m. this was complete. Battalion HQ was in a house on the Tilleul Road. D Company was in the front line on the right and B Company was on the left, A Company was in support in Pollard Support.[9] C Company was in reserve with D Company of 1/Cameronians providing an additional reserve. Supplies were to be brought up via a trench tramway from Wulverghem to Bethlehem Farm. The next few days were uneventful and on the evening of 7–8 November the battalion was due to be relieved by 1/Queen's; the relief commenced at 3:00 p.m. However, at 8:30 p.m., Post No 11, the left-most post of the left-hand company (B Company) was raided by the enemy. Hostile artillery fire wounded six men of 1/Queen's in Pollard Support. Two men and a Lewis gun were missing after this raid. The 33rd Division General Staff (GS) war diary stated; 'A court of enquiry was held on the rest of the post as the whole episode was unsatisfactory. The court found that sufficient resistance was not put up by the rest of the post nor had a proper look out been kept.'[10] The relief was complete by 10:30 p.m. 20/R Fus then marched to Kortepyp 'A' Camp. 8 November was spent in cleaning up and likely making up losses in equipment. The next five days were spent in conducting training whilst in camp. On the 13th, Second Lieutenant Geoffrey Colbourne took over as the Battalion Adjutant; the former incumbent, Captain Cyril Catchpole, took command of A Company.

6   Second Lieutenant Richard Hugh Worley, formerly of 16/KRRC, was gazetted into the Royal Fusiliers in April 1917.
7   Of these casualties GS/63986 Private Thomas Livesey was killed in action on 11 October 1917; he had previously served with the Yorkshire Regiment. He was buried in Bedford House Cemetery. GS/63980 Private Alfred Taylor, aged 26, died of his wounds on 14 October 1917. He was buried in Nine Elms British Cemetery. Both men had joined the battalion on 13 September.
8   Captain Cecil Butler had previously served as a sergeant in 22/R Fus in France for five months during 1916. He was gazetted in May 1917 and five months later was commanding a company.
9   This may have been dug by Major Hollingsworth's works party.
10 TNA: WO 95/2406: 33rd Division GS War Diary.

The 20/R Fus battalion band. Three members of specific interest are (centre) Ernest Brierley the Sergeant Drummer or 'Drum Major'. He had served in 20/R Fus since 1914 and was steadily promoted during early 1916 to leading the band. He later served with 13/R Fus and was gazetted to a commission in March 1919 with the Manchester Regiment. The battalion CO, Lieutenant Colonel Modera, is seated to the left of Brierley. He took over on 11 October 1917 having risen from subaltern to lieutenant colonel in two-and-a-half years. Modera has notably aged since his earlier photograph. On the right of Brierley is Second Lieutenant Geoffrey Colbourne who became adjutant in November 1917 thus dating the image to after that event. (Reproduced with the permission of Angela Jennings)

On 14 November the Battalion marched to Strazeele via Waterloo Road, Ravelsberg and Bailleul. 20/R Fus was billeted in camps to the west and southwest of the village. The 15th and 16th were spent conducting more training in camp. Next day the battalion departed early in the morning, on buses, and were dropped off at a point on the Ypres-Dickebusch Road. They then marched to bivouacs just north of the White Chateau on the Menin Road. On 18 November the Battalion was at rest and cleaned up their camp and conducted a reconnaissance of 'H' Track and the Zonnebeke– Potijze Road. Next day the battalion changed camp to a point on the Menin Road slightly nearer Ypres.

Between 20 and 23 November 20/R Fus provided working parties to conduct tasks in the forward areas. Donald Price recalled the death of one comrade, likely on 23 November:

> There was always one man dedicated as the post man, he used to go to the headquarters and get the bag of post for the battalion and then it was his job to distribute the letters and parcels ... our company fellow was a fellow named Hill. Sid and I were in this dugout and we were getting what we called fifteen inch shells regularly ... they came on like a train

… 'Bill' Hill [sic] was coming along with this bag of mail for us and all of a sudden these fellows [shells] came over and landed. There was no Hill, he'd gone, and all the letters and things were all over the place. The shell had blown him, and everything else, to pieces and eventually I know he was picked up and sort of gathered together, pieces of him, and that was poor 'Bill' Hill. A shocking business…[11]

Price potentially witnessed the death of GS/63325 Lance Corporal Edward Hill, who was from London. He had formerly served with 1/Londons and had joined 20/R Fus on 11 July 1916. His longevity with the battalion might be explained by his role as Post NCO. Hill was buried in Menin Road South Military Cemetery along with GS/52099 Private William Thomas Jamieson and GS/63495 Lance Corporal Albert Edward Duke, who were killed on the same day.[12]

Passchendaele – 'The whole place had been pock-marked with shell holes …'

On the 24th they moved up the Zonnebeke–Potijze Road, then 'H' and 'J' Tracks, and took over the right support position with the Battalion HQ at Seine. They were in the area to the southwest of Passchendaele, an infamous site at the very tip of the Ypres Salient which had been heavily fought over. The next two days saw more work and carrying parties brought ammunition and supplies to the front line. According to Donald Price, who was remembering this approximate period of the war, conditions in this sector were awful:

There was no land there at all, there was no earth, it had all been moved and jumbled about … the whole place had been pock-marked with shell holes and it was a case of water and earth … no trees, no vegetation, no nothing. … All those fellows up there in the front line had got to be fed, had got to be kept alive. One of the first things we had to do, I remember, we had to carry these duck boards up to put on the earth to make a path up to the line. … You'd be amazed, we must have taken up thousands of these things … over the course of the battle, to get anywhere. … It was quite true, in many cases, that if you fell off, and you weren't helped back on [that] you were a gonner.[13]

The only indications as to where men were, came from signposts.

Meantime, the forward positions were reconnoitred in preparation of them being occupied by the Battalion in the future. There had been a small number of fatalities suffered over this period; one man was killed on 19 November, three on 23 November, two on 25 November and another man was lost on the 26th.[14] This slow, steady, trickle of casualties slowly added up to death by a

---

11 Price may have been confused with Lieutenant 'Bill' Hill whom he had served as batman at Arras. IWM: Donald Price Sound Recording 10168, Reel 12.
12 Albert Duke and William Jamieson were buried in Menin Road South Military Cemetery.
13 IWM: Donald Price Sound Recording 10168, Reel 12.
14 GS/68556 Private Charles Nelson, aged 19, died of his wounds on 19 November and was buried in Ypres Reservoir Cemetery. GS/68917 Private Charles Ashton, aged 29, died on 20 November and was buried at Lijssenthoek. PS/11820 Lance Corporal Herbert Edward Victor Brown, aged 20, was killed on 25 November as was GS/63981 Private William Tucker, aged 29. G/50142 Private Harold Brown, aged 23, was killed next day. Herbert Brown, Tucker and Harold Brown were buried in

thousand cuts and was typical of battalions holding the line in such conditions. To this figure must have been added numerous men wounded, evacuated sick or exhausted.

On 27 November the Battalion went forward and took over the right subsector from 5/6 Scot Rif; 20/R Fus oversaw the sector by 11:20 p.m. Price described the nature of the front line at this time:

> … there were no trenches as such, there were no continuous trenches, you made a trench and it was blown up the next minute. … eventually, when things settled down a bit they did start trenching a bit … as far as I remember there were no trenches. … I was lying in a shell hole for two days with water to your knees … it was a job to skirt shell holes which were full of water.[15]

Now the Battalion was back in the front line the casualties mounted. On 28 November there was a heavy German bombardment on the 20/R Fus front line positions. An SOS signal went up on the right flank of the battalion. During this heavy shelling the battalion suffered numerous casualties which were likely exacerbated by the rudimentary nature of the shelter from shellfire the positions offered. The final death toll on 28 November was twenty-one men.[16] In total, Second Lieutenant Louis Solomon and approximately fifty men became casualties. Solomon had enlisted with 19/R Fus and been gazetted to a commission in the RFC after serving in the trenches over the winter of 1915–1916. He later returned to the infantry and joined 28/R Fus in April 1917 and joined 20/R Fus on 3 June. He recovered from his injury and rejoined the battalion in January 1918. One of those killed was PS/5173 Lance Corporal Leonard Sandham Knott, of A Company, No 1 Platoon; he was an 'original' who had travelled to France with 20/R Fus on 14 November 1915; by this stage there were relatively few men of his vintage left with the battalion.[17] Another six men died of their wounds over the next two days.[18]

    Dochy Farm New British Cemetery.
15 IWM: Donald Price Sound Recording 10168, Reel 12.
16 The following eleven men were buried in Tyne Cot Cemetery: SR/9296 Private Chester Abbett had gone to France in November 1914; G/34815 Private Harold Goodwin, aged 27; GS/66479 Private John Graham, aged 19, was the son of a railway surface layer; GS/66416 Private Leonard Hanchett worked for a ruler manufacturer; GS/53123 Private Albert Kinson; GS/63652 Private William Rawle; GS/53124 Private Henry Sheerman, aged 31; GS/66494 Private Robert Thurston was an errand boy for a grocery store; GS/63917 Private James Spencer; GS/63964 Private Harry Orme, aged 28, a butcher's assistant; GS/63971 Private Henry Robinson, aged 19; GS/63978 Private Joseph Taylor, aged 25. Taylor died of his wounds on 28 November and was buried at Lijssenthoek. Orme and Spencer were buried in Tyne Cot Cemetery. Robinson is commemorated on the Tyne Cot Memorial. GS/6197 Corporal George Lipscombe, aged 23, a farm labourer and GS/66443 Private Matthew O'Reilly, aged 33, are also named on the Tyne Cot Memorial. G/46563 Private George Beale, aged 34, and 228046 Private Joseph Eke, aged 22, an errand boy, died on 28 November and were buried in Passchendaele New British Cemetery. PS/9093 Private Sidney Arthur Bendell Wightman, aged 21, from Felixstowe, and GS/63910 Private Jack Levy, aged 24, a Jewish soldier who was born in Whitechapel, both died on 28 November and were buried in Dochy Farm New British Cemetery. GS/63476 Lance Corporal George Bates, aged 26, and GS/66469 Private David Graham, died of their wounds and were buried in Nine Elms Cemetery.
17 Knott, aged 22, and from Airdenshaw, was buried in Tyne Cot Cemetery.
18 GS/53205 Private Isaac Harrison, aged 28, and GS/63694 Private Percy Holman, aged 35, both died on 29 November and were buried at Dochy Farm. GS/63574 Private Alfred Clancy, aged 29, died of his wounds on 29 November 1917 and was buried in Nine Elms Cemetery. GS/52596 Private Alfred

Next day Major D.W. Hollingsworth took over command of 20/R Fus after Lieutenant Colonel Modera went sick and departed to hospital. At 7:15 a.m. on 30 November the SOS signal was again sent up by the right front of the battalion. German troops were seen to advance against the unit on the right flank of 20/R Fus on the Keiberg Spur. That evening the battalion was relieved by 2/Worcs R and once this was complete 20/R Fus moved to billets on the Ypres-Menin Road. Next day the Battalion marched from billets near the Menin Road to St Jean Station via Irish Farm; after a short train ride they disembarked at Brandhoek and marched to Toronto Camp where they arrived at 12:00 p.m. During November 1917 the battalion had lost one officer wounded, 29 men killed, 69 wounded and two missing.

Major General Pinney returned to command 33rd Division on 28 November. Major General Wood returned to England. According to Captain Dunn, Wood had allegedly 'strafed' his Brigade Commanders for not being active enough.[19] However, Pinney had apparently used his influence to get back into command. This was helped by Wood supposedly having fallen out with the corps commander.

Further training took place until 6 December on which day Lieutenant Colonel Modera took over command again. After this short rest the Battalion marched back towards Ypres and arrived at Sapper Camp on the Ypres–Potijze Road where they relieved 1/9 HLI. Further training took place for the next three days. On the 11th the Battalion entrained at St Jean and travelled to Abeele and marched to billets in various farms near Watou. On 12 and 13 December the Battalion cleaned and improved their billets and spent another five days training. On 20 December Lieutenant Colonel Modera again departed; this time having been injured in an accident. Major Hollingsworth was away on leave so Captain G.E.R. de Miremont of 2/RWF took over temporary command.

On 21 December the Battalion moved to Poperinghe where the battalion was billeted. From the 22nd until Christmas Eve the Battalion provided three companies to the Commander RE of VIII Corps every day. These companies worked at erecting barbed wire defences for the Corps Line. There were no working parties on Christmas Day and the battalion was presumably able to celebrate for a brief time. Boxing Day and the three subsequent days were also spent on working parties conducting wiring. Lieutenant Colonel Modera returned from hospital and assumed command. No sooner than he had taken over, Modera departed on leave leaving Hollingsworth back in charge.

At 11:00 a.m. on 29 December the Battalion was relieved by 4/King's. B Company and Battalion HQ marched to their old billets near Watou. The other companies were carried by train from where they were working back to Abeele; they marched from there to billets near Watou. There was a lecture to all battalion officers and sergeants in 33rd Division on New Year's Eve by the Corps Commander, Lieutenant General Sir Aylmer Hunter Weston (VIII Corps), at Steenvoorde. He was followed by Lieutenant Colonel Levy DSO who lectured on training the individual soldier without training areas. Training took place for the rest of the battalion.

Wilson, aged 35, died of his wounds on 29 November and was buried in Divisional Collecting Post Cemetery and Extension. GS/69061 Private Harry Wood, aged 32, died on 30 November and was also buried in Nine Elms Cemetery. GS/63558 Private Frederick Cattell, aged 37, also died on 30 November and was buried at Lijssenthoek.

19 Dunn, *War the Infantry Knew*, p.413.

According to the war diary there was only one man wounded during December 1917. GS/63485 Private Alfred James Chapman, from Cambridge Heath, London, died of his wounds at Etaples on 8 December 1917. 228424 Private Thomas Clarke, aged 37, from Longton, also died of his wounds on 20 December 1917 and was buried in Boulogne Eastern Cemetery. A couple of men returned from hospital, including Second Lieutenant Solomon who had been wounded near Passchendaele. Several junior officers also went on training courses concerning musketry, gas, bombing and Lewis guns.

Summary of 1917

Having been decimated at High Wood and having also lost men in subsequent fights and to commissions 1916 had ended badly for 20/R Fus. Getting through the winter 1916-1917 had been a matter of endurance but the Spring had been an opportunity for a renaissance for 20/R Fus. Officers and men had been able to 'bed in' and training had enabled the battalion to reach a better level of combat efficiency. The Battle of Arras was an indication of their performance but results were mixed; a costly failure on 16 April was matched with a partial success on 20 May. But these were operations when better battalions than 20/R Fus had also not succeeded. The summer by the seaside had not been a rest cure but had been preferable to insertion into the line at Ypres and the casualties other battalions had suffered. Further training and replacements might also have created optimism prior to the Third battle of Ypres. In September the battalion had reached a low ebb having lost its commanding officer, and much of Battalion HQ, to gas. Its ignominious removal from the line without even seeing action had been a low moment. It is difficult to see how the unfortunate loss of these officers could have been avoided but the other battalions in 19th Brigade had to pick up the slack. This likely reinforced the low opinions held by the critics of 20/R Fus and few might hold a good opinion of the battalion. This was followed with labouring tasks and holding the line; this was vital work but it further bled the battalion of men and did not allow much time for training. The rest of the autumn and the winter was another period that had to be endured.

On a positive note, a UPS officer was now in command and there were still UPS men in the ranks to hold up the previous traditions of the battalion. Moreover, the previous replacements had grown experienced and added their own character to the battalion. It was likely hoped that 1918 might bring a further revival in the fortunes of 20/R Fus and the year would be an opportunity for further glory once the Battalion was given a chance to rest.

30

Early 1918 in the Ypres Salient and Disbandment

We were divided

'We were called out onto the parade ground ... we were all lined up and ... we were divided.[1]

The newcomers were nearly as strong as ourselves ...[2]

New Years Day saw the battalion still in Watou, a holiday enjoyed by all ranks. The respite was short, training was renewed the next day. On the 3rd the battalion travelled by railway from Abeele Station to Brandhoek before marching to Toronto East Camp which was a kilometre southwest of Brandhoek. The battalion transport travelled by road. At some point on 3 January PS/5516 Private Peter Stanley Reynolds of 20/R Fus, died of his wounds. He was attached to 19th Brigade Trench Mortar Battery when he died.[3] Next morning the Battalion moved nearer the front by bus arriving at Ypres; it marched onwards to Whitby Camp which was 400 metres beyond the Menin Gate on the Potijze road. It arrived at 8:30 a.m. The Transport and QM's stores moved to Potijze, arriving at 9:00 a.m. Only the fighting strength of the battalion was at Whitby Camp – only 80 men per company had gone forward. The remainder of the battalion, minus transport and QM's stores, remained near Brandhoek (near Shelac Farm). A demonstration platoon was formed under Second Lieutenant William Gould; this was presumably to aid the training of the remainder.[4] The rest at Whitby Camp was only temporary, prior to the battalion moving into the line that evening. The battalion departed at 3:30 p.m. and marched via Potijze and Zonnebeke to Seine where 1/4th Yorks was relieved. The next day 20/R Fus went forward and took over the right-hand sector of the front line from 9/DLI. The relief was made difficult by heavy enemy shelling and was slow. The battalion was situated south of Passchendaele between the Moorslede Road and Tiber Copse with one company in support and battalion HQ at 'Hamburg'. On the night of 8–9 January an officer's patrol of men from

---

1   IWM: Donald Price Sound Recording 10168.
2   Chapman, *Passionate Prodigality*, p.223.
3   PS/5516 Private Peter Stanley Reynolds, aged about 24, was an original member of the battalion who had gone to France in November 1915. He was buried in Nine Elms British Cemetery.
4   Gazetted in 1917, Second Lieutenant William Thomas Gould, aged 22, from Newmarket, had formerly been a sergeant in the Royal Fusiliers who had served in France during 1915.

20/R Fus encountered an enemy covering party 400 yards to the east. It was surmised that these Germans were a covering party protecting the erection of some barbed wire. On 9 January a German wandered into the line held by 20/R Fus. However, his capture did not identify any new German regiments in the vicinity.

20/R Fus maintained their positions until they were relieved on 9 January. 1/Queen's RWS had previously been in support and at 5:45 p.m. commenced taking over the front line; by 12:30 a.m. on 10 January this was complete. On relief 20/R Fus returned to Whitby Camp; considerably more tired than when they had left.[5] At 4:30 a.m. on 10 January the battalion moved to Brandhoek by light railway and then marched to Toronto Camp East where the men spent three days cleaning themselves up and training before returning to Ypres by light railway where it was billeted. From here 20/R Fus provided parties for working nearer the front line from 14 to 20 January 1918. Due to having already served four days in the line in this way and another battalion being missing from 19th Brigade and the poor weather it was decided to pool battalions within the division and to cycle them through the different brigades holding the front line for 48-hour stints. Each brigade would hold the line for four days. On 21 January the battalion marched via Potijze and Judah Track to supporting positions at Seine again where 1/Mx was relieved. Here they remained until 25 January when 9/HLI relieved them. 20/R Fus marched to Alnwick Camp where it rested and refitted for a day. 27 January saw a change in pattern for 20/R Fus; it entrained at St Jean and moved to St Omer before marching to St-Martin-au-Laert where it conducted training. According to the 20/R Fus War Diary three men were killed in January with another dozen wounded and one missing.[6]

'Our Chocolate Soldiers are disbanded'

According to Donald Price, one of the few long-serving survivors left, the public school aspect of the battalion was 'finished'. There were very few original UPS members left. These consisted of only one or two men left on the transport to which Price was attached early in 1918.

In the Spring of 1918, there was a manpower crisis within the British Army in France. Political pressure was placed on Field Marshal Sir Douglas Haig by Prime Minister David Lloyd George to reduce casualties by holding replacement soldiers in the UK. Depleted divisions were therefore unable to adequately return to fighting strength through external replacements. The only solution was to break up infantry battalions to ensure others were replenished whilst not reducing the total number of infantry divisions. Divisional and brigade commanders were given some power in selecting three battalions from under their command for disbandment amongst some deliberate moves of regular battalions between divisions to retain national identities for certain divisions. There was undoubtedly a desire to cut off dead wood. Though 20/R Fus was

---

5   During this stint in the line GS/46136 Private Percy Newman, aged about 35, was killed in action on 7 January 1918. His grave was found near German lines in 1919. On 8 January GS/43602 Private George Corday died of his wounds. Also killed that day was GS/62798 Private Charles Pilbeam, aged 23. All three were buried in Tyne Cot Cemetery.

6   The missing man was likely GS/63914 Private Ernest Richards. He was recorded as being killed on 9 January but there are no records of his burial or commemoration. TNA: WO 95/2423: 20/R Fus War Diary. The 19th Brigade Diary recorded one man killed, eight wounded and one missing. TNA: WO 95/2421: 19th Brigade War Diary.

disbanded, this was not necessarily because of inefficiency. Whether 20/R Fus was as bad as was opined, amongst 19th Brigade, it was the only possible candidate. No Regular Army battalions, nor Territorial Force battalions, were disbanded (though first- and second-line Territorial Force Battalions were amalgamated). Disbandment had to come from the New Army which lost 30 percent of its battalions.[7] The 33rd Division pioneer battalion, 18/Middlesex could not be removed and this left comparison between 20/R Fus and 16/KRRC; the latter was retained. However, Major General Pinney presumably made his decision based on battalion performance rather than seniority, regimental considerations or personal affiliations; Pinney had originally joined the Royal Fusiliers in 1884. However, the residue of 20/R Fus were not of poor quality.

According to the 19th Brigade War Diary on 1 February 20/R Fus commenced disbandment except for the battalion transport. The 20/R Fus War Diary stated; 'Orders received for disbandment of the Battalion, personnel to be posted to the 2nd, 4th and 13th Battalions Royal Fusiliers.'[8] Whilst men of 2/RWF might have taken pleasure in the demise of the last Public School Battalion, that they had long derided, they too departed 19th Brigade on 4 February destined for 38th Division. In their place 1/Queen's RWS was transferred from 100th Brigade to join the two Scottish battalions.

The battalion played the final of its inter-platoon football tournament between No 9 Platoon and No 13 Platoon; the former was awarded the trophy by Lieutenant Colonel Modera. Training took place for three days and on 5 February the battalion was inspected by Major General Pinney and Brigadier General Mayne. No sooner than this was completed the first drafts to other battalions commenced departing on the 5th with larger drafts departing on 9 February. According to Captain Dunn:

> 'Our Chocolate Soldiers are disbanded. The 2nd Royal Fusiliers, getting a draft of over 300 of them, sent a party to pick and choose. They took the Band, football team, and "Ruffles" – the Concert Party, and after that anyone for makeweight. All the rest of the Division tried, and failed, to get some of their horses, which go to Remounts.'[9]

Dunn paid an indirect compliment to 20/R Fus. The battalion did not create an enviable band, football team, concert troupe and well-horsed transport section by accident. 270 men would go to 2/R Fus along with Captain Cyril Catchpole, Lieutenants Thomas Harrison and Louis Solomon, and Second Lieutenants William Claughton, William Gould, Mark Whyte, George Cook, John Smith, William Stokes, Richard Darker, Douglas Neish and George Kenny-Sivewright. 222 men and six officers left for 2/R Fus on 8 February; Gould joined on 16 February, Catchpole and Darker on the 21st, Harrison on the 22nd, and Solomon on 1 March; presumably they all arrived from leave or courses.

On 10 February 41 men arrived with 4/R Fus along with Second Lieutenants Patrick Freeland, Edward Sutro and Stanley Honeywill. Sutro was sent to hospital four days after arriving. A further 20/R Fus draft of 28 men joined on 14 February. Second Lieutenant Arthur Charles joined on the 23rd and a further draft of 19 men presumably joined later to bring the

---

7   Martin Middlebrook, *Your Country Needs You* (London: Leo Cooper, 2000), p.149.
8   TNA: WO 95/2423: 20/R Fus War Diary.
9   Dunn, *War the Infantry Knew*, p.441.

total number to 88.[10] 4/R Fus also received large drafts from 8/R Fus which had also been disbanded.

Donald Price remembered little of the disbandment of 20/R Fus after several decades:

> We had no idea we were going to be disbanded. We were called out onto the parade ground … we were all lined up and … we were divided. So many of us went to so and so or some other battalion. I happened to be sent to 13th Battalion Royal Fusiliers … they'd had a hell of a knock as well; we were sort of building them up.[11]

Price went with a draft of seven officers and 214 men for 13/R Fus on 9 February.[12] He hoped to end up in the transport section of 13/R Fus which he saw as a potentially comfortable job. Arriving in 13/R Fus was a new experience. Almost all officers and men were strange to him, he confessed that; 'I lost sight of everybody'.[13] Eight officers and 213 men joined 13/R Fus on 9 February; the officers were Captains George Ziegler and Frederic Bower, Lieutenant Bryan Marshall, Second Lieutenants Arthur Floyd, Harold Kirk DCM, Arthur Taylor, Charles Vickers MM, Richard Worley. A further batch of officers joined 13/R Fus on 17 February; Second Lieutenants Andrew Blain, Ernest Bowler and Harold Pye; these three were only posted to 20/R Fus on 13 February 1918. Second Lieutenants William Miles, Ronald Barker and Roy Keller, joined on the 22nd and 25th February and 5 March respectively; they had presumably been detached on courses or leave. In all 271 other ranks and fifteen officers went from 20/R Fus to 13/R Fus. Percival Tompkinson also went to 13/R Fus. Though Dunn would suggest 13/R Fus got the leftovers from 2/R Fus they were complimentary of their new draft.

Some officers were found jobs conducive to their rank and experience; Major Hollingsworth left on 10 February to take command of the VIII Corps Reinforcement camp at Poperinghe. Captain John Templar proceeded to the Senior Officers Course in Aldershot, presumably as a precursor to future command of a battalion.

After the main drafts had departed some warrant officers, sergeants and signallers were surplus to requirements and, on 14 February, they were despatched to Hollingsworth's VIII Corps Reinforcement Camp for further posting. On 15 February the battalion transport of 20/R Fus, Captain George Jones-Williams and thirty-four men, departed to be attached to 19th Brigade. According to a 19th Brigade movement order dated 16 February they were accommodated with the Details and QM Stores of 20/R Fus at Camp 32 south of Brandhoek. If Dr Dunn is to be believed their horses were returned to 'Remounts' and the men were presumably absorbed into the transport of other brigade units.

By 15 February there remained unposted only the commanding officer, Lieutenant Colonel Frederick Modera MC; the Adjutant, Captain Geoffrey Colbourne; Quartermaster, Honorary Lieutenant and QM Percy Stokes and Captain Daniel Pigache. Other officers, not recorded here, were away on leave or courses and were posted to new units on their return. On 16 February Modera finally closed the war diary for 20th Battalion Royal Fusiliers and the battalion ceased to officially exist. In April 1918 Modera took over 1/LF. Pigache had been largely employed as

---

10   TNA: WO 95/1421: 4/R Fus War Diary.
11   IWM: Donald Price Sound Recording 10168.
12   Price's service records state he was posted to 13/R Fus on 30 January 1918.
13   IWM: Donald Price Sound Recording 10168.

a canteens officer attached to divisional HQ but had likely remaining on the strength of 20/R Fus. He was described as 'always accessible, unwearying in energy, and gave to us a canteen anywhere and everywhere.'[14] He departed to 7/R Fus.

However, large contingents of this, and previous UPS battalions, were still serving in other battalions. In addition, numerous officers around the Army had served in these battalions or who had been gazetted from their ranks. The UPS battalions had disappeared from the Army order of battle but the men who comprised them still had a long war to fight and more would still make the ultimate sacrifice in the fighting to come. The likes of Dunn, eternally-critical of the UPS battalions, likely felt that disbandment was a just desert for 20/R Fus. Others, like Seton Hutchison remarked, 'on the very great regret of the whole [33rd] Division in losing such fine fighting battalions …[including 20/R Fus]…'[15]

This could be a neat point at which to end this story of the UPS; the disbandment of the last UPS battalion would wrap up the story of these unusual units. However, this would negate the blood, sweat and tears that the surviving members of the six UPS battalions would still endure to help bring the Great War to a victorious conclusion. The stories and fates of these men must be covered for three reasons. First, to determine whether they upheld the reputations of their previous units. Second, to highlight the wider effectiveness of the UPS battalions as a source of effective platoon commanders. Third, the bravery and sacrifice of some of these men during their subsequent service must be recorded to round-off their stories and the history of the battalions that had helped mould them. Before doing this, the experience of the men going to 13/R Fus will be examined to see how well they were incorporated into their new battalion and how well they performed in their first battle.

13/R Fus draft

Other units gained from the influx of manpower from the disbandment of 20/R Fus. The initial activities of 13/R Fus after receiving the 20/R Fus draft are of note to help understand what these newly arrived men experienced. According to Guy Chapman of 13/R Fus; 'We were heartened by the arrival of some two hundred officers and men from our 20th Battalion.'[16] Chapman described the difficulties of absorbing the new draft from 20/R Fus, 'The newcomers were nearly as strong as ourselves, and though they were willing enough to conform to our tenets and customs, they were naturally reluctant to part with those of their disbanded battalion … Some things we did they considered silly: some they believed in we thought rank heresy.'[17] A non-UPS officer later recalled arriving with 13/R Fus; No nicer crowd could have been found than the 13th Battalion … They were now nearly at the end of a long war and casualties had been replaced time after time with men of the New Armies. Even so, something of the old friendly spirit of the territorial remained and I had every reason to be pleased with my welcome.'[18] The men of 13/R Fus would have presumably welcomed both old UPS-men and

---

14 Hutchison, *Thirty-Third Division*, p.174.
15 Hutchison, *Thirty-Third Division*, p.78.
16 Guy Chapman, *A Passionate Prodigality* (London: Ivor Nicholson & Watson, 1933), p.221.
17 Chapman, *Passionate Prodigality*, p.223.
18 E.W. Parker, *Into Battle* (London: Leo Cooper, 1994), p.93.

more recently joined members of 20/R Fus. The position of RSM of 13/R Fus became vacant after the arrival of the 20/R Fus draft. Armour took over as RSM of 13/R Fus in March 1918 and made a strong impression, 'Armour, as fine a soldier as ever I have seen. A tall, lean, trim-waisted martial figure with a long Scots jaw and a fair moustache, he was always the cleanest thing in the battalion. Not only the cleanest but the most efficient; for he knew everything about the army and the war that was worth knowing.'[19] Chapman also noted that Armour was teetotal and seldom smoked. Whilst men of 13/R Fus and 20/R Fus might co-exist this was not possible in other regiments. On the piping out of 2/RWF from 19th Brigade, the Pipe Major of 1/Cameronians stated, within earshot of his sister battalion; 'I don't know why they didn't take that lot [referring to 5/6 Scot Rif]', instead of the RWF.[20]

'A seething, splashing turmoil'

Within a month the 13/R Fus draft were in action and did not have long to assimilate with their new comrades. GS/63311 Private George Allen joined 13/R Fus on 30 January but within three weeks he was killed in action on 22 February. Others of his draft would soon follow; shortly afterwards 13/R Fus was involved in heavy fighting on 8 March 1918. An attack was expected and the British artillery was stood-by with a pre-programmed defensive bombardment. Guy Chapman recalled; 'The S.O.S. rocket went floating up. Fifteen seconds later there was a roar over our heads … and a vast billow of smoke and flame broke along the German front. Our gunners had been standing-to for hours, waiting for the signal. When it came they put down the most devastating defensive barrage … a seething splashing turmoil.'[21] In spite of this counter-barrage, at about 5:45 p.m., there were infantrymen on the receiving end of the heavy German artillery fire and they had suffered severe casualties. According to Donald Price the German artillery:

> … pulverised us, bloody big stuff was coming over, shells, whizz-bangs, the lot. Eventually everybody in the trench but me was either wounded or killed and the trench started caving in …eventually I found one of our fellows had a Lewis gun … and I got hold of this machine gun and I thought that if anything happens at least I've got a gun, I can do something if something happens … it really was very terrible because [we had] no water no nothing; there was a hole at the back, a shell hole … I got a drink [from it] but it was covered in blood … and I couldn't drink it.'[22]

Price was not the only man left as he could hear other men engaging a German aircraft with rifles. A runner arrived in Battalion HQ from No 3 Company requesting stretchers to be sent; the company commander, Captain Bower, was dead and three other officers were casualties leaving a sergeant in charge. The battalion lost 145 men killed and wounded and stretchers

---

19 Chapman, *Passionate Prodigality*, p.235.
20 Dunn, *War the Infantry Knew*, p.442.
21 Chapman, *Passionate Prodigality*, p.228.
22 IWM: Donald Price Sound Recording 10168, Reel 13.

passed battalion HQ all evening.[23] Bower was a former 'original' member of 20/R Fus and had fought bravely at High Wood.

Though the 13/R Fus war diary, and Chapman, suggest the Germans did not enter the No 3 Company trenches there was likely some incursion. GS/66391 Private John Brown, of No 3 Company, was captured, unwounded, on 8 March.[24] However, most of the 13/R Fus casualties were caused by the Germans having smothered the battalion with shells to allow an attack to the left of their positions. These losses were Captain Bower and one officer and seven men killed; five officers and 33 men wounded.[25] This included Lieutenant Bryan Marshall and Second Lieutenant Andrew Blain wounded but the latter remained at duty. In addition at least three former 20/R Fus men were killed.[26]

There were a few lighter moments on this terrible day. A few days before going into the line Lieutenant Marshall had brewed an alcoholic punch that was the envy of all the members of the 13/R Fus officers' mess. The MO jokingly refused to allow a wounded Lieutenant Marshall to leave his regimental aid post until he had imparted the formula for the concoction. Marshall stated that his batman knew the recipe and departed down the line; much to the MO's chagrin the batman in question later arrived wounded and unconscious and he never learnt this secret.

---

23 Chapman, *Passionate Prodigality*, p.229.
24 Brown was aged 28 and from Camberwell. He was held in Friedrichsfeld and Limburg.
25 Captain Bower was buried in Hooge Crater Cemetery. He had formerly served as PS/4508 Lance Sergeant Frederic William Bower, aged 31, from Timperley, Cheshire. He was educated at Manchester Grammar School and subsequently worked for an engineering firm, and played hockey for a local club, before enlisting in September 1914. He was promoted to sergeant and in December 1916 was gazetted to a commission in 20/R Fus; he did a short course attached to No 42 Squadron RFC in January 1918 before being posted to 13/R Fus.
26 GS/63665 Private Walter Brown and GS/66454 Private Walter Shrubsole, aged 25, were buried in Hooge Crater Cemetery; G/29272 Private Harry Keeble is commemorated on the Tyne Cot Memorial. SPTS/4910 Private Evan Hansbury, aged 26, a postman from Manchester, is also commemorated on the Tyne Cot Memorial.

31

Unternehmen Michael, March 1918

My God they are here

> *There was a colossal crash as thousands of projectiles rent the air on their mission of death and destruction …*[1]

> *Col Stuart Wortley … was telling me the bad state they were in and that he was being very heavily attacked … when he said "My God they are here …"*[2]

By the time the German offensives in March 1918 occurred the 20th Battalion Royal Fusiliers had gone, alongside the other University and Public Schools battalions. There were many ex-UPS officers and other ranks who were still serving and fought throughout the Army in France and further afield. Many of them would contribute to the decisive battles of 1918. Some would pay the ultimate price and many were wounded or captured in what were some of the hardest battles of the Great War. This section will follow a sample of UPS men during the German 'Michael' offensive in March 1918.

Before the offensive, on 18 March 1918, a battery of German 77mm guns fired on a party of officers from 39th Division who were conducting an appreciation of the terrain over which they would later fight. This shellfire killed a Brigadier General and the nose fuse of a shell hit Captain Lionel Whitby, a company commander with 39th MG Battalion, and amputated his right leg. This left Edmund Page as the last of the three Bromsgrovians from his section in No 4 Company, 21/R Fus.

The German Spring offensive commenced on 21 March 1918; though the attack had been anticipated, the ferocity of the German artillery fire, when it came, was unprecedented. The German artillery fired 3.5 million shells in the first five hours of the bombardment which softened-up the British positions with high explosive and gas shells before the German infantry appeared. Basil Peacock, an ex-UPS man, now serving as an officer with 22/NF (34th Division), remembered, '[We]…flattened ourselves against the forward walls of the dug-outs, we could only communicate by signs, for most of the time we could not hear ourselves speak.

---

1   IWM: N.A. Pease Papers 8230, p.45.
2   TNA: CAB45/187: Letter from Captain Freeman-Tailyour.

Fortunately most of the projectiles hit the far bank of the cutting, and at first we escaped with few casualties.'[3] After a couple of hours, 'our bodies began quivering and we were all gasping for breath, and beginning to feel drowsy when we heard a gas alarm, followed by some duller explosions. We clawed at our masks, revolted by their smell, and within moments the eyepieces clouded up so that we were partially blinded.'[4]

The defences of Fifth Army and the neighbouring Third Army were based on mutually supporting redoubts. These positions should have been able to mutually support each other with machine gun fire and employ their fields of observation to direct artillery fire onto German infantry. A thick morning ground mist on 21 March negated many defensive features and gave limited pre-warning of the German attacks which were often led by specially trained Storm Troops with infantry divisions following behind to mop up. Additionally, many of the redoubts were unfinished. Captain Harold Saxon of 9/R Sussex (24th Division) recalled 'Trinket Redoubt' near Hesbécourt:[5]

> The Redoubt had been made from part of an existing partly dug trench line. ... It allowed a field of fire to the immediate front of about a dozen yards beyond the wire ... It was very strongly wired in front ... There was a partly completed "single apron" [of wire] on the right front but no wire on the remainder of the right or on the rear or left flank. Only in one small portion was the trench so much as a yard deep. There was a supply of S.A.A., bombs, rifle grenades, iron rations and water.[6]

Saxon's men rapidly commenced digging.

Under these conditions ex-UPS officers fought for their lives all along the front; from the hard-pressed 34th Division in the north with Third Army to the right flank of Fifth Army. Some units fought harder than others and not every British soldier fought to the 'last man'. One case study for UPS-men in this offensive is provided by 18th Division at the southern end of the line.

Fort Vendeuil was a pre-existing fortress originally built to protect the approaches to la Fère; it was re-used in March 1918 to protect a company of 7/E Kent. The garrison included Lieutenant Sydney Harvey who had earlier been wounded on 18 November 1916.[7] The defences were bombarded at 4:20 a.m.; Harvey recalled:

> The enemy used gas shells and shells of all calibres and in a very short time all wires were cut and the fort was completely isolated. ... Between 10 a.m. and 11 a.m. the enemy, under

---

3   Peacock, *Tinkers' Mufti*, p.73.
4   Peacock, *Tinkers' Mufti*, p.73.
5   Formerly PS/864 Corporal Harold Saxon, of 19/R Fus, was aged 37 and was from Altringham, Cheshire. He was gazetted into 3/R Sussex in August 1916 and was awarded the MC in July 1917 for good work during an attack.
6   TNA: WO 95/2219: 9/R Sussex War Diary.
7   Formerly PS/2829 Private Sydney Arthur Harvey of No 4 Company, 21/R Fus, from Hornchurch. He had been educated at Romford High School and King's College London and was a bank clerk, before enlisting in November 1914. After being gazetted in August 1916 he joined 7/E Kent. He was wounded in November 1916 in the leg and right buttock. On recovery he returned to 7/E Kents. He was repatriated in December 1918.

cover of a thick mist, advanced [and] occupied the village of Vendeuil and surrounded the fort. Their main attack on the fort developed at around 11 a.m. and was beaten off with heavy losses … In the afternoon the mist lifted and the enemy concentrated a very large number of guns of all calibres on the fort. During the night … patrols were sent out but … they were unable to get in touch with any of our forces.[8]

Next morning there was a further bombardment, but the fort had been bypassed on both sides. Harvey continued, 'Communication with our forces appeared to be impossible … In addition the supply of ammunition was rapidly running short. A conference of the officers was therefore called … at about 4:00 p.m. on the 22nd and it was decided to surrender the fort.'[9] Lieutenant Frederick Winter of 7/E Kents was also captured.[10]

One lucky officer was Captain Nicholas Pease; a few days before the battle he was nominated to attend a training course. Pease recalled that 'At 4:40 a.m. there was a colossal crash as thousands of projectiles rent the air on their mission of death and destruction. Dawn broke to reveal a thick mist with visibility down to less than 50 yards.'[11] However, Pease was nonetheless packed-off on course at 9:00 a.m.; despite requesting to stay. His course was cancelled, and he later rejoined the battalion, 'My company, which I had left on the 21st March 120 strong, was down to a handful.'[12] Pease was lucky. Captain Ernest Mecey of 8/R Berks was injured when a dug-out beam fell on him on 20 March.[13] He was getting treated when the barrage commenced next morning; his replacement was wounded and brought news back of heavy casualties. Mecey was at battalion HQ when the Germans finally attacked later in the morning, 'Here we were able to make a stand for some time but finally, having no more bombs and ammunition running out, I decided to make my way to posts manned by the support company … On my way … I and the few men with me were surrounded by a large number of enemy, who appeared suddenly out of the fog…'[14] Mecey, like many officers, blamed the fog for his capture, 'We were unable to see our own trench mortar SOS and totally unable to distinguish any object until almost upon it while the fog seemed to deaden all minor sounds.'[15]

On the left of 18th Division was 14th (Light) Division which had several infantry battalions overwhelmed in the forward areas. One example was 6/SomLI. The bombardment fell at 4:30 a.m. with high explosive and mustard gas. At about 10:00 a.m. the German infantry attacked in massed ranks through the mist and surprised and overwhelmed the outpost line. Captain Wilfrid Hensley was killed on 21 March 1918; he is commemorated on the Pozieres Memorial. Others were more fortunate; Second Lieutenant Theodore Twist was captured and survived.[16]

8   TNA: WO 339/58995: Officer File S. Harvey.
9   TNA: WO 339/58995: Officer File S. Harvey.
10 Formerly PS/6782 Frederick Charles Winter of 18/R Fus, from Stratford-upon-Avon, was repatriated on 29 November 1918.
11 IWM: N.A. Pease Papers 8230, p.45.
12 IWM: N.A. Pease Papers 8230, p.46.
13 Formerly PS/1183 Lance Corporal Edward John Mecey, C Company, 19/R Fus. He was from Thatcham and aged 23 when he enlisted in November 1914. He was a law student and articled clerk having been educated at Reading School. He was wounded in the head on 16 January 1916 but returned to duty two weeks after. He was gazetted to a commission in the R Berks in August 1916.
14 TNA: WO 339/58882: E.J. Mecey Officer File.
15 Mecey was repatriated in December 1918 and was demobilised in March 1919.
16 Formerly PS/9316 Private Theodore Farquhar Twist, of 28 or 29/R Fus. He was gazetted into 6/

Left: Captain Wilfrid Henry Hensley (formerly PS/471, 19/R Fus) was from Bourton-on-the-Water, Gloucestershire, and attended Greyfriars School. He was gazetted in July 1916 and served with 6/SomLI. (Reproduced with permission of Sutton Archives)

Right: PS/978 Sergeant Noel Shipley Thornton originally served with 19/R Fus. He was a 34-year-old from East Malling, Kent, who had been educated at Charterhouse School and Trinity College, Cambridge. Thornton attended an OCB in March 1916 and was gazetted to the RB in July 1916. He had already been awarded the MC and been MiD twice before he received the DSO and was killed with 7/RB. Thornton was buried in Abbeville Communal Cemetery Extension. (Reproduced with permission of Sutton Archives)

Only the adjutant and a handful of men escaped from twenty officers and 540 men. 7/RB was also overwhelmed losing Lieutenant John Hall who was missing by the end of the day.[17] Major Noel Thornton of 7/RB was awarded the DSO for holding the line of the Crozat Canal at Jussy. His citation read, 'He stopped stragglers and organized them into formed bodies and defended a most exposed position for three hours under heavy artillery fire and machine gun barrage.' Major Thornton was in action with his battalion in front of Villers-Bretonneux on 3 April 1918 and was badly wounded. His DSO citation continued, 'Some days later he saved some of his company from being cut off. He continued to give a magnificent example of courage and leadership until badly wounded.' He died of wounds on 10 April.

Lieutenant Edward Batty, 41st Brigade intelligence officer, subsequently recollected:

> On the evening of the 23rd March I was told to go out and find out as much as I could about the dispositions of the enemy who were surrounding Cugny. I took my runner with me.

SomLI on 25 April 1917 and was repatriated on 11 December 1918.
17 PS/428 Private John Smith Hall, aged 33, from Overzeal, Ashby de la Zouche. He had previously served in the ranks of 19/R Fus. He is commemorated on the Pozieres Memorial.

> Obtaining some important information I sent my runner back with a message to the G.O.C. 41st Brigade. I then went on myself … As I was going through one of the streets of Cugny I was suddenly surrounded by about 50 of the enemy who were patrolling the village.[18]

Similar examples of bravery and loss concerning UPS men occurred all along the line and this work will identify examples to illustrate their stories in lieu of a comprehensive account.

Before noon, Harold Saxon and his men could see the Germans advancing unhindered by other British troops near le Verguier. He described the attack as it intensified, 'There was much dead and broken ground on the right and rear and the bosche were obviously getting much closer and were now firing from our right rear and were in much greater numbers. They made several little rushes …[but] … couldn't time the rushes so that they all came together and our shooting was very good.'[19] British artillery fire also landed on Trinket Redoubt. Later in the day, with the last of the ammunition distributed and 50 Germans and two MGs within 130 yards of Saxon's men:

> We refused a summons to surrender brought by an NCO but it was useless. The Boche came on strongly on their right and rear and with the greater part of the men held to the trench by these protected guns, no weight of fire could meet them… The men on the right were simply overwhelmed … This was the end of C Coy's effort…[20]

At about 4:00 p.m. Saxon and his men were finally captured.[21]

Second Lieutenant Tom Bickerstaffe, a section commander in 66th MG Battalion (66th (2nd East Lancashire) Division), remembered:[22]

> … I was in charge of the machine guns in Hargicourt Village with a section of Hd Qs [Headquarters] slightly in front of the guns. After the enemy barrage had lifted, I endeavoured, accompanied by a Sergeant, a runner & batman, to work our way back to the two guns I had in the village. There was a dense fog at the time & in addition we were wearing gas helmets owing to the gas. We had proceeded about 200 yds when we were surrounded by a party of about 30 of the enemy. Unable to escape we were compelled to surrender, as any resistance there would have been useless.[23]

Lieutenant Reginald Shann of 4th Battalion East Lancashire Regiment was also near Hargicourt. He had recovered from wounds received in April 1917 but was reported missing on 21 March and was last seen lying dead in a trench.[24]

---

18 TNA: WO 339/60056: E.A.F. Batty Officer File.
19 TNA: WO 95/2219: 9/R Sussex War Diary.
20 TNA: WO 95/2219: 9/R Sussex War Diary.
21 Saxon was held at Rastatt and was repatriated on 11 December 1918.
22 Formerly PS/6957 Private Tom Stanley Bickerstaffe a Lewis gunner from A Company, 18/R Fus. He was aged 22, from Blackpool, and a theatrical proprietor; he enlisted in May 1915 and was gazetted to a commission in the MGC in September 1916. He was awarded the MC for good work in an attack on 26 November 1917.
23 Bickerstaffe was repatriated on 30 November 1918. TNA: WO 339/65085: T.S. Bickerstaffe Officer File.
24 PS/2050 Sergeant Reginald Arthur Shann of B Company, 18/R Fus, was educated at Tonbridge School and the Royal Agricultural College, Cirencester, and was employed as a land agent. He had enlisted, aged 26, in September 1914. He is commemorated on the Pozieres Memorial. TNA: WO

21st Division initially frustrated German attempts of cut off the Flesquieres Salient from the south; 110th Brigade blunted the initial German attacks near Epehy but at significant cost. Second Lieutenant William Norton of 7/Leicesters, a contributor to The Gasper, was attached to the 21st Divisional Signal School. On hearing of the German offensive he returned to his battalion with a detachment from the school in time to fight. He was killed on 23 March near Moislains and Clery. Another ex-UPS officer of 7/Leicesters was Second Lieutenant Frank Henry Carr; he had been on traffic control duties and rejoined hurriedly but was mortally wounded and died on 23 March.[25] There were other UPS-men spread around 21st Division. Second Lieutenant Frank Fosbery from Brighton was with 21st MG Battalion but was killed on 21 March.[26] The next day Lieutenant Robert Thorp MC, attached to 64th Brigade Trench Mortar Battery, was reported killed in action and Lieutenant John Henderson of 14/NF was wounded in the spine and captured on 22 March 1918 but succumbed to his wounds on 26 April 1918.[27] Lieutenant Urban Stephenson of 1st Battalion Lincolnshire Regiment (1/Lincs) was killed in action on 23 March.[28]

Crucial to the performance of 47th (London) Division, which held the Flesquieres Salient, near Cambrai, was its resilience to German poison gas. The Divisional Gas Officer was Captain Douglas Smith, formerly of 18/R Fus, who had plenty of experience from service in the RE Special Brigade. Though the division suffered many gas casualties on 21 March he remained in his role after this battle which suggested that his advice had proved vital in protecting the division from the German gas onslaught.[29] North of this salient was 17th Division. Two officers of 12/Manch R, Lieutenants Reginald Hamer and Malcolm Liggett, were both ex-UPS men. Both men were wounded and captured on 21 March.[30] Another ex-UPS casualty with 17th Division was Captain James Wood; he was killed on 26 March whilst 10/LF were withdrawing across the Somme battlefields.[31] His body was later identi-

---

      339/61148: R.A. Shann Officer File.
25 Formerly PS/2521 Private Frank Henry Carr (21/R Fus) was one of the Derby contingent recruited in September 1914. He served in France with 19/R Fus until he was sent home in March 1916. He returned in July 1916 and was posted to 20/R Fus until sent back to the UK for a commission in September 1916. He was gazetted into the Leicestershire Regiment in December 1916 and was sent to 7/Leics R. He was killed, aged 31, and was buried in Roye New British Cemetery.
26 Formerly PS/7578 Private Frank Sidney Fosbery, 19/R Fus; he is commemorated on the Poziéres Memorial
27 Previously PS/3805 Private John Easton Henderson of 18/R Fus. He was buried in Le Cateau Military Cemetery. PS/3294 Lance Sergeant Robert Oakley Vavasour Thorp, of 21/R Fus, was aged 39 and from Chathill, Northumberland. He was gazetted into 1/NF in July 1916 and was awarded the MC in November 1916 for a trench raid. He was buried in Saulcourt Churchyard Extension.
28 Previously PS/6892 Private Urban Arnold Stephenson, 19/R Fus, who was from Althorpe, Doncaster. He was buried in Peronne Communal Cemetery Extension.
29 Smith was killed in a road accident on 9 September 1918. He was buried in St Pol British Cemetery.
30 These two officers were formerly: Private Reginald Barnes Hamer, 20/R Fus, was an articled clerk from Bolton and enlisted in September 1914. He was gazetted in to 12/Manch R on 21 December 1914 and was wounded on 20 July 1916. He was captured on 21 March 1918. PS/7695 Private Malcolm Liggett, 21/R Fus, from Fallowfield, had gone to France with that battalion before returning to the UK to be gazetted into 12/Manch R in August 1916. He was captured on 24 March 1918. Both were repatriated in December 1918.
31 Formerly PS/2266 Private James Buckley Wood of 18/R Fus. He was aged 24 and from Bramhall, Cheshire. He was gazetted into 21/LF on 6 July 1916.

fied by a cigarette case and a small black cat mascot he must have carried for good luck.[32] As the British withdrew across the Somme Battlefields men of the Tank Corps (Tank C) were deployed as dismounted Lewis gun teams. On 23 March, somewhere near Moislains, 1/Tank C was thus employed; Lieutenant William Moss Wardell MC, formerly of 19/ R Fus, was killed in trying to stem the German tide.[33]

Further north were 51st and 6th Divisions. The former had a strong reputation as an attack division but like many in the line on 21 March it was likely experiencing German set-piece offensive firepower for the first time. Second Lieutenant William Beck-Savage, one of the original artists for The Pow-Wow and The Gasper, was, by 1918, an officer with B Company, 51st MG Battalion.[34] The battalion was in the Beaumetz sector spread in depth across the divisional area. Beck-Savage was wounded on 21 March and though his wounds were dressed in the field, he had to be left behind by the medical orderly who treated him.

A few days later Lieutenant George Melville Cooper of 4th Battalion Seaforth Highlanders (51st (Highland) Division) was captured on 23 March 1918.[35] The battalion was holding a position near Beaumetz-les-Cambrai:

> At about 2 p.m. my Commanding Officer … & myself started off from Batt HQ to see how our right flank was fairing. We had only gone a few hundred yards when Capt Gray was slightly wounded by a MG bullet in the head and was unable to continue at once. I succeeded in reaching the right flank of the Battalion, which was intact at the time, but was a few moments afterwards hit in the left thigh, also by an MG bullet. I was put on a stretcher but only succeeded in reaching a sunken road… when one of the stretcher bearers was hit.[36]

Cooper was unconscious when he was captured.[37] Serving with 2/Notts&Derby (6th Division) was Lieutenant Herbert Taylor; he was captured on 21 March 1918 after his position was outflanked by Germans attacking on his right:

> It was evident that our line had been broken and a trench block was made. Meanwhile a heavy TM bombardment was put on our immediate front which knocked the trench in very badly causing heavy casualties to men, arms and ammunition. A steady fire was maintained by the survivors of the two platoons on the enemy, and further ammunition collected from all sources. At (I believe) about 11 a.m., an attack by overwhelming numbers was made

32 Wood was initially buried in Moislains General Cemetery; he was later interred in Heath Cemetery, Harbonnieres.
33 Wardell is commemorated on the Pozieres Memorial.
34 PS/862 Private William Beck-Savage, of No 2 Company, 19/R Fus, was aged 33 and from London. He was educated at City of London School (1894-1899) and the Royal Academy School and was an artist by trade. He enlisted in September 1914 and was posted to 28/R Fus just prior to the departure of 19/R Fus to France. He went to France with the first reinforcement draft in December 1915. Beck-Savage was gazetted to the MGC in October 1916 and served with 153rd MG Company. Beck-Savage is commemorated on the Arras Memorial.
35 Formerly PS/3631 Private George Melville Cooper, of A Company, 19/R Fus. He was educated at King Edward's High School, Birmingham, and his profession was in the theatre. He enlisted, aged 20, on 14 December 1914. He was gazetted to a commission on 4 September 1916.
36 TNA: WO 339/60980: G.M. Cooper officer file.
37 Cooper was repatriated on 9 December 1918.

on us from our right flank and my right post was lost. The survivors maintained a position near the road for some time longer but this was rushed by the enemy and two or three of us rushed further to the left. In doing so I fell into a shell hole, became entangled in some wire and was captured before I could rise.[38]

"My God they are here"

On the left of 6th Division the 59th (2nd North Midland) Division was largely overwhelmed. After being removed from command of 21/R Fus in January 1916 Lieutenant Colonel John Stuart-Wortley had served as the 2IC of 2/6 S Staffs (59th Division) and later took command from March 1917. On 21 March 1918 the battalion positions were placed under a heavy bombardment. According to one source, 'The German bombardment then lifted, in order to avoid its own troops. Battalion Hqrs. were rushed, and the CO. (Colonel Stuart Wortley) was killed.'[39] John Buchan summarised his death:

> … his battalion was holding the forward zone at Bullecourt, and in the misty dawn it took the shock of the German onslaught. We know how at that dark hour of our fortunes the thin outpost troops stood their ground from Arras to the Oise, and by their sacrifice enabled our armies to fight that great battle against odds which in the long run broke the enemy's power. Jack was of the race that defends forlorn hopes, and his men under his command did not yield. They were overwhelmed, and some time about nine o'clock their leader fell …[40]

Buchan's lyrical account perpetuates myths about the German Spring Offensive that require countering; though many British soldier did fight to the death, large numbers surrendered. According to an officer at 176th Brigade: 'The actual last message I received on the telephone was from Col Stuart Wortley himself shortly after 12 noon, in which he was telling me the bad state they were in and that he was being very heavily attacked … when he said "My God they are here". After that no more telephonic communication was maintained…'[41] Another eyewitness was the battalion adjutant, 'it was necessary to evacuate the position and while doing this I was close to the Colonel I saw he was hit and I saw him fall, as far as I could tell he was shot through the head by machine gun fire, there was a very heavy barrage from enemy M.G.s…'[42] Stuart-Wortley was believed to have been killed in London Reserve trench; he is commemorated on the Arras Memorial.

---

38   Formerly PS/6297 Private Herbert Gordon Taylor, of 21/R Fus, from Sutton in Ashfield, Nottinghamshire. He later served in 26/R Fus before being gazetted in May 1917. He was repatriated on 29 November 1918. TNA: WO339/83819: H.G. Taylor officer file.
39   A Committee of Officers, *The War History of the Sixth Battalion the South Staffordshire Regt* (London: Heinemann, 1925), p.201.
40   John Buchan, *These for Remembrance* (London: Enright, 1919), pp.58–59.
41   TNA: CAB 45/187: Letter from Captain Freeman-Tailyour.
42   TNA: WO 374/76958: J.S. Wortley officer file.

59th Battalion MGC was an integral part of their defence. Second Lieutenant Eric Nuttall was one of many members of the battalion missing after the fighting.[43] Another UPS man with 59/MG Battalion was Second Lieutenant Edward Friend:

> The attack commenced, all communication was cut early on, and information was very scarce. I could not get to know what was really happening. Until I saw a large body of troops advancing, who altho' being extremely difficult to distinguish even thro' glasses. I came to the conclusion they must be Bosche because they were in such strength, the range would be 900x. I opened fire & continued to fire until I was surrounded, by which time both guns were put out of action by the enemy. When my position was reached several bombs were thrown at us and I had no alternative but to give in.[44]

Also in action with 59/MG Battalion on 21 March 1918 was Second Lieutenant Wilfrid Brackenridge who was awarded the Military Cross for repelling numerous attacks using his machine guns.[45] Other ex-UPS men serving with 59/MG Battalion were Sinclair Womersley of 19/R Fus and Harry Day of 21/R Fus.

The first that the battalions of 34th Division knew of the enemy advance on their right was when members of 59th Division withdrew down their trenches in a state of panic followed closely by German bombing parties. During this fighting Basil Peacock temporarily held the Germans by throwing grenades over a railway cutting:

> I had just thrown a grenade and had turned to pick up another from the box; when I straightened up I saw a dozen Germans a few feet away in the cutting. I was so appalled that I stood paralysed, until a black-bearded *feldwebel* pointed a Luger pistol at my stomach and remarked threateningly, *'Sie werfen Granaten'*. Fortunately, I had not drawn the pin out of the grenade I was holding, so I dropped it. … a German officer … pushed away the NCO's pistol …[46]

Lieutenant Sydney Whitaker was a section commander in 40th MG Battalion (40th Division) having only joined the unit two weeks before; he was killed on 22 March 1918.[47] Another UPS

---

43 Second Lieutenant Eric John Nuttall, aged 24, from Gainsborough, Lincolnshire, had formerly served as PS/6757 with 19/R Fus and was gazetted into the 3/W Yorks in August 1916; he was later attached to 12/W Yorks before transferring to the MGC. He is commemorated on the Arras Memorial.
44 Formerly PS/357 Lance Corporal Edward John Friend of B Company, 19/R Fus. He was from Sheffield, educated at West End School, Aldershot, and was employed as a civil servant in the audit department; his father was a tailor in Aldershot. Friend enlisted in September 1914, aged 22. He was gazetted into the MGC in September 1916 and served in France as an officer from February 1917. Friend was repatriated in December 1918. TNA: WO339/62933: Officer File for E.J. Friend.
45 PS/8245 Private Wilfred Brackenridge from Chorlton-cum-Hardy had served with 21/R Fus. He was gazetted into the MGC in September 1916. The London Gazette, 26 July 1918, p.8779.
46 Peacock, *Tinker's Mufti*, p.75.
47 Formerly PS/7364 Private Sydney George Whitaker of 19/R Fus, from Lincoln. He was gazetted into the Leicestershire Regiment and embarked for France in July 1917. He was appointed to 40/MG Battalion on 8 March 1918. Whitaker is commemorated on the Arras Memorial

man with 40th MG Battalion was Lieutenant Tom Bowker of D Company.[48] He was constantly in action from 22 to 25 March 1918 and was recommended for the Military Cross for keeping his guns in action to cover the withdrawing infantry and later taking over a party of infantrymen who were without an officer.'[49] His actions enabled the line to be stabilised as did those of Lieutenant Claude Dunn, also of 40/MG Battalion, who helped defend Henin Hill on 22 March. Another notable member of 40/MG Battalion was Second Lieutenant Hugh Spurrell who was wounded on 22 March 1918 and he too received the Military Cross, '…In the evening the enemy attacked in masses, and he was wounded, but remained with his guns, and fired one gun himself when his team was knocked out…'

Near to this fighting was PS/10387 Lance Corporal Archie Brown, of 4/R Fus (3rd Division).[50] On 23 March this unit moved to trenches on Henin Hill; these trenches were originally part of the Hindenburg Line. On 24 March he recollected:

> Later in the afternoon the Germans on our front secured a footing in an old communication trench leading to our lines and the proceeded to bomb their way along the front line trenches. The left platoon of our company was overrun … At the same time a small enemy attack was occurring on my front, but we managed to contain this… Realising the danger to our position from the left, I worked my way along the trench to see where the Germans were … after proceeding some 15 yards I heard German voices. I threw my bomb, and we withdrew to my section post, where I quickly ordered the section to block the trench and man it with a Lewis gun … Things then quietened down, but by evening of the 24th March the Germans had gained a footing in other parts of the front line … Consequently, it was decided to withdraw…[51]

Brown was wounded in the chest by a piece of shrapnel the next day whilst in support of an attack. He was later awarded the MM.[52] The commander of 4/R Fus was Lieutenant Colonel John Cabourn Hartley who had been a major in the UPS. His leadership was likely a key part of the steadfastness by 4/R Fus which was, in turn, crucial as the defence by 3rd Division which formed a buttress limiting the Germans expanding the offensive northwards.

50th (Northumbrian) Division came into action after the first day. 50th MG Battalion took up positions at Caulaincourt in the Green Line where they were attacked at 2:30 p.m. A rearguard action was fought but during the withdrawal Second Lieutenant Tom Wainwright of A Company was captured.[53] Lieutenant Hopkin Maddock, also with 50/MG Battalion, was later awarded the MC, 'This officer, during a withdrawal, held on unsupported by infantry until

---

48 Formerly PS/3985 Private Thomas Bowker of 21/R Fus. He was aged 25 and from Old Trafford, Manchester. He was gazetted into the MGC on 25 September 1916. He had formerly served with 244th MG Company.
49 The London Gazette, 16 September 1918, p.10924.
50 Alfred Thomas Archie Brown, aged 18, and from Blackwood, Monmouthshire, had enlisted in January 1916 and joined 28/R Fus in Oxford. In July 1916 he was sent to France and joined 4/R Fus and was later appointed as a lance corporal.
51 Brown, *Destiny*, p.28.
52 *The London Gazette*, 12 July 1918.
53 Formerly PS/3373 Corporal Tom Wainwright, of 21/R Fus, aged 28, from Hurstbourne Priors. He was gazetted to the MGC in September 1916. He was repatriated on 28 November 1918.

almost surrounded, when he withdrew in good order, doing great execution as he retired. He was the last to cross a bridge before its destruction.'[54] He was wounded in the knee but remained at duty.

"Retire and leave me boys"

26/R Fus was heavily involved countering this offensive and was commanded by Lieutenant Colonel Henry Tuite who had already succeeded a former UPS officer in command. Tuite was not to survive this campaign as one soldier recorded:

> On Sunday March 24th on the Somme at a place not far from Bapaume, about 5 p.m. I saw our Colonel get wounded in the groin by a bullet. When he was hit I was only three feet away. He dropped and I tried with a stretcher bearer to get him away but could not. We then tried to pull him along on a waterproof sheet, but again failed. The enemy were close on us, and our Colonel saw we could do nothing, so he said "retire and leave me boys"…[55]

Lieutenant Colonel Tuite succumbed to his injuries and is commemorated on the Arras Memorial. Similarly, Captain Francis Steinthal, formerly of 21/R Fus and a rugby international with German ancestry, was also with 26/R Fus. He was wounded in the head on 24 March which fractured his temporal bone; he never returned to the trenches. Meanwhile, 17/R Fus (2nd Division) was fighting throughout this period and two ex-UPS men were amongst their officers. Captain John Aylmer (formerly 19/R Fus) had already been wounded once.[56] Also with 17/R Fus was Lieutenant Sydney Antill, formerly of 19/R Fus, had been gazetted into 20/R Fus in August 1916. Aylmer's A Company, 17/R Fus, was holding the line on 25 March near Pys. According to the war diary:

> Shortly after 2.30 p.m. the enemy concentrated a heavy barrage on the ridge, enfiladed the position with heavy machine gun fire and advanced. For half an hour we held him in check. But trouble began when first one flank and then the other began to give way. All the officers by rushing along the top of the ridge with their revolvers and exposing themselves without fear, magnificently rallied the men and the line held.[57]

Aylmer was wounded in the right knee during this action and suffered from gas poisoning. Antill was also injured. There were several ex-UPS enlisted men with 17/R Fus. PS/7019 Sergeant Arnold Uttley MM was one of the Hebden Bridge contingent of 19/R Fus; he was killed on 24 March along with four other ex-UPS men. Meanwhile, Hugh Alan Panting, formerly of 21/R

---

54 *The London Gazette*, 16 September 1918, p.3419.
55 TNA: WO 339/6304: H.M. Tuite officer file.
56 John Aylmer, formerly Private PS/3426, had enlisted in 19/R Fus in September 1914. He was aged 26, from Fincham in Norfolk, educated at Hurstpierpoint School and was an electrical engineer by trade. He was gazetted into 19/R Fus on 27 October 1914 and had served with 28/R Fus, 26/R Fus, was wounded on 7 October 1916 near Flers, and attached to both the Queen's RWS and 13/Essex, prior to joining 17/R Fus.
57 TNA: WO 95/1363: 17/R Fus War Diary.

Fus, was awarded the MC for good work during a withdrawal when he rallied his small band of men five times after they had suffered severe casualties.[58]

Members of the UPS Brigade played a significant part blunting the German Spring Offensive in March 1918. The number of ex-UPS officers and men involved, many more than have been mentioned, show how important and widespread the officer replacements provided by the UPS were in sustaining the leadership of the Army during this critical period. However, there were still many more ex-UPS officers and men still in France and Flanders and they would still face two large German offensives in April and May 1918 as well as the 'Hundred Days' which led to the Allied victory. However, the stand near Bucquoy by 13/R Fus demonstrates the large number of ex-20/R Fus men who were in that battalion's ranks.

58 *The London Gazette*, 26 July 1918.

32

Battle of the Lys, April 1918

To our surprise & disgust we found ourselves surrounded

> *I ... was shouting that we were British officers when to our surprise & disgust we found ourselves surrounded by Germans who disarmed us.*[1]

> *We had to run across an open ploughed field ... The machine gun bullets rained all round, the noise making one continual whistle & finding many of the boys on the way.*[2]

Though April 1918 is predominantly considered to concern the heavy fighting near Ypres during the Battle of the Lys there were several other offensives further south which were sequels to the German March 1918 attacks. Ex-UPS officers and men were involved throughout this period and along the whole length of the front. Many fought hard, were killed, wounded or captured and many were decorated. This section is only a snapshot of their many stories to give an appreciation of the nature of the fighting and how ubiquitous these officers and men were.

'A bad, muddled business' – Bucquoy and Aveluy Wood

Before the German offensive on the Lys 'kicked-off' there were further actions involving ex-UPS men further south. On 1 April 1918 13/R Fus moved into the front line near Bucquoy and at 7:15 a.m. they had to repel a German party that attempted to raid one of their bombing posts; two officers and twelve men were wounded. GS/63363 Private Alexander Hayward, formerly of 20/R Fus and GS/63623 Private George Kitchener both died of their wounds.[3]

---

1. Formerly PS/4232 Private Richard Leicester Statham of 21/R Fus, was aged 29, born in Sao Paulo, Brazil and lived in Eccles where he was a bank clerk. He was gazetted to 3/Dorsets on 25 September 1916 and was awarded the MC for bravery during the summer of 1917. He was repatriated in November 1918. TNA: WO339/64078: Statham Officer File.
2. IWM: Stoneley Papers, 7716, handwritten account, pp.2–3.
3. Hayward is buried in Nine Elms British Cemetery. Kitchener died next day and is buried in Doullens Communal Cemetery Extension No 1. He had formerly attended Dartford School.

On 5 April at 5:30 a.m. the Germans commenced bombarding British battery positions and an hour later they commenced shelling Bucquoy and the 13/R Fus front line positions south and east of the village. The trenches of the two forward companies were obliterated but deep dugouts reduced the casualties amongst 13/R Fus. At about 8:45 a.m. the barrage lifted, and German infantry attacked across the open and with bombing parties down two trenches. The rifle and Lewis gun fire from 13/R Fus stopped the attack in the open and one bombing party was repulsed. A second pushed the Fusiliers company back to company HQ before Second Lieutenant Kirk DCM and PS/4674 Sergeant H.R. Bowden, two former 20/R Fus stalwarts, led a counterattack to restore their positions. Several German dead were found. By 10:00 a.m. the situation was rectified. However, on the left 1/8 LF (42nd (East Lancashire) Division), had withdrawn, leaving the flank of 13/R Fus unprotected. The CO of the Fusiliers left to try and contact the HQ of 1/8 LF and found them unaware of the situation. The commanding officer and 2IC launched a small counterattack but both were killed. The adjutant of the 13th, Guy Chapman, on his own initiative, withdrew the forward companies to avoid them being cut off and to face the threat from their left. Some of the withdrawal was covered by Lieutenant Charles Vickers MM (formerly 20/R Fus). Something went wrong around the HQ of one company where several casualties occurred and several men of Company HQ were surrounded and captured as they tried to dig out some men buried in a collapsed dugout. Second Lieutenant Roy Keller, formerly 20/R Fus, now attached 112th Trench Mortar Battery, held an advanced position with some of his own men and a mixed bag of 13/R Fus and 1/8 LF personnel.

Having captured Bucquoy the Germans did not press on further. Eighteen ex-UPS men or ex-20/R Fus men were killed or died of their wounds. They included PS/6830 Private Thomas Mitchell who had been wounded with the 20th in March 1916; GS/63323 Private Sidney Taverner, a Gallipoli veteran; GS/66425 Private James Lester, and GS/68140 Private Woolf Kitchenoff. The whole event left a bad taste in the mouth for 13/R Fus; 'After mutual recrimination, the matter fizzled out. It had been a bad, muddled business without a single brightness to redeem it.'[4]

'A general offensive had started all along the line …'

The massed German attack during the Battle of the Lys fell on where the line was perceived to be at its weakest; along the front held by the Portuguese Corps. However, it is to be doubted whether any British formation, in the same situation, would have fared differently. Lieutenant Alfred Standage undoubtedly looked the part of a Western Front veteran by sporting a gap-toothed smile having lost two teeth to a previous facial wound. He was attached as an interpreter and liaison officer to the HQ of 23rd Portuguese Battalion at Lansdowne House, near Windy Corner, in the Richebourg sector. He recalled what happened:

> At 4 a.m. on the 9th April a gas alarm was sounded and a heavy bombardment commenced. Our telephone wires were cut so I sent a pigeon message to … Brigade HQ, which was at le Touret. My duty was to remain at Battalion HQ and to endeavour to keep in touch with Brigade HQ and to call in case of emergency – I asked for artillery retaliation, but

4   Chapman, *Passionate Prodigality*, p.247.

this was useless as a general offensive had started all along the line. At 9 a.m. the Germans had taken our three lines ... and their barrage lifted again shortly after 9 a.m., when the German infantry surrounded our HQ, and captured us.[5]

Lieutenant Richard Statham MC of 5/Dorsets was also attached to the Portuguese, as an interpreter. Having been gazetted from 21/R Fus he had spent a year in France and had been both decorated for bravery and acquitted at a general court martial for drunkenness. Statham had got back from leave just before the German attack and went forward to find out what was happening on 9 April 1918:

> We had nearly reached Brigade HQ through an inferno of shellfire when we caught sight of a party in grey, pouring down the road to our left, it was a misty morning & we thought they were the Portuguese bolting from the line. ... We rushed across to try & drive them back & they opened fire ... we thought they had lost their heads. Just then I was bowled over with a M.G. bullet through my head... I came into the open and was shouting that we were British officers when to our surprise & disgust we found ourselves surrounded by Germans who disarmed us.[6]

The 'Portuguese' that Statham had seen were Germans who had broken through.

On the left of the Portuguese was 40th Division who were in this sector to recuperate after heavy fighting further south back in March. They were hit hard by the German artillery but it was the penetration of the Portuguese lines further south that significantly endangered the 40th Division position. A member of 40th MG Battalion recorded for 9 April:

> 4 a.m. Sudden opening of enemy barrage, plenty of gas shells. We take up position and do our firing in great discomfort with masks on – 4.50 to 5.5.a.m. Then take up position in support line with sect. of C Compy. About 11 a.m. we hear that enemy has broken thro' Portuguese on right, surrounded the MID's [Middlesex Regiment] and is now flanking us...[7]

Still serving with 40/MG Battalion was Lieutenant Tom Bowker and he likely experienced similar confusion first-hand on 9 April 1918 in trying to stave off the German strike. He was killed in action never having known he was awarded the MC for his earlier good work in March near Vaulx-Vraucourt.[8] Lieutenant Claude Pridmore Dunn, also with 40/MG Battalion, was also unaware of his MC when he was captured on 9 April 1918 near Windy Corner. His section of four guns was last heard fighting at 1:30 p.m. though practically surrounded.

---

5   Standage was held at Karlsruhe and was repatriated on 29 November 1918. TNA: WO 339/63325: A.K. Standage Officer File.
6   Formerly PS/4232 Private Richard Leicester Statham of 21/R Fus, was aged 29, born in Sao Paulo, Brazil and lived in Eccles where he was employed as a bank clerk. He was gazetted to 3/Dorsets on 25 September 1916 and was awarded the MC for bravery during the summer of 1917. He was repatriated in November 1918. TNA: WO 339/64078: Statham Officer File.
7   IWM: A.E. Mackrell Papers, 4797.
8   He is buried in Rue-Petillon Military Cemetery, Fleurbaix.

Second Day, 10 April 1918

Hopkin Thomas Maddock, the former rugby international, rejoined 50th Battalion MGC on 6 April 1918 after his wounding the previous week. He was again wounded on 10 April, this time more seriously in the neck and thigh, and was evacuated for treatment. Three days later another ex-UPS man of the same battalion, Hubert Parsons, ex-18/R Fus, was killed.

There was much air activity during this fighting. After his brother's death on the Somme in November 1916 Kelvin Crawford went on to score four more victories, three were shared. He went back to France for a third time in March 1918 and joined No 60 Squadron. He was last seen near Bucquoy engaged in an aerial combat with two enemy aircraft on 11 April 1918 but was shot down and killed.[9]

2/R Fus and the Battle of the Lys

From 10 April to 13 April 2/R Fus was in action facing the German Lys Offensive and suffered many casualties amongst ex-UPS and 20/R Fus soldiers and officers over this period. Their story will be covered separately to the general narrative. At 6:30 a.m. on 10 April buses dropped 2/R Fus at Vieux Berquin and they marched to Merville where they were to initially support 151st Brigade (50th Division) near Estaires. Next day the battalion took over positions near Doulieu. During this time the battalion was under intermittent shellfire and on 12 April it was holding posts along the line of a road to the west of Doulieu facing southeast. PS/5705 Ernest Stoneley was a former member of 20/R Fus and a High Wood survivor; he was serving with Y Company, 2/R Fus. During these trying days Stoneley remembered one vignette:

> Before morning we were out in extended order for open fighting, here we held on under a very heavy & systematic trench mortar shelling & machine gun fire, which sent many a good lad west & thinned our line considerably, the German aeroplanes troubled us too. At last the line of Germans came over the crest backed up by a terrific machine gun fire & by this time, in our section, we were very few, & to attempt to hold on was hopeless, so, being left to our own judgement, we decided … we would try to get back to a stronger line that was some distance in our rear … we had to run across an open ploughed field making ourselves a sure target for the Germans. The machine gun bullets rained all round, the noise making one continual whistle & finding many of the boys on the way.[10]

At 9:00 a.m. the Germans commenced pushing forward skirmishers and after the division on the left withdrew, 2/R Fus had to give ground. 86th Brigade staged an orderly withdrawal to the northwest and a line was held along the road running eastwards from Vieux Berquin. On the morning of the 13th the Germans launched a heavy attack which was held. However, the troops on both the left and right withdrew leaving the Fusiliers to throw out defensive flanks to cover these vulnerabilities. The Germans, unable to advance, dug in 800 yards away, though they had

---

9   Captain Kelvin Crawford, aged 22, is commemorated on the Arras Flying Services Memorial along with his brother.
10 IWM: Stoneley Papers,7716, handwritten account, pp.2–3.

PS/5705 Ernest Stoneley was a 27-year-old artist from Broughton, Manchester. He was a former student at the Manchester School of Art and one of the original illustrators for The Pow-Wow. He had served pre-war with the 4th Manchester Volunteer Battalion and enlisted in September 1914. Following repatriation, he was released from the Army in February 1919. (Reproduced with the permission of Janet Wood)

troops in Vieux Berquin which threatened the British right flank. After dusk 2/R Fus withdrew through troops who were forming a further line to the northwest. However, Ernest Stoneley recalled the moment he was captured:

> From my position I could watch the enemy moving about, so I made a sniper of myself from the side of my head-cover, & so held on. Very soon the first line of Germans made a fresh rush, by this time our numbers were few for the boys had been knocked out right & left … When the first line reached us many were killed & others were taken prisoners, myself being one to be taken. The [German] men I had to face seemed to be men who would play the game & they allowed me to pass through.[11]

One of the officers killed on 11 April was Second Lieutenant Horace Mepham from X Company. He was a 22-year-old who had previously served with 18/R Fus; his body was not found and he named on the Ploegsteert Memorial.[12] With W Company were Second Lieutenants Darker

---

11 Aged 27, and from Moss Side, Manchester, Stoneley had been a pre-war member of the 4th Manchester Volunteer Battalion for 4 years. He served with 20/R Fus with little break (though was for a time attached to HQ 33rd Division) and was posted to 4/R Fus in January 1918 and moved again to 2/R Fus in March 1918. He was captured between 11 and 13 April 1918 and was a PoW in Dulmen. IWM: Stoneley Papers,7716, handwritten account, pp.3–4.

12 He was formerly PS/6399 Private Horace Leslie Mepham, from Aldershot. He served with 18/R Fus until April 1916 and then thence with 2/R Fus until he returned to the UK for officer training on

Sketch by Ernest Stoneley depicting his last moments before capture. Other ex-UPS men underwent similar experiences during the German offensive. (Reproduced with the permission of Janet Wood)

and Solomon. Louis Solomon, one of several UPS poets, was killed on 12 April 1918. Second Lieutenant William Gould, with Y Company, was captured on 12 April having been wounded in the side during the fighting. Gould's company was in support and he observed battalions on either flank withdrawing:

> As both our flanks had gone we were now suffering from heavy rifle, MG and light field gun fire from either flank. The order was given to retire and my platoon and myself was to stay and provided covering fire. We did this and then began to dribble my men back. I was with the last four to leave and after running about 50x a man went down wounded. I slung him over my shoulders and continued but was soon joined by 2Lt Darker of my Bn who helped me carry the man another few yards and Mr Darker was shot through the arm and after a few more yards felt very faint…[13]

Gould left the two men but was shot in the side by a sniper as he withdrew; he continued and was later half buried by a shell before being captured by a German field gun crew. Richard Darker was missing after this engagement and his name is on the Ploegsteert Memorial. Gould, Darker and Solomon had all transferred from 20/R Fus when the battalion was disbanded.

February 1917. He was gazetted to 6/R Fus in June 1917 and was attached to 2/R Fus.
13   TNA: WO339/70575: W.T. Gould Officer File.

The 2/R Fus lost eight officers killed, five wounded and two missing along with 23 men killed, 156 wounded and 145 missing. The number of missing, many of whom were POWs, suggested that the withdrawals were not as well ordered as the war diary suggested. About twenty ex-20/R Fus other ranks were captured during these operations and ten were killed.[14] The experience of 2/R Fus was possibly so bad that it drove PS/5133 Sergeant Vernon Jepson to try to end his own life on 15 April 1918; he died on 26 April from these self-inflicted wounds. It is unknown what drove him to such desperate behaviour.[15] The deaths did not stop; another four ex-20/R Fus men succumbed over the next few days.[16]

'Easy targets' – 33rd Division and the Battle of the Lys

There were several UPS officers and men still with 33rd Division during the Lys operations. Lieutenant Thomas Crompton had served with 19/R Fus before being gazetted to 1/Queen's RWS in September 1916. Lieutenant Geoffrey Stevenson (formerly 18/R Fus), Second Lieutenant Hugh Bertram Denny (ex 20/R Fus) and Lieutenant Kenric Murray East (formerly 19/R Fus) were also with the battalion. On 12 April the battalion was in billets at Meteren when orders were received to take up fighting positions south of the village; the camp was cleared in fifteen minutes and the companies were in position within four hours near Hoegenacker Mill and Belle Croix Farm; Crompton was with B Company near the latter; Stevenson was with C Company near the former; Denny was with D Company but was commanding a road block further west; East was with battalion HQ southwest of Meteren. At 8:45 a.m. next morning a German assault overran three posts of B Company which were commanded by Crompton. The garrisons were reported missing and no further information on the fortunes of these men was available. Crompton was later reported as having been killed on 13 April.[17] Denny was wounded shortly afterwards. Stevenson and East survived this action. On the night on 14/15 April 1/Queen's was relieved by 4/King's having lost thirteen officers and 357 men.

The 100th Brigade (2/Worcs R, 16/KRRC and 1/9 HLI) was fighting further to the east near Neuve Eglise having been detached to 49th Division. The latter had been hit hard by

---

14 G/50267 Private Edward May, aged 32; GS/63401 Private Edward Apling, aged 28; GS/63494 Private Sidney Davis; GS/63911 Private William McNells, aged 22; GS/63913 Private Leonard Moules; GS/63927 Private Joseph Bound; GS/63982 Private Leonard Tonkin, aged 28; GS/66935 Private Jack Conquest, are all commemorated on the Ploegsteert Memorial. PS/10881 Private William Thomas Allen, aged 37 and GS/63508 Private Alexander Frankis, aged 26, were buried in Merville Communal Cemetery Extension. G/51768 Private James Irvine is buried in Cabaret-Rouge British Cemetery.
15 Jepson was formerly a sergeant in 20/R Fus, was aged 32, from Heaton Mersey, and was educated at Dunstable. He worked for the Manchester Corporation City Surveyors Department. He was a noted amateur actor and lacrosse player and is buried in Ebblinghem Military Cemetery
16 228036 Private Leonard Alfred Bussey, aged 33, was formerly of 20/R Fus. He died of his wounds on 19 April 1918, having served with 2/R Fus, and is buried in Etaples Military Cemetery. PS/5101 Private John Henry Hurst, aged 26, formerly of 20/R Fus, died of his wounds on 22 April 1918 and is buried at Wimereux Communal Cemetery. GS/63939 Private John Hampson, formerly of 20/R Fus, died of his wounds on 24 April in Etaples and is buried there. G/50421 Herbert Webber, aged 23, formerly of 18/R Fus, was wounded in mid-April 1918 and died of his wounds and is buried near Swansea.
17 Crompton is commemorated on the Ploegsteert Memorial.

the German thrusts. On 12 April 1918 Lieutenant Clarence Uttley (formerly 18/R Fus), was serving with 1/4th Battalion York and Lancaster Regiment (49th Division). The battalion was involved in the fighting near Neuve Eglise and in the process Uttley was wounded and captured. Lieutenant Henry Kenyon Bagshaw was an ASC officer who was attached to 1/7 DWR. Bagshaw's new battalion was engaged near la Leuthe and in the process he lost his life on 13 April 1918. Bagshaw was educated at Hunstanton, and later Oundle School, and was employed as a land agent.

The 100th Brigade arrived to help bolster the defences on the right of 49th Division. Around this time Albert Knighton (formerly 18/R Fus), now a corporal with 9/HLI, was entrenched at Neuve Eglise. He recalled:

> You could see them [the German infantry] coming down [a forward slope], no artillery firing at them, our artillery had gone west, and when the Germans had got to within, say, three or four hundred yards of the Portuguese we all fired. The Portuguese ran away, so we fired into the Portuguese as well as the Germans. Well the Germans advanced to within 25 yards of our trench but they were easy targets …[18]

By the end, there were only about thirty men left of Knighton's company.

The rest of 100th Brigade was hard-pressed around this time. The situation was so dire that the brigade commander ordered his three battalions to bring up their 'minimum reserve' personnel. Their commitment to battle shows the level of desperation. 16/KRRC was also engaged in this fighting. Second Lieutenant Herbert Arthur Cram was with the 'minimum reserve' and was called forward. He had served as a sergeant with 19/R Fus. Throughout 16 April the Germans bombarded the KRRC positions and Cram was severely wounded by shellfire; he succumbed next day to his injuries and was buried at Klein-Vierstraat British Cemetery.

The 2/Worcs R was fighting to defend Neuve Eglise. The German attack was relentless and by 6:00 p.m. on 13 April the battalion had only two platoons and battalion HQ to hold the Mairie in the village. The remaining companies had already been overrun. Second Lieutenant John Turley, formerly 19/R Fus, organised a rifle grenade bombardment from the garden in front of the Mairie to knock out German MGs which were placing the defenders under pressure. For his good work he was awarded the MC with the following citation; 'For conspicuous gallantry … in the defence of a building which was repeatedly attacked by the enemy. He organised parties of rifle grenadiers to eject the enemy from nearby houses, and when the enemy brought up trench mortars and bombarded the building, he showed conspicuous coolness in assisting the defence.'[19] The Worcesters held out for as long as possible before withdrawing from Neuve Eglise in the afternoon of 14 April. Only about six officers and 100 men from the battalion escaped. Turley was wounded on 16 April and suffered a broken arm. Also on 14 April, Lieutenant Harold Clarke, of 49th Battalion MGC (formerly 18/R Fus), was captured during this fighting. By 18 April the situation at Neuve Eglise had been stabilised.

---

18   The battalion war diary does not mention withdrawing Portuguese troops. IWM: Knighton Sound Recording, 10263, Reel 1.
19   *Edinburgh Gazette*, 18 September 1918, p.3462.

Further to the west, back at Meteren, Second Lieutenant Joseph Francis Marrion was the Lewis gun officer with 4/King's. The battalion was in the line near Meteren trying to hold back the German offensive. He left the following account:

> On April 16th 1918 at 7 a.m. the Germans succeeded in taking the positions of A, C and D Coys of the battalion frontage. About 30 men were all that could be gathered together of these three companies and they were placed under the command of Lieut Pack who was D Coy commander … Captain Warburton who was commanding the battalion in the line came across during the afternoon to inform us that we were being relieved the evening of the 17th … I went out about 3 minutes later and found the headquarters already surrounded by the enemy. We held them off for about half an hour … Lieut Pack and myself were taken prisoner.[20]

Captain George Harold Edmondson Warburton MC was the Adjutant. He was aged 22 and from Warrington; he had served in France with 18/R Fus before being gazetted in September 1916. He later joined 4/King's. Warburton attempted to escape the German encirclement but was killed.[21]

This was a high-water mark in the German attack. They had suffered heavy casualties and captured significant swathes of France and Flanders; they needed an operational pause to regroup and prepare for their next blow.

'I considered further resistance useless …' – Villers-Bretonneux

Another decisive battle occurred in April 1918 at Villers-Bretonneux. This was a German attempt near Amiens to capitalise on the earlier March offensives. On the morning of 24 April, a section of Tanks of A Company, 1st Battalion Tank Corps (1/Tank C), went into action. The two MG-armed 'females' and one 6-pounder gun-armed 'male' found themselves up against German A7V *Sturmpanzerwagen*. This was the worlds' first tank-vs-tank engagement. A tank officer, Frank Mitchell, was in the 'male' tank; his account of this affair glossed over the other two tanks:

> Nearing the village of Cachy, I saw to my astonishment that the two female tanks were slowly limping away to the rear. They had both been hit by shells almost immediately on their arrival and had great holes in their sides. As their Lewis guns were useless against the heavy armour-plate of the enemy and their gaping sides no longer afforded them any defence against machine-gun bullets, they had nothing to do but withdraw from action.[22]

Though Mitchell's tank knocked out, an A7V, the war diary report stated that both female tanks returned MG fire and Mitchell alludes that this exchange between them and the A7V took place before he was in action. Though he had only arrived with 1st Tank Battalion two

---

20 He had undergone enlisted service with 29/R Fus. TNA: WO 339/91583: J.F. Marrion Officer File.
21 Warburton is buried Abeele Aerodrome Military Cemetery.
22 F. Mitchell, When Tank Fought Tank, *True World War I Stories* (London: Robinson, 1997), p.234.

PS/4845 Private Walter John Francis was aged 39 and was educated at Marling School, Stroud, and Kings College London. Employed as a schoolmaster, Francis enlisted in December 1914 and joined A Company, 19/R Fus. Wounded and captured near Villers-Bretonneux, he was repatriated on 1 December 1918. (Reproduced with permission of Sutton Archives)

days before, Lieutenant John Webber, formerly of 19/R Fus, commanded one of the 'female' tanks. Though his tank was outgunned by the German vehicle, Webber's crew engaged the German tank in this one-sided battle. As the initial retaliation against Mitchell's tank was only from MG fire it is likely that the two females diverted the attention of the German main armament and may have given Mitchell's gunner additional time to hit the A7V. Webber survived this action but suffered from the effects of mustard gas.[23]

The British Army had 'enjoyed' the sole ownership of tanks and most infantry units were unlikely prepared for encountering German armoured vehicles. Dug in east of Villers Bretonneux, Walter Francis was attached to 2/Mx:

> In the early morning … the enemy commenced a heavy bombardment of our lines and his infantry later advanced to within 60 yards of our position where we held him. Information then reached me that enemy tanks had broken through on our right and later the enemy were seen occupying our trenches in that direction. I sent a report to Coy HQ by runner who did not return. About an hour later I saw three enemy tanks moving from our left rear, one of which approached us along [the] rear of [the] trench. Being engaged on our front and right flank, with one tank in a position to enfilade us on our left, I considered further resistance useless…[24]

23 Formerly PS/1123 Private John Webber, aged 21, from Hartlepool, was educated at Henry Smith School Hartlepool and University College, Reading, and was a schoolmaster. He enlisted in 19/R Fus in October 1914, served with the battalion in France. Webber was sent to No. 5 OCB in April 1916 where he took part in rowing, hockey and football and gazetted to the MGC in September 1916. In August 1917 he transferred to the Tank Corps. After training Webber served with 9/Tank C from October 1917 to April 1918.
24 TNA: WO 339/63254: W.J. Francis Officer File.

Francis had been wounded in the right hand and left arm by shrapnel. Second Lieutenant Charles Hastings Reeves, of 8th MG Battalion, 300 yards in front of Villers-Bretonneux, was also captured.

* * *

Spring 1918 had been bloody. The sterling work of ex-UPS men has already been seen during the March 1918 German offensives. Similar officers and men further demonstrated their steadfastness and sacrifice during further German attacks in late April near Ypres and in late May 1918 on the Aisne. The ethos and training provided whilst with the UPS, combined with the OCBs and their new regiments, helped these men fight to the best of their ability and contribute to holding the line. However, from late-May 1918 onwards, the British and Allied armies were looking for opportunities for offensive action which would bear fruit during the autumn; the so-called 'Hundred Days' offensive.

33

UPS Men and the Advance to Victory

Keeping up a relentless push forward

> *It is rather like taking the footer field with a very raw team, to play against somebody who has whacked you earlier in the season and whom half of your team think are experts…*[1]

> *Here we went over [the top] again – keeping up a relentless push forward. It was really open warfare … We kept on going forward …*[2]

The German offensives proved indecisive and led to a further stalemate. Meantime, the British Army held the line and conducted a series of local offensives. This short-term return to position warfare would see the balance of power on the Western Front alter as the Allies increased in strength whilst the German Army, having lost the strategic initiative, dug in again.

On the evening of 6 June 17/R Fus (2nd Division) paraded ready for the front. After a ride on a light railway they marched to the trenches to take over a sector near Ayette. They were heavily shelled and lost an officer and nine men killed and ten men wounded. One of the latter was GS/58980 Lance Corporal William Martin Green, formerly of 29/R Fus; 'I picked up my 'Blighty one' the night of 6th June, 1918. The worst wound was in my left knee, and with my left leg firmly encased in a Thomas's splint I travelled in stages uncomfortably across France.'[3] Despite the severity of his injury, Green would survive the war. Having already joined up under-age, he received his 'official' call-up papers for conscription whilst he was recovering from his wounds!

The Tide Turns – August 1918

The great Allied counter-offensive on 8 August 1918 was the start of the Allied offensive that would win the war. Though largely spearheaded by Australian and Canadian divisions on 8

---

1   Everard Wyrall, *The 17th (S) Battalion Royal Fusiliers 1914-1919*, London: Methuen, 1930, pp.251–252.
2   IWM: Donald Price Sound Recording 10168, Reel 14.
3   Green, *Here's a How-De-Do*, p.19.

August, British divisions and ex-UPS officers and men were involved. The 9/R Fus was still with 12th Division; they were actively involved and lost several men. Lieutenant William Ernest 'Bill' Hill had been the scout officer for 20/R Fus and Donald Price's platoon commander. He was killed near Morlancourt on 8 August 1918, aged 24, and was buried in Beacon Cemetery, Sailly-Laurette. He was educated at Leigh Grammar School, attended the Manchester Athenaeum for 2 years and was a shipping clerk.[4] Buried in the same cemetery was PS/8789 Private Thomas Nimrod Stephens (ex-19/R Fus), aged 23, from Merthyr Tydfil who had attended Westminster Training College. PS/8600 Private Alfred Worthington, aged 21, from Salford, and PS/11296 Private Matthew Henry Brown, from Chiswick, were also killed on 8 August with 9/R Fus.[5]

7/R Sussex (12th Division) was involved in a series of attacks from the 8th to the 14th. Second Lieutenant Reginald Clements was a platoon commander at the forefront of these operations. He had attended Hereford Cathedral School and was a university theological student. He enlisted in September 1914 and served with D Company, 19/R Fus. He had been awarded the MC for bravery during patrols and a raid on 8 March 1918. He again performed gallantly from 8–13 August but was killed in action 13–14 August 1918 by a stray bullet as the battalion was withdrawn from action. One of the other company commanders in 7/R Sussex was Lieutenant Colin Clayton (formerly 18/R Fus) who was wounded on 7 August. Ernest Clifton Gorringe MC (formerly 19/R Fus), Lieutenant Norman McCracken (formerly 18/R Fus) and George Alfred Phipps (formerly 19/R Fus) were also with the battalion. Phipps was wounded on 26 August but rejoined in October 1918. Gorringe, aged 32, was killed on 5 September and was buried in Peronne Communal Cemetery Extension. McCracken was wounded on 21 September.

British tanks actively supported the offensive. Walter Edwin Stockley, aged 24, had served with the UPS before being gazetted into the Middlesex Regiment in 1915 and had already been wounded. He was a tank officer with 14/Tank C but was killed when his tank was knocked out on 9 August.[6]

On 19 August 1918 2/R Fus attempted some local attacks south of Ypres. They successfully captured several farms between Merris and Vieux Berquin and captured 111 prisoners and ten MGs. This attack cost two officers and eighteen men their lives, including Lieutenant Mark Gilchrist Whyte, aged 20. He had previously served with the Artists Rifles and joined 20/R Fus in March 1917; he moved to 2/R Fus in February 1918. Whyte was buried in Borre British Cemetery. Four other ex-UPS or 20/R Fus men were killed.[7]

---

4   Hill enlisted in September and served with A Company, 20/R Fus. He had several brushes with military discipline during 1915 but in 1916 he was steadily promoted. As a sergeant he was placed in charge of the battalion observers and later led a platoon. Colonel Garnett having recommended him for a commission, Hill was gazetted in December 1916 and remained with 20/R Fus as an officer. He attended a sniping course and the Fourth Army telescopic sight course prior to gas poisoning on 26 September 1917. Returned to France in April 1918, he joined 9/R Fus the following May.
5   Worthington and Brown are commemorated on the Vis-en-Artois Memorial.
6   Stockley had attended the City of London School (1909–1911) and had been a member of the OTC. He is buried in Vrely Communal Cemetery Extension.
7   An 'original' UPS man was PS/6918 Private William Ratcliffe Mason, formerly of 18/R Fus, he was aged 23 and was from Haslingden, Lancashire. Three further men were formerly members of 20/R Fus; GS/575 Corporal Reginald Hickox, GS/62800 Private Alfred Searle, and GS/63384 Private Frederick Nelson, aged 29. Searle, Hickox, and Mason are commemorated on the Ploegsteert Memorial. Nelson is buried in Outtersteene Communal Cemetery Extension.

There were further counter-offensives by the British south of Arras to further press the Germans. On 21 August 1918 Lieutenant Colonel John Hartley, commanding 4/R Fus, was badly wounded in the chest by a shrapnel bullet which penetrated a lung. He would not return to the front but was awarded the DSO in January 1919. His divisional commander, Cyril Deverell, a future Chief of the General Staff, considered him 'to be a hardworking, careful, painstaking C.O. who can be trusted to get anything done…'[8] Two other ex-20/R Fus men were killed in this fight.

These attacks spread southwards to the old Somme battlefields. In late August 1918 38th Division fought its way across the River Ancre near Albert. Second Lieutenant Robert Humphrey Davies, formerly 21/R Fus, had transferred to the RE Special Brigade in 1915 and had been gazetted into the RE; by August 1918 he was fighting with 13/RWF in 113th Brigade. He was killed on 23 August and his body lies in Bouzincourt Ridge Cemetery. Davies was aged 30 and from Aberystwyth. 2/RWF, now part of 115th Brigade, was involved in efforts by the division to break through on the old Somme Battlefields. On 25 August, Second Lieutenant Edwin Ledbury of 2/RWF (ex-19/R Fus) was wounded when his company encountered a German strongpoint near Bazentin-le-Petit Wood. Lieutenant Cyril Thomas Osmond, aged 21, from Canton, Cardiff (formerly 21/R Fus), was with 14/Welsh (38th Division). He died of his wounds on 25 August 1918 and lies in Bagneux British Cemetery, Gezaincourt.

As part of these offensive operations 13/R Fus, with its former 20/R Fus contingent, was actively involved. Donald Price, now a veteran, recalled what occurred on 21 August:

> A massive bombardment started – our barrage was certainly terrific. It was quite a foggy morning and when we went over to the attack, it was quite impossible to keep in touch. It seemed so extraordinary to me as there was practically no opposition. We had gone quite some distance and were ordered to dig in where we were. Germans started coming through and giving themselves up, and when the fog cleared and the sun came up, we had realised we had taken our 1st objective and gone considerably farther.[9]

Casualties were light in this attack. Price continued:

> We were withdrawn after a day for a rest of about three days but were ordered in again at Achiet-le-Petit. Here we went over [the top] again – keeping up a relentless push forward. It was really open warfare. Taking shelter for a spell – A young German was in the shell hole with me. … I had no desire to kill him, but gave him a cigarette and when the shelling had slackened, sent him off to the rear on his own … We kept on going forward…[10]

This second attack on 23 August was harder and more lives were lost. Two attacks were made during the day. One officer, Second Lieutenant Alexander McCarthy, and three men of the battalion with 20/R Fus or UPS connections, were killed in action and were buried in Achiet-le-Grand Communal Cemetery Extension.[11] PS/4638 Corporal Albert Clegg, aged 26, formerly

---

8 TNA: WO 374/31620: J.C. Hartley Officer File.
9 IWM: Donald Price Sound Recording 10168, Reel 14.
10 IWM: Donald Price Sound Recording 10168, Reel 14.
11 PS/9547 Private Alexander McCarthy had briefly served with 19/R Fus before joining 11/R Fus

a stalwart of the victorious 20/R Fus football team, was killed the next day and was buried in the same cemetery. Several ex-20/R Fus officers also suffered; Harold Kirk DCM was badly wounded in the head and Charles Vickers MM was also injured. 13/R Fus was a battalion in the ascendant and it may have been aided by having received its draft from 20/R Fus. They further benefited from receiving a new Quartermaster in Leonard Raven in May 1918; he was the recently commissioned ex-RSM of 20/R Fus.

'An utter disregard for his own personal safety …'

Second Lieutenant Cecil Harold Sewell, formerly of 21/R Fus, was a Tank Corps officer with 3rd Light Tank Battalion. His bravery near Fremicourt on 29 August was recorded in the London Gazette:

> When in command of a section of Whippet Light Tanks in action this officer displayed most conspicuous bravery and initiative in getting out of his own Tank and crossing open ground under heavy shell and machine-gun fire to rescue the crew of another Whippet of his section which had side slipped into a large shell-hole, overturned and taken fire. The door of the Tank having become jammed against the side of the shell-hole, Lt. Sewell, by his own unaided efforts, dug away the entrance to the door and released the crew. In so doing he undoubtedly saved the lives of the officer and men inside the Tank as they could not have got out without his assistance. After having extricated the crew, seeing one of his own crew lying wounded behind his Tank, he again dashed across the open ground to his assistance. He was hit in doing so but succeeded in reaching the Tank when a few minutes later he was again hit, fatally, in the act of dressing his wounded driver. During the whole of this period he was within full view at short range of the enemy machine guns and rifle-pits, and throughout, by his prompt and heroic action, showed an utter disregard for his own personal safety.[12]

Sewell died that day, aged 23, and was buried in Vaulx Hill Cemetery. He was awarded a posthumous Victoria Cross.

September 1918

As part of the continuation of these offensives 13/R Fus (37th Division) was in action again on 4 September 1918 north of the Canal du Nord near Hermies. Five men, formerly of 20/R Fus, were killed out of twenty fatal casualties.[13] Days before this action Donald Price had been

---

and was later gazetted to a commission in the Royal Fusiliers in July 1917. GS/53273 Private Bertie Webb; GS/63335 Sergeant Stanley Mehrtens, aged 21, a Lewis Gunner in No. 9 Platoon, 20/R Fus, and GS/66710 Private Richard Wilkes.

12 *The London Gazette*, 29 October 1918, pp.12801–12802.
13 GS/63581 Private William Richard Siggers was awarded the Military Medal just days before his death; G/46713 Private John Steadman, aged 26; and GS/63621 Private Henry Smith. All three are commemorated on the Vis-en-Artois Memorial. GS/62813 Private William Brigden and GS/66938

promoted from private to corporal in the space of a single day, presumably to backfill NCO casualties. According to Guy Chapman, replacement soldiers had barely fourteen weeks service and had never fired a rifle or thrown a grenade.

As the advance was pressed forward there were risks of German retaliation. On the morning of 19 September a German counterattack retook African Trench, west of Gouzeaucourt, from 14/RWF. They counterattacked with two companies and recaptured the positions but by the end of the day only 25 men of these companies returned. Second Lieutenant Colin Parker was wounded and captured. Despite medical problems, and discharge from the UPS, he had re-enlisted and been gazetted to a commission in a garrison battalion and had transferred to 14/RWF nineteen days before. He died from sepsis as a prisoner on 25 October.[14]

'The spiteful clatter of Boche machine-guns' – 17/R Fus and the Crossing the Meuse-Escaut Canal

It is impossible to mention every ex-UPS officer and man during this campaign. However, focussing on a few notable events involving certain Royal Fusilier battalions will highlight the ongoing contributions of many UPS men. The morale and combat effectiveness of different battalions was an unknown quantity during the 'Hundred Days'. Units had suffered heavy casualties during the Spring and losses had been replaced by young soldiers. The commander of 17/R Fus described the condition of his battalion in the Autumn of 1918:

> I have two damned fine company commanders in Gibson and Ashwell, both with experience. Panting is as keen as mustard, too young and will probably lose his head. … The four company sergeant-majors are all good, proved men in action. The R.S.M could not be bettered. Some of the junior officers are a bit shaky … The men are an unknown quantity. A good many have never been in a show and others only know the March fiasco: they probably think the Bosche is a damned sight finer fellow than he will prove to be … I foresee a lack of "guts" in our men: it is not entirely their fault. They would mostly have been rejected earlier in the War … I think it will depend very largely on how we kick off; if a good beginning is made it may mean a huge success … It is rather like taking the footer field with a very raw team, to play against somebody who has whacked you earlier in the season and whom half of your team think are experts…[15]

Junior officer leadership would be vital. 'Panting' was Hugh Alan Panting, formerly of 21/R Fus.[16] The 17th Battalion Royal Fusiliers still comprised a handful of ex-UPS men and they would be put to the test in late September 1918. On 28 September, 6th Brigade attacked the Meuse-Escaut Canal near Noyelles but did not cross successfully. 17/R Fus was originally

---

Private John Farmer were buried in Hermies Hill British Cemetery.
14   Parker is buried in Hautmont Communal Cemetery.
15   Everard Wyrall, *The 17th (S) Battalion Royal Fusiliers 1914-1919*, London: Methuen, 1930, pp.251–252.
16   Hugh Panting was from Yelverton, Devon, and was educated at Worksop College. He went to France with the battalion in November 1915 but returned to the UK shortly after, wounded. Panting was gazetted to 5/R Fus. Returning to France to join 17/R Fus, he was awarded the MC on 26 July 1918 for bravery in March 1918.

providing flank protection but at 11:00 a.m. on 28 September the battalion was ordered to cross the canal and capture the German trenches beyond. Second Lieutenant Frederick Waters (formerly C Company 18/R Fus) and Lance Corporal Harry Harvey (formerly 18/R Fus) were sent forward to reconnoitre a crossing site.[17] They reached the lock gates and both swam through a tunnel carrying the river Scheldt beneath the canal. Having reconnoitred this method of crossing the canal, both men returned to report. They devised a plan for getting a Lewis gun, and gunner, through the tunnel on a raft to provide covering fire to allow more men to reach the other side. D Company was assigned to cross the canal but after attempts at raft-building failed, and with zero hour (7:00 p.m.) approaching, a decision was made:

> The Captain [Panting] was keenly disappointed. He glanced anxiously at his watch. "Lowe, this is useless; we must try across the Lock. I'll try and dash it. Sergeant-Major, see they follow one at a time; no bunching!" Before Lieutenant Lowe could expostulate, the gallant O.C. "D" Company had sprung up on to the towpath, and was dashing along the length of the bridge, some 25 yards, greeted by the spiteful clatter of Boche machine-guns. The double lock gates were at the end of the bridge. Stepping on to a narrow footboard which ran along the top of the gates, and regardless of the bullets which spent themselves on the iron-work and the unresponsive granite about him, he strode – some would have said "scudded" – across; doubled along the tow-path on the opposite side for a familiar distance, to drop panting under the friendly lee of the now vacated lock houses. … Gamely, his lieutenant, and at due intervals, his men, followed his example. Anxiously each watched his predecessor set out on his run for life…[18]

Once Panting's company was across they gathered themselves and rushed the German trench. When they entered it the two platoons diverged; one cleared left and the other right. The right-hand platoon was likely Waters'; they captured two MG posts. The left platoon encountered strong resistance but captured their section of trench.

The Germans later counter attacked and after fierce fighting managed to push the British out, but they retained their foothold across the canal. Though Harvey underplayed his part in the operation his citation for the DCM added 'during an enemy counter-attack, he showed the greatest coolness under severe machinegun fire, and immediately led forward his section to counter-attack. Though wounded, Harvey remained at duty for four hours until ordered back by his platoon commander.'[19]

Panting, Waters and Harvey had directly contributed to the success of their adopted battalion, 17/R Fus. Panting and Waters were wounded; Waters was awarded the MC, Harvey the DCM.[20] Panting had already been awarded the MC for bravery in March 1918. Second

---

17 Harvey referred to Waters as 'Lieutenant Lowe' and records himself as Waters' un-named companion. PS/7327 Private Frederick George Waters, from Hitchin, Hertfordshire, had enlisted in 18/R Fus in 1915 and went to France with the battalion. He was promoted to corporal and transferred to 2/R Fus in late April 1916. Waters returned to an OCB in early 1917 and was gazetted into the RF on 29 May 1917. He later joined 17/R Fus. See Harvey, *Battle-Line Narratives*, pp.230–231.
18 Harvey, *Battle-Line Narratives*, p.234.
19 *The London Gazette*, 2 December 1919, p.14848.
20 Waters was also 'in charge of the leading wave of the attack and led his men with great courage and determination against two machine guns, killing both crews. Later, when the enemy counter-attacked, he rallied his men and led them forward, remaining at duty after being wounded.'

Lieutenant Hugh Etheridge MM and Bar, also formerly of C Company, 18/R Fus, was not a promising officer according to his OCB report; 'Lacks self-confidence and is rather a slow sleepy character but has shewn very distinct improvement is conscientious and anxious to please. May yet make an efficient but never brilliant Platoon Commander.'[21] Etheridge had earned the MC at Beaumetz on 3 September 1918 and was severely wounded in the thigh during the German counterattack near Noyelles. He later succumbed to his injuries.[22] From his hospital bed Waters wrote to Etheridge's family; 'I have just seen in the Times … that your son has died of wounds … I met your son for the second time about two weeks ago and recognised him as one of the old C Coy 18th Royal Fusiliers men in which I had the honour of being in the ranks. It is always so nice to meet someone of the old battn…'[23] It is also likely that Waters chose Harvey to accompany him based on their previous service with the UPS.

October 1918

Second Lieutenant Harold Mumby, one of the contributors to The Pow-Wow and The Gasper', joined 5/Tank C on 24 September 1918. Having already been wounded twice he could easily be considered to have done his 'bit'. Just over a week later he was in one of seventeen tanks that attacked the Beaurevoir Line supporting 32nd and 46th Divisions on 3 October. Eight of the tanks were knocked out, Mumby was killed when his tank was hit by a shell on arriving back in the lying-up point after the action.[24] Also in action on 3 October was Second Lieutenant T.F. Mitchell, of D Company, 1/8 Notts&Derby. His platoon was involved in the successful capture of Wiancort but he was mortally wounded leading his men on Ramicourt shortly afterwards; he died next day. They had been aided by tanks; likely of 5/Tank C.

On 5 October 1918 the commanding officer of 13/DLI was wounded leaving Captain Leonard Greenwood DSO MC* (formerly 18/R Fus) to command the battalion in the next action. He was gassed whilst leading his men during the capture of the village of St Benin. Though he continued in command he was admitted to hospital three days later and died on 17 October of pneumonia. Greenwood was one of the more highly decorated ex-UPS officers.[25]

21   TNA: WO 374/22949: H.D. Etheridge Officer File.
22   PS/1462 Private Hugh Dimsdale Etheridge, aged 30, was from Pyrford, Woking, and was a farmer. He went to France with 18/R Fus in November 1915 but returned to the UK wounded or sick on 8 February 1916. He later returned to France and joined 7/R Fus on 24 July until 2 December 1916. He served with 12/R Fus 24 April to 16 November 1917 before returning to the UK for officer training. Etheridge was gazetted on 25 June 1918 and later joined 17/R Fus. He died on 2 October 1918 and is buried in Grevillers British Cemetery.
23   FML: RFM.ARC.2482.47: Documents of H.D. Etheridge, Waters letter 7 October 1918.
24   PS/5366 Private Harold Cheffings Mumby was 27, had been born in Manchester, and was educated at Manchester Central High School. He later attended the Manchester School of Art. He enlisted in September 1914 with 20/R Fus. He was slightly wounded on 29 April 1916 and severely wounded in the head and shoulder on 20 July 1916 by a hand grenade. He was in hospital until November 1916 and on recovery was posted to 26/R Fus before returning to the UK as a MGC officer candidate in May 1917. Gazetted into the Tank Corps in November 1917, Mumby was killed in action on 3 October 1918 and is buried in Bellicourt British Cemetery.
25   Captain Leonard Montague Greenwood, aged 25, was from Streatham Common, London, and was educated at Dulwich College. He had enlisted in 18/R Fus in September 1914 and had served with C Company and played brigade rugby. Gazetted into the DLI in January 1915, he embarked for

13/R Fus in October 1918

Donald Price was in various battles during in the Autumn of 1918. He recalled constantly leap-frogging forward:

> We just attacked you see. It was so common then [to be told] you're going over in the morning and you'd go over; and then the officer said … when we get to a certain line we'd stop. It was common to do two miles or three miles and then rest and wait two days whilst someone else came through you. There was very little shelling going on … It was so easy after Ypres or the Somme. There were machine guns firing at us but no such thing as ten machine guns firing at you; they [the Germans] were oddments.[26]

Price was now commanding a section and he remembered that his men knew that the war was being won; morale was high and there was little caution. Conversely, an officer of 13/R Fus had scant memories of this period which were far from rose-tinted; 'Those last months [of 1918] hang cloudily in my mind. The richness of detail which make the early years vivid is lacking. … Probably I was more fatigued than I knew … The next six weeks remain in my memory a mere set of disconnected pictures with periods of complete blankness.'[27] However, the 13/R Fus War Diary recorded that there was still hard fighting. On 8 October 13/R Fus, supported by a tank, were to attack Hurtebise Farm which lay to the southwest of Caudry. Zero Hour was 4:30 a.m. and the battalion had to fight its way to the start line which had not been secured by another battalion. The battalion was delayed by long-range MG fire originating from beyond the extent of the British barrage. Price's section was waiting in a sunken road to 'go over':

> A tank came along and 'Chick' Edgar, a friend of mine, and I, went [forward] behind this tank. It was easy, we were walking and hardly anything was happening. There was a great big valley and on the other side we could see three Germans and they'd just come out of this trench. Chick … put his rifle up and had a go … and one of these fellows dropped. We went on for a bit and just before I got to the top of this valley I got one … this bullet went straight through my leg and blew a great big hole in the back of my leg. It didn't hurt but I was really thankful I'd got it.[28]

Though Price had seen several Germans retreating, some had remained firing from a trench. He was carried off the field by six German PoWs and recovered from his wound just in time for the Armistice. Price was one of the lucky ones. One company was fired at from the rear by a German strongpoint. This precarious situation took time to resolve which further delayed the battalion. Later that morning another battalion advanced to the next objective. Guy Chapman recalled his sparse memories of this period (8–10 October): 'the sight of men walking and falling in alternate

---

France with 13/DLI in August of that year. Promoted to captain in February 1917 and acting major in April 1918, Greenwood was awarded the MC in June 1917; the DSO in January 1917, and MiD in May 1918. Following his death, Greenwood was awarded the MC and was again MiD. He is buried in St Sever Cemetery Extension.

26 IWM: Donald Price Sound Recording 10168, Reel 14.
27 Chapman, *Passionate Prodigality*, p.267.
28 IWM: Donald Price Sound Recording 10168, Reel 14.

groups up a slope towards a line of wood ... of a small party lying round a strong point while a tank crawls towards it ... next morning at Ligny[-Haucourt], with the battalion silhouetted above a ridge and then dodging forward in a spatter of bullets across a patch of cabbage ...'[29] On 10 October 13/R Fus was to form part of a western pincer to establish posts beyond Caudry whilst another battalion advanced from the east; the operation was successful and the village was liberated. This successful three-day operation cost 13/R Fus twelve officer casualties and 104 casualties to the rank and file. An ex-UPS man, Second Lieutenant Eric Rees, was killed in this action.[30] Another officer wounded with 13/R Fus was Lieutenant Reginald Young Daniel.[31]

Some men returned to the front for the dying days of the war and still saw action. Second Lieutenant Horace James Widgery, of A Company, 11/Essex, was on patrol with two men on the night of 12 October 1918 having joined his battalion six days beforehand:

> It was a dark night, and I had to use my compass as no stars were visible. We reached a trench which the enemy had been holding and I entered same, finding that it was unoccupied, then, as we were making for a copse, which the enemy had held during the previous day, we encountered a German patrol of about eight men, which we had great difficulty in shaking off. We entered a wood, and took cover, but, unfortunately, my compass had broken and was useless. After a time, some men could be seen in the distance, as, I believe my own men with instructions that, if they heard any firing, or, if I failed to return, they were to try and get back to our own lines, I went forward to investigate. I managed to get quite close ... but was suddenly challenged and found myself ambushed. To have offered resistance would have implicated my men, and, as the wood was held by the enemy there would have been no chance of any of us getting back, so I surrendered.[32]

Widgery's patrol returned to British lines. Next day, near a village named Saulzoir, Second Lieutenant Rowland Bruce Norman (19/LF) took part in an attack; 'He led his platoon with great courage and determination through a very heavy machine gun fire. Although wounded in the head, he continued to direct the fire of the platoon until again severely wounded, when he returned to battalion headquarters, and fully reported the situation.'[33] Norman was formerly PS/10402, from Stonehouse, Gloucestershire. His battalion was commanded by Lieutenant Colonel Herbert Fenn, formerly of 21/R Fus.

Thistle Robinson, having recovered from the accidental wound from the rifle grenade explosion in February 1917, had been sent to 26/R Fus in October 1917. Despite hospitalisation for his nerves in late 1917, and stints as an instructor, he saw out the campaign with 26/R Fus. On 25 October 1918 the battalion engaged in its last offensive action of the war when it attacked Ooteghem near the Scheldt. According to the regimental history, 'The German barrage was

---

29 Chapman, *Passionate Prodigality*, p.267.
30 PS/6208 Private Eric Montague Rees, formerly 21/R Fus, had been sent to 11/R Fus on 27 April 1916 but was sent on to an OCB shortly after joining his new battalion. Gazetted into 6/R Fus, he was attached to 13/R Fus. Rees is commemorated on the Vis en Artois Memorial.
31 Likely wounded at High Wood, PS/4742 Private Reginald Young Daniel went with 20/R Fus to France, returning to the UK in July 1916. Gazetted in January 1917, he was subsequently attached to 13/R Fus.
32 TNA: WO 374/74187: H.J. Widgery Officer File.
33 *The London Gazette*, 19 December 1919.

PS/4669 Sergeant Frank Eaden Cook, aged 28, was a commercial traveller in wool and woollen waste from Denby Dale near Huddersfield. He enlisted in 20/R Fus in 1914 and is buried in Belle Vue British Cemetery, Briastre. (Reproduced with permission of Sutton Archives)

very heavy, and the machinegun fire so intense that the whole line was held up on the west of Ooteghem ... An attempt was made to rush the windmill on the ridge southwest of Ooteghem. Lieutenant T. Robinson, of A Company, was killed in a first gallant dash; but it was eventually captured.'[34] After heavy losses 26/R Fus dug in; they saw no further action.[35]

Another man not destined to survive the war was Lieutenant Frank Cook whose brother John was killed at High Wood. Frank had already been twice-wounded. He earned the MC at Riencourt on 30 August 1918: 'This officer, finding that the platoon on his left was held up, after reconnoitring the hostile position, successfully pushed on with his platoon, and with great gallantry and skill drove the enemy from his position, thus enabling the platoon on his left to gain its objective ...' Sadly, his luck ran out on 20 October 1918 when he was killed near Briastre whilst serving with 1/10 Manchesters.

British prisoners in Germany had little idea of the progress of the Allied armies and they still succeeded in escaping. One man was PS/7193 Private Max Goodman, 21/R Fus, who had been captured during Whittington's raid in January 1916. He managed to cross the frontier into Holland on 18 October and was held in quarantine in Sittard until he could be repatriated.[36] Many others remained in German camps and some would die of illnesses before they could return. Walter George Craig, formerly C Company, 19/R Fus, died, aged 27, on 5 November 1918. He was from Gateshead and had been captured on 22 March 1918 with 11/DLI.

On 24 October 13/R Fus was to launch another attack; this time near Ghissignies. In the process Captain Ziegler earned the MC with the following citation:

> He led his company forward over ... some four thousand yards, his right flank being completely unprotected and captured the village of Ghissignies. He then, under heavy fire,

---

34 H.C. O'Neill, *The Royal Fusiliers in the Great War* (London: Heinemann, 1922), p.327.
35 Robinson is buried in Heesteert Military Cemetery.
36 TNA: FO 383/381: POW Escape, Private Max Goodman.

consolidated and maintained his position for forty-eight hours. Over a hundred prisoners, twenty machine guns and two heavy field guns were captured and many of the enemy killed.[37]

The battalion captured a crossroads and an orchard beyond the village. The platoon defending the latter were all killed but the remainder of the objective was held, despite the positions being saturated with gas.[38]

Influenza

That autumn many men, tired and weakened by exertion or injuries, were struck down with what became known as 'Spanish Flu'. Some of the bravest and the best succumbed to what was a world-wide pandemic that did not respect rank or length of service. Since March 1918 Captain George Moor VC had been serving as the ADC to Major General Williams of 30th Division. Despite Moor's bravery and front-line service he was admitted to 28th Field Ambulance and died of influenza and pneumonia on 3 November 1918.[39] Lieutenant Francis Maurice had served in France with the Grenadier Guards from December 1917 to May 1918 before being demobilised to allow for him to resume his medical studies. He might have thought himself safe from the war in late 1918 but he died of influenza and pneumonia on 29 October 1918.[40] His brother, Charles, had died on 24 January 1917. In August 1918 Henry Julian Day of the MGC was sent to the GHQ MG School in Camiers and relative safety; in September he returned to the UK, to another 'safe' instructional post at Grantham, for six months. He was still stalked by death; though he survived to the Armistice he caught influenza at Belton Park and died on 3 December 1918.[41]

The Last to Die

Though the 'Hundred Days' comprised a series of offensive victories, the morale of many soldiers was low. There was still heavy fighting occurring, and many men were reluctant to take chances. In the words of Godfrey Skelton; 'As the end of the war came nearer one's personal anxiety increase and naturally one tried to avoid any unnecessary risks of being killed or wounded.'[42] Tragically this proved to be the case for some soldiers. On 4 November Lieutenant Robert Poole of 13/KRRC was mortally wounded by MG fire whilst leading his platoon in an attack

---

37 *London Gazette*, 4 October 1919, p.12351.
38 Chapman, *Passionate Prodigality*, pp.268–269.
39 Moor is buried in Y Farm Cemetery, Bois Grenier.
40 PS/649 Private Francis Thomas Maurice, from Reading, was educated at Blundell's School and St John's College, Oxford, before joining 19/R Fus. He served in France with the battalion before being gazetted into the Grenadier Guards in July 1916. Maurice was studying at St Thomas' Hospital in London before his death.
41 Day is buried at Great Yarmouth New Cemetery.
42 IWM: G. Skelton Papers, 13966, p.73.

on Louvignies near le Quesnoy.[43] His commanding officer wrote of him; 'Of his qualities as an officer I cannot speak too highly. He was brave in action, generous in the treatment of his men, and ever courteous … His loss is very keenly felt by all ranks, and leaves a great gap.'[44] However, many ex-UPS men were still full of 'fight'. Just days before the Armistice Lieutenant William James Henry Clare Lovett from 32nd Battalion MGC was awarded the MC near Avesnes on 8th November, where he made a reconnaissance, identified the locations of enemy MG positions and pushed his own guns forward to silence the opposition.[45]

Though the war had days left to run, this was not known at the time. In early November 1918 13/R Fus was preparing to make one last effort in the attack on the next objective, the Forêt de Mormal. They had suffered heavy casualties in a previous attack in late October on Ghissignies and were badly understrength. The battalion reformed into two rifle companies of a hundred men each; A Company was commanded by Captain George Ziegler. The battalion advanced in support of the Essex Regiment on 4 November. Ziegler recorded the attack:

> … dug in on a road far side of Railway Line. Finally the Essex went through us, we followed going E[ast] towards Ferme de l'hopital where we helped the Essex mop up a pocket of Bosch who had been left. After this we went on to Jolimetz where we rested for some time continually going through the Essex at the western edge of the forest. At 4 p.m. we hadn't got far before it was dark and we met some machine guns on the railway …[illegible]… centre of wood, these were got rid of, and we established a line halfway through at a crossroads … we established a headquarters in the cottage at the level crossing. The Somersets getting into touch on our right. The 5th Div went through us in the morning.[46]

During this attack GS/8021 Sergeant William Green (formerly of 20/R Fus) was commanding a platoon. He was praised for pushing his men forward, through the wood, and was the only force to reach the objective at 6:30 p.m. Green held on until morning, completely unsupported, and even patrolled the ground for 1,000 yards to the east. The battalion was withdrawn the next day. Though the UPS battalions were long gone, their men were still serving in large numbers, and fighting hard with other Fusilier battalions like 13/R Fus. On the 9th, George Ziegler had returned to their battle site at Ghissignies 'walked over to railway cutting in Ghissignies to see the Bosch positions which held us up and where we had so many casualties.'[47] Here, his sporadically kept personal diary ended.

Captain Duncan MacKay rejoined No 55 Squadron RAF on 26 June 1918. In four months he took part in twenty-two bombing raids and three photographic reconnaissance missions, leading seven of them. He must have felt the same as Skelton, but kept flying missions, nonetheless. On 10 November 1918 he took part in a bombing raid on Cologne, along with his observer. Their aircraft was shot down and Mackay was mortally wounded; he died next day. Whilst at

---

43 Lieutenant Robert Evelyn Sandford Poole, aged 22, from Harlow, was educated at Marlborough College and had passed the entrance examinations for Pembroke College, Cambridge when war broke out. Enlisting in 19/R Fus, he was commissioned into 6/KRRC in June 1915. He is buried in Beaurain British Cemetery.
44 Obituary, *The Times*, 13 November 1918.
45 *The London Gazette*, 10 December 1919, p.15351.
46 Fusiliers Museum: RFM.ARC.3015.2: Diary of Captain Ziegler.
47 Fusiliers Museum: RFM.ARC.3015.2: Diary of Captain Ziegler.

No 4 OCB, William Bentley shared a room with MacKay and described him as a 'nice fellow' who was studious but easy to get along with.[48] Duncan MacKay was the last ex-UPS man to be killed in action in the Great War; however, he was not the last victim of the conflict. He was awarded the Distinguished Flying Cross on 3 December 1918 and was MiD in July 1919.[49]

Armistice

Godfrey Skelton was still with an RE field company with 35th Division when; 'On November 11th, 1918 strong rumours of the acceptance and signing of an armistice were circulating and later in the day these were confirmed officially, and so the four and a half years of fighting and slaughter ended. It seemed unbelievable that the chances of survival existed, that I should come through alive and not even having been wounded.'[50] For many the Armistice was underwhelming. It was not spent in merriment by 13/R Fus; just beforehand they had buried the dead from 24 October 1918 near Ghissignies. Many of these were of seriously wounded men who had allegedly been executed and mutilated by the retreating enemy.

All UPS men and survivors of 20/R Fus were scattered by this stage. Vinden was instructing at an OCB in the UK, near Bath; Edward Chapman, having lost an eye, was with a Reserve Battalion. Nicholas Pease was convalescing in England and was part of the festivities in London, he recalled; 'The West End had spontaneously gone mad. To participate in the crazy rejoicings was a never-to-be-forgotten experience.'[51] Martyn Green was undergoing a massage as part of his convalescence when the news of the Armistice filtered through; he and other injured men rushed home to enjoy the celebrations. Ashley Gibson had returned from Africa and somehow secured a role as an assistant curator at the Imperial War Museum. He too was in London to enjoy the Armistice. After convalescing Donald Price found himself at Etaples when the Armistice occurred, he recalled that; 'It was terribly cold, even the Bully Beef was frozen when it came up for breakfast.'[52] A blistered heel prevented him from joining the celebrations but he enjoyed a bottle of wine. In a PoW camp in Germany Basil Peacock's reminiscences of the Armistice were different: 'Then came the news of the Armistice, which the Germans called *Waffenstillstand*. There was little cheering as the word went round, for we had nothing to celebrate with ...'[53]

48　Liddle Archives: WW1/GS/0125: Papers of W.G. Bentley, p.17.
49　PS/631 Lance Corporal Duncan Ronald Gordon MacKay, aged 23, was a student at Cheltenham College from 1909–1914 before enlisting with 19/R Fus in September 1914. Gazetted to 13/A&SH in June 1916, he joined the RFC and served with No. 55 Squadron in France from 17 March 1917. Wounded on 23 April 1917, he returned to 55 Squadron on 22 June 1918. MacKay is buried in Joeuf Communal Cemetery. See *The London Gazette*, 3 December 1918, p.14323.
50　IWM: G. Skelton Papers, 13966, p.73.
51　IWM: N.A. Pease Papers, 8230, p.56.
52　LA: D.J. Price Papers, WW1/GS/1299, p.44.
53　Peacock, *Tinkers' Mufti*, p.80.

The flu rages on

The Spanish Flu and other illnesses and injuries killed further ex-members of the UPS. The former QM of 18/R Fus, John Gray, was serving as the QM of 1/R Fus in November 1918. Six days after the Armistice he succumbed to bronchitis and influenza.[54] Lieutenant Wilfred Hulme of the King's Regiment (formerly 19/R Fus), died in the UK on 14 November 1918. He was aged 22, and from Erdington, Birmingham. PS/5855 Private Hugh Brabham was a signaller with 19/R Fus. He was discharged in August 1916 from 28/R Fus due to neurasthenia, weak muscles, gastritis and being under-age. He took up munitions work and later contracted tuberculosis. He died in January 1919, aged 24. Captain John Franklin Hopwood Templar, one of the 20/R Fus stalwarts, had been a rising star having attended the Senior Officers' School and been one of the youngest captains in the British Army. He died on 8 February at No 14 General Hospital, from influenza. Though a veteran officer who had enlisted in 1914, he died a day before his 26th birthday. Henry Stanley Way, aged 22, of the Tank Corps, was accidentally killed on 5 May 1919 whilst preparing tanks for entrainment. He had attended Oxford Prep School and Blundells School, Tiverton, and was formerly of A Company, 19/R Fus.[55] Furthermore, PS/2448 Private Richard Atherton Brook served with 21/R Fus having been a bank sub manager. He had served with 26/R Fus from May 1916 before transferring to the Labour Corps in mid-1917. He was discharged in February 1919 after an accident fractured his left leg. He was in a nursing home in Bexhill and died in June 1919, aged 38. These are only a few of the many who did not live long after the Armistice.

---

54   Gray is buried in Solesmes British Cemetery.
55   Templar is buried in St. Pol British Cemetery, St. Pol-Sur-Ternoise. He had also served in Palestine.

34

'Side-Show' Theatres

From the snows of Russia to the dust of Mesopotamia

The First World War was a conflict which saw British soldiers serve and die in far-flung regions across the globe. Though the UPS Brigade was rumoured to be being sent to some exotic locations in 1915 they never left the Western Front. However, ex-UPS men were every bit involved across the world; from the snows of Russia to the dust of Mesopotamia in the many 'side-shows' away from the main campaign in France and Belgium. However, these many and varied theatres and campaigns were every bit as dangerous and taxing for the participants. They included the Salonika and Italian campaigns; Mesopotamia and Palestine; West and East Africa; India, the Northwest Frontier and Afghanistan; Persia, the Caucasus and Russia.[1]

Italian Front 1917-1918

With the successful Austro-German offensive at Caporetto, a French and BEF force, the latter consisting of five divisions (5th, 7th, 23rd, 41st and 48th (South Midland)), were dispatched to Italy. Each British formation had ex-UPS men with them. Cecil Isom, now with 26/R Fus, recorded aspects of the long train journey to Italy; 'Nice 5.30 a.m. Lovely scenery all morning. Monte Carlo most beautiful I have seen. Passed through Monarco and Menton 7 a.m. Crossed Franco-Italian border about 7.15 a.m.'[2] He also recorded eating plenty of grapes. The likelihood of a German offensive in March 1918 saw 41st Division, including 32/R Fus and 26/R Fus, return to France. It is unlikely that the return journey was enjoyed by these men. 5th Division followed shortly after. Harold Oakley, the illustrator, briefly served with 8/Yorks, 23rd Division, in Italy before moving back to France in June 1918 to be an Aide de Camp.[3]

---

1 Gallipoli has already been chronicled.
2 IWM: C.H. Isom Papers, 16336.
3 Oakley was appointed as an Aide-de-Camp for the commander of 32nd Division in June 1918. He was demobilised in April 1919. After the war he offered to do silhouettes for members of the 'Old 21st' club.

Much of the Italian Campaign was relatively quiet with raids or local operations and few major offensives. An exception was the response to a local attack on 15 June 1918 when the enemy had broken through 11/NF. Harold Cowling, as the Lewis gun officer, formed a defensive position having dug out two Lewis guns. He was awarded the MC for this. PS/10996 Private Ridsdale Henry Grant, aged 18, of 29/R Fus, later served as a subaltern with 1/7 R Warwicks in Italy. He was shot and wounded in the shoulder during a British raid on the Asiago Plateau on the night of 8–9 August 1918 and was evacuated back to the UK. The raid was successful with one hundred prisoners and two MGs captured for the loss of one man killed and seventeen wounded. Second Lieutenant Vincent Edwards of 10/DWR (formerly 19/R Fus) participated in a raid at Canove in August 1918 to capture a prisoner. Edwards' platoon successfully entered the target positions but found the garrison had withdrawn into dugouts and refused to come up. Rather than throw down grenades and kill these potential prisoners Edwards went down with an electric torch and a revolver to persuade the Austrians to the surface. Casualties were light and the raid was a success.[4]

The campaign heated up in the Autumn of 1918. In a sequel to the raid above, during the crossing of the Piave in October, Edwards reached the far bank with only twelve men of his platoon. There he found several disorganised Austrian soldiers with more in dugouts nearby. Again, Edwards and his men were lucky; 57 enemy soldiers were persuaded to surrender.[5] Weeks before the Armistice two ex-UPS men; Lieutenant Alan Brown of 10/NF and Captain Harold Roberts of 9/S Staffs; were both were killed on 27 October 1918 when the 23rd Division crossed the River Piave.[6] Furthermore, Captain George Richardson MC of 22/Manch R (7th Division) was killed on 29 October 1918 during the death throes of the Italian Campaign. Meanwhile, Captain Harold Edward Cowling, the Adjutant of 11/NF, was awarded the DSO for bravery on 27 October when he reorganised his battalion after the CO and 2IC were incapacitated.[7] Charles Peter Yapp, formerly of 18/R Fus, was attached to 12th Battery MMGC for the final advance in Italy. He was wounded severely in the left side on 30 October 1918, at Sacile. The end in Italy, when it occurred, was quick. Vincent Edwards' battalion was formed up on parade by his commanding officer and had the terms of the Armistice read to them; their war was over.

The Salonika Campaign – 'The heat continues to increase daily. What it will be like in the summer, I hardly like to think …'[8]

In summer and autumn of 1916 numerous men of the UPS reserve battalions might have been dismayed at being sent to 3rd Battalion Royal Fusiliers (3/R Fus) to serve on the Salonika front, a seemingly malaria-riddled side show. Many enlisted ex-UPS men had a hard war in

---

4   LA: WW1/GS/0505: Papers of V. Edwards, p.5.
5   LA: WW1/GS/0505: Edwards Papers, p.6.
6   Formerly PS/2431 Private Alan George Brown, aged 24, from Darlington and PS/3188 Private Harold Roberts, aged 25, from West Bromwich, both formerly 21/R Fus. Roberts was awarded the MC and Brown was MiD; both were buried in Giavera British Cemetery, Arcade.
7   He was formerly with 18/R Fus (as PS/1386) having enlisted in September 1914 and risen to acting CSM, aged only 21, by April 1916. He was formerly of Emanuel School, Wandsworth Common (where he was part of the school OTC) and was a stockbrokers clerk.
8   IWM: P.F.F. Spaull Papers, 15453, letter 29 May 1917.

Macedonia but a higher proportion survived than of those sent to France. One ex-UPS man who served as an officer at Salonika was Second Lieutenant Philip Spaull (12/Hamps R, 26th Division).[9] He recorded some impressions of the climate and trenches in a letter home:

> You seem to have had some terrible weather in England. Here, we occasionally have a thunderstorm with rain which hits the ground and rebounds like smoke but fails to clear the air. The heat continues to increase daily. What it will be like in the summer, I hardly like to think. I saw my first snake yesterday and another one today. Someone had killed the latter in the bottom of a trench. One thing, we have plenty of lumps of rock handy for them.[10]

Some were unlucky and succumbed to debilitating diseases such as malaria. Also with 12/Hamps R was Geoffrey Martin Hubbuck who was MiD three times during the campaign between November 1915 and April 1919.[11]

The 26th Division is an example case study for ex-UPS officers who fought in this theatre. Some battalions received several ex-UPS officers; 10/Devons got at least six. On 10 February 1917 the battalion raided the Petit Couronné to destroy Bulgarian trench mortar emplacements. Second Lieutenants Edward Hudson, Howard Mercer, Sidney Smith and William Wilson, all formerly of 19/R Fus, and James Kirby, formerly 21/R Fus, were with the battalion and were engaged in this attack. The raid cost 10/Devons five officers and 32 men killed, four officers and 96 men wounded but the operation was believed to have inflicted many casualties on the Bulgarians. Smith and Kirby were missing but were later reported killed; both are commemorated on the Doiran Memorial.[12] Hudson died of his wounds on the 13th and is buried in Sarigol Military Cemetery.[13] Mercer survived and was awarded the MC in July 1917; '… Though severely wounded he carried on … to the objective. Here he rendered valuable assistance and set a splendid example to all.'[14] Some officers and men lost their lives just holding the front line. William Eccles (7/OBLI) was killed on 28 February 1917 whilst his battalion was in the trenches at Doldzeli. Little happened except for some Bulgarian shelling in the morning and a blizzard in the evening.[15]

---

9 PS/3592 Corporal Philip Spaull enlisted in B Company, 19/R Fus in September 1914. He attended King Edward VI School, Norwich. He lived in Paris 1910–1914 and was employed as a purser with Cunard. After leaving 19/R Fus, he was gazetted to the 10/Norfolks and was attached to 12/Hamps. Spaull contracted malaria in July 1917 and returned to the UK in January 1918.
10 IWM: P.F.F. Spaull Papers, 15453, letter 29 May 1917.
11 Private Geoffrey Martin Hubbuck, formerly of Tonbridge School, enlisted with 18/R Fus in September 1914 and was gazetted to 12/Hamps in November. He served with the battalion in France and Salonika and was promoted captain in early 1917.
12 PS/2892 Private James Sabey Kirby, aged 24, was from Bishop Auckland, County Durham. PS/899 Private Sidney John Howard Smith, like Kirby, had returned to the UK in March 1916 and gazetted to 11/Devons in July/August 1916.
13 PS/510 Private Edward Stanley Hudson, 19/R Fus, was from Ilfracombe and was educated at Victoria College Jersey (1906–1911) and Exeter College, Oxford (1911–1914) where he graduated BA and 1st Class Honours in modern languages. He was gazetted in May 1915 to 11/Devons. He left for Egypt on 22 November 1915 and served at Lemnos, Egypt, and Salonika.
14 Mercer was from Frinton on Sea and was educated at Bedford School. He was a sergeant with 19/R Fus before being gazetted to 3/Devons in September 1916. *The London Gazette*, 26 July 1917.
15 PS/6693 Private William Eccles, aged 30, was from High Wycombe and had formerly served with 19/R Fus. Eccles is buried in Karasouli Military Cemetery.

The first major assault was at Doiran on the evening of 24 April; this would also be the first battle into which various ex-UPS officers would be committed. On the evening of 24 April 10/Devons were to attack the Petit Couronné again, this time with the view to capturing and holding it. Second Lieutenant William James Wilson was another ex-UPS man who had survived the previous raid. 7/Wilts and 12/Hamps R would also attack. Further left 11/Worcs R and 7/R Berks would attack between Hill 380 and the Petit Couronné; 8th Battalion King's Shropshire Light Infantry (8/KSLI) attacked the Mamelon and 13/Manch R went for Pip Ridge. From the start 10/Devons was struck by artillery fire as it advanced along a narrow ravine and suffered heavy casualties; the survivors still managed to capture part of the position but had to withdraw later in the night. Wilson was missing after the attack and his name is on the Doiran Memorial.[16] Meanwhile, 7/O&BLI were employed providing carrying parties for the attacking troops. Second Lieutenant Herbert Pickford had been educated at The Grammar School, Burford, and was formerly of 18/R Fus. He was also killed on 25 April and is named on the Doiran Memorial.[17] Second Lieutenant Colin Arthur Hutchins was also with 7/O&BLI and was in action in May 1917 during the next attack on the Petit Couronné which also failed; Hutchins suffered a shrapnel wound to his left thigh.[18] Second Lieutenant Norman Parker, 10/Devons, attached 7/O&BLI, was awarded the MC in the London Gazette in July 1917; '… Although wounded, he took command and held on to the captured trench, though continuously counter-attacked, and when both machine guns were out of action, and he had run out of bombs…'[19]

September 1918 saw the beginning of the end in Salonika and a significant increase in activity both in fighting and ex-UPS casualties. Robert Edwards Childs, formerly of 21/R Fus, served with 7/R Berks in Salonika from July of that year. He was wounded on 25 September 1918 as the battalion pursued the Bulgarians near Izlis. He died next day.[20] Lieutenant Percy Cottrell was employed as a staff officer with 78th Brigade. He was mortally wounded in the chest, near Snevce. He was awarded the MC, Greek Order of the Redeemer, Greek MC and was MiD.[21]

16 PS/6178 Private William James Wilson, aged 23, from Carlisle, had been enlisted with 19/R Fus.
17 PS/6594 Private Herbert Thomas Reade Pickford, aged 20, was from Highworth, Wiltshire. He was educated at The Grammar School, Burford, Oxon. He enlisted with 18/R Fus and served with No. 2 Platoon, A Company, before being gazetted in September 1916.
18 PS/6988 Private Colin Arthur Hutchins had been aged 19 on enlistment in 19/R Fus in May 1915. He has been educated at Aylesbury Grammar School and Oxford City Technical School and had become a teacher. He has been gazetted in September 1916 and joined 7/O&BLI in November. After being wounded he also suffered from malaria in September 1917 and later served with 2/4 O&BLI and with an armoured car company in Persia.
19 Formerly PS/6087 Private Norman Parker of 19/R Fus was gazetted to 11/Devons and later served with the MGC.
20 G/47651 Private Robert Edwards Childs, aged 27, had served with 21/R Fus in France from November 1915 to May 1916 and was MiD. Childs is buried in Karasouli Military Cemetery.
21 PS/547 Company Sergeant Major Percy Baglietto Cottrell, aged 25, had been educated at King's School Canterbury and Worcester College, Oxford. He had returned from Athens to enlist but poor eyesight kept him from being gazetted; he enlisted with 19/R Fus in September 1914 and was finally gazetted in January 1915 with 19/R Fus. He served with the Intelligence Corps attached to 78th Brigade in France in 1915 and later Salonika 1916–1918. His MC citation stated, 'On 18th September, 1918, he was mortally wounded whilst gallantly attempting to carry information under very heavy machine-gun and rifle fire…' The London Gazette, 1 February 1919. Cottrell is buried in Sarigol Military Cemetery.

Not every notable ex-UPS officer served in the front line or with the infantry. One of the most famous ex-UPS men to serve at Salonika was Isaac Leslie Hore-Belisha. He had enlisted in 1914 and was later gazetted to a commission in the ASC. After service in France in a staff role in HQ Third Army he was sent to General Milne's Salonika HQ in early 1917. He was promoted to major in the ASC and was 'Mentioned' but was sent home in March 1918 suffering from malaria.[22]

Mesopotamia

A large proportion of the ex-UPS men who served in the Mesopotamian campaign did so with 13th Division. As this was a 'New Army' division raised within Kitchener's first Hundred Thousand it is unsurprising to find ex-UPS representatives in almost every infantry battalion in the division. Taking 5/Wilts as an example, PS/1399 Private Donald Edward Cruickshank, aged 28, formerly of 18/R Fus, was gazetted to the Border Regiment but was attached to 5/Wilts having embarked for the MEF on 27 December 1915. John Eric Binns, formerly 19/R Fus, attended Bedford Grammar School and was a student at the Institute of Civil Engineering. He was gazetted in June 1915 and served with 5/Wilts in Mesopotamia. Early in April 1916 13th Division was to attempt to breakthrough the Turkish defences to relieve besieged Kut-al-Amara; 5/Wilts captured the Turkish trenches at Hannah but were held up beyond them in the open countryside. They were withdrawn to rest but attacked before dawn next day, 9 April 1916; their objective was Sanna-i-Yat. Direction was lost and though a few men reached the Turkish trenches the battalion was only able to reach a position 500 yards short of the Turkish lines. Cruickshank was reported missing and Binns was killed. PS/5698 Private Robert Oliver Stanley (formerly 20/R Fus) attended the Royal Belfast Academic Institution. After being gazetted he arrived in Mesopotamia on 28 February 1916 and likely served with 8/RWF (40th Brigade) alongside 5/Wilts. He was also killed on 9 April 1916. PS/1935 Private Benjamin James Polack, previously of 18/R Fus, was serving with 9/Worcs R (39th Brigade). This battalion also attacked at Sanna-i-Yat and lost nine officers, including Polack. who was a modern languages master at Battersea Grammar School. Giles Vellacott Daunt attended St George's College and the City of London School (1907–1911) and was a secretary with the London City and Midland Bank. He was attached to 6/S Lancs, 38th Brigade and was yet another ex-UPS man killed on 9 April 1916. Daunt, like the four others, is commemorated on the Basra Memorial. Also with 38th Brigade, PS/3115 Private Roy Sinclair Perry (ex-21/R Fus) was killed four days before hand with 6/LNL.

The 13th Division also saw action in early 1917. 38th Brigade had two ex-UPS officers in the same battalion killed on the same day. John Savill Tatham, aged 27, was educated at Winchester College and was a clerk for a merchant banking firm. In June 1915 he was gazetted to 15/KRRC and was wounded in February 1916 whilst attached to 9/KRRC. Tatham joined 6th Battalion King's Own Royal Lancaster Regiment (6/King's Own) in Mesopotamia on recovery. He was killed on 9 February 1917 as his commanding officer related:

---

22 Isaac Leslie Hore-Belisha had attended Clifton College, Heidelberg, the Sorbonne and St John's College, Oxford and enlisted after returning from a holiday in Germany.

He met his death on Friday in the most glorious way imaginable, fighting most desperately at one of our bomb-heads in the Turkish trenches we had captured… Your little brother gave his life most gallantly in defending this, and it's quite impossible to say what the splendid behaviour of all ranks at this point may have meant to the whole operation… The poor boy was killed at about midnight, February 9th–10th, having been hit in the head by a bomb. He could have felt nothing and was quite unconscious…[23]

PS/2068 Private Gerald Coldwell Siordet, formerly 18/R Fus, was gazetted to 13/RB and was attached to 6/King's Own. He was awarded the MC in September 1916; '…After his company commander had been killed he rallied the company under heavy fire, and consolidated the position gained. When the order to withdraw was given he brought the battalion back to our trenches, remaining on duty until wounded himself.'[24] Siordet was killed on 9 February 1917 and his name is inscribed on the Basra Memorial along with Tatham's.

Palestine

Relatively few ex-UPS officers and men served with Murray or Allenby in the Palestine Campaign. However, those that did underwent a hard campaign. Evelyn Charles Bradley Wodehouse, aged 32, had formerly served with 18/R Fus. He was gazetted to the Bedfordshire Regiment and went to Egypt in February 1916 and was attached to 1/5 Bedf R. In early October the battalion was in the coastal sector near Hereford Ridge. On 4 October 1917 a fighting patrol of three officers and forty-six men went out from Subket Post. The patrol was fired upon and Wodehouse was fatally injured; six others were badly wounded.

Gilbert Alfred Bencher, aged 24, originally served with 20/R Fus (PS/4465). In March 1916 he was gazetted and joined 2/19 Londons (60th (2/2nd London) Division) in Palestine. He was mortally wounded and died on 9 December 1917 and is buried in Jerusalem War Cemetery. Philip Graham Egerton, formerly of

Private Evelyn Charles Bradley Wodehouse, aged 32, was from Hertford and enlisted with 18/R Fus and is interred in Gaza War Cemetery. (Reproduced with permission of Sutton Archives)

---

23 *Winchester College Roll of Honour* <https://www.winchestercollegeatwar.com/RollofHonour> (accessed 5 February 2023).
24 *The London Gazette*, 22 September 1916, p.9282.

Tonbridge School, also served with 2/19 Londons and was shot in the knee in May 18. He later succumbed to his injuries.

East Africa – 'Every man present in this engagement must have fought like a hero …'

Several UPS men served in East Africa as part of the active campaigning against the skeleton German forces there. The focus for this section must fall on 25th (Service) Battalion Royal Fusiliers (Frontiersmen) who were formed in early 1915 by the Legion of Frontiersmen. Several UPS men served with that battalion or were sent to Africa as replacements. Captain Arthur Wynell Lloyd had written the UPS Song Book as a private in 19/R Fus but had left in August 1915 to take a commission.[25] Lloyd was the adjutant for 25/R Fus, and later a company commander. He was the only one of forty officers to be present at all battalion engagements at Ziwani, Tandamuti, Narunu and Nyngao. He was awarded the MC for setting; 'A good example throughout all the hardships that Battalion went through …', and; 'Conspicuous bravery in all these actions and most soldierly conduct during the retirement at Tandamuti.'[26] Lloyd's commanding officer recorded the activities of 25/R Fus on 17 October, at Nyngao:

> An attack was being delivered by several battalions against the enemy holding a lone line. No impression could be made. The Fusiliers, now sadly reduced after long service, consisting of about 40 rifles, with Lewis and Machine Guns, total strength about 106, were sent forward. They got into the enemy positions but were not supported; it would appear that the others retired. The Fusiliers, isolated, held their own and fought with their usual courage for several hours being attacked in front and flank. After sustaining some 70 casualties, they were relieved by the Nigerians. All the officers … were either killed or wounded, Capt. Lloyd being shot through the head… Every man present in this engagement must have fought like a hero…[27]

A bullet struck him approximately three inches behind his left ear and exited between the same ear and an eye. Several doctors considered this was a fatal wound; he survived but it affected his memory, partially paralysed his face and partially-deafened him. Several other ex-UPS or 20/R Fus men served with 25/R Fus.[28]

Other ex-UPS men served in Africa. PS/127 Sergeant Herbert Julian Brooke was born in 1888 and was from St Leonards, Sussex. He attended Gresham's School, Holt and in 1913 he

---

25 PS/618 Private Arthur Wynell Lloyd of D Company, 19/R Fus, was aged 31 and was from Hartney Wintney. He was educated at Rugby School and was a *Punch* artist. He enlisted in September 1914. Lloyd served in the East African Campaign and contracted malaria in August 1916 and in January–February 1917. He was later employed by the Ministry of Labour.
26 TNA: WO 339/36825: A.W. Lloyd Officer File.
27 TNA: WO 339/36825: A.W. Lloyd Officer File.
28 GS/49293 Private William Saunders, aged 22, was MiD and was promoted to sergeant. He is buried in Dodoma Cemetery. G/47362 Private William Charlesworth and G/24748 Private Edward Box were likely wounded at High Wood. They were later shipped to 25/R Fus in East Africa in February 1917. PS/7868 Private Walter Codling also survived High Wood and went on to join 25/R Fus in July 1917.

oversaw a farm in South Rhodesia. He was a sergeant in 19/R Fus by February 1915 and was sent to an OCB in March 1916. He was gazetted to the King's African Rifles in June 1916 and served with 2/2 KAR. In September 1917 he saw action at Mpingo Ridge, East Africa, and was awarded the MC for carrying in a badly wounded N.C.O. whilst under heavy fire.[29] Ashley Gibson also served in Africa. On recovery Gibson applied to join the West African Field Force ('the Waffs') because he could speak a little Hausa. However, due to an administrative error he was transferred to the King's African Rifles which operated in East Africa and required knowledge of Swahili.

Persia, Azerbaijan, Georgia

Robert Leach Petty was rejected five times due to eyesight before joining 19/R Fus (PS/7626). He served as an officer in Mesopotamia. In early 1918 Major General Lionel Dunsterville was sent to train local armies in the Caucasus to stop advancing German and Turkish forces. Dunsterville's 'Dunsterforce' required reinforcements to get to Baku where it could start its task. 7/N Staffords, to which Petty was attached, fought the Turks at Baku to protect Dunsterforce whilst it trained local levies. The force was evacuated but not before Petty was killed in action on 31 August 1918. He had attended Dollar Academy; 'His fellow-pupils will remember him as a boy of gentle, lovable disposition; but … his gallant conduct during the war showed that there was a deep stratum of manly grit beneath his mild appearance.'[30] Philip Haye was another ex-UPS man who found his way to a far-off military garrison. A former pupil of Wellington College, Rossall School and Edmonton University, Canada, Haye had served with 19/R Fus. He was an officer with 82nd MG Company at Salonika and was part of the garrison of Tiflis in Georgia. His horse was attacked by dogs when he was riding on 2 April 1919 and he was thrown off. Haye died of his injuries and is commemorated on the Haider Pasha Memorial. There were also UPS-men amongst the small British forces operating in Persia. PS/2541 Sergeant Donald Nevill Carr's parents were members of the Church Missionary Society at Isfahan, Persia. Carr was gazetted to 21/R Fus in March 1915 and was a captain with 4/Border. He was later attached to the South Persia Rifles. He died, aged 23, on 26 November 1918 and is commemorated on the Tehran Memorial; he was awarded the MC in March 1919. L.G.H. Heaver, formerly of 18/R Fus, had attended Tonbridge School; he served in Egypt and Palestine and was seconded to the Indian Army in July 1917. Whilst attached to the 1/41st Dogras he served on the Northwest Frontier until July 1918 after which he operated in South Persia. E.C. Johnson, formerly a private in 19/R Fus, was also educated at Tonbridge. After being wounded at High Wood with 2/R Sussex he had transferred to the Indian Army. Whilst attached to the 1/2nd Gurkhas he received a gunshot wound in the right hand from an enemy MG at Menzil in North Persia on 4 August 1920.

---

29 *The London Gazette*, 16 July 1918, p.8453.
30 Petty is commemorated on the Tehran Memorial.

Russia – 'there was no trace of any bodies ...'

The end of the Great War saw ongoing civil war in Russia with France, Britain and the United States supporting the White Russians against the Bolsheviks. Three main Allied interventionist military forces (Crimea, North Russia and Archangel) were deployed to different parts of Russian territory to either keep order or support the fight against Bolshevism. One ex-UPS man in Russia was Lieutenant George de Coundouroff who was the interpreter for a party examining the Indo-European telegraph system between Ekaterinodar and Jubka. They were ambushed near Defanskaya on 25 July 1919 and George was reported missing. De Coundouroff's young wife as a friend of Vera Brittain who recorded the news at home:

> The blow which was to strike the Cottingham household barely awaited Winifred's return from France ... news came that George de Coundouroff was missing in Russia. Since Edith [de Coundouroff] was helplessly tied to the baby daughter born a month after he left, Winifred went up to London and wrestled with every Government department which undertook to answer inquiries about officers serving in Russia.[31]

A letter from the War Office eventually reported that:

> ... the party ... was ascending a very steep hill ... when about 300 yards from the top, it was suddenly attacked by rifle fire from both sides of the road at point blank range... it was found that your husband with others was missing. Two search parties were subsequently formed and the dead horse of Lt. de C. was discovered ... but there was no trace of any bodies...[32]

His young wife was left to bring up their daughter alone. De Coundouroff was one of several officers and men of the UPS who died in far-off lands, and in curious conflicts, despite the Armistice having been signed.[33] From Lindi to Archangel, from Salonika to Peshawar; the UPS had contributed much to these so called 'sideshows'. For every ex-UPS man who died during these campaigns there were further men who served and survived. The UPS, as a unit that germinated officer candidates, had a disproportionate effect on the Great War as both a European and a global conflict.

---

31 Vera Brittain, *Testament of Experience* (London: Harper Collins, 1980).
32 TNA: WO 374/19052: De Coundouroff officer file.
33 De Coundouroff is commemorated on the Haider Pasha Memorial.

35

The University and Public Schools Brigade Legacy

Remnants now of those who survived

It was like a burial at sea. His grave was the whole earth, as the grave of a dead sailor is the infinite ocean.[1]

Free from fear and those bloody lice ... and sleeping in a bed. Was there ever such bliss![2]

This indestructible bond between us ... we were remnants of a battalion that suffered heavily; and remnants now of those who survived...[3]

Physical and mental scars

Many men suffered life-changing injuries and would never recover their previous good-health because of their service. Despite having had two close shaves during 1917 Second Lieutenant Joseph Finch's severe wound in September 1917 severely affected him. His application for a wound gratuity highlighted his plight: 'I was wounded by shrapnel on Sept 25th 1917, losing my left eye and large portion of my forehead; since that date I have been in hospital, have undergone large operations, one operation being to take bone from my ribs to build up my head; my headaches and dizziness still continue and I find myself greatly handicapped in many ways.'[4] Some men were plagued by their wartime injuries and never recovered from them. Hopkin Maddock was unable to return to his pre-war rugby career or play for Wales. Though he was released in 1919 he died in 1921 of the injury he sustained in 1916, aged 40.

According to Jim Peacock, Basil's brother, the battles of the Great War 'are cemented into the very fabric of our lives.'[5] Some men were changed in more subtle ways. Sidney Platt, like

---

1   Buchan, *These for Remembrance*, p.59.
2   David Clarke, *Great War Memories* (Blackburn: THCL Books, 1987), p.53.
3   Hodson, *Return to the Wood*, pp.15–16.
4   TNA: WO 339/66635: Officer file Second Lieutenant J.H. Finch.
5   Peacock, *Tinkers' Mufti*, p.88.

many other 'temporary gentlemen' coveted his Sam Browne belt long after the war. Later in life he would never touch whiskey, preferring rum, as he had seen too many officers die in action having got drunk on whiskey.[6] Donald Price remembered his feelings on leaving the Army; 'I am now eighty-six and look back at my demob. I had the option of five pounds or an ill-fitting suit. I took the five pounds and brought an overcoat … I sought no reward, no award, I was home, paradise itself. Free from fear and those bloody lice … and sleeping in a bed. Was there ever such bliss!'[7] The Great War changed his outlook to hardship; everything thereafter would be seemingly easy by comparison.

Some ex-UPS men used their experiences of the war and new-found skills to move on to achieve greater things. Arthur Whitten Brown, formerly of 20/R Fus, flew across the Atlantic in just over sixteen hours on 14 June 1919 which was a boundary-pushing feat at the time. He was awarded a £10,000 prize by the *Daily Mail* and a KBE by The King. George Eyston, formerly of 19/R Fus, became addicted to speed and raced cars at Brooklands; he later set the world land speed record three times in the 1930s.

Some went back to relatively normal lives. Harold Tyson survived High Wood and was later gazetted and joined the RFC. After the war he married in 1919 and returned to the Westminster Bank. He worked in branches in Tyldesley, Radcliffe and Blackburn before retiring as a chief clerk in Crewe in 1954. He played lacrosse and cricket; was part of an amateur concert party and was a freemason. He died in 1967.

Post-war service

Though a number of ex-UPS men served in different operational theatres straight after the war, in Russia especially, relatively few chose to make the Army their career. Though a few ex-UPS men were employed as majors or lieutenant colonels during the war very few held higher ranks. Many returned to pre-war jobs or returned to education to start or finish the degrees they had been due to undertake in 1914. Some men did choose to transfer into the Regular Army and served in the inter-war period. Many of those who continued into peacetime soldiering had to revert to being subalterns. When the Second World War came there were few ex-UPS men in senior leadership roles. Hubert Vinden served on and operated in Ireland during the post-war conflict as did Terence Doherty whose loss of an eye did not affect his Army career. Another UPS officer who was active in 'intelligence' out there was Captain Frederick Shove, formerly of 19/R Fus. Vinden also served in Colchester, Gibraltar and Shanghai with the Suffolk Regiment.

Another case study was Basil Peacock who became a Territorial Army officer with 4/NF in 1921 and trained to be a dentist. Money was tight for Peacock in the early 1920s despite being a war veteran, a part-time officer, *and* a medical professional. The same would be true for many ex-servicemen and many 'temporary gentlemen' were also impoverished. For Peacock the Territorial Army was valuable as a way to let off steam and when he moved to London he transferred to 4/Queen's RWS and later to the 44th Division RASC which was still horse-drawn throughout the 1920s.

6   IWM: S. Platt Papers, 17681.
7   David Clarke, *Great War Memories* (Blackburn: THCL Books, 1987), p.53.

Old Comrades

Many UPS men remained in touch both during and after the hostilities. Like many units they maintained their close ties through old comrades' groups and met up for sociable dinners and events. These organisations also attempted to help others out during the financial difficulties of the inter-war period. In 1930 the Old 21st Association, the old comrades' association for 21/R Fus, applied to place a plaque in Ashtead Church in memory of the battalion. Graham Dawbarn, a member of the association, acted as the architect with a Mrs Petrie as the artist and sculptor. Donations were requested of one guinea from members, but payments of any size were welcomed. However, the committee stated, we have no intention of limiting the subscription of one guinea because we wish to devote the surplus, if any, to the relief of ex-service men in distressed circumstances, resident of the Parish of Ashtead. We feel that this will be a suitable, if totally inadequate return for the kindness and hospitality of Ashtead residents in 1914–15'.[8] This highlighted the bond between the battalion and the village that nurtured their initial existence.

The silk Union Jack Flag representing the King's Colour for 21/R Fus had been laid up in the Garrison Church in the Royal Fusiliers Depot at Hounslow. After permission was granted by the Secretary of State for War this Colour was moved to Ashtead Parish Church where it was to be permanently stored. This installation was marked by a church service where the association paraded, wearing medals, on 22 March 1931; the memorial was also unveiled. A Colour Party from the Royal Fusiliers Depot received the King's Colour from the altar of the Garrison Church from the Depot Chaplain. A guard of honour was provided by the Depot courtesy of the OC, Major H.H. Cripps.

The service was conducted by Reverend E.J. Austin the Rector of Ashtead; the act of laying the Colour on the altar was marked with the following from Rev Evans: 'We now lay up within Thy House if Prayer this emblem of Thy Grace and Blessing. May all who look upon it be reminded of their duty to Thee and to their King, their Country, and their School.'[9] This was followed by a rousing; 'God Save the King'.

The memorial was unveiled by Brigadier General Sir Robert Gilmour CB CVO DSO. He was well into his 70s by this time and even suggested he may need to doze in the car on the way from the Guards Club to Ashtead. Gilmour was accompanied by the head of the association, R.L. Watson, to pick him up and introduce him to members he may not know. Brigadier General A.H. Leggett CMG DSO, the chairman of the Ashtead Branch of the British Legion, also took an interest in these events, when he was well enough to attend, until he died suddenly in 1936. A 20/R Fus association was also still going in the early 1930s with annual dinners taking place in Manchester.[10] According to Hodson, Colonel Bennett wept at a reunion dinner for not releasing more men to commissions.[11]

Hodson recalled of his comrades in the 1950s:

> We were not the men we were, of course. We are grey or bald and we creak at the joints. We said we should be disguised as ancient monuments and nobody would know us. But

---

8   IWM: R.L. Watson Papers, 4255.
9   IWM: R.L. Watson Papers, 4255.
10 IWM: R.L. Watson Papers, 4255.
11 Hodson, *Return to the Wood*, p.44.

An elderly Brigadier General Gordon Gilmour (top hat) inspects the ranks of the 'Old 21st Association' who are parading on The Common south of Ashtead Station. The man to his left sporting the DSO, MC and umbrella, is likely Colonel Denison who took over 21/R Fus when Stuart-Wortley was relieved of command. (Image from the papers of R.L. Watson)

some of us knew one another right enough; not the same men, and yet, in a sense, just the same. When I think of them I see them more often as they were thirty-eight years ago; to that small extent, immortal.[12]

Sadly the paper shortage during the Second World War put paid to 21/R Fus old comrades association publishing newsletters.

Return to the Western Front

In June 1921 the task of finding and exhuming bodies was still taking place. On 27 June 1921 an exhumation party was despatched to attempt to find Claude Briggs' body based on a map provided to his family by another soldier. Briggs, of 21/R Fus, had been killed in action on 20 September 1917 whilst serving with 26/R Fus. He was believed to have been buried where he fell somewhere near Shrewsbury Forest. A body was found buried in a waterproof sheet and identified as Briggs. He was buried in Bedford House Cemetery the next day. His family visited

12   Hodson, *Return to the Wood*, p.14.

France and Belgium almost annually to see his grave and were some of the many battlefield tourists to the region.

The French and Belgian farmland continued to disgorge missing bodies of UPS men. In 1926 a farmer ploughing near Givenchy accidentally found the remains of a dugout which contained the body of Captain Price-Edwards which allowed him to be properly buried in Cabaret-Rouge British Cemetery, Souchez. Many bodies were never found. The eminent author John Buchan wrote of Jack Stuart-Wortley:

> He had always a horror of being laid in a trim cemetery under a neat cross. On that grim March [1918] morning the guns of both sides made the place where he fell a tormented wilderness, and his body was never recovered. It was like a burial at sea. His grave was the whole earth, as the grave of a dead sailor is the infinite ocean.[13]

His words are apt to describe many of those of the Great War whose bodies were consigned to the earth without formal burial.

Harry Harvey visited the battlefields in 1921 *en route* from the French Riviera. He remembered the Somme:

> The "old front line," that zone of glory of July, 1916, like the barbed-wire has disappeared! Just here and there a trace of upturned chalk and the "bent, battered but unbroken" Line which stretched away to the sea. On every hand are to be seen countless circular patches, the old wounds of 5-9's; a few rusty piles of "duds" and neglected stacks of corroded wire, cut and bundled for the furnace, on the roadside. Here and there a shell-battered, half-hid "tin-hat" – where he that wore it lies buried! The few dug-outs on the roadside are sunken, collapsed or almost hidden beneath mud thrown from the broad, straight, though now deserted highway. No song of the bullet – no wail of the shell![14]

To him it was a lonely place; 'What memories of our trudges to the Line of fire and furies! Yet mournful now without "the boys"…'[15] Many other men chose to return to France to revisit their former battlefields to both remember the good and the bad times from their service. James Hodson took the opportunity to visit his former battlefields in 1939:

> High Wood, which has deep memories for me, and which I last saw torn and defiled, has sprung up afresh and within its borders a large farm is busy… Mametz Wood, where twenty-three years back, we lay in holes, was under a cloud of shadow, but a few minutes later the sun caught the vast coloured expanse and turned it to glory.[16]

According to James Hodson the old comrades of 20/R Fus were still meeting in the 1950s and decided to conduct a battlefield visit to High Wood; 'I should not have begun to set down this record had I not attended within the past year a reunion dinner of my old battalion. High and

13   Buchan, *These for Remembrance*, p.59.
14   Harvey, *Battle-Line Narratives*, p.250.
15   Harvey, *Battle-Line Narratives*, pp.245–250.
16   Hodson, *Somewhere in France*, p.105.

thoughtless good spirits were intermingled with patches of melancholy. Affection was abroad, wine was in the blood, and some were reluctant to think another year would have gone by before we assembled again.'[17] They also visited the front-line further north; 'Cambrin Church looks as it did in 1915, except that the unexploded shell stuck in the wall had gone. Round the corner in those day was the entrance to Moulin Rouge Alley, and across the road in a battered house with parts broken by shell-fire you could still get eggs and chips … We wandered into the churchyard …'[18] There, Hodson saw several graves of fallen comrades.

As the fiftieth anniversary of the war brought it back into focus in the 1960s more men chose to visit. The many war graves cemeteries gave passing visitors the chance to reflect on old comrades. Basil Peacock and his brother visited the battlefields in 1965. They stopped at the cemetery in Annequin where just twelve Commonwealth War Graves Commission burials were situated. His brother knelt to read the first headstone on the left; 'Jim … then straightened up saying gravely, "Good heavens here is poor Corporal Pepperday, one of our first fatal casualties. I was present when he was hit and he died as I was attending to him." My brother stood for a few moments bareheaded profoundly moved and I feel sure that he said a little prayer…'[19] Skelton made a pilgrimage tour to Ypres and the surrounding area with the British Legion in October 1967.

Hodson, described the dwindling group of 20/R Fus veterans; 'This indestructible bond between us … we were remnants of a battalion that suffered heavily; and remnants now of those who survived; for Time takes his toll. "Do you know what's happened to So-and-so?" – "Died a year or two back." That was common. Made us feel old; and I suspect we're at the dangerous age for men who had a good many narrow escapes …[20]

The UPS and the Second World War

The 1930s were a troubled time with war looming. Morgan Williams, then a reverend with the Presbyterian Church of Wales recalled in 1937 that; 'The twenty-three years that have rolled by since August 1914, have exacted their price, yet if I had them all back, I would gladly pay it again.'[21] Sadly he died in 1938.

Many ex-UPS men were on the upper age margins for service by 1939. A minority served during the Second World War. The final newsletter for the 'Old 21st' Association observed:

> … we think it advisable to suspend activities of the Association for the period of the war. In addition to the fact that Meetings are so difficult, members on reserve have been called up and many have joined up. Others, unfortunately, have fallen on evil times owing to the dislocation of their work. As soon as the war is over, we intend to carry on … We wish you good luck wherever you are, whatever you are doing, and every blessing for your sons and daughters, particularly if they are serving with His Majesty's Forces.[22]

17 Hodson, *Return to the Wood*, p.11.
18 Hodson, *Return to the Woo*, p.26.
19 Peacock, *Tinkers' Mufti*, pp.201–202.
20 Hodson, *Return to the Wood*, pp.15–16.
21 Williams, *From Khaki to Cloth*, p.180.
22 IWM: R.L. Watson Papers, 4255, newsletter 14 March 1940.

Herbert Vinden, a 'Regular', was involved in the British Army decision to adopt the Bren light machine gun in the 1930s (replacing the Lewis gun). He was also involved with the formation of the School of Infantry and served in Singapore from 1937 with the intelligence bureau. His memoirs highlight the lack of preparedness for war; attempts to improve the situation were not greeted with interest. Other ex-UPS men would bear the suffering of these errors. Curiously, another ex-UPS officer, Lieutenant Colonel George Cecil Shipster MC (formerly OC of D Company, 19/R Fus), was employed as the Chief Inspector of Small Arms from 1934 onwards and may have also had a part in the introduction of the Bren gun. He had served with the Royal Tank Corps and worked in the War Office (1930–1934). He was a colonel on the General Staff in 1940 but died in 1941, aged 55.[23] Vinden served with the second British Expeditionary Force in 1940 as GSO2 at 1st Division under General Alexander. Unfortunately, he went sick with suspected Dengue fever in May 1940 and was evacuated to the UK just before the German *Blitzkrieg*. He later helped set up the War Office Selection Boards (WOSB) which likely incorporated some of his experience from service in the ranks of the UPS and attending an OTC in 1916.

One of the most prominent ex-UPS men was Isaac Leslie Hore-Belisha who was the Secretary of State for War 1937–1940. He feuded with senior generals in the run up to the Second World War and was against committing the British Army to a land war in Europe. He was dismissed in January 1940.

In the Second World War the Royal Fusiliers were again expanded to form war-raised battalions. The second incarnations of the 18th, 19th, 20th and 21st Battalions were not raised with links to the UPS and none of them saw significant fighting.

1939 and 1940 saw many ex-UPS officers apply for Emergency Commissions but most were denied. James Hodson wrote that; 'This new war had become very much my concern. If the Emergency Officers' Reserve had no place for me (and, so far, it had not), perhaps this was the next best thing to do. Or a better thing.'[24] That 'better thing' for Hodson was to join the BEF in 1939 as a war correspondent and he chronicled the 'Phoney War' whilst also observing some of his former battlefields. His memory of the first wartime Armistice Day was:

> Not before has a solemn Armistice Day ceremony been observed … when Lord Gort and the Duke of Gloucester stood, without greatcoats, and in a downpour of rain, to pay their respects to the dead of France and Britain. This situation, terrible in its irony, of still grieving for those fallen in the last war on an old battlefield that may soon be a new one and by men who may themselves fall in a renewal of the old struggle, must have struck at many of us.[25]

Many former UPS men were only employed with second-tier units, supporting arms or in non-combatant roles. Some joined the Local Defence Volunteers or, later, the Home Guard. Alfred Frank Jolly, formerly a second lieutenant in 19/R Fus, joined the Home Guard and served with 5th Dorset Battalion. He was killed in his car during a German bombing raid on Weymouth in April 1942. Keith Radford also served with the Leeds Home Guard until 1944. In 1939

---

23 He was buried in Highgate Cemetery.
24 Hodson, *Somewhere in France*, foreword.
25 Hodson, *Somewhere in France*, p.107.

Daniel Pigache decided to serve in the ranks again, this time as a private in the RASC. Wilfrid Kinton Clarke also served with the RASC and was gazetted to a commission with the Corps in 1941. Several men, including Robert Davies, Reginald Knight, Stanley Sparey, Thomas Storrs, Alexander Schurig, Alexander Pilling and Norman Dunn served with the Auxiliary Military Pioneer Corps or Pioneer Corps during the war. Knight became a Warrant Officer, Dunn became a Captain and camp commandant and Schurig became an adjutant. Thomas Entwistle Storrs served with several Pioneer Corps Smoke Companies as a company 2IC. Some found other ways of serving; William Bloomer, formerly of 21/R Fus, was a Metropolitan Police special constable in 1939. Ellis Fell, the former 20/R Fus footballer wounded at High Wood, served as the Chief Warden with his local Air Raid Precautions detachment. Kenneth Norman also served with the ARP. Others took on industrial or ministry roles like George Eyston who worked for the Ministry of Production. He was made an OBE in 1948. Guy Mackarness, the '…empty headed boy from Charterhouse', had become a schoolmaster and head teacher between the wars. He served with the Ministry of Economic Warfare, the Ministry of Home Security and the Ministry of Information throughout the war. Another man who indirectly assisted the war effort was John Bingham Morton (known as 'Beachcomber') who improved morale in Britain with his newspaper column which ridiculed Nazi propaganda and British 'red tape'. A High Wood survivor who had an impact was Charles Cundall who became a well-noted official war artist of some memorable paintings.

Other ex-UPS men joined the ever-expanding RAF. Cator Barclay Holland, formerly of 18/R Fus, and Thomas Archibald Metford Stuart Lewis (of 19/R Fus) were re-employed in the forces as an Acting Pilot Officers in the Training Branch or in training schools of the RAF Volunteer Reserve. Alfred Eltringham was a squadron leader with the RAFVR and received the Air Force Cross in 1946. John Lloyd Williams resumed his service in September 1939 with the RAF Volunteer Reserve. He died on duty in a road accident and was one of many uncounted casualties of the blackout. Arthur Whitten Brown joined the Home Guard but later resigned to join the RAF and worked at Training Command. Martyn Green entertained the troops with ENSA and in April 1941 was gazetted to a commission in the RAFVR. He was an administrative officer for a British pilot training unit eighty miles from Los Angeles.

Archie Brown had a curious second war. He was commissioned into the Territorial Army in the 1920s and was mobilised again in 1939 to train soldiers. Likely seen as being too old for active service he was appointed to investigate cases of self-inflicted wounds in Normandy; these were on the rise due to worsening morale during the Normandy breakout. He subsequently became the Town Major of Ypres during the winter of 1944–1945. The winter was so cold, and supplies so scarce for the population, that the trees on the ramparts around the Menin Gate were cut down for firewood despite orders for their preservation.

Several other ex-UPS men reached higher ranks or had prominent roles during the second conflict. Lionel Whitby, formerly of 21/R Fus, having lost his leg, went into medicine at Downing College Cambridge. He worked at Middlesex Hospital as a haematologist. In the Second World War he joined the RAMC and was a brigadier by March 1942. He was Knighted in 1945.[26] Herbert Christopher Hatton-Hall left the 19th Battalion in November 1914 to attend the Royal Military College. He had a fruitful career and ended the Second World War as a GSO1 in the Department of the CIGS, but died in November 1945, aged 53. Roger Morton,

---

26 Lionel Whitby died in 1956.

who left 20/R Fus in 1915 to attend Sandhurst, ended up as a brigadier having commanded several formations in the UK. A former officer with 19/R Fus, Philip Ingleson, was also quite high-profile during the Second World War and worked in Sudan with the Sudanese Political Service. He was made CMG in 1945 and received an MBE.

Very few of the 'Ups' saw real action in the second war. Leslie Wendt, formerly 18/R Fus, was a lieutenant colonel in the Union Defence Forces at Tobruk in 1942. He was captured at Tobruk whilst in command of the 2nd Battalion Royal Durban Light Infantry. After repatriation from being a POW he was awarded the DSO for his conduct at Tobruk.[27] Terence Doherty was the Adjutant of 7/Essex in the mid-1930s and was a major in 1938. In 1940, despite having only one eye (and being nicknamed 'Cyclops'), he was the Second in Command of 2/Essex in France, aged 44. He escaped from Dunkirk with his battalion and later commanded an Infantry Training Centre at Warley, Essex, and afterward served with the RAF Regiment at Debden. After the war he arranged the repatriation of German troops from Norway and commanded an Italian labour battalion.

Basil Peacock, Erroll Shearn, Arthur Percival and Benjamin Silly were all prisoners of the Japanese. Peacock, having become a POW in March 1918 was extremely unlucky to suffer a worse fate in the Second World War. In 1936, he became a company commander for a new searchlight unit which he formed from scratch and was mobilised with them in 1939. He later took a draft to Singapore which formed a part-Malay, part-British searchlight regiment. He was captured by the Japanese and had to survive years of captivity in appalling conditions. Shearn was with the ARP in Singapore and survived Japanese bombing of Kuala Lumpur before being enrolled as a pilot officer in the RAF; he escaped Singapore on a ship reserved for RAF personnel and reached Java but was caught up in the Japanese invasion there. Though he put some of his Great War infantry training to use instructing RAF aircraftsmen he nonetheless had to march into captivity. He recalled; 'Among others in the HQ I found an Air Commodore Silly (his actual name). We discovered that we had both been members of the U.P.S. Brigade in World War 1.' Silly and the higher-ranking officers departed to another location and they were separated. Shearn and Peacock were the lucky ones; they survived until 1945 and were repatriated. Air Commodore Silly died in captivity in 1943; his son was killed in action in France in October 1944. Other ex-UPS men were affected by their children serving during the Second World War. Colonel Garnett's son was killed in May 1940. Major Abbay's son died in an accident near Cirencester in June 1941 having had a difficult time escaping from Dunkirk. Arthur Brown's only son, known as 'Buster', served in the RAF but was killed in June 1944 when his Mosquito crashed.

Though relatively few became casualties during the Second World War, the number of UPS veterans dwindled during the latter half of the century. Many had passed away by the middle of the 1980s. Alongside many of his comrades Philip Wright reached 1980; Vernon Bartlett lived until 1983; Gordon Jacob died in 1984; Philip Ingleson and Hugh Spurrell passed away in 1985; Harold Wilson died in 1986; Claude Dunn and James Dykes lasted until 1989. One of the last members of the UPS was Don Price. He revisited the battlefields in 1983, aged 85, including High Wood and Fontaine-les-Croisilles.[28] He was still able to visit High Wood on a tour in May 1987 and planted a sapling English Oak to commemorate the 20th Battalion Royal

27 *The London Gazette*, 19 December 1946.
28 *Stand To!*, Issue 9, Winter 1983, pp.4–7.

Fusiliers.²⁹ He passed away in Nottingham on 16 February 1989, aged 90. The final survivor was likely Harold Augustus Aldridge, formerly of 19/R Fus, and the Tank Corps, who died in November 1994 in Brighton at the age of 97.

Final thoughts

Donald Price spoke later about his experiences:

> People will read this no doubt and say how terrible it must have been … no one can possibly describe it and no one can possibly understand it. Can one adequately describe his feeling on finding that he was sleeping, leaning on a dead man, or having to pick his pal up in pieces and put his remains into a sandbag after he was blown into smithereens … So many instances of complete ghastliness appeared – but the mind seemed to accept it all. … I do wonder why I was able to survive, and yet I feel a sense of pride somehow that I was able to have been there.³⁰

Hugh Spurrell recalled; 'I expressed a desire to enlist as a private. This desire of mine was very unpopular at home [having been offered a commission]. I managed to carry out my intention, however, and have seldom regretted it. The experience I gained as a Tommy has been invaluable, although it cost me two years seniority.'³¹ William Bentley learned many life lessons from his time with the UPS; to put himself second when achieving common goals, to persevere through adversity and the value of a team ethos.³² Arguably, much of the credit received by the UPS Brigade must be passed back to the public schools, and universities, and the OTCs they ran. The ethos, values, and basic training they imparted were invaluable to the UPS-men who served.

The last word should go to the 'official' history of the UPS Brigade which summed up the 'old' brigade:

> No finer body of men was ever assembled, and it was apparent in the early days in Hyde Park that the nation possessed in the Brigade a wonderful ready-made source of supply of officers. No member of the old U.P.S. need regret his connection with it, and probably there are few who are not proud of having once been members of the famous Brigade, and who will not without pleasure and regret look back on those happy days spent at Epsom, Ashtead, and Leatherhead. Nor will they ever forget the hospitality and kindness of the inhabitants of these towns, which made the many weeks spent there pass so happily.³³

---

29 David Clarke, *Great War Memories* (Blackburn: THCL Books, 1987), p.54.
30 IWM: Donald Price Sound Recording 10168.
31 IWM: H.W. Spurrell 2138, pp.2–3.
32 LA: WW1/GS/0125: Papers of W.G. Bentley, p.21.
33 Anon. Author, *University & Public Schools Brigade*, p.63.

Between the hedges of the centuries,
A thousand phantom armies go and come.
While Reason whispers as each marches past,
*'This is the last of wars, – this is the last!'*34

Lieutenant Gilbert Waterhouse, 2/Essex (formerly 18th Battalion Royal Fusiliers (1st Public Schools)

34 Gilbert Waterhouse, *Rail-head: and other poems* (London: E. MacDonald, 1916).

Conclusion

We are essentially amateurs. But we are not too bad

I believe what happened was that the Public Schools Battalion 20th RF came away at dark; and so did most of the Scotsmen ...

Robert Graves

Despite most heavy casualties, both from shell and machine gun fire, the Royal Fusiliers ... with great steadiness and courage, and sticking to their task, in keeping with the Albuhera tradition, fought on until the whole wood was in our hands.

G.S. Hutchison

Before assessing the performance of 20/R Fus and the UPS in general there are outstanding criticisms of these units concerning High Wood that must be addressed. Robert Graves, in *Goodbye to All That*, had little positive to say about 20/R Fus to the extent of providing an imbalanced and biased account. His most critical comments concerned High Wood. Graves recounted a pre-battle briefing by Major Crawshay, commanding 2/RWF, that he predicted that the Royal Fusiliers, Cameronians and Scottish Rifles would run away.[1] Graves also regurgitated a statement he had heard from a wounded staff officer; 'They [2/RWF] hung on to near the end. I believe what happened was that the Public Schools Battalion 20th RF came away at dark; and so did most of the Scotsmen. Your chaps were left there more or less alone ... They steadied themselves by singing...'[2] Graves came under considerable fire for the veracity of this account, which was damaging to the reputation of the Cameronians, Scottish Rifles and survivors of 20/R Fus.[3]

The official documents and personal accounts employed in previous chapters counter the assertion that the RWF fought on entirely alone during the latter stages of 20 July 1916. According to Brian Bond, 'Graves frankly admitted in 1930 that he had deliberately mixed and spiced up all the military incidents he could think of to produce a bestselling book because he desperately needed the money'.[4] Others have been unkind to Graves as a writer; Paul Fussell described him

---

1 Graves, *Goodbye to All That*, pp.179–180.
2 Graves, *Goodbye to All That*, p.183.
3 Mr Graves Withdraws Retreat Story, *Edinburgh Evening News*, 3 November 1930, p.3.
4 Brian Bond, *The Unquiet Western Front: Britain's Role in Literature and History* (Cambridge: Cambridge University Press, 2002), p.31.

as; '… a tongue-in-cheek neurasthenic farceur whose material is "facts."' Fussell asserted that Graves wrote an exaggerated account to make 'a lump of money'; Graves also made little effort to hide this and revelled in the controversy.[5] Dunn knew Graves ('Graves had reputedly the largest feet in the Army, and a genius for putting them both in everything.')[6] However, he did not recognise him as a contributor in *The War the Infantry Knew* and especially not the chapter concerning High Wood.[7] This suggests that Dunn either did not trust Graves' writing or, more likely, he recognised that Graves was not an eyewitness.

Contrary to Graves' account the 33rd Division historian, Lt Col. Graham Seton Hutchison stated:

> I hold no special brief either for the 20th Royal Fusiliers … or for the Scottish Rifles; but I am an intimate witness of the whole of these operations … I believe Robert Graves' memory to be at fault when he alleges that his Colonel made such a remark, and if the Colonel did so it was one of prejudice of which he should be heartily ashamed … It passes my comprehension how a brigade major, except possibly that of the 19th, Graves' own brigade, could have observed the 2nd R.W.F.s 'shaking out into artillery formation, and remaining singing in the wood, whilst the Manchester boys of the Royal Fusiliers and the Jocks legged it.'

Members of 20/R Fus, whose battalion had been; 'Generally sniped at throughout the book…' also countered Graves' memoirs.[8] James Hodson's accounts in 1929 and 1955 were partly drafted to rebut some of what had been written by Graves as well as doing so in newspaper articles. Graves withdrew the comments against the two Scottish battalions in the introduction to '*But Still it Goes On*' but maintained that; '…the only battalion that disappeared was the Public Schools Battalion of the Royal Fusiliers attached to our brigade …'[9] James Hodson responded to this in the press highlighting that 20/R Fus melted away through severe casualties rather than its men absconding in panic.[10] Harold Tyson wrote a detailed rebuttal. 'you state that Dr. Dunn said the Public Schools Battalion … disappeared. In the first place we were decimated and without officers of non–coms., and we carried on in isolated groups … I am sorry Dr. Dunn has got the impression that we 'cut off,' for this was not the case.'[11] Tyson, also took the opportunity to highlight his admiration for Dunn. Graves' responded: 'I believe Mr Fyson [sic]; though the melting away of numbers of the battalion is beyond dispute. Monsignor McShane has recently written to the Scottish Press confirming this part of the story.'[12] Despite this answer, Tyson's letter got the following added to future editions of Graves' book: 'This was not altogether accurate. I know now that some men of the Public Schools Battalion, without

---

5   Paul Fussell, *The Great War and Modern Memory* (Oxford: Oxford University Press, 1975), pp.204-206. Graves, *But Still it Goes On*, pp.13-16.
6   Dunn, *War the Infantry Knew*, p.166.
7   Dunn, *War the Infantry Knew*, p.xv, p.222.
8   Norman, *The Hell they Called High Wood*, p.241.
9   War Poets Apology, *News Chronicle*, 3 November 1930.
10 It is highly probable that all four battalions suffered from some stragglers.
11 Graves, *But Still it Goes On*, p.37.
12 Graves, *But Still it Goes On*, p.38.

officers or NCOs, maintained their positions in the left centre of the wood, where they stayed until relieved … twenty-two hours later.'[13]

Dunn was also not generally positive about 20/R Fus, though unlike Graves, Dunn's work is far less stilted and would give credit where it was due. For example, concerning 16 April 1917; 'This was another dud show. The Chocolate Soldiers went well, and suffered for it…'[14] However, such comments can be ambiguous.[15] He also accidentally praised aspects of 20/R Fus such as the quality of its band, transport and football team. Dunn did employ certain members of the battalion as sources (Lieutenant Colonel Modera, RSM Armour and 'Captain H.H. Fyson' (Corporal Tyson)) which suggested that he approved of their character and trusted their testimony. It is unknown whether this was to capture the other side of the argument, to provide a more comprehensive account or to enhance the reputation of 2/RWF. Though 2/RWF was a strong, cohesive infantry battalion they, and Dunn, seemed to evaluate other battalions with a hauteur which few other units could satisfy (even the other Regular and long-serving Territorial battalions in 19th Brigade or 33rd Division). External observers like Hutchison had a different view: 'I do not consider that the Royal Welch Fusiliers was by any means its [33rd Division's] best battalion… Neither do I judge that the 1st Cameronians … lived up to their historical traditions.'[16] Hutchison could, on occasion, be a vocal advocate for 20/R Fus but gave a misleading a speech to the Royal Fusiliers Depot:

> Despite most heavy casualties … the Royal Fusiliers, supported by the 2nd Battalion Royal Welch Fusiliers, with great steadiness and courage, and sticking to their task … fought on until the whole wood was in our hands … Thus did the 20th Battalion … respond to the finest characteristics of Englishmen and sustain the traditions of your regiment…[17]

Hutchison's hyperbole is an exaggeration too far in the opposite direction due to his audience and because his former commander, and patron, Major General Pinney, was a former Royal Fusilier.

The 20th Battalion Royal Fusiliers performed as well as could be expected in the carnage at High Wood. Their ability to succeed was drastically reduced by their heavy casualties and mixing with the other battalions. The losses in officers and NCOs in 20/R Fus was a sign of gallant leadership. Their casualties amongst the rank and file showed the men had high morale and were highly motivated to fight with dash in the initial attacks and doggedly in defence.[18] To draw a line under the debate concerning High Wood, men of 20/R Fus were probably aware of the 'chocolate soldiers' epithet the RWF applied to them. According to one 20/R Fus

---

13 Norman, *The Hell they Called High Wood*, p.241. Graves, *Goodbye to All That*, pp.183–184.
14 Dunn, *War the Infantry Knew*, p.329.
15 Dunn's words might have been crystal clear to those who had experienced these events. But, over a century later, some of his comments can be misinterpreted or have different meanings attached to them.
16 Hutchison was naturally biased *towards* the battalions of 100th Brigade, to which his MG Company was attached. He also had reason to be biased *against* battalions of the Regular Army having been sent home from 2/A&SH in 1915 after being falsely accused of impropriety. TNA: WO 374/36057: G.S. Hutchison Officer File; Graham Hutchison, *Footslogger* (London: Hutchinson, 1931), p.178.
17 Hutchison, *Footslogger*, p.177.
18 Norman, *The Hell they Called High Wood*, pp.141–142, p.147; IWM: Donald Price Sound Recording 10168, Reel 9.

officer who survived High Wood: 'We have suffered rather heavily ... As one sergeant of the Regulars tersely put it, "I've called 'em chocolate soldiers scores of times, but I'll take my words back now."'[19]

Alternatively, Morgan Williams observed that 21/R Fus enjoyed better relationships with its Regular counterparts; '...we [21/R Fus] lived and worked on terms of friendship and understanding with the 1st Middlesex and the 2nd Argyles and Sutherlands, two battalions of the very division to which the 2nd Royal Welch Fusiliers belonged.'[20] However, that relationship involved some banter with 21/R Fus. In an *estaminet* when two UPS men were trying to talk in schoolboy French to the waitress; 'Suddenly from the back of the room a gruff Scots voice chipped in with the remark, "We're a fine brigade the noo; two battalions of fechters and two of interpreters."'[21]

On paper, the UPS fulfilled some of the criteria expected of a cohesive military organization. Though limited training resources and expertise; camp construction work; wastage due to commissions; and labour demands delayed training, these were problems experienced by many 'New Army' battalions. However, for the UPS the excessive drain from large numbers of commissions, and the manpower 'churn' replacing them, exacerbated these problems and had a considerable impact on coherent training. Officer-man and NCO-man relations were generally good; sometimes too good in the latter case, with more friendly, rather than disciplined, relationships between NCOs and men. Discipline was more informal as the primary means of resolving problems.

The performance of the UPS and 20/R Fus during periods of trench warfare was not noteworthy but was not worthy of undue criticism. Paddy Griffith suggested some battalions 'were known for their easy-going ways ... and their general failure to make an impact ...'[22] Tony Ashworth's idea of less effective and aggressive battalions fostering a live-and-let-live mentality during trench warfare is also relevant.[23] Placing an emphasis on the negative traits of these battalions might place the UPS battalions into this bracket. However, patrolling and raiding was planned and carried out by the UPS and 20/R Fus, albeit sometimes unsuccessfully. Clear attempts were made to counter the live-and-let-live system of trench warfare. This was not continuous, but few units will have agitated their enemies at every moment. However, these efforts did not achieve dominance of No man's land. Having looked at 20/R Fus in detail the battalion does not wholly fit within the brackets described by Griffith and Ashworth but it was also not an outright aggressive battalion. It's performance also fluctuated over time, as did many battalions, due to changes in leadership, composition or under different campaigning conditions. The Winter of 1916–1917 saw the battalion both surviving but also building itself back up after a disastrous Somme campaign; a reduction in aggression during this period could be understood. However, no sooner had it regained its self-confidence than it faced a near impossible attack at Arras and suffered heavily for little gain. These losses left the battalion less cohesive and less effective as its character and ethos had been diluted over time.

19 *Manchester Evening News*, 27 July 1916.
20 Williams, *From Khaki to Cloth*, p.61.
21 Williams, *From Khaki to Cloth*, p.61.
22 Paddy Griffith, *Battle Tactics on the Western Front* (London: Yale University Press, 1994), p.79.
23 Tony Ashworth, *Trench Warfare 1914–1918, The Live and Let Live System* (London: Pan, 2004), passim.

The combat performance of the UPS battalions is only one yardstick by which their effectiveness should be measured. In supplying a wave of officer candidates who had experienced trench conditions, the UPS Brigade made a considerable contribution to the whole BEF. Such officer candidates would likely make better platoon commanders who might survive for longer and keep their men alive for longer. Ex-UPS men, gazetted *en masse* during the summer of 1916, alongside those who had departed earlier, filled the gaps left by the Somme campaign. For the latter stages of that battle, and for the operations at Arras and Ypres, these battle-inoculated ex-UPS officers proved their worth in numerous engagements. Likewise, the Royal Fusiliers battalions that inherited the remaining men from the three disbanded battalions received experienced volunteers rather than more reluctant 'Derby men' or conscripts. Frustratingly, most of the men trained by the two UPS reserve battalions did not go to 20/R Fus but instead rebuilt other battalions. When 20/R Fus was finally disbanded the men it sent to other units still fought until the Armistice and many maintained the reputation of their former unit. Therefore, the three disbanded UPS battalions, and 20/R Fus, provided a disproportionate benefit to the Western Front, and other theatres, which cannot be overestimated.

Though not elite units, history should not leave the four battalions of the UPS, and especially 20/R Fus, as the subject of ridicule. Whilst the UPS battalions and 20/R Fus were not outstanding their effectiveness was underplayed by commentators. Likewise, some criticisms of the UPS Brigade were exaggerated to suit the whims of their accusers. Whilst the UPS were initially founded on an 'elitist' concept with high martial standards, the UPS battalions, especially 20/R Fus, had no delusions of being 'elite' by 1916–1917. One officer wrote in early 1917; 'The battalion is going strong. It is a better battalion now than it was 6 or 7 months ago. We are not a first-class battalion by any means – we are essentially amateurs. But we are not too bad.'[24] For some of the externally imposed handicaps the UPS had to endure '…not too bad …' was an understated, but adequate, assessment of their performance.

24 IWM: E.F. Chapman Papers, 1799, p.56.

Appendix I

Victoria Cross Awards to UPS Brigade Members

1. Second Lieutenant G.R.D. Moor VC, 3/Hamps R

George Raymond Dallas Moor was born in Melbourne, Australia, and was educated privately at Little Appleby on the Isle of Wight and later at Cheltenham College. He enlisted in 21/R Fus, aged 19, on 18 September 1914 at Exeter. Moor did not spend long with the UPS and was gazetted into the Hampshire Regiment on 17 October 1914. He joined 2/Hamps and embarked with the battalion on 20 March 1915 for Egypt. He landed at Gallipoli and was wounded with a bullet wound to his right shoulder on 28 April 1915. He recovered from this injury and re-joined his battalion. In a subsequent attack his behaviour was observed for displaying the utmost gallantry and he was recommended for the Victoria Cross. He left the field sick with exhaustion on 2 July and after a spell on a hospital ship and at Malta he recovered and re-joined his men on 7 July 1915. On 16 September 1915 he was admitted to hospital with diarrhoea and was sent back to the UK on 22 September suffering from dysentery and jaundice. In January 1916 he was considered too young for a staff appointment though he was clearly brave and battle experienced. One patron recorded; 'I have made several efforts to get him such a place, as he is very able and agreeable, besides being tall and good looking, but I have not so far had any success.'[1] Never one for a quiet life he was posted at Fort George where he suffered cuts and bruises after flying accident on 27 February 1916. He served for a week with 14/Hamps and joined 1/Hamps on 3 October 1916. On 30 October Moor was promoted to lieutenant and he served in France until he was hospitalised for a month with tonsillitis on 1 March 1917. He enjoyed the opportunity of leave in August and returned to 1/Hamps R on 6 October 1917. He went on leave again from 1–15 November 1917. He was wounded in action near Monchy-le-Preux on 23 November 1917 suffering a gun-shot wound to the left arm.

After recovery, Moor was an aide-de-camp to the GOC of 30th Division, Major General Williams, from 18 March 1918. He served almost until the Armistice until he was admitted to a field ambulance suffering from influenza. He died on 3 November 1918 of pneumonia with the end of the war in sight. He was buried at Y Farm Cemetery, Bois Grenier. Though young, tall and fine-looking, he was the antithesis of the stereotypical effete and clueless ADC that is predominant in literature on the Great War. He also earned the Military Cross prior to his death which was gazetted on 2 December 1918; 'For conspicuous gallantry and skill. He carried

---

1 TNA: WO 339/34862: G.R.D. Moor Officer File.

out a daylight reconnaissance all along the divisional front in face of heavy machine-gun fire at close range, in many places well in front of our foremost posts.'[2] He was awarded a bar to this MC on 15 February 1919.

2. Second Lieutenant G.S.M. Insall VC, No. 11 Squadron, RFC.

Gilbert Stuart Martin Insall was born in Paris on 14 May 1894. He lived in Wellington, Surrey, and was educated at the Anglo Saxon School, Paris. Pre-war he was employed as a student in dental surgery before he enlisted as PS/1667 Private G.S.M. Insall, 18/R Fus, on 15 September 1914 in Westminster. On 13 March 1915 Insall was gazetted to a commission in the RFC and became a flying officer on 16 July 1915. He joined No. 11 Squadron in France and flew a Vickers biplane. On 13 December 1915 Insall and Donald were on another patrol and they saw a German aircraft near Arras which turned southwards; after a short dog-fight they were lured into the fire of German anti-aircraft guns and were wounded before conducting a forced landing behind German lines. Whilst Insall and Donald were trying to set fire to their aircraft a German came up on a bicycle and covered them with his pistol. Insall had a 450g shell fragment removed from his hip by a German surgeon which he was given as a souvenir. He would spend over a year in PoW camps in Cologne and Heidelberg whilst he recovered from both his wounds and appendicitis. Once recovered, Insall was able to escape to Switzerland and returned to the UK on 11 September 1917. He was made a flight commander on 19 March 1918 and served as an instructor with No 86 Squadron at RAF Northolt. He was later promoted major.

3. Temporary Lieutenant G.St.G. Shillington Cather VC, 9/RIrFus

Geoffrey St. George Cather was born on 11 October 1890 in Streatham, London. He lived at St Johns Wood and was educated at Rugby School. Prior to the Great War he was in the tea trade and was a member of 28/Londons (1909–1911). His business took him to America and he bought himself out. By 1914 he had returned and he enlisted in 19/R Fus on 3 September 1914 in Westminster and served as PS/1269 Private Cather of B Company.

On 22 May 1915 Cather was gazetted to a commission in 9th Battalion Royal Irish Fusiliers (36th Division). He went to France with his battalion in October 1915. By July 1916 he was a lieutenant and was the adjutant of his battalion. His battalion attacked on 1 July 1916 and suffered heavy casualties; sixteen officers and 518 men. By the end of the day the military objectives had not been obtained and activity changed to trying to find and save the wounded in no man's land. That evening Cather worked in no man's land looking for, and extracting, casualties. The after-action report in the battalion recorded that; 'Several parties were organized to search NO MAN'S LAND to bring in casualties, their search was continued by parties sent up to HAMEL … On July 2 while carrying out this duty Lieut & Adjt G. CATHER was killed.'[3] Cather's work was of the utmost gallantry but he was killed by MG fire on 2 July 1916. Though

---

2 *The London Gazette*, 2 December 1918.
3 TNA: WO 95/2505: 9/R Irish Fus War Diary.

he was buried where he fell, his grave was later lost and his name is recorded on the Thiepval Memorial. He was gazetted to the Victoria Cross on 8 September 1916.

4. Second Lieutenant F.B. Wearne VC, 3/Essex, attached 11/Essex

Frank Bernard Wearne was born in Fulham on 1 March 1894 and was the son of a wine merchant. He was educated at Bromsgrove School until 1912 and attended Corpus Christi College, Oxford. Wearne was in the OTC both at school and university and he left Oxford in 1914 to enlist. He had already been rejected as unfit by another battalion, likely due to short-sightedness, when he enlisted in 18/R Fus on 3 September 1914 and served in B Company.[4] On 22 November 1914 PS/2214 Private F.B. Wearne was appointed lance corporal but was reduced on 20 December. On 15 May 1915 he was gazetted to a commission in the 3rd Battalion Essex Regiment. He was sent to France on 13 December 1915 and, on 5 June, whilst with 10/Essex, the war diary recorded his small patrol capturing a German prisoner in No man's land. He was described as a 'plucky and resourceful officer' and was wounded near Montauban on 3 July 1916 in four places in his right arm.[5]

On recovery he was posted to 11/Essex in France in May 1917. On 21 May Frank's brother, Captain Keith Wearne, 1/Essex, was killed by shell fire near Orange Hill, Arras. On the night of 28 June 11/Essex conducted a raid on the German trenches near Hill 70, near Loos. Second Lieutenant Frank Wearne was killed in action on 28 June 1917. A report on the raid attached to the battalion war diary recorded that Wearne was killed whilst leading his section in the bombing attack over open ground.[6] His body was not recovered and he is recorded on the Loos Memorial to the missing. His VC was gazetted on 2 August 1917.

5. Lieutenant C.H. Sewell VC, late Queen's RWK, attached Tank Corps

Cecil Harold Sewell was born on 27 January 1895 in Greenwich. He was educated at Dulwich College, where he was a member of the OTC. He was employed as an articled clerk, having been a law student when he enlisted in 21/R Fus on 23 November 1914. Sewell was a member of the MG Section and went to France with the battalion in November 1915 and was sent to No1 OCB for officer training on 24 March 1916. In August he was gazetted to a commission in the Royal West Kent Regiment. He joined 1/Queen's RWK in September 1916 but transferred to the Heavy Branch of the MGC (later the Tank Corps) on 23 December. He initially served with C Battalion before joining the 3rd Brigade Signals Section in May 1917 followed by joining 3/Light Tank Bn. On 29 August his battalion went into action near Fremicourt; Sewell was killed, aged 23, and was buried in Vaulx Hill Cemetery.

4   TNA: WO 339/50235: F.B. Wearne Officer File.
5   T.M. Banks and R.A. Chell, *With the 10th Essex In France* (London: Gay and Hancock, 1924), p.121.
6   TNA: WO 95/1616: 11/Essex War Diary, 29 June 1917.

Appendix II

20/R Fus. Roll of Honour, High Wood, 20 July 1916

Thiepval Memorial

PS/4370 Private Esmond George Akehurst, aged 18, was from Palmers Green.

PS/4393 Private George Oswald Ashby, aged 21, was from Oldham and was an accountant.

PS/7067 Private Archibald Bailey was aged 26 and was from Stockport.

G/24661 Private Herbert Frederick Balls, aged 24, was from Northolt.

PS/7759 Private Augustus Barford, aged 20, was born in Newmarket and was an office boy.

PS/4423 Private Geoffrey Barker, aged 21, was from Didsbury, Manchester, attended Manchester Grammar School and worked for a bank. He enlisted in September 1914 and served in C Company.

PS/4594 Private Victor Abraham Bass was born in Wrexham and lived in Buxton. He attended Repton School and Clare College, Cambridge, and was connected to the Bass brewing family.

PS/4524 Private Cyril Bowden, aged 18, was from Whitfield, Glossop

PS/10002 Private Harry Boxold, aged 21, was from Banbury.

PS/4408 Lance Corporal Kingsley Brown, aged 21, was born in Bolton.

PS/4421 Private Frederick Walter Brownridge was born in 1893 and lived in Whalley Range. He worked in the treasurer's office of the Manchester Corporation. He enlisted in September 1914.

PS/6005 Private David Buchanan, aged 27, was from Manchester and worked for the Home Office.

PS/4584 Lance Corporal Huyton Ernest Ulric Budge, aged 23, was from Fallowfield, Manchester. He attended Hulme Grammar School, Princess St Training College and Manchester University. He graduated with a BSc in 1914. He was a Signaller in A Company.

PS/7760 Private Lionel Hewett Cansdale was born in 1897 in Ashton-in-Makerfield.

L/16126 Private William George Capper, aged 19, was from St Pancras and attended Manchester Street School, London.

PS/4610 Lance Sergeant Edward Charles Cass was born in 1895 and lived in Colne Lancashire. He was an apprentice with Levinsteins, in Manchester. He attended Colne Secondary School and at Manchester University where he was a chemistry student and a member of the OTC. He enlisted in September 1914.

PS/4618 Private John Willis Challoner, aged 32, was from Burnley. He attended the Mechanics Institute having been awarded a science scholarship and worked for the Scottish Sea Fisheries Board. He enlisted in November 1914.

PS/9076 Private Herbert Enderby Chambers, aged 23, was born in Suttersby, Lincolnshire. He was an outfitters assistant for a draper and outfitters company.

G/24641 Private Albert Thomas Chandler, aged 29, from Hounslow, was a tin cutter at a gunpowder mill.

G/24723 Private Charles Frederick Green Church, aged 28, had lived in Ealing. He was a ticket examiner on the Great Western Railway.

PS/4640 Private John Stainton Clark was born in Crosthwaite and lived in Keswick

PS/4634 Private Roland Clayton, aged 26, was from Poulton-le-Fylde

PS/4637 Private Agnew Maurice Clegg was from Higher Crumpsall and attended Manchester Grammar School

PS/4671 Private John Eadon Cook.

PS/4660 Private Charles Henry Coop, aged 24, was from Oldham.

GS/22282 Private Thomas Edward Corlett, aged 36, was born and enlisted in East Ham.

PS/8702 Private Hector Crews, of A Company, was aged 19 and from Kings Norton, Worcestershire.

PS/4713 Private Richard Crossland was born in Todmorden and lived in Bolton.

PS/8887 Private John Harris Davies, aged 21, was from Newark Farm.

PS/4752 Sergeant George Hume Dennis, of C Company, was aged 33, from Cookham Rise, Berkshire, and attended the Royal Belfast Academic Institution.

PS/4780 Private Wilfred Edwards, aged 24, was from Colwyn Bay.

PS/8249 Lance Corporal George Ashton Elliott was from Urmston.

PS/4833 Private Sydney Fox, aged 21, was from Stockport. He attended Manchester University.

PS/4850 Private Alfred Garne lived in Oswaldtwistle. He attended Manchester University where he read engineering and was a member of the OTC. He enlisted in September 1914.

PS/4878 Lance Corporal Thomas Goodier, aged 26, was from Stockport.

PS/8322 Lance Corporal Reginald Wilfred Green was originally from Bury St Edmunds.

PS/4904 Private Julius Gregory, from Levenshulme, attended Manchester Grammar School.

PS/4921 Sergeant Frank Harris, aged 30, was from Rusholme, Manchester.

PS/8724 Private Walter Bernard Harrison, aged 20, was from Manchester. He attended Kingswood School, Bath, where he won an open Classical Exhibition to Merton College Oxford awarded by Lancashire County Council. He was a member of Oxford OTC. He enlisted from university in July 1915.

PS/4991 Private Hugh Heald, aged 25, was from Chorley, Lancashire.

PS/4985 Private Leonard John Heywood was from Crumpsall, Lancashire, where he was an articled clerk. He enlisted in September 1914.

PS/5004 Private Henry Hill was from Prestwich, Manchester.

PS/5007 Lance Corporal Thomas Eric Hill was killed on his 24th birthday. He was from Withington, Manchester, and attended Manchester Grammar School.

PS/5037 Private John Howard, aged 18, was from Manchester.

PS/5040 Private Norman Howard, aged 32, was from Bolton. He served with A Company.

PS/5127 Private Wallace Jackson, aged 19, was born in Stretford and lived in Manchester.

PS/5132 Private Leslie Newlyn Jeffrey, aged 24, from Moss Side, was employed with the Commercial Union Assurance Company. He was; 'a young man of great promise in business'. He served with B Company.

PS/8180 Private Thomas Judson, aged 18, was from Newark, Nottinghamshire.

PS/5152 Private Thomas Kelly was born in Horwich, Lancashire, and lived in Chorlton-on-Medlock.

PS/5151 Lance Corporal Richard George Kelsall, was born in Pendleton, Manchester and lived in Swinton, Lancashire.

PS/5156 Private John Kershaw, aged 23, was from Droylesden and was a manager for John Kershaw and Co. He served with C Company.

PS/5176 Private Alan Joures Knudson, aged 22, was from Chorlton-cum-Hardy.

PS/5194 Private Leslie Stewart Lapraik, of C Company, was aged 18, and lived in West London.

PS/5204 Corporal (Joseph) Henry Leather, of B Company, was aged 31, from Moss Side, and was a journalist for the Manchester Guardian.

PS/5238 Private Frederick Lucas was from Fulwood where he was an accountant.

PS/8054 Private Walter Mason, aged 23, was a clerk to a veterinary medicine provider in Crewe. His father was a labourer in a steel forge for a railway company. He enlisted in July 1915. His brother was PS/8052 Acting WO1 Arthur Mason.

PS/5287 Private Charles Maynard was born in Stoke Newington and lived in Stamford Hill.

PS/8199 Private John Alfred McMillan, aged 26, from Heaton Moor, attended Manchester Grammar School.

PS/5312 Private Edward Eric Mellor, aged 23, was from Oldham. He attended Hulme Grammar School and studied law at Manchester University before becoming an articled clerk. He enlisted in September 1914.

PS/5322 Private Carl Henry Francis Miede, aged 24, was from Manchester and attended Manchester Municipal Secondary School and Manchester University. He was awarded a BSc in Sanitary Engineering in 1912. He served in the OTC. Miede was a surveyor in the town planning department of the Manchester Corporation.

PS/5372 Lance Corporal James Higginson Norbury was from Knutsford, Cheshire, and was a bank clerk.

PS/5384 Lance Corporal Charles Nuttall, of C Company, was aged 26 and from Farnworth, Lancashire.

PS/5450 Sergeant Harry Pemberton was born and lived in Glossop and was a clerk with the CoOp wholesale society.

PS/8202 Private Harry Lawrence Picking, aged 18, was from East Hanningfield. He was the captain of King Edward VI Grammar School, Chelmsford and served in the school OTC. He served with C Company.

PS/8650 Private James Albert Pigram, was from Swansea where he was a 'cutter' for a ladies and gents tailors.

PS/5514 Lance Sergeant James Henry Reason was from Manchester. He was last seen wounded in a shell hole with a comrade.

PS/5529 Private Henry Rider, aged about 23, was a clerk in the cotton trade, from Salford.

PS/5528 Private Longford Brooke Risk, aged 22, lived in Old Trafford where he was a town hall clerk for the Manchester Corporation. He served in C Company.

PS/5546 Private Joseph Robinson, aged 24, lived in Sale, Manchester. He attended Crewe Secondary School and Crewe Training College for Teachers. He served in B Company.

PS/5609 Lance Corporal Thomas Wilson Sedgley, aged 23, was from Withington; he attended Manchester University and was a labour exchange clerk. He served with C Company.

PS/5620 Private Clifford Sharples, aged 21, was from Rhodes, Lancashire; he attended Cheetham Central School. Sharples was wounded on 18 July but went back to front line; he fell in the repulse of a desperate counterattack.

PS/5626 Sergeant Ernest Shorrocks, aged 41, was from Chorlton-cum-Hardy. He attended Hulme Grammar school and Victoria University, Manchester, and was awarded an MSc in 1900. He was a schoolmaster at Hulme Grammar School, Queen's College, Taunton, and Huish Grammar School, Taunton. He enlisted in September 1914. Shorrocks played football for B Company and played cricket and rugby for Somerset.

PS/8688 Acting Corporal Henry Duncan Ross, aged 40, was born in Bellary, Madras, and was a schoolmaster in Hampstead. He studied at Rugby School and Queens College, Oxford. He enlisted in Newcastle in September 1915.

PS/5630 Private Arthur Simpson, aged 21, was from Macclesfield, Cheshire.

PS/5646 Private William Kenneth Smethurst, aged 21, lived in Levenshulme, Manchester; he attended Victoria Park School. He enlisted in October 1914.

PS/7865 Private William Sowby, aged 32, was from Boston.

PS/5685 Lance Corporal Hubert Sansom Staines, aged 24, was from Ruthin, Denbigh.

PS/5710 Private Harold Claude Stockdale, aged 21, was from Oldham.

PS/5720 Private Joseph Edwin Sunderland lived in Todmorden, Lancashire. He was a wholesale millinery apprentice and he served in D Company, No 13 Platoon.

PS/5726 Sergeant Ernest Swire, aged 26, was from Colne. He attended Nelson Secondary School and Owen's College and was employed in the offices of Andrew Swire, cotton manufacturers. His brother, Percy Swire, survived High Wood.

PS/5738 Private William Knapton Taylor, aged 21, lived in Burnley where he attended Burnley Grammar School. He served in A Company as a signaller.

PS/5736 Private Alfred Thomason was from Urmston.

PS/5764 Private William Maurice Tripp, aged 21, was born in Chicago and lived in Altrincham.

PS/8203 Private Kenneth Leopold Raymond Trussler, aged 25, was from Pinner.

PS/5751 Private James Tweedale was from Rochdale. He served in D Company, No 13 Platoon.

PS/8501 Private John Waldron, aged 24, was also from Rochdale.

PS/8901 Private William Waugh, aged 20, lived in Willington-on-Tyne.

PS/9335 Private John Albert Weeks, aged 21, was from Hitchin and was a clerk for an oil company.

PS/7871 Private William Edgar Williamson, aged 20, was a warehouse assistant from Derby.

PS/8853 Private Arnold Wilson, aged 19, was from Rochdale. He served with C Company, No 12 Platoon and had trained with James Dykes.

PS/5967 Lance Corporal Frank Wilkinson, aged 40, was born and lived in Brighton. He attended Hurstpierpoint College, Burgess Hill, Sussex and was a partner in a family auctioneer and estate agent business in Brighton.

PS/5946 Lance Corporal Harry Winterbottom, aged 31, was from Manchester. He served with A Company.

PS/5972 Private Arthur Worthington.

PS/5981 Corporal Basset Wright.

GS/114 Private Thomas Wright, aged 21, was from Islington. He had formerly served in France with 9/R Fus in 1915.

PS/5997 Sergeant Joseph Eric Young, aged 30, lived in Disley, Cheshire, and attended Manchester Grammar School.

Caterpillar Valley Cemetery, Longueval

PS/4445 Private Leslie Ball, of C Company, No 4 Platoon, was aged 21 and was from Chorlton.
G/24680 Private Frederick George Bates lived and enlisted in Staines.
PS/4509 Lance Corporal Herbert Woodward Boumphrey was from Crumpsall and played football for C Company.
PS/8878 Private George Alan Bradley, aged 19, was from St Helens. He attended Bury Grammar School, Liverpool College and Liverpool University.
PS/4657 Lance Corporal Harry Cecil Cox was born in Kingstown Ireland and lived in Prestwick
PS/4868 Sergeant George Douglas Gillmore, aged 36, was from Camberwell. He served in D Company, No 14 Platoon.
PS/7846 Private Edward Francis Gould, aged 32, was from Leicester and served with D Company.
PS/4903 Private Frederick John Grant, aged 30, was from Manchester.
PS/5212 Private Nissim Lisbona, aged 34, lived in Manchester and attended Manchester Grammar School and Manchester University. He was called to the Bar in 1908. He served with C Company.
PS/5302 Private Ernest Matthews, aged 43, lived in Piccadilly, Manchester. Ernest attended Manchester Grammar School. He enlisted in September 1914 and served with A Company, 20/R Fus. He was; '… always very keen and was an excellent soldier.'
PS/8849 Private Ralph Fernando Morton, was a material clerk for an engineering contractor in Blackpool.
PS/5432 Lance Corporal Leonard Pateman, aged 19, was from Accrington. He attended Accrington Secondary School and graduated with honours in maths from Manchester University.
PS/8983 Private Gordon Wesley Powell was from Bath. He attended the University of London and became a schoolmaster at Kingswood School, Bath. He was one of Dykes' draft and joined No 9 Platoon, C Company. He was pedantic, calm, philosophical and never got ruffled.
PS/5480 Private Harold Harvey Proudfoot was from Broughton Park, Manchester. He attended Manchester Grammar School and was a warehouse assistant for a cotton manufacturer.
PS/5540 Lance Corporal William Roberts lived in Salford; he attended Manchester Grammar School.
PS/5560 Sergeant Alexander Ross was born in 1892 in Northampton and lived in Chorlton-cum-Hardy. He attended Manchester Municipal Secondary School and studied for a BSc in electrical engineering at Manchester University. He enlisted in September 1914.
PS/5670 Private Harold Smith was from Higher Broughton. He attended Colne Secondary School and Mill Hill Public School and assisted his fathers' cotton manufacturing business. He had been nominated for a commission in the LNL.
PS/5744 Lance Corporal Fred Taylor, aged 26, was from Birmingham. He served with A Company.
PS/5943 Sergeant Ernest Williamson was from Rusholme and attended Manchester University.

PS/5885 Private Herbert Waterhouse was born in Eccles and lived in Lostock, Bolton.
PS/5958 Private Frederick James Wood, aged 21, was from Newton Heath. He attended Manchester Grammar School and the School of Technology, Manchester.
PS/5971 Corporal Cecil Worthington served with A Company.

Serre Road Cemetery No 2

PS/4917 Harold Grindrod, aged 22, was from Bolton. His body was found in High Wood and was identified from a spoon marked 'H G'.
PS/5043 Lance Corporal Leonard Alexander Hoole, aged 23, lived in Reigate, Surrey. He attended the Old Hall School, Wellington, Salop and had private tutors. He enlisted in London in September 1914 and served with the Stretcher Bearers. His brother E.H. Hoole was also killed. Leonard's body was found in High Wood in 1928 and was identified by his height and by a spoon and pocketbook.
PS/5634 Private Mason Simonton, aged 21, was born in Ballymacarrett and lived in Belfast
PS/5920 Lance Corporal Cedric Whittaker, aged 26, was a salesman from Prestwich, Manchester. He served in C Company with James Hodson. He had married in Epsom in December 1914. His remains were found near one of the rides in High Wood and were identified by four false teeth and the remains of a named photograph.

Bouzincourt Ridge Cemetery, Albert

PS/4503 Warrant Officer Class 2 Robert Blackstock was the Company Sergeant Major of A Company. He was aged 35 and lived in Whalley Range; he attended Manchester Grammar School. Captain Templar, commanding A Company, wrote: 'During the preliminary bombardment we lay side-by-side, and he chatted away quite unconcernedly under a perfect canopy of shrapnel. Then at the word 'Go' we went forward together, and he was invaluable in keeping the line, rallying stragglers and generally organising his immediate portion of the attack. It was fine to hear his voice ringing out and giving orders just as if we were on parade … he was among the first to go – a shell hit him and he died at once.' His body was later identified by a watch.
PS/7429 Private Elijah Laban was from Walsall.
PS/4572 Lance Corporal Herbert Buckland, aged 30, lived in Old Trafford and served in A Company.

Ovillers Military Cemetery

PS/7435 Private Frederick Moffat, aged 20, was from Macclesfield. He was a battalion signaller.
PS/5568 Private William Stanley Russell, aged 23, from Manchester, attended Manchester Grammar School. Moffatt and Russell were buried in the same un-marked grave in High Wood and were re-interred in 1933.

Delville Wood Cemetery, Longueval

PS/5333 Private George Henry Moore, aged 22, lived in Higher Broughton, Manchester. He was employed as a junior clerk for a cotton goods shipping company. He served in C Company, No 10 Platoon, along with his brother James.

PS/5509 Private Peter Lloyd Rees, aged 21, lived in St Anne's on Sea. He was a member of C Company.

The following men died of their wounds and are buried in Heilly Station Cemetery

PS/5982 Private Donald Hodgson Wright, aged 21, was from Brooklands, Cheshire, where he was an articled clerk. He enlisted in September 1914 and died of wounds on 20 July.

SR/2296 Private Joseph Barham lived in Lambeth. He died on 21 July.

PS/4519 Private Donald Rumney Boddan, aged 20, was born in Manchester and attended Marlborough College. He enlisted in September 1914 and served as a signaller. He died on 22 July.

PS/5769 Sergeant Basil Graham Taunton, aged 31, was the Platoon Sergeant for No 10 Platoon, C Company. He had recently returned having been wounded in March 1916. He died on 22 July, aged 31, of wounds received in action.

PS/5267 Sergeant Thomas McCreath, from Wapping, died on 24 July.

PS/5363 Private Frederick Alexander Murdoch, aged 25, lived in Barton-on-Irwell, near Manchester. He was an elementary school teacher. He succumbed on 21 July.

PS/4677 Private Vernon Jack Cooper, aged 23, was from Oxford. He died on 28 July.

GS/16103 Private Alfred Burden, aged 18, was from Paddington. He had formerly served with 4/R Fus. He died on 29 July.

The following men died of wounds and are buried elsewhere

PS/7457 Private George 'Midge' Alcock, aged 24, was from Prestwick and attended Manchester Grammar School. He was missing and died as a PoW near Velu on 31 July. He was buried in Achiet-le-Grand Communal Cemetery Extension.

PS/4943 Private Cyril Hartington, aged 29, lived in Castledon. He died on 22 July and was buried in Bois Guillaume Communal Cemetery.

PS/8964 Private Harold Robert Lumley was born in Barnard Castle, Durham. He served in C Company, No 11 Platoon, having joined in January 1916. He was described by Dykes as a man; '...alive and full of energy'. He died on 30 July.

PS/5070 Private Harry Holdsworth was from Hazelhurst. He was in the yarn business with his brother. He was badly wounded and left in a shell hole in High Wood by Alan Proctor. He was of fine physique and was awaiting a commission but died on 27 July having succumbed to his injuries whilst in captivity. Lumley and Holdsworth were buried in Caudry Old Communal Cemetery.

PS/5565 Lance Corporal Dudley Rogers, aged 26, lived in Upper Chorlton. He served with B Company and died on 23 July. He was buried in Etaples Military Cemetery.

PS/5650 Private Edward Baxter Southern, aged 24, was from Southport. He attended Charterhouse School (1906–1911) and Oriel College, Oxford. He enlisted in September 1914. He died on 21 July and was buried in Mericourt-L'Abbe Communal Cemetery Extension.

PS/5149 Private Ernest Cecil Kay was from Middleton, Lancashire, and attended Manchester Grammar School. He died on 27 July and was buried in Middleton.

PS/7261 Private Frederick William Patman, aged 28, was from Whitehaven. He was a bank clerk. He died of wounds on 28 July and was buried in Netley Military Cemetery.

PS/5209 Lance Corporal Sydney George Lewis, aged 29, was born in Tottenham and lived in Chingford. He died on 27 July and was buried in Puchevillers British Cemetery.

PS/4695 Private George Coles, aged 22, was born and lived in Beswick. He died on 23 July.

PS/4823 Private Stephen Ambrose Fisher, aged 24, lived in Seaton, Devon. He was the son of an Inspector of Schools and attended Manchester University. He died 23 July.

PS/7439 Private William Owen Jones, aged 28, was from Ferry, near Chester. He died on 31 July. Coles, Fisher and Jones were buried in St Sever Cemetery Extension.

PS/4795 Private Frederick William Ekin, aged 23, from Belfast, served in No 13 Platoon, D Company. He attended the Royal Belfast Academic Institution and Queen's University where he was a member of the OTC. He was grazed by a whizz-bang on 17 May but was severely wounded on 20 July and died on 18 August; he was buried in Belfast.

Appendix III

Key Documents

Document 1.

Letter by 'Eight Unattached' published in The Times on 26 August 1914:[1]

'To the Editor of "The Times"

Sir,

We attended the recent meeting at the Hotel Cecil to consider the formation of a corps of past public schools men, and found that the organizers only required grey-haired, spare-time veterans.[2] We are between thirty and thirty-five, absolutely fit and game for active service. The meeting showed that there must be hundreds of men in the same position as we are, who, between the years 1898 and 1903, were marksmen, and attended the Bisley musketry camps and Aldershot training camps with school or university corps. We have applied for commissions in the new Regulars, but find we are too old. We have offered our services as musketry instructors, and are informed that we are too young, and that none under thirty-five are selected.

After endless inquiries there seems only one way in which our services are acceptable, and that is by joining the ranks. Many advantages would result if we all joined the same regiment, and all public school men of similar age and qualifications are invited to attend a formal meeting on Thursday next, the 27th instant, at the address below, between 8 p.m. and 9 p.m., to discuss the formation of a " Legion of Marksmen " with a view to offering its services en bloc to one of the new battalions now being formed.

We should be obliged if you would insert this invitation in your columns.

Yours,

Eight Unattached

59A BROOK STREET,
      GROSVENOR SQUARE, W.'

---

1   Anon. Author, *University & Public Schools Brigade*, p.15.
2   This meeting had concerned recruiting for home defence.

Document 2.

Initial criteria proposed by Mr Boon on 27 August 1914 for the raising of the UPS:[3]

1. Inasmuch as it is known that the applications for temporary Commissions in His Majesty's Army have been greatly in excess of the number required, and further it is felt that there are a great number of old public school boys who are anxious to serve their country, but at the same time are somewhat chary of joining the regular army with the ordinary run of recruits, it is proposed to raise a Regiment of old public school boys.
2. For the purpose of determining what schools are public schools, all past members of these schools the names of which appear in the current number of The Public Schools Year-book shall be eligible for admission to this Regiment.
3. The number of boys at present members of the above schools is in excess of 30,000, and it is, therefore, felt that there should be little difficulty in raising a Regiment 5,000 strong of five battalions.
4. The promoters of the Regiment to obtain the sanction of the War Office to the formation of the Regiment, which shall form part of Lord Kitchener's New Army.
5. In order to save the War Office as much work as possible, it is suggested that the recruiting should be carried out by a Committee to be appointed, and that the Committee shall undertake the medical examination of all recruits in accordance with War Office instructions.
6. As it is a matter of national importance that recruits should be obtained as soon as possible, a period shall be fixed, say fourteen days, within which every effort shall be made to recruit the 5,000 men required.
7. The Central Organizing Office shall be in London, and recruiting offices shall also be opened throughout the country.
8. The following shall be asked to render service free:
   (a) Doctors, medical examination of recruits.
   (b) Newspapers-free displayed appeals for recruits.
   (c) Landowners-training-grounds.
9. If found practicable, the men of various schools shall be recruited in companies-i.e. a Harrow Company, a Charterhouse Company, and so on.
10. The War Office shall be asked to sanction the recruiting of men between the ages of twenty-one and thirty-five, and possibly up to thirty-eight or forty, and youths between eighteen and twenty-one, with parents' permission.
11. The Regiment to be one recruited for active service abroad.
12. When the Regiment is at full strength an effort shall be made to obtain the Royal Assent to its being named "Princess Mary's Own."
13. The question as to whether the members of the Regiment should pay for their uniforms and equipment, other than rifles and bayonets, shall be considered at a later date, and the promoters will not make any statement as to this when calling for recruits.

---

3   Anon. Author, *University & Public Schools Brigade*, pp.16–18.

Document 3.

Letter to Lord Kitchener 28 August 1914.

To the Secretary of State for War, War Office, S.W.

Sir,

At a representative meeting of old public school and university men held last night at Claridge's Hotel it was unanimously resolved that an effort should be made to raise a force of least 5,000 men to form part of the new armies now being raised.

A committee was appointed for the purpose of obtaining the sanction of the War Office to the raising of such a force, and it was further resolved that, in the event of such sanction being obtained, that such Committee should undertake the whole work of publicity, enrolment, and medical examination of recruits in accordance with the instructions of the War Office.

The Committee are of opinion that the formation of such a force would result in a very large number of men joining the New Army who would not otherwise do so. The applications for commissions have been greatly in excess of the vacant positions, and it is thought that those who have not yet obtained commissions would join the rank and file of this new force.

The Committee intend to make a widespread appeal to old public school and university men, and will open offices throughout the country for recruiting purposes, and, as the result of

inquiries made, are confident that they will succeed in raising the whole of the 5,000 men and propose to set themselves a fortnight within which to complete this work.

In making this appeal the Committee do not wish to in any way claim any privileges for the members of this force, and desire to make it clear that the men will be recruited on exactly the same terms and conditions as those laid down for the new regular army.

We have the honour to remain, sir,

Your obedient servants,

Document 4.

Letter sent to Lord Mayors, Provosts, and Mayors of the chief cities and towns in Great Britain:

THE PUBLIC SCHOOLS AND UNIVERSITY MEN'S COMMITTEE,
66 VICTORIA STREET, WESTMINSTER, S.W.

August 28, 1914.

To His Worship the Mayor of -------,
EX-PUBLIC SCHOOLS AND UNIVERSITIES FORCE

Dear Sir,

I am instructed by my Committee to send you the enclosed particulars relating to the above force, the raising of which has been officially sanctioned by the War Office, and to say that they would esteem it a favour if you would, on receipt of this letter, use your best endeavours to obtain the services of some ex-Public School or University man of good standing in your town to immediately undertake the work of providing a recruiting office.

It has been decided that all provincial recruiting officers be asked to give their services free and to place at the disposal of the Committee suitable offices.

The procedure to be adopted by all recruiting officers in accepting this position is as follows:
(1) Open recruiting offices and to place the posters which will be issued by the Committee in prominent positions.
(2) Issue to the local press a notice informing intending recruits of the address of the recruiting office and the hours between which the same will be open.
(3) Notify the Secretary of the Committee at the above address, by telegram if possible, of the address of the recruiting office and telephone number that can be used in connection with same.
(4) Secure the services of a magistrate or other qualified persons for the purpose of swearing in recruits.
(5) Obtain the services of qualified medical practitioners for the purpose of carrying out the medical examination of recruits.

Thanking you in anticipation, I have the honour to remain,

Your Lordship's obedient servant,

Sec. to Committee.

Document 5.

Letter to Lord Kitchener:[4]

September 8, 1914.

The Right Hon. the Earl Kitchener of Khartoum, P.M., K.P.,
Secretary of State for War, War Office, S.W.

My Lord,

We, the undersigned members of the Committee of the Old Public Schools and University Men's Force, wish to call your Lordship's urgent attention to a most serious state of affairs existing at the present moment.

Acting under your Lordship's direct approbation and with your cordial sanction, we have worked night and day and have at the time of writing succeeded in enrolling 4,000 men-the most splendid material in the eyes of competent observers who have witnessed their drills and route marches.

At the moment we, the Committee, after many promises, verbal and written, are unable to get the necessary military formalities concluded. We appeal to your Lordship to take this matter in hand yourself immediately and insist that the necessary formalities be concluded in order that the men whom we have enrolled shall be medically examined and attested at once, thus making them part of the regular army.

We, the Committee, are prepared and are only too willing to do all in our power to assist the War Office in this matter, and, further, we have offered to arrange for the organization, clothing, housing, and feeding the men.

We trust that this urgent appeal made on behalf of 4,000 men, whose only desire is to serve their country, will receive your earnest personal attention.

We beg to remain,

Your Lordship's obedient servants

---

4   Anon. Author, *University & Public Schools Brigade*, p.27.

Document 6.

UPS Brigade Mobilisation Order dated 15 September 1914:

Old Public School and University Men's Force

66, Victoria Street,
Westminster, S.W.

15th September, 1914

Mobilisation Order

You are hereby instructed to report yourself at 12 o'clock noon on Friday next the 18th, inst., in Hyde Park, opposite Knightsbridge Barracks, and bring with you the articles specified on the attached list.

Battalions will proceed in the course of the afternoon to Epsom for training.

<u>Failure to report yourself in accordance with these instructions will result in orders being issued for your immediate arrest as a deserter.</u>

Parade instructions.

Only one small bag allowed per man. An overcoat, preferably waterproof, must be brought. The enclosed label must be addressed and attached to bag. Only caps must be worn. No umbrellas. Bags must be brought to the parade ground and deposited in accordance with the instruction of the N.C.O. on the ground.

Document 7.

UPS Brigade Mobilisation Kit List dated 15 September 1914:

Old Public School and University Men's Force

66, Victoria Street,
Westminster, S.W.

15th September, 1914

List of articles which every Recruit must bring to camp with him, and which must be packed in one bag, which should be as small as possible:-

Pair of Boots (spare)
2 Soft Collars
Braces
Hair Brush
Shaving Brush
Tooth Brush
Comb
Razor
2 Flannel Shirts
5 pairs Socks
2 Towels
Knife
Fork
Spoon
Housewife

Optional – Extra pair of Trousers, or a spare suit.

Document 8.

Letter to the Press to refute criticisms as to the reluctance to release enlisted men for commissions dated April 1915:

PUBLIC SCHOOLS BRIGADE
REPLY TO CRITICISMS[5]
To the Editor of -------

Sir,

Our attention has been called to correspondence which has appeared in the Press with regard to the selection of members of the Public Schools Brigade for commissions in the Army, and, in view of the erroneous statements which have been made, we think that the true position of affairs should be made known.

In September last the War Office authorized the raising of a Brigade, to consist in all of 5,400 public school and university men.

Recruiting was energetically carried on, and we reached a total of very little short of that number. The need for officers for the new Army then began to make itself apparent, and, as was natural, in a Brigade composed practically entirely of public school and university men, large numbers began to be taken from our ranks to receive commissions in other regiments.

This went on without any check until early in the year, when the Brigadier General and officers of the Brigade and the men under them began to fear that the Brigade had practically been turned into an Officers' Training Corps, and approached our Committee with a view to ascertaining the exact position. We therefore made inquiries from those in authority at the War Office, and received assurances that the Brigade was intended to continue to exist as a unit, and not as an Officers' Training Corps.

Further, we were assured that there was no intention of drawing upon the Brigade for more officers except in special cases, of which there would only be a small number.

At that time 1,700 men had already been recommended for commissions. Since then further need for officers had arisen. We, realizing this, agreed to more men being taken, and when this new demand on us has been satisfied a total of not less than 3,083 men will have been taken altogether out of our Brigade.

We think it is only fair to the Brigade itself and to ourselves as a Committee to make public these facts and figures, which speak for themselves, and surely afford a conclusive answer to the criticisms that have appeared to the effect that the Brigadier General and his officers, and we as a Committee, have put obstacles in the way of men obtaining commissions.

All the men who are fit for and desirous of commissions have now been recommended for appointment.

ARTHUR STANLEY, Chairman.
LURGAN, Vice-Chairman.
H. J. BOON. / J. W. ORDE.

Committee of the Public School Brigade, Royal Fusiliers,
Committee Room 65, 83 Pall Mall, S.W., April 15.

5   Anon. Author, *University & Public Schools Brigade*, pp.61–62.

Document 9.

Letter from the War Office concerning the 16th Battalion Middlesex Regiment in May 1916:

War Office Letter to Municipal Authorities Dated 24 May 1916[6]

War Office
London, SW
20/General Number/4257 (AGI)
24 May 1916

Sir,

I am commanded by the Army Council to inform you that, owing to the increase in the Army Abroad in the various theatres of war in which operations are taking place, the question of the supply of reinforcements has been engaging their strenuous attention.

The Council fully appreciate the generous and spontaneous offers of assistance, put forward by many communities and individuals in the early days of the War, which have added so many battalions to the Army, nearly all of which have now taken their place in the field. They further recognise the desire of those responsible for the raising of these battalion not only that the men should serve together but that they should be reinforced by men obtained from the same locality of the same class. They have also noted with satisfaction the esprit de corps which animates local battalions.

Whenever it is found possible, men of local reserve battalions will be sent as reinforcements to their own service battalions, but the Council reserve themselves the power to transfer trained men wherever and whenever it may be necessary, in order to meet the pressing demands of the forces in the field, and thus ensure the successful conduct of the war.

I am,
Sir,
Your obedient Servant
B.B. Cubitt

To Municipal Authorities and Gentlemen responsible for raising Local Battalions.
Copy to General Officers
Commanding-in-Chief.
Officers i/c Records.'

---

6   A similar letter was likely circulated to the UPS. TNA: WO 32/11343: 'Proposal to raise a Public School Battalion', May 1916.

Document 10.

Instruction sent from GHQ to the HQ of I Corps in late February 1916:

CONFIDENTIAL   1st Army No: G.S. 285

Subject – Withdrawal of battalions for use as O.T.C.

1st Army.

With reference to your *C/48/A.M.S. of the 21st January, it has been decided to withdraw the undermentioned battalions to G.H.Q. area as a temporary measure: –

18th Royal Fusiliers
19th Royal Fusiliers
21st Royal Fusiliers
16th Middlesex

As the withdrawal of the large number of N.C.Os and men recommended for commissions will, for the time being, cripple these battalions as fighting units, it is proposed to reconstitute them, and to form two or three battalions from the residue. These reconstituted battalions will probably be available eventually to rejoin their division. Meanwhile they will be replaced by four T.F. battalions under arrangements detailed in G.H.Q letter number O.B./1370 of 17th February.

G.H.Q.   (Sd) R.BUTLER, Major General,
17th February 1916.   Deputy C.G.S.

Appendix IV

UPS Brigade Poets

PS/1258 Lance Corporal Robert Beckh – Robert Harold Beckh of Haileybury and Jesus College, Cambridge, enlisted in 18/R Fus in Westminster on 2 September 1914, aged 20, along with some other Cambridge friends. Though he had intended to become a clergyman Beckh believed 'that the best training for an officer was through the ranks.' In April 1915 Beckh was appointed as a lance corporal but his application for a commission was already in motion. On 26 May 1915 he departed to a 'Pals' Battalion of the East Yorkshire Regiment and joined on 1 July 1916.[1] On 15 August 1916 the war diary of 12/E Yorks recorded that Beckh was killed by MG fire during a patrol to Ferme Cour d'Avoué.[2] Beckh was later reburied in Cabaret-Rouge British Cemetery. The commanding officer wrote to Beckh's family:

> The death of your son has been, and is, a very great blow and sorrow to me and to all the others in this battalion. He was most deservedly popular with all ranks. ... He had one more than one occasion done exceedingly good work on patrol duty, once bringing back useful information from a German trench. On the night of the 15th instant (August 1916) he was employed in the same duty, and was unfortunately killed after passing back through the enemy's wire... Your son always did his duty, and died doing so...[3]

Beckh was an aspiring war poet who left several war poems which were published in 'Swallows in Storm and Sunlight' as a memorial after his death. He wrote two other poems just two days before his death.

PS/2496 Private Andrew Buxton, 21/R Fus – Andrew Richard Buxton was aged 35, from Easneye, Ware, Herts, and lived at St Johns Wood, London. He was an old Harrovian who had attended Trinity College, Cambridge. Buxton was both a local director of the Westminster Branch of Barclay and Co (1909–1914) and the winner of the gundog retriever trials in 1910. He enlisted in 21/R Fus on 15 September 1914 in Westminster. On 23 January 1915 he was gazetted to a commission in 6/RB and went to France in July 1915 to join 3/RB. He was

---

1   Beckh, *Swallows in Storm*. <https://www.jesus.cam.ac.uk/robert-harold-beckh-2nd-lieutenant-east-yorkshire-regiment> (accessed 20 May 2018).
2   TNA WO 95/2357: 12/E Yorks War Diary.
3   Beckh, *Swallows in Storm*, pp.13–14.

clearly a promising and effective officer who, as a captain, was employed as a staff learner with 73rd Brigade. Buxton was killed on 7 June 1917, aged 37, in the attacks on Messines Ridge. Lieutenant Colonel Pigot DSO MC, OC of Buxton's Battalion, wrote of him:

> He was just coming back from the front line after an attack yesterday when he was hit by a bullet and died almost at once. I can't tell you how much I deplore his loss. He had been with us a long time and on ever so many occasions had shown himself a very brave man. Everyone loved him, and all the men of his Company will, I know, regret his loss. He was always doing his best to make his men comfortable, and I can assure you he will be a very great loss to us all.[4]

He is buried in Oosttaverne Wood Cemetery.

PS/9798 Lance Corporal Arthur Choyce – Arthur Newberry Choyce, from Hugglescote, Leicestershire, enlisted in 28/R Fus in January 1916, aged 22. He joined D Company. He had attended Market Bosworth Grammar School and Nottingham University College and was an elementary school teacher. Choyce was promoted to corporal in August 1916 and served with 104th Training Reserve Battalion (TR/10/60342). He attended No 7 OCB in September 1916 and was gazetted to the Leicestershire Regiment; he joined 9/Leics R in France. Choyce was badly wounded in the left arm and abdomen whilst advancing with his men on 3 May 1917. He lay out in No man's land for twenty hours before darkness allowed him to painfully crawl 600 yards to where he was collected by a stretcher party. When Choyce recovered from his wounds he was sent to the US for a lecturing tour as part of the Department of Information. He saw no further active service and had *Memory Poems of War and Love* published in 1918. Released from service in April 1919, he died in 1937. His poem, 'The Boy Leader' highlights his survivors' guilt after wounding:

> *There down by Arras with his thirty men*
> *He boldly dared to play a leader's part;*
> *And steeled himself to stifle every sign*
> *Of smallest tremor in his boyish heart.*
>
> *By glorious days of constancy displayed;*
> *By little things that proved his soldier worth,*
> *He caught and held the perfect trust and love*
> *Of those who shared the grandest task on earth.*
>
> *Then came a day when wild and wounded pain*
> *Destroyed the happy triumph in his eyes,*
> *And rushed to overwhelm him in a place*
> *Where thirty soldiers suffered 'neath the skies.*
>
> *And down by Arras there are thirty graves*

---

4   Woods, *Andrew Buxton*, p.289.

Unnoticed quite in War's exacting plan.
But one boy sobs through many an English night
And pleads for courage still to play the man.[5]

PS/192 Private Reginald Clements – Assistant Editor of The Pow-Wow ('Bow-Wow') – Reginald Francis Clements was from Oban, Argyll, and was a former student of Hereford Cathedral School and was a university theological student. He enlisted, aged 22, on 4 September 1914 in Westminster and joined D Company, 19/R Fus. He went to France with the Battalion in November 1915 and was with one of the earliest drafts of officer candidates on 22 March 1916 to No 2 OCB. He was gazetted in August to the Royal Sussex Regiment. He unluckily tripped over some barbed wire whilst on patrol with 7/R Suss on 13 November 1916 and suffered a badly twisted left knee keeping him out of action for a year. A brave and resourceful officer, he was awarded the MC for a raid in March 1918. He was killed on 14 August 1918 and his body lies in Morlancourt British Cemetery No 2.[6] Clements' poems were published in 1917 in a book called 'Salisbury Plain'.

PS/907 Private Louis Solomon – Louis Bernard Solomon was born in Oakland, California in February 1896. He lived in Upper Norwood and his parents also lived in Blaby Hill, Leicester. He was educated at Dulwich College (1909–1911) and passed the matriculation for London University. He described himself as an experimental motor engineer who worked for the Standard Motor Company in Coventry. Solomon enlisted, aged 18, with 19/R Fus on 15 September 1914 and was appointed as a signaller with D Company. His training was uneventful and he went to France in November 1915. After enduring the winter trenches he attended No 1 OCB in March 1916 with one of the early batches of officer candidates. He was selected for the RFC in mid-June 1916. After being made a Flying Officer in September he was injured whilst training in the UK but went to France with No 6 Squadron at the end of October 1916. He was in hospital within a month and ultimately returned to the infantry, and was posted to 28/R Fus, in April 1917. In early June he joined 20/R Fus. He was wounded in the scalp on 28 November 1917 but returned from hospital on 11 December 1917. He attended a signalling course at the VIII Corps School and was posted to W Company, 2/R Fus, on 30 January 1918 on the disbandment of 20/R Fus. Solomon was killed in action on 12 April 1918 near Bailleul, aged only 22; he is buried in Outtersteene Communal Cemetery Extension, Bailleul. After his death a collection of his verse was printed under the title *Wooden Crosses and Other Poems*.[7]

PS/1021 Private Robert Vernede – Robert Ernest Vernede was born in 1875 and lived Standon, near Ware, Herts. He attended St Pauls School (where he joined the OTC) and St Johns College, Oxford. He enlisted, aged 39, on 4 September in Westminster and joined B Company, 19/R Fus. On 15 May 1915 he was gazetted to a commission in 5/RB and served with 12/RB in France. He wrote an up-beat letter to his wife on 8 April 1917 but next day he was involved in the Battle of Arras during which he was mortally wounded, aged 41. He is

---

5   Choyce, *Memory Poems*, p.31.
6   TNA: WO 339/60321: R.F. Clements Officer File.
7   TNA: WO 339/60873: L.B. Solomon Officer File.

buried in Lebucquiere Communal Cemetery Extension. He was a noted author and poet and a collection of his poems titled *War Poems, And Other Verses* was published in 1917.[8]

PS/2207 Private Gilbert Waterhouse – Gilbert Waterhouse was born in Chatham on 22 January 1883. He lived in Thames Ditton, Surrey and had attended Bancroft's School, Woodford, and the University of London. He was an architect and a member of the Royal Institute of British Architects. He enlisted with 18/R Fus on 15 September 1914 in Westminster and joined B Company. He left the battalion early with a commission to 3/Essex on 8 May 1915. In October one of his poems, a sonnet, was published. He was sent to France on 4 December 1915 and later joined 2/Essex R. He was killed, aged 33, on 1 July 1916 and his body could not be recovered. In August 1917 his body was found and identified by a set of ID discs which were recovered along with an automatic pistol and a satchel. He was interred in Serre Road Cemetery No 2.[9] Some of his poems (pre-war works inclusive) were published posthumously in *Rail-Head and Other Poems*.[10]

8   TNA: WO 339/54474: R.E. Vernede Officer File.
9   TNA: WO 339/50234: G. Waterhouse Officer File.
10 Waterhouse, *Rail-head*.

Select Bibliography

Primary Sources

Imperial War Museum (IWM)
IWM: Documents 21795: L.K. Briggs Papers
IWM: Documents 1799: Captain E.F. Chapman Papers
IWM: Documents 18911: W.K. Clarke Papers
IWM: Documents 12961, 2003-12-22: T.O'C. Doherty Papers
IWM: Documents 7378: J.N. Dykes Papers
IWM: Documents 11046, 1980-12: F. Henley Papers
IWM: Documents 12059: G.W.G. Hughes Papers
IWM: Documents 14441, 2006-01-31: Lieutenant L.W.C. Ireland Papers
IWM: Documents 16336, 2008-02-26: C.H. Isom Papers
IWM: Sound 10263: A.E.L. Knighton Sound Recording
IWM: Documents 2783, 1986-07-04: J.H. Leather Papers
IWM: Documents 10271, 1976-03: *The UPS Song Book* by A.W. Lloyd
IWM: Documents 6756, 1978-02: Reverend E. Mannering Papers
IWM: Documents 1708, 1985-11-05: W.B. Medlicott Papers
IWM: Documents 16924: E. Page Papers
IWM: Documents 8230: N.A. Pease Papers
IWM: Documents 17681: S. Platt and V. Platt Papers
IWM: Sound 10168: Donald Price Sound Recording
IWM: Documents 21103: L.E. Salter Papers
IWM: Documents 2033, 1992-08: E.D. Shearn Papers
IWM: Document 1385: F.J. Shield Papers
IWM: Documents 13966, 2005-12-20: G. Skelton Papers
IWM: Documents 15453, 2007-01-16: P.F.F. Spaull Papers
IWM: Documents 2138, 1992-11: H.W. Spurrell Papers
IWM: Documents.7716: E. Stoneley Papers
IWM: Documents 17109: G.K. Twiss Papers
IWM: Documents 5565: F.H. Vinden Papers
IWM: Documents 4255, 1983-10: R.L. Watson Papers
IWM: Documents 3606, 1985-07-17: H. Wilson Papers
IWM: Documents 17675: S.D. Wright Papers

Liddle Archives (LA)
LA: WW1/WF/01/C10: E. Connold Papers
LA: WW1/WF/01/C11: G.K. Cooke Papers
LA: WW1/WF/01/C12: C. Coom Papers
LA: WW1/GS/1188: K.V. Norman Papers
LA: WW1/GALL/001: D.A. Addams-Williams Papers
LA: WW1/GS/0125: W.G. Bentley Papers
LA: WW1/GS/0505: V. Edwards Papers
LA: WW1/GS/0831: P. Ingleson Papers
LA: WW1/GS/0841: G.P.S. Jacob Papers
LA: WW1/TR/03/92: R.C.B. Jones Sound Recording
LA: WW1/GS/1299: D.J. Price Papers
LA: WW1/GS/1316: K.C.K. Radford Papers
LA: WW1/GA/LOV/4: G. Worthington Papers

Royal Sussex Regiment Museum
Add Mss 30070
34666: R. Tudor Correspondence

Lancashire Infantry Museum
Lancashire Infantry Museum: Documents, E.A. Holden Correspondence

Fusiliers Museum London (FML)
FML: RFM.ARC.2482.47: Documents of H.D. Etheridge
FML: RFM.2013.8.1/2: P.N. Wright Diary
FML: RFM.ARC.3015.2: G.G. Ziegler Diary

The National Archives (TNA)
TNA: WO95/2405–2407: 33rd Division GS War Diary
TNA: WO95/2408: 33rd Division A&Q War Diary
TNA: WO95/2420–2421: 19th Brigade War Diary
TNA: WO95/2424–2425: 98th Brigade War Diary
TNA: WO95/2423: 18/R Fus War Diary
TNA: WO95/2427: 19/R Fus War Diary
TNA: WO95/2423: 20/R Fus War Diary
TNA: WO95/2427: 21/R Fus War Diary
TNA: WO95/2423: 2/RWF War Diary
TNA: WO95/2422: 1/Cameronians War Diary
TNA: WO95/2422: 5/6 Scot Rif War Diary
TNA: WO32/11343: Proposal to raise a Public School Battalion (16/Mx)
TNA: WO158/344: Notes and Lessons on 1916 Operations, Fifth Army

Formation/unit war diaries (WO 95), Historical Section letters (CAB 45) and commissioned officer files (WO 339 and WO 374) from the National Archives have been acknowledged in footnote citations but are too numerous to list.

UPS Personal Accounts and Autobiographies

Bartlett, Vernon, *I Know What I Liked* (London: Chatto and Windus, 1974)
Bartlett, Vernon, *This is My Life* (London: Evergreen Books, 1941)
Brown, Archie, *Destiny: A Man of Two World Wars* (Bognor Regis: New Horizon, 1979)
Brown, Sir Arthur Whitten, *Flying the Atlantic in Sixteen Hours* (New York: Frederick Stokes, 1920)
Woods, Edward (ed.), *Andrew. Andrew Buxton: A Memoir* (London: Robert Scott, 1918)
Clayton, Albert, *Long Before Daybreak* (Unknown location: M.J. Duckworth, 2020)
Eyston, George, *Safety Last* (London: Vincent, 1975)
Gibson, Ashley, *Postscript to Adventure* (London: Dent & Sons, 1930)
Green, Martyn, *Here's a How-De-Do* (London: Max Reinhardt, 1952)
Harvey, H.E., *Battle-Line Narratives 1915-1918* (London: Brentano's, 1928)
Hiscock, Eric, *The Bells of Hell Go Ting-A-Ling-A-Ling* (London: Arlington, 1976)
Hodson, James, *The Soul of a Soldier* (London: George Routledge, 1918)
Hodson, James, *Grey Dawn-Red Night* (London: Victor Gollancz, 1929)
Hodson, James, *Return to the Wood* (London: Victor Gollancz, 1955)
Kelly, D.V., *39 Months with the "Tigers"* (London: Ernest Benn, 1930)
Kelly, Sir David, *The Ruling Few* (London: Hollis & Carter, 1952)
Morton, J.B., *The Barber of Putney* (New York: Penguin, 1939)
Peacock, Basil, *Tinkers' Mufti* (London: Seeley Service, 1974)
Private No 940 (Sturges, Robert), *On the Remainder of Our Front* (London: Harrison and Sons, 1917)
Ravenscroft, P.D., *Unversed in Arms* (London: Crowood Press, 1990)
Williams, Morgan, *From Khaki to Cloth* (Unknown: Western Mail and Echo, 1949)

Other Personal Accounts

Blackham, Robert J., *Scalpel, Sword and Stretcher* (London: Sampson Low, Marston and Co; 1931)
Davidson, Andrew, *The Invisible Cross* (London: Heron Books, 2016)
Eyre, Giles, *Somme Harvest: Memoirs of a P.B.I. in the Summer of 1916* (London: Stamp Exchange, 1991)
Graves, Robert, *Goodbye to All That* (London: Penguin, 1957)
Hart-Davis, Rupert (ed.), *Siegfried Sassoon Diaries* (London: Faber and Faber, 1983)
Hodson, James Lansdale, *Somewhere in France* (London: Cherry Tree, 1940)
Hutchison, Graham Seton, *Footslogger* (London: Hutchinson, 1931)
Hutchison, Graham Seton, *Warrior* (London: Hutchinson, 1932)
Hutchison, Graham Seton, *Pilgrimage* (London: Rich & Cowan, 1935)
Maclean, Hugh and Baynes, John, *A Tale of Two Captains* (Edinburgh: Pentland Press, 1990)
Richards, Frank, *Old Soldiers Never Die* (Eastbourne: Anthony Rowe, 1933)
Sassoon, Siegfried, *Memoirs of an Infantry Officer* (London: Faber and Faber, 1993)
Stewart, Alexander, *A Very Unimportant Officer* (London: Hodder, 2009)

Poetry

Beckh, Robert, *Swallows in Storm and Sunlight* (London: Westminster Press, 1917)
Choyce, Arthur Newberry, *Memory Poems of War and Love* (London: John Lane, 1918)
Clements, R.F., *Salisbury Plain and other poems* (Salisbury: Bennett Brothers, 1917)
Waterhouse, Gilbert, *Rail-head and Other Poems* (London: E. MacDonald, 1916)

Formation/Unit Histories

Anon. Author, *The History of the Royal Fusiliers University & Public Schools Brigade (Formation and Training)* (London: The Times, 1917)
Dunn, Captain J.C., *The War the Infantry Knew 1914-1919* (London: Jane's, 1938 republished 1987)
Fox, Martin, *With the Special Brigade RE: A Brief History of 186 Company RE and C Special Company RE 1915-1919* (Toronto: Edgar Cross, 1957)
Hutchison, Graham Seton, *History and Memoir of the 33rd Battalion Machine Gun Corps* (London: Waterlow Bros, 1919)
Hutchison, Graham Seton, *The Thirty-Third Division in France and Flanders 1915-1919* (Uckfield: N&MP, originally 1920)
Martin, Major David (ed.), *The Fifth Battalion, the Cameronians (Scottish Rifles) 1914-1919* (Glasgow: Jackson, Son & Co, 1936)
O'Neill, H.C., *The Royal Fusiliers in the Great War* (London: William Heinemann, 1922)
Story, Colonel H.H., *History of the Cameronians (Scottish Rifles) 1910-1933* (Aylesbury: Hazell Watson & Viney, 1961)

Biographies and Autobiographies (Alphabetical by Subject)

Buchan, John, *These for Remembrance* (London: Buchan and Enright, 1919)
Hill, Prudence, *To Know the Sky: The Life of Air Chief Marshal Sir Roderic Hill* (London: William Kimber, 1962)
Namier, Julia, *Lewis Namier, A Biography* (London: OUP, 1971)
Rendell, Jerry, *Profiles of the First World War: The Silhouettes of Captain H.L. Oakley* (The History Press, 2013).

General

Behrmann, Franz, *Die Osterschlacht bei Arras 1917. II Teil* (Berlin: Oldenburg, 1929)
Edmonds, Brigadier General Sir James, *Military Operations France and Belgium 1918 Vol. I: The German March Offensive and its Preliminaries* (London: HMSO, 1935)
Edmonds, Brigadier General Sir James (ed.), *Military Operations France and Belgium 1918 Vol. IV: The Franco-British Offensive* (London: HMSO, 1947)

Falls, Cyril, *Military Operations France and Belgium 1917: The German Retreat to the Hindenburg Line and the Battles of Arras* (London: Macmillan, 1940)
Fussell, Paul, *The Great War and Modern Memory* (Oxford: Oxford University Press, 1975)
Hurry, C. (ed.) *The Pow-Wo No. 1–No. 20, 18 November 1914* to *16 April 1915* (Epsom: Birch and Whittington, 1914)
Lewis-Stempel, John, *Six Weeks: The Short and Gallant Life of the British Officer in the First World War* (London: Weidenfield and Nicholson, 2010)
Middlebrook, Martin, *The Kaiser's Battle, 21 March 1918: The First Day of the German Spring Offensive* (London: Allen Lane, 1978)
Middlebrook, Martin, *Your Country Needs You: Expansion of the British Army Infantry Divisions* (London: Leo Cooper, 2000)
Moore-Bick, Christopher, *Playing the Game: The British Junior Infantry Officer on the Western Front 1914-18* (Solihull: Helion & Company, 2011)
Nicholls, Jonathan, *Cheerful Sacrifice: The Battle of Arras 1917* (Barnsley: Pen & Sword, 1990)
Norman, Terry, *The Hell they Called High Wood* (Wellingborough: Patrick Stephens, 1984)
Parker, Peter, *The Old Lie: The Great War and the Public School Ethos* (London: Constable, 1987)
Pigeon, Trevor, *The Tanks at Flers* (Cobham: Fairmile Books, 1995)
Purdom, C.B., *Everyman at War* (London: Dent and Sons, 1930)
Ruvigny, Marquis De, *The Roll of Honour* (London: Standard Art Book Company, 1916)
Seldon, Anthony & Walsh, David, *Public Schools and the Great War* (Barnsley: Pen & Sword, 2013)
Warner, Rex, *English Public Schools* (London: Collins, 1945)

Index

PEOPLE

Abbay, Major Bryan, 283–284, 490.
Addams-Williams, Donald, 34, 44, 57–58, 109,
Armour, RSM Robert, 256, 283, 285, 324, 354, 434, 495.

Badenoch, Second Lieutenant Ian, 361.
Bartlett, Charles ('Vernon'), 15, 21, 53, 57, 106, 490
Beal, C.F., 17.
Beck-Savage, Second Lieutenant William, 118, 155, 442
Beckh, Second Lieutenant Robert, 313, 519.
Bennett, Lieutenant Colonel Charles, 41, 57, 60, 89, 123, 207–209, 232, 263, 280, 281, 283, 291, 294, 296, 317–319, 351–353, 484.
Bentley, Second Lieutenant William, 31, 40, 43–44, 48, 50, 52, 60, 72, 82, 135, 217, 410, 471, 491.
Blackstock, Sergeant Major Robert, 237, 279, 506
Boon, Hector ('The Spectre'), 17, 19, 38–39, 63, 510, 516.
Bower, Captain Frederic, 288, 432, 434–435.
Brierley, Drum Major Ernest, 165, 198, 231, 243, 424.
Briggs, Private Claude, 221, 412, 485.
Brown, Lance Corporal Archie, 365, 413, 445, 489.
Brown, Lieutenant Arthur Whitten, 15–16, 22, 26, 32, 114, 483, 489.
Buxton, Captain Andrew, 47–48, 50–51, 55, 57, 64, 314, 407, 408, 520.

Catchpole, Captain Cyril, 423, 431.
Cather, Captain Geoffrey, VC, 248, 499–500.

Chapman, Lieutenant Edward, 306, 323–324, 327–330, 333, 349–350, 352–353, 356–362, 374–375, 381, 384–387, 390, 392–393, 471, 497.
Chell, Randolph, 19, 35.
Choyce, Second Lieutenant Arthur, 363, 372, 520.
Clark, Captain Henry, 351, 381, 394–395, 416, 418.
Clarke, Second Lieutenant Wilfrid, 338–339, 489.
Clayton, Private Albert, 312, 340, 373.
Clements, Second Lieutenant Reginald, 92, 218, 460, 521.
Coggin, Lieutenant Algernon, 202, 240–241, 267, 294, 296, 324–325, 328–330, 361
Colbourne, Captain Geoffrey, 396, 406, 423–424, 432,
Connold, Eric., 217.
Cooke, Private George, 311.
Crawshay, Major, 241, 284–285, 287–288, 355–356, 493.

Dale, Second Lieutenant Eric ('Said Fazil'), 365.
Darker, Second Lieutenant Richard, 422, 431, 452–453.
De Coundouroff, Private George, 481.
De Miremont, Captain G.E.R., 427.
Denison, Lieutenant Colonel E.B., 182, 188–189, 222–226, 485.
Doherty, Terence, 23, 37, 50, 63, 65, 77, 79–84, 89, 130, 132, 134, 150, 152–153, 177, 344–345, 483, 490.
Dunn, Second Lieutenant Claude, 445, 450, 490.

Dunn, Captain J C, xii, xiii, 231, 241, 242, 269, 272, 280, 281, 283, 284, 285, 286, 287, 288, 356, 359, 360, 361, 377, 378, 380, 384, 396, 398, 405, 415, 420, 427, 431, 432, 433, 434, 494, 495.
Dykes, Private James, 66, 202, 203, 227, 228, 230–234, 236–237, 240–242, 253–256, 258–267, 270, 272–274, 276–277, 279, 281–282, 300–301, 306, 490, 504–505, 507.

Edwards, Second Lieutenant Vincent, 24, 167, 409, 474.
Etheridge, Second Lieutenant Hugh, 97, 338, 407, 408, 409, 411, 465.
Eyston, Captain George, 21, 22, 26, 27, 32, 34, 36, 38, 53, 112, 372, 483, 489.

Fenn, Major Herbert, 16, 17, 30, 58–59, 128, 138, 140, 170, 178, 181, 183, 188, 212, 467.
Figg, Lieutenant Colonel Donald, 358, 359, 360, 361.
Finch, Second Lieutenant Joseph, 395, 400, 417, 418, 482.

Garnett, Lieutenant Colonel William, 318, 328–329, 331, 351–352, 354, 357–358, 374, 460, 490.
Gibson, Second Lieutenant Ashley, 39, 42, 48, 58–59, 76, 79, 86, 138–139, 151, 227, 244, 272–273, 276–277, 295–296, 307, 320–325, 327, 330–333, 337, 350, 353, 471, 480.
Gluckstein, S., 17, 38, 47, 69.
Goodman, Private Max, 184, 185, 468.
Gordon-Gilmour, Brigadier General Sir Robert, 39–40, 51, 90, 123, 484–485.
Gould, Second Lieutenant William, 406, 429, 431, 453.
Graves, Lieutenant Robert, xii, 283–284, 287, 306, 355–356, 493–495.
Gray, Lieutenant and QM John, 472.
Green, William ('Martyn'), 459, 471, 489.

Haig, Field Marshal Sir Douglas, 371, 376, 378, 384, 388, 407, 420, 430.
Hallett, Captain George, 17.
Hammond, Lieutenant Colonel Robert, 412.
Hartley, Major John, 130, 411, 445, 461.
Harvey, Lance Corporal Harry, 145–147, 159, 223, 311, 464–465, 486,

Hele-Shaw, Dr, 17.
Henderson or Westwood-Henderson, Major James, 17, 24,
Henley, Second Lieutenant F, 328, 329, 330.
Hill, Second Lieutenant Roderic, 23, 111–112.
Hill, Lieutenant William 'Bill', 388, 391, 418, 425, 460.
Hiscock, Private Eric, 407, 414.
Hodgson-Jones, Captain Douglas, 237, 323, 329, 392, 394.
Hodson, Sergeant James, 30–31, 43, 68, 136, 159, 174, 201, 208–209, 230, 252, 254–256, 258–259, 268, 270, 277, 279, 281, 286, 290, 299, 302–304, 306, 310, 318–320, 323–326, 328, 331–332, 337, 351, 482, 484–488, 494, 506.
Holden, Sergeant Ernest, iii, 81–82, 84, 86–89, 93, 95, 99–100, 117–118, 228, 233, 237, 289, 298, 344.
Holliday, Second Lieutenant John, 369–370.
Hollingsworth, Major David, 294, 335, 360, 394, 395, 418, 420, 423, 427, 432.
Hore-Belisha, Captain Isaac, 23, 477, 488.
Howell, H., 17, 38.
Hughes, Colonel G.W.G., 195, 320.
Hume, Second Lieutenant Ronald, 294, 367.
Hutchison, Lieutenant Colonel Graham, 257, 358, 421, 433, 493, 494, 495.

Ingleson, Second Lieutenant Philip, 222, 490.
Insall, Lieutenant Gilbert, VC, 115, 116, 499.
Ireland, Second Lieutenant Leslie, 72, 78–79, 81–82, 84, 87–89, 93, 121, 216, 218–219, 223, 363–364.
Isom, Private Cecil, 307–308, 319, 322–323, 325, 336, 350, 353–355, 359–360, 375–376, 381–382, 387, 392, 412, 473.

Jacob, Second Lieutenant Gordon, 31, 44, 98, 103, 194, 369–370, 490.
Jones, Second Lieutenant Robert, 100, 140–141, 224.
Jones-Williams, Captain George, 397–398, 400, 432.

Kelly, Captain David, 20–21, 25, 33–34, 57, 106, 113, 339, 371–372.
Kirk, Second Lieutenant Harold, 398, 421, 432, 449, 462.

Kitchener, Lord, xii, xiii, 16–17, 30, 34, 37–38, 51–52, 55, 81, 108, 141, 209, 477, 510–511, 513.
Knighton, Private Albert, 36, 61, 455.
Knighton, Major Gerald, 366.

Leader, Lieutenant Colonel Lionel, 360–361, 381, 393, 404, 418.
Leather, Lance Corporal Joseph, 83, 300, 503.
Lloyd, Captain Arthur, 32, 39, 105, 479.
Lurgan, Lord, 38, 516.

Mackarness, Captain Guy, 37, 98, 221, 223, 489
MacKay, Captain Duncan, 470–471.
Mannering, Reverend Ernest, 230–231, 286, 300, 304, 350, 418.
Maxwell, Captain Allen, 232, 263–264, 294, 325, 411.
McShane, Father, 283–284, 494.
McConnell, Captain W, 337, 360, 416, 418.
Medlicott, Second Lieutenant Walter, 15, 25, 40, 60, 79–81, 87, 89, 124, 126, 130, 147, 158–159, 211, 214–217, 219–221, 225, 309, 346–347.
Modera, Lieutenant Colonel Frederick, 59, 60, 119, 137, 325, 329, 331, 354, 357–358, 394, 404, 420, 422, 424, 427, 431, 432, 495.
Moor, Captain George, VC, 108, 469, 498.
Morton, Lieutenant John ('Beachcomber'), 44–45, 68, 489.
Murgatroyd, Lieutenant Arthur, 337, 351, 354, 376, 399, 416.

Namier, Private Lewis, 22–23, 38, 58.
Norman, Second Lieutenant Kenneth, 19–20, 26, 32–34, 36, 38, 42–43, 45, 48–49, 60, 63, 65, 76, 78, 82–84, 89, 93–94, 103, 119, 121, 126, 130–131, 146–148, 193, 197, 209–211, 218, 338, 345–346, 370.

Oakley, Captain Harold, 47–48, 473.

Page, Captain Edmund, 23, 33, 35, 37, 42, 55–57, 106, 113, 314, 436.
Peacock, Second Lieutenant Basil, 179, 341, 436–437, 471, 482–483, 487, 490.
Pease, Second Lieutenant Nicholas, 23, 34, 43, 45, 120, 130, 134, 204, 349, 372, 414, 436, 438, 471,

Pigache, Lieutenant Daniel, 432, 489.
Pinney, Major General Reginald, 354, 374, 388, 398, 415, 427, 431, 495.
Platt, Sidney and Vincent, 138, 148, 160, 179, 199–201, 203, 208, 229, 231, 237–240, 243–244, 256, 261, 266, 277, 280, 299, 305, 306, 482–483.
Price, Donald, 44–46, 60–61, 69, 95, 119, 124, 194, 198–200, 255–256, 266, 272, 278, 285, 288, 296, 302, 304, 306, 365, 374–375, 378, 391, 397, 400–401, 405, 415, 417, 418, 421, 424 - 426, 429–430, 432, 434, 436, 459 - 462, 466, 471, 483, 490–491, 495.
Price-Edwards, Captain Owen, 158, 211, 225, 242, 486.

Radford, Keith, 488.
Raven, RSM Leonard, 256, 354, 394, 462.
Ravenscroft, Lieutenant Pelham, 246, 310.
Richards, Private Frank, xii, 356, 380.
Robinson, Captain Thistle, 352, 357, 359, 467–468.

Salter, Leonard, 71–75, 77–78, 86–88, 90, 93, 97–98, 100, 103, 123, 125, 128, 132, 134, 142, 144–145, 152, 154, 156, 162–163, 167–169.
Sassoon, Siegfried, 374, 377–378, 380, 384.
Saxon, Captain Harold, 217, 437, 440.
Selfridge, Gordon, 38, 79.
Sewell, Second Lieutenant Cecil, VC, 218, 462, 500.
Shearn, Erroll, 15–16, 21, 25, 34, 36, 43–45, 57, 63, 208, 245, 247, 490.
Shield, Frederick, 232–234, 236, 238–240, 243, 252–255, 259–260, 262, 306.
Solomon, Second Lieutenant Louis, 218, 426, 428, 431, 453, 521.
Spaull, Philip, 98, 103, 105, 113, 117–118, 121, 125, 132, 144–145, 149, 151, 153, 162, 169, 176, 190, 206, 211–212, 216, 474–475.
Spurrell, Second Lieutenant Hugh, 23, 29–30, 34–35, 37, 42, 45, 52, 64–65, 69, 71, 74–75, 77–79, 81–86, 90, 96–98, 118, 120–121, 124, 126, 139–141, 145, 151, 153, 161–162, 170–173, 179, 181–183, 185, 188–189, 191, 206, 212–214, 216, 221, 223, 225, 368, 445, 491.
Standage, Lieutenant Alfred, 449, 450.

Index 531

Statham, Lieutenant Richard, 448, 450,
Stoneley, Private Ernest, 33, 50, 69, 71, 80, 448, 451, 452, 453.
Stuart-Clarke, Chaplain, 83, 118.
Stuart-Wortley, Lieutenant Colonel 'Jack', 41–42, 52, 55, 74–75, 110, 118, 170, 181–182, 225, 443, 485, 486.
Stuart, Major Edward, 16–17,
Sturges, Second Lieutenant Robert, 93, 103–104, 119, 123, 125, 126, 128, 132, 165, 170, 174–177, 190, 210, 214–215.

Taunton, Sergeant Basil, 228, 263–265, 267, 507.
Templar, Captain John, 28, 237, 279, 291, 354, 418, 432, 472, 506.
Thompson, J.P., 17.
Tudor, Private Robert, 340, 341,
Tuite, Lieutenant Colonel Henry, 223, 224, 412, 446.
Twiss, Major Edward, 131, 132, 158, 160, 192, 193, 195, 199, 396.
Tyson, Private Harold, iii, 199, 256, 273–274, 286, 288–291, 302, 383, 483, 494–495.

Vernede, Second Lieutenant Robert, 314, 363, 364, 521.
Vinden, Captain Herbert, 16, 21, 22, 24, 25, 41, 47, 135, 138, 197, 198, 204, 208, 209, 347, 366, 367, 471, 483, 488.

Warner-Abbatt, Major Frank, 17, 24.
Waterhouse, Second Lieutenant Gilbert, 247, 492, 522.
Watson, Lieutenant R.L., 484–485.
Wearne, Second Lieutenant Frank, VC, 410, 500.
Whitby, Lionel, 436, 489.
Williams, Lieutenant Morgan, 21–22, 34–35, 43, 48–49, 57, 61–62, 68, 79, 83–84, 100, 103, 117–118, 120–121, 126, 129, 138–140, 143–144, 208, 212, 218–219, 221, 347–348, 367, 487, 496.
Wilson, Second Lieutenant Harold, 119–120, 123, 125, 128, 130, 132–134, 142, 145, 150–151, 153–154, 164–165, 174, 177, 179–181, 190, 213, 218, 490.
Wolrige-Gordon, Lieutenant Colonel Walter, 40–41, 107, 118, 169, 177, 213, 215, 224–225.
Wright, Second Lieutenant Donald, 156, 178, 210–211, 348, 368–369.
Wright, Captain Eric, 229–231.
Wright, Private Philip, 120, 123, 126, 128, 182, 185–186, 188, 190, 206, 208, 212–214, 217, 221–222, 245, 250–251, 308–309, 339, 490.
Wylie, Captain, 254, 259–260, 324, 333, 352.

Ziegler, Captain George, 203, 266, 294, 296, 397, 403, 416, 418, 432, 470.

PLACES

Albert, 256, 461.
Arras, 107, 115, 362–373, 374–396, 407, 416, 428, 443, 461, 496–497, 499–500, 520–521.
Arras Memorial, 108, 365, 370, 376, 377, 381, 383, 387, 392–394, 396, 442–444, 446.
Ashtead, 26, 30, 32–34, 43, 51, 58, 65, 76, 94, 237, 484–485, 491.

Bazentin-le-Petit, 260–261, 263, 265–266, 268–270, 283, 461.
Bedlam Buildings, 92, 95–97, 102, 129, 133,
Béthune, 127–132, 135, 137–140, 143, 148–150, 153–156, 159, 162, 164, 167–171, 177–178, 181, 185, 189–191, 194, 197–198, 200, 209, 210, 227–228, 231, 233, 234, 237–238, 245, 349.

Béthune Town Cemetery, 131, 137, 138, 156, 167, 169, 181, 185, 190, 191, 194, 197, 198, 228, 231, 238, 239, 241–243, 253.
Beuvry, 132, 138–140, 171, 195, 198, 203, 239, 349.
Boulogne, 119–121, 123, 125, 218, 224, 415, 417, 428.
Bromsgrove School, 107, 370, 500.

Cambridge University, 21, 26, 43, 57, 58, 89, 106–107, 111, 168, 216, 220, 225, 246, 266, 309–310, 313, 328, 337, 341, 366, 407, 428, 439, 470, 489, 501, 519.
Cambrin, 155, 160, 165, 175, 179, 193–194, 197–198, 228, 254, 487.

Cambrin Churchyard Cemetery, 131, 172–173, 175–179, 189–191, 194, 196, 199–201, 203, 227–228, 230–231, 234, 238.
Charterhouse School, 37, 43, 131, 194, 259, 318, 341, 439, 489, 508, 510.
Cheltenham College, 29, 169, 471, 498.
Clery, 355, 358–360, 441.
Clipstone Camp, 77–91, 93, 94, 101, 132, 374.
Croisilles, 369, 371, 376–378, 381, 383, 388, 390, 394–396, 407,
Cuinchy, 110, 157, 161–163, 166, 168, 179, 193–194, 228, 234, 236, 252.

Dulwich College, 44, 61, 218, 223, 224, 465, 500, 521.

East Africa, 168, 473, 479–480.
Epsom, 21, 23, 26, 27, 30, 32–36, 38, 39, 42–46, 51–52, 58–60, 62, 65–67, 72–79, 82, 84, 86, 89, 94, 102, 132, 150, 299, 330, 374, 395, 491, 506, 514.
Etna Crater, 229.
Eton College, 21, 39–41, 224, 264.
Ewell, 26–27.

Fontaine-les-Croisilles, 369, 371–372, 376–378, 385, 388, 390, 394–395, 490.

Gallipoli, 94, 105, 108–109, 111, 206, 304, 326, 329, 336, 356, 359, 360, 383, 392, 406, 449, 473, 498.
Gibbons Crater, 229.
Givenchy, 110, 132, 134–135, 145, 148, 158, 181, 184, 193, 198, 239, 244, 252, 359, 486.

Harrow School, 43, 68, 246, 510.
High Wood, iii, xiii, 15, 257, 260, 262, 264, 266–307, 310, 316, 317, 319–320, 322–323, 330–331, 336–337, 342, 344, 349, 355, 356, 360, 367, 377, 396, 411, 428, 435, 451, 467, 468, 479, 480, 483, 486, 489–490, 493–496, 501–508.
Hindenburg Line see *Siegfriedstellung*
Hooge Crater, 411, 435,
Hyde Park, 20–21, 24–27, 29, 30, 88–89, 491, 514.

Italy, 473–474.

Le Transloy, 327, 341–342.
Leatherhead, 26, 29, 32–34, 37, 45–46, 60, 65, 76, 86, 94, 491.
Lesboeufs, 325–329, 331–334, 336–337, 342, 349–351.
Lombartzyde, 401–403.
Loos, 110–113, 130, 191, 410, 500.
Loos Memorial, 111, 113, 147, 181, 410, 500.
Lump Lane, 395.
Lys, 448–456.

Manchester, 26–28, 30, 56, 60, 72, 83, 107, 131, 137, 165, 170, 172–174, 178, 181, 194, 198, 200–201, 207–208, 223–225, 227–228–229, 235–236, 239–240, 242, 253, 262, 265–266, 272, 277, 280, 288, 295, 298–299, 304, 310, 321–322, 330–331, 336, 357, 381, 403, 435, 445, 452, 454, 460, 465, 484, 494, 496, 501–508.
Manchester Grammar School, 28, 59, 165, 176, 200, 225, 228–229, 235, 237, 241–242, 265–266, 274, 288, 305, 310, 322, 329, 335, 387, 435, 501–503, 505–508.
Manchester University, 28, 113, 225, 305, 341, 362, 375, 383, 501–503, 505, 508.
Marlborough College, 54, 109, 113, 167, 196, 216, 225, 228, 300, 339, 341, 470, 507.
Menin Gate Memorial, 429, 489.
Menin Road, 411, 414, 416, 424, 425, 427.
Mesopotamia, 361, 473, 477–478, 480.
Messines Ridge, 407–408, 422, 520.

Nieupoort (Nieuwpoort), 397, 400–406.

Oldenburg Lane, 390.
Orchard Redoubt/Keep, 156, 240, 241
Oxford, 216–217, 223, 261, 353, 395, 445, 472, 476.
Oxford University, 21, 23, 26, 30, 37, 39–40, 43, 59, 68, 100, 108, 113, 121, 148, 168, 176, 194, 211, 215, 218, 228, 230, 246, 259, 264, 300, 311, 323, 339, 341–342, 352, 365, 371, 407, 469, 475–477, 500, 502, 504, 507–508, 521.

Palestine, 472, 473, 478–480.
Passchendaele, 414, 425–429.
Perham Down, 90, 93, 104, 120.
Plum Trench, 394.

Red Dragon Crater, 239, 243, 253.
Renescure, 210, 218, 222.
Rugby School, 111, 179, 248, 294, 479, 499, 504.

Salonika, 473–477, 480–481.
Siegfriedstellung, 363, 369, 371, 375–378, 380, 384–386, 388, 390, 394–395, 445.
Somme, 146, 191, 244, 245–251, 252, 255–348, 349, 351, 354–355, 362, 370, 386, 410, 441–442, 446, 461, 486, 496–497.
St Omer, 204, 209, 210–212, 215, 218–220, 253, 309, 344, 415, 420, 430.
Switch Line/Trench, 268–270, 273, 276–277, 285, 289, 310.

Thiepval Memorial, 87, 221, 223, 248–248, 251, 261–262, 265–266, 280, 286, 291–292, 294–295, 300, 305, 309–310, 321, 326, 329, 335–336, 339,
Tidworth, 93, 96–97, 100–104, 118, 119, 132, 181, 374.
Tunnel Trench, 371, 388–391, 393–395.
Tyne Cot, 412, 416, 426, 430, 435.

Victoria Steet (no 66), London, 19, 512, 514, 515.

West Africa, 480.
Winchester College, 109, 225, 309, 354, 477, 478.

Ypres and Ypres Salient, 106, 110, 113, 366, 388, 399–400, 406, 407, 410–411, 414–416, 419, 421–422, 424–425, 427–430, 448, 458, 460, 466, 487, 489, 497.

FORMATIONS & UNITS

Divisions
Guards Division, 315.
1st Division, 94, 111, 124, 210, 488.
2nd Division, 130, 138, 140, 155, 160, 223, 311, 347, 446, 459.
3rd Division, 347, 445.
4th Division, 247, 325, 365.
7th Division, 130, 138, 250, 268, 269, 473–474.
8th Division, 249, 250.
9th (Scottish) Division, 94, 111, 113, 365.
11th (Northern) Division, 47, 109.
12th (Eastern) Division, 113, 222, 233, 308, 340, 373, 460.
13th (Western) Division, 94, 109, 477.
14th (Light) Division, 316, 366, 415, 438.
15th (Scottish) Division, 365.
16th (Irish) Division, 181, 409.
17th (Northern) Division, 131, 250, 322, 332–333, 441.
18th (Eastern) Division, 222, 250, 372, 437–438.
21st Division, 111–112, 250, 371–372, 375–377, 384, 394, 441.
23rd Division, 311, 409, 416, 418, 420, 473, 474.
24th Division, 111–112, 149, 220, 223, 321, 407, 411, 437.
26th Division, 475.
29th Division, 226, 247.
30th Division, 375–376, 381, 385, 469, 498.
32nd Division, 248, 400, 465, 473.
33rd Division, xii, 86, 90, 94–95, 97, 99–100, 103, 109, 113, 124, 131, 141, 151, 160, 176, 178, 195, 197–198, 208, 210–211, 223, 234, 251, 254, 257, 259, 266–269, 285, 291–292, 303, 307, 321, 345, 349–350, 352, 355, 358, 362, 369, 371, 374–377, 381, 385–388, 390, 393–396, 398–400, 415–418, 423, 427, 431, 433, 452, 454, 494–495.
34th Division, 436–437, 444.
35th (Bantam) Division, 354, 471.
36th (Ulster) Division, 248, 499.
37th Division, 113, 462.
38th (Welsh) Division, 203, 228, 431, 461.
39th Division, 234, 436.
40th Division, 360, 444, 450,
41st Division, 340, 411–412, 473.
42nd (East Lancashire) Division, 449.
46th (North Midland) Division, 322–323, 465.
47th (London) Division, 111, 316, 441.
48th (South Midland) Division, 473.
49th (West Riding) Division, 400, 454–455.
50th (Northumbrian) Division, 445, 451.
51st (Highland) Division, 442.

56th (London) Division, 316, 342, 377, 380–381.
59th (North Midland) Division, 443–444.
60th (London) Division, 478.
62nd (West Riding) Division, 371.

Brigades
19th Brigade, xii, 131, 138, 141, 145, 148,
  157, 160–161, 192, 194–195, 234, 236, 239,
  244, 257, 259, 262, 268–270, 275, 278, 281,
  284–285, 288–289, 292, 300, 303, 305–306,
  318–321, 326–327, 329, 331, 333, 335, 345,
  352, 354–357, 359–360, 369, 375–378,
  384–386, 390, 394, 396, 400, 415–418,
  420–423, 428–432, 434, 495.
54th Brigade, 251.
64th Brigade, 371, 441.
76th Brigade, 366–367.
98th Brigade, 18, 88, 90, 96–97, 99, 101, 118,
  122, 124–125, 131, 137, 141–145, 150–157,
  170–191, 208, 210, 259, 269, 292, 319–320,
  354, 357, 369–370, 384–386, 388, 390,
  417–418.
99th Brigade, 90, 96–97, 223.
100th Brigade, 96–97, 259, 417,
110th Brigade, 113, 371–372, 441.
118th Brigade, 18, 89, 90.
UPS Brigade
  Attestation, 19, 20, 24, 38, 202, 513.
  Billeting, 32–36, 38, 43, 46–47, 62.
  Commissions, 16, 19, 22, 23, 29, 44, 52,
    53–61, 71, 88–89, 100, 105.
  Committee, 17, 19, 20, 24, 26, 28, 38–40, 53.
  Embarkation, 117–121.
  Recruiting, 19–29.
  Mobilisation, 26–32.
  Training,
    Bayonet Training, 40, 81, 86, 94, 97,
      99–103, 126, 157, 194, 198, 200, 276.
    Running, 36, 46, 109.
    Drill, 22–26, 32–33, 35–36, 45–46, 72,
      214, 260, 416.
    Grenade / Bomb throwing, 99, 103, 140,
      148, 151, 152, 198, 361, 463.
    Route Marching, 20, 24, 35–36, 44, 48,
      70, 84–86, 89, 94, 99, 103, 123–124,
      126, 145, 150, 152, 155, 157, 159, 210,
      213–214, 221, 232, 397, 513.
    Trench digging, 49–51, 57, 65.
    Trench warfare, 95–96, 102.

Corps, Regiments and Battalions
Royal Field Artillery (RFA), 31, 35, 94, 97.
Royal Engineers (RE), 21, 99, 220, 269,
  272; Field Companies; 59, 471; 11th Fd
  Coy, 269, 277–278, 284–285, 288; Special
  Brigade (RE), 110, 178, 194, 223, 441, 461;
  Tunnelling Companies (RE), 134, 181, 191,
  195, 233, 234, 239, 241, 243, 321.
Grenadier Guards (GG), 39, 223–224, 469.
Scots Guards (SG), 295.
Royal Scots (Lothian Regiment) (R Scots),
 Queen's (Royal West Surrey Regiment)
  (Queen's RWS), 1st Bn, 139–140, 171, 181,
  194, 237, 288, 320, 345, 369, 386, 396,
  430–431, 454.
Buffs (East Kent Regiment) (E Kents), 6th Bn,
  113; 7th Bn, 437–438; 8th Bn, 411.
King's Own (Royal Lancaster Regiment)
  (King's Own), 325, 367; 6th Bn, 477–478.
Northumberland Fusiliers (NF), 10th NF, 474;
  11th Bn, 474; 12th Bn, 111, 13th Bn, 394;
  14th Bn, 441; 16th Bn, 249; 21st Bn, 311;
  22nd Bn, 436.
Royal Warwickshire Regiment (R Warwicks),
  180; 1st Bn, 326; 2nd Bn, 138; 2/5th Bn,
  239; 1/7th Bn, 474.
Royal Fusiliers (City of London Regiment) (R
  Fus),
  1st Bn, 383, 412, 472.
  2nd Bn, 247–248, 431–432, 451–454, 460,
    521.
  3rd Bn, 106, 474.
  4th Bn, 354, 365, 413, 431–432, 445, 461.
  7th Bn, 241, 296, 433, 465.
  8th Bn, 308–309, 312, 340–341, 373, 432.
  9th Bn, 309, 312, 340, 460.
  10th Bn, 229, 398.
  11th Bn, 222, 250–251, 309, 339.
  12th Bn, 112–113, 223, 408, 411.
  13th Bn, 148, 424, 432–435, 447–449,
    461–462, 466–468, 470–471.
  17th Bn, 156, 223, 347, 446, 459, 463–464.
  18th Bn, xii, 36, 40, 44–45, 48, 60, 64–65,
    67, 78, 89–90, 93, 97, 99–104, 107,
    110–111, 114–115, 119, 123, 125–127,
    129–131, 141, 145–148, 157–161, 165,
    171, 181, 192–198, 200, 204, 209–211,
    214–218, 220, 222–225, 242, 245,
    246, 248–249, 311, 315, 316, 338–339,

342–343, 365–366, 372, 393, 407–410, 441, 451–452, 454–456, 460, 464–465, 472, 474, 476–478, 480, 489–490, 499–500, 519, 522.
19th Bn, xii, 21, 31, 37, 40–41, 43, 48, 50, 60, 65, 67, 72–73, 77–78, 86, 89–90, 93, 95, 98, 100, 103, 107, 109–110, 112, 119–121, 123–128, 130, 132, 134–135, 142, 144–145, 150–153, 155–157, 161–169, 171, 173–181, 190, 192, 196–198, 200, 206, 210–215, 217–225, 245, 251, 316, 340, 342, 347, 349, 364–365, 368, 407, 409–413, 426, 439, 442, 444, 446, 454–455, 457, 460–461, 468, 472, 474–475, 477, 479–480, 483, 488–491, 499, 521.
20th Bn, xii - xiv, 29, 36, 38, 41, 44–45, 54, 57–59, 64–67, 82, 86, 88–89, 93, 95–96, 104–105, 107, 109–110, 113, 119, 122–125, 127–128, 130, 135–138, 141, 147–149, 158–161, 165, 171, 173, 179, 192, 194–198, 200–204, 206–209, 224, 226–244, 245, 247, 251–307, 310–311, 316–337, 340–341, 344–345, 349–362, 369, 371, 373, 374–406, 407, 411–412, 414, 415–435, 446–449, 451–454, 460–462, 470–472, 477–479, 483–484, 487–488, 490–491, 493–497, 501–508, 521.
21st Bn, xii, 34–35, 41–42, 44–45, 47–49, 54, 57–59, 65–67, 72–74, 78, 81, 86, 90, 93, 96, 98–102, 104, 106–108, 110–111, 114, 120–121, 123, 126 - 128, 130, 138–140, 142–145, 148, 150–157, 161–162, 167, 170–173, 178, 181–184, 186–191, 195–196, 198, 204, 206–208, 210, 212–214, 216–218, 221–225, 244–245, 251, 309, 311, 314, 339, 346–348, 355, 367–368, 372, 375, 411–412, 436, 443–446, 450, 461–463, 467–468, 472, 475–477, 480, 484–485, 489, 496, 498, 500, 519.
22nd Bn, 223, 311, 347.
23rd Bn, 206, 221–223, 225, 311, 347.
24th Bn, 90, 222–223, 346–347.
25th Bn, 479.
26th Bn, 221, 304, 325, 342, 411–412, 414, 446, 467–468, 472–473, 485,
28th Bn, 102, 121, 217, 340, 343, 372, 426, 472, 520, 521.

29th Bn, 96, 202, 232, 311–312, 325, 340–341, 373, 412, 459, 474.
32nd Bn, 340–341, 375, 412, 472.
King's (Liverpool Regiment) (King's), 1st Bn, 130; 4th Bn, 239, 265, 319, 361, 390, 394, 427, 454, 456.
Norfolk Regiment (Norfolks), 6th Bn, 86; 10th Bn, 475.
Lincolnshire Regiment (Lincs), 162; 1st Bn, 441.
Devonshire Regiment (Devons), 218, 371; 8th Bn, 413; 10th Bn, 475–476.
Suffolk Regiment (Suff R), 2nd Bn, 347, 367; 1/4th Bn, 319, 355, 384–386; 7th Bn, 198, 233; 13th Bn, 215.
Prince Albert's (Somerset Light Infantry) (SomLI), 470; 1st Bn, 365; 6th Bn, 438–439.
West Yorkshire Regiment (W Yorks), 1st Bn, 325.
East Yorkshire Regiment (E Yorks), 7th Bn, 332; 12th Bn, 313, 519.
Bedfordshire Regiment (Bedf R), 216; 1st Bn, 338; 2nd Bn, 132; 1/5th Bn, 478; 8th Bn, 368.
Leicestershire Regiment (Leic R), 218; 6th Bn, 57, 372; 7th Bn, 441; 9th Bn, 371–372, 520.
Yorkshire Regiment (Green Howards or Yorks), 1st Bn, 146; 2nd Bn, 132; 1/4th Bn, 429; 7th Bn, 250; 8th Bn, 47, 473.
Lancashire Fusiliers (LF), 224; 1st Bn, 432; 2nd Bn, 342; 10th Bn, 441; 19th Bn, 467; 20th Bn, 257.
Royal Scots Fusiliers (RSF), 347; 2nd Bn, 132; 6/7th Bn, 312.
Cheshire Regiment (Ches R), 23, 225; 10th Bn, 309.
Royal Welsh Fusiliers (RWF), 208; 2nd Bn, xii, xiii, 136, 148, 160, 192, 197–198, 200, 203, 224, 231, 233–234, 236–237, 243, 253, 269, 280, 283–285, 287, 292, 305, 318, 322–323, 326, 328, 332, 335, 352, 355–360, 374, 377–378, 380, 384, 386–387, 394, 398, 404–405, 417–418, 420, 427, 431, 434, 493, 495; 8th Bn, 206, 477; 10th Bn, 218, 347–348, 367; 13th Bn, 203, 461.
South Wales Borderers (SWB), 4th Bn, 57, 109.
Cameronians (Scottish Rifles) (1/Cameronians, 5/Scot Rif), 1st Bn, xii, 145, 198, 201, 231, 233–234, 236, 238–239, 244, 252, 256–258,

260, 269–270, 273, 276, 283–286, 288, 292, 306, 319–320, 322–323, 325, 329, 345, 352, 355, 356, 358, 376–378, 380–381, 384, 386, 388, 390–392, 394–396, 404–405, 415, 417, 420, 422–423, 434, 493, 495; 1/5th or 5/6th Bn, xii, 136, 145, 148, 196, 202–203, 228, 233, 252, 257, 263, 269, 272–273, 282, 284, 292, 319, 358, 376–378, 384, 386, 388, 390, 404–405, 417–418, 420, 426, 434.

Gloucestershire Regiment (Glosters), 8th Bn, 310.

Worcestershire Regiment (Worcesters or Worc R), 2nd Bn, 140, 161, 170, 203, 232, 269–270, 390, 393, 396, 427, 454–455.

East Lancashire Regiment (E Lancs), 4th Bn, 440.

East Surrey Regiment (E Surrey), 8th Bn, 251, 372, 414; 9th Bn, 149.

Duke of Wellington's (West Riding Regiment) (DWR), 1/4th Bn, 403; 1/7th Bn, 455; 10th Bn, 409, 421, 474.

Border Regiment (Border), 477, 480; 8th Bn, 397.

Royal Sussex Regiment (R Sussex), 217; 7th Bn, 144, 460, 521; 9th Bn, 437, 440.

Hampshire Regiment (Hamps R), 1st Bn, 208, 247, 498; 2nd Bn, 108, 498; 12th Bn, 475–476; 14th Bn, 498.

South Staffordshire Regiment (S Staffs), 1st Bn, 138; 2nd Bn, 130, 140–141; 2/6th Bn, 443; 7th Bn, 109; 8th Bn, 322; 9th Bn, 474.

Dorsetshire Regiment (Dorsets), 21, 53; 1st Bn, 106, 132, 406; 5th Bn, 450.

South Lancashire Regiment (S Lancs), 188, 216; 2nd Bn, 249; 6th Bn, 477.

Welsh Regiment (Welsh), 10th Bn, 228; 14th Bn, 461; 17th Bn, 360.

Black Watch (Royal Highlanders) (Black Watch or BW), 40, 111, 365.

Oxfordshire and Buckinghamshire Light Infantry (O&BLI), 21, 138, 476.

Essex Regiment (Essex), 283; 1st Bn, 368, 500; 2nd Essex, 247, 492, 522; 11th Bn, 344, 410, 467, 500; 17/Essex, 90.

Sherwood Foresters (Nottinghamshire & Derbyshire Regiment) (Sherwood Foresters or Notts&Derby), 2nd Bn, 442; 1/8th Bn, 465; 16th Bn, 223; 17th Bn, 223.

Loyal North Lancashire Regiment (LNL), 216, 505; 1st Bn, 111; 6th Bn, 477; 8th Bn, 344.

Northamptonshire Regiment (Northants), 397; 1st Bn, 111; 2nd Bn, 410.

Royal Berkshire Regiment (Berkshires or R Berks), 1st Bn, 105, 112, 130; 2nd Bn, 249; 7th Bn, 476; 8th Bn, 438.

Queen's Own (Royal West Kent Regiment) (Queen's RWK), 500; 1st Bn, 414; 4th Bn, 332; 11th Bn, 412–413.

King's Own Yorkshire Light Infantry) (KOYLI), 265, 365; 2nd Bn, 248, 249; 7th Bn, 364; 8th Bn, 421; 10th Bn, 387.

King's (Shropshire Light Infantry) (KSLI), 8th Bn, 476.

Middlesex Regiment (Middlesex or Mx), 240, 309, 318, 336, 384, 450, 460; 1st Bn, 131, 141–142, 144–145, 150, 152, 161–162, 171, 180, 189, 197, 208, 231, 237, 326, 384, 423, 430, 496; 2nd Bn, 457; 12th Mx, 339, 16th Bn, xii, 17, 29, 90, 198, 209, 226, 422, 517–518; 18th Bn, 90, 97, 175, 243, 269, 287, 431.

King's Royal Rifle Corps (KRRC), 30, 57, 182; 1st Bn, 130; 2nd Bn, 246, 310; 7th Bn, 106, 113, 314; 16th Bn, 90, 162, 164, 177, 189, 196, 198, 203, 239, 252, 260–261, 288, 330, 369–370, 390–391, 394, 423, 431, 454–455, 469.

Wiltshire Regiment (Wiltshires or Wilts), 2nd Bn, 132, 134; 5th Bn, 109, 477; 7th Bn, 476.

Manchester Regiment (Manchesters or Manch R), 223, 363, 424; 1st Bn, 124; 9th Bn, 298; 1/10th Bn, 468; 12th Bn, 363, 441; 13th Bn, 476; 18th Bn, 375–376; 22nd Bn, 474.

North Staffordshire Regiment (N Staffs), 7th Bn, 480; 9th Bn, 410.

Durham Light Infantry (DLI), 179; 9th Bn, 429; 11th Bn, 468; 13th Bn, 311, 465–466.

Highland Light Infantry (HLI), 257, 260; 2nd Bn, 138–139; 1/9th Bn, 226. 303, 352, 388, 390, 394, 427, 430, 454–455.

Seaforth Highlanders (Seaforths), 4th Bn, 442.

Gordon Highlanders (Gordons), 351; 1st Bn, 106; 6th Bn, 142.

Queen's Own Cameron Highlanders (Camerons), 1/4th Bn, 146.

Royal Irish Fusiliers (1/RIrFus), 9th Bn, 248, 499.

Connaught Rangers (Conn R), 361.
Argyll and Sutherland Highlanders (Argylls or A&SH), 361; 2nd Bn, 131, 136, 173, 190, 203, 208, 234, 319, 353, 357, 384, 386, 421, 495, 496.
Leinster Regiment (Royal Canadians) (Leinsters), 2nd Bn, 220; 5th Bn, 108; 7th Bn, 181, 189, 197, 201.
Royal Munster Fusiliers (Munsters or RMF), 243–244; 1st Bn, 409; 6th Bn, 203.
Royal Dublin Fusiliers (Dublins or RDF), 218; 2nd Bn, 326.
Rifle Brigade (Prince Consort's Own) (RB), 30, 57, 216, 3rd Bn, 314, 407–408, 411, 519; 5th Bn, 521; 6th Bn, 57, 519; 7th RB, 439; 12th Bn, 342, 364, 521; 13th Bn, 478.
London Regiment (Londons or 1/Lond R), 304, 336; 1st Bn, 329, 343, 392, 425; 2nd Bn, 352; 7th Bn, 111; 10th Bn, 326, 355–356, 381, 405; 19th Bn, 478–479; 24th Bn, 359.
Machine Gun Corps (MGC),
  Battalions: 8th MG Bn, 458; 21st MG Bn, 441; 32nd MG Bn, 470; 39th MG Bn, 436; 40th MG Bn, 444–445, 450; 49th MG Bn, 455; 50th MG Bn, 445, 451; 51st MG Bn, 442; 59th MG Bn, 444; 66th MG Bn, 440.
  Companies: 19th MG Company, 269, 292; 96th MG Company, 249; 100th MG Company, 390, 495; 208th MG Company, 371; 217th MG Company, 411.
Army Service Corps (ASC), 79, 158, 200, 210, 234, 375, 383, 393, 415, 416, 455, 477.
Royal Army Medical Corps (RAMC), 19, 151, 181, 194, 280, 311, 337, 406, 420, 489.
Labour Corps (Lab Corps), 165, 180, 229, 324, 452.
Army Ordnance Corps (AOC), 395.
Tank Corps (Tank C), 216, 221, 315–316, 442, 472, 488, 491; 1st Bn, Tank Corps, 442, 456–457; 3rd Light Tank Bn, 462, 500; 5th Bn, 465; 14th Bn, 460.
Army Cyclist Corps (ACC),
  33rd Divisional Cyclist Company, 86, 176, 195, 292, 345.
Royal Flying Corps:
  No 11 Squadron, 115, 499.
  No 20 Squadron, 367.
  No 24 Squadron, 315.
  No 48 Squadron, 367–368.
Regular Army, iii, xii, 16, 25, 44, 60, 100, 108, 111, 132, 134–136, 141, 152, 160, 182, 199, 208, 220, 255–256, 273, 317, 359, 430–431, 483, 488, 495–496, 509–511, 513.
Territorial Force, 16–17, 19, 22–23, 37, 111, 295, 359, 361, 431.
'Pals' Battalions, xiii, 519.
Officer Cadet Battalions, 131, 181, 204, 219, 221, 224, 317, 458.

Miscellaneous
Armistice, The, xii, 466, 469–472, 474, 481, 488, 497, 498.
'Chocolate Soldiers', xii, 208, 224, 374, 380, 430–431, 495–496.
'Gentleman rankers', xii, xiii, 17, 44, 317.
Publications:
  The Gasper, xiii, 87, 100, 118, 125, 127, 133, 137, 144, 152–155, 205, 216, 218, 226, 307, 441–442, 465.
  The Pow Wow, xiii, 33, 47, 50, 55, 67, 75, 78, 80–82, 85–87, 90, 236, 309, 367, 442, 452, 465, 521.
  The UPS Song Book, 32, 39, 105, 479.
Public Schools Yearbook, xii, 510.
'Temporary Gentlemen', xii, xiii, 483,
Victoria Cross, 115, 243, 248–249, 410, 462, 498–500.
War Office, 17, 24–26, 38, 48, 56, 58, 62, 73, 84, 103, 181, 204, 207–208, 211, 219, 224, 341, 365, 481, 488, 510–513, 516–517.
Weapons:
  Grenades ('bombs'), 138, 143, 188.
  Lewis machine gun, 99, 165, 240, 277, 303, 311, 324, 326, 328, 339, 362, 381, 385, 390, 394, 423, 428, 434, 440, 442, 445, 449, 456, 462, 464, 474, 488.
  Rifle grenade, 132, 140, 172, 177, 180, 194–195, 198–199, 201, 227–230, 233, 252, 356–357, 362, 385, 437, 455, 467,
  Snipers, 95, 102, 140, 148, 157, 172, 175, 197, 253, 288, 319, 388, 405,
  Tanks, 315–316, 339, 365, 372, 456–457, 460, 462, 465–467, 500.
  Trench mortars, 132, 172, 180, 185, 228, 234, 237, 247, 252, 276, 284, 356, 455.
  Vickers machine gun, 172, 285.